T £27.00

A Specialist Periodical Report

Surface and Defect Properties of Solids

Volume 6

A Review of the Recent Literature published up to mid-1976

Senior Reporters
M. W. Roberts, *Department of Chemistry, University of Bradford*
J. M. Thomas, *Edward Davies Chemical Laboratories, University College of Wales, Aberystwyth*

Reporters
J. Corish, *University College Belfield, Dublin, Eire*
R. M. Hooper, *University of Exeter*
J. L. Hutchison, *University College of Wales, Aberystwyth*
P. W. M. Jacobs, *University of Western Ontario, Canada*
D. A. Jefferson, *University College of Wales, Aberystwyth*
G. G. Kleiman, *Universidad Nacional Autonoma de Mexico, Mexico*
Z. Knor, *Czechoslovak Academy of Sciences, Prague, Czechoslovakia*
U. Landman, *University of Rochester, New York, U.S.A.*
D. R. Lloyd, *University of Birmingham*
C. M. Quinn, *University of Birmingham*
S. Radhakrishna, *Indian Institute of Technology, Madras, India*
N. V. Richardson, *University of Birmingham*
J. N. Sherwood, *University of Strathclyde, Glasgow*
B. K. Tanner, *University of Durham*
M. J. Tricker, *Heriot-Watt University, Edinburgh*

The Chemical Society
Burlington House, London W1V 0BN

ISBN: 0 85186 300 0
ISSN: 0305-3873
Library of Congress Catalog Card No. 72-78528

Copyright © 1977
The Chemical Society

All Rights Reserved. No part of this book may be reproduced or transmitted in any form or by any means – graphic, electronic, including photocopying, recording, taping or information storage and retrieval systems – without written permission from The Chemical Society.

Printed in Great Britain by
Billing & Sons Limited, Guildford, London and Worcester

Preface

For the first time in this series we have considered physisorption, one of the pillars of contemporary surface chemistry. Landman and Kleiman have adopted a detailed microscopic approach covering both theoretical and experimental aspects of physisorption and drawing widely from diverse areas of science to provide a thorough analysis of the current state of knowledge. An interesting contrast is evident in the articles of Knor on the one hand and Lloyd, Quinn, and Richardson on the other. Knor has chosen to consider in an essentially phenomenological fashion the interplay between theory and experiment with relation to the surface chemistry of metals. He poses problems for the theoretician and asks questions of the experimentalist. Lloyd, Quinn, and Richardson consider the particular advantage of one recent experimental development, namely angle-resolved ultraviolet photoelectron spectroscopy, in relation to theoretical aspects of metal surfaces *per se* and also adsorption on these surfaces. The general area of photoemission from metal surfaces has been considered in previous volumes (Roberts, Brundle, Spicer *et al.*, Mason *et al.*) but this is the first time we have discussed in detail the advantages that accrue from angular studies. A new departure in this volume is a review of the application of conversion electron Mössbauer spectroscopy to surface studies, by Tricker, who describes the experimental approach, discusses various aspects of the sensitivity of the technique, and outlines some successful applications.

Tanner's contribution also represents a new departure for this series. His succinct account, illustrated with examples from disparate fields, of how crystal perfection may be assessed by X-ray topographic methods is particularly timely, since it is very likely that considerable reliance will in future be placed on non-destructive topographic methods with the continued advances in X-ray optics and with the increasing availability of synchrotron radiation. The comprehensive chapter by Corish, Jacobs, and Radhakrishna returns to the topic of point defects in ionic crystals which was the subject of a major review (by Corish and Jacobs) in Volume 2. Ionic solids continue to elicit the interest of a wide spectrum of researchers, from the computational physicists on the one hand to those engaged in technological developments associated with fuel cells, solar energy converters, electrocatalysts, and the like on the other. It is appropriate, therefore, that as well as dealing with intrinsic and impurity (ionic and molecular) defect parameters in ionic crystals, fast ion conduction is also discussed.

The two remaining chapters differ markedly from one another. Sherwood's summarizing contribution focuses attention upon highly plastic molecular crystals where interesting advances have been made, many by Sherwood himself, in the past three years. In this context the dislocation is of paramount importance. In the final chapter Hutchison, Jefferson, and Thomas survey the current scene in the solid-state structural chemistry of a range of minerals. Enormous progress, chiefly through the agency of high-resolution microscopes (the principles of which were reviewed by Anderson and Tilley in Volume 3), has been made in this area within the past few years. In this particular context, the importance and ubiquity of planar faults,

Wadsley defects, recurrent twinning, coherent intergrowths (all of which have been touched upon in previous volumes of these reports) are strikingly apparent. Necessarily, several classes of minerals, notably the feldspars, have had to be omitted in this survey. We plan to repair this omission in a future report.

April 1977

M. W. Roberts
J. M. Thomas

Contents

Chapter 1 Microscopic Approaches to Physisorption: Theoretical and Experimental Aspects   1
By U. Landman and G. G. Kleiman

  1 Introduction   1

  2 Theoretical Approaches for the Study of Physisorption   2
    Introduction   2
    van der Waals Forces and Physical Adsorption   5
    Semi-empirical Calculations   19
    Fundamental Calculations of Physisorption Binding Energies   21

  3 Microscopic Experimental Methods for the Study of Physisorption   33
    Introduction   33
      General Introductory Remarks   33
      On the Use of UPS in Characterization of the Physisorption State   34
    Experimental Techniques for the Study of Atomic Arrangement in Physisorption Systems   39
      LEED: Background Remarks   40
      LEED Studies of Physisorption Systems   43
      Neutron Scattering: Background Remarks   56
      The Adsorption of N_2 and ^{36}Ar on Graphite   60
      The Study of Thermal Vibrations using LEED   62
      Low Energy Molecular (Atomic) Beam Scattering (LEMS)   66
      Experimental Observations. Scattering Regimes   68
      Experimental Observations. Diffraction of Rare Gases from Metals   70
      Experimental Observations. Selective Adsorption   74
      On Interaction Potentials used in Theoretical Analysis of LEMS   77
    Experimental Techniques for the Study of the Electronic Structure of Physisorbed Systems   82
      Ultraviolet Photoemission Spectroscopy (UPS)   82
      X-Ray Photoemission Spectroscopy (XPS)   85
      Electron Energy Loss Spectroscopy (EELS)   90
      Field Emission Techniques   90
      Field Emission and the Probe-hole Techniques   92
      Field Emission. Theoretical Considerations   94

	Measurement of Adsorption Isotherms *via* Electron Spectroscopy	97
	Auger Electron Emission in Adsorption Isotherm Studies	100
4	Conclusion	103

Chapter 2 57-Iron Conversion Electron Mössbauer Spectroscopy 106
By M. J. Tricker

1	Introduction	106
2	Principles of the Mössbauer Effect	106
3	Interaction of the Nucleus with its Environment	
	Chemical Isomer Shift	108
	Quadrupole Interaction	109
	Magnetic Interaction	109
	Spectral Areas	110
4	Surface Studies	111
5	Internal Conversion	113
6	Theory of the Attenuation of ^{57}Fe Conversion Electrons	115
7	^{57}Fe CEMS Experimental Considerations	116
8	Probing Depth and Sensitivity	120
9	Applications	127
	Corrosion and Oxidation Studies	127
	Surface Phase Analyses of Steels	130
	Surface Stress Measurements	131
	Ion Implantation Studies	131
	Applications in Geochemistry	134
10	Depth Selective ^{57}Fe CEMS	137
11	Conclusions	138

Chapter 3 The Interplay of Theory and Experiment in the Field of Surface Phenomena on Metals 139
By Z. Knor

1	Introduction	139
2	Bare Metal Surfaces	139
	Theoretical Models of Metal Surfaces	139
	Comparison of Theoretical and Experimental Results	151

3 Chemisorption of Gas Molecules on Transition Metal Surfaces 160
Model Representation of the Gas–Metal System 160
CFSO–BEBO Approach 162
The Localized Free Electron Interplay Model 163
The Comparison of Theoretical and Experimental Results 167

4 Conclusion 178

Chapter 4 Angle-resolved Ultraviolet Photoelectron Spectroscopy of Clean Surfaces and Surfaces with Adsorbed Layers 179
By D. R. Lloyd, C. M. Quinn, and N. V. Richardson

1 Introduction 179

2 Photoemission from Clean Surfaces 180
Theory 180
Comparisons with Angle-resolved Spectra 189

3 Angle-resolved Spectra of Adsorbed Species 205
Theory 205
Comparison with Experiment 211

Chapter 5 Point Defects in Ionic Crystals 218
By J. Corish, P. W. M. Jacobs, and S. Radhakrishna

1 Introduction 218

2 Intrinsic Defect Parameters 219
Self-diffusion 219
Ionic Conductivity 221

3 Impurity Defect Parameters 227
Diffusion of Impurities 227
Impurity–Vacancy Complexes 228

4 Theoretical Calculation of the Properties of Point Defects 234
Defect Energies: Methods 234
Defect Energies: Results 235
Defect Interactions 243
Rare Gas Diffusion in Ionic Crystals 245
Defect Entropies 245

5 Fast Ion Conduction 246
Ag^+ and Cu^+ as Fast Ion Conductors 247
Fast Conduction by Alkali Ions 249
Anionic Fast Conduction 251

6 Paraelectric Impurities — 252
  Static and Dynamic Behaviour of PEC — 252
  Devonshire Model — 253
  Multi-well Potential Tunnelling Model — 253
  Experimental Techniques and Data — 254
    Electro-optical Technique — 254
    Electrocaloric Method — 255
    Specific Heat Measurements — 255
    Paraelectric Resonance — 255
    The Kerr Effect Method — 255

7 Molecular Impurities — 258
  Diatomic Molecular Ions — 258
  Triatomic Molecular Ions — 262
  Planar Molecular Ions — 262
  Molecular Ions of Tetrahedral Symmetry — 263
  Complex Molecular Ions — 266

8 Impurity Ions with the S^2 Configuration — 269
  Emission — 276

Chapter 6 Assessment of Crystal Perfection by X-Ray Topography — 280
By B. K. Tanner

1 Introduction — 280

2 Techniques of X-Ray Topography — 281
  Laboratory X-Radiation Techniques — 281
    The Berg-Barrett Topography — 281
    The Lang Technique — 282
    The Double Crystal Method — 284
  X-Ray Topography using Synchrotron Radiation — 288
  Direct Viewing of X-Ray Topographs — 289

3 Contrast on X-Ray Topographs — 290
  Perfect Crystal Phenomena – Pendellösung — 290
  Contrast of Dislocations — 294
  Contrast of Stacking Faults — 298
  Contrast of Twins and Sub-grains — 300

4 Applications to Crystal Defect Studies — 300
  Crystal Growth — 300
  Metallurgy and Plastic Deformation — 303
  Mineralogy — 304
  Control of Electronic Devices — 305
  Oxidation Studies — 307

Contents

Chapter 7 The Plasticity of Highly Plastic Molecular Crystals — 308
By R. M. Hooper and J. N. Sherwood

1 Introduction — 308

2 High Temperature/Low Stress Deformation – Self-diffusion Control — 309

3 Low Temperature/High Stress Deformation – Dislocation Slip — 314
  Evaluation of Dislocation Energies — 317

4 Conclusions — 319

Chapter 8 The Ultrastructure of Minerals as Revealed by High Resolution Electron Microscopy — 320
By J. L. Hutchison, D. A. Jefferson, and J. M. Thomas

1 Introduction — 320

2 The Etch-decoration Technique and its Use in the Detection of Stacking Faults — 323

3 Pyroxene Minerals — 326
  Structures — 327
  Enstatite Polymorphism — 328
  Orthopyroxenes — 330
  Clinopyroxenes — 331

4 Pyroxenoids — 333
  Wollastonite — 333
    Planar Defects in Wollastonite — 334
    Linear Defects and other Anomalous Features in Wollastonite — 335
  Other Pyroxenoids — 336

5 Amphiboles — 337
  Exsolution in Amphiboles — 339

6 Serpentine Materials — 340
  Chrysotile — 340
  Antigorite — 341
  Lizardite — 342
  Polygonal Serpentine — 342
  Garnierites — 342

7 Silicates based on the Mica Type of Layer — 343
  Mica Group — 343
  Chlorite Group — 346
  Zussmanite — 347
  Stilpnomelane — 348

8 Orthosilicates 349
Chloritoid 349
Mullite 352

9 Graphite and its Intercalates 353

10 Sulphides 354
$4C$ Pyrrhotite 354
Intermediate (nC) Pyrrhotite 355
nA Pyrrhotite 356
Smythite, Fe_3S_4 356
Troilite, FeS 356

11 Fluorocarbonates 357

Author Index 359

1
Microscopic Approaches to Physisorption: Theoretical and Experimental Aspects

BY U. LANDMAN AND G. G. KLEIMAN

1 Introduction

The writing of scientific reviews can be regarded as a service to the scientific community as well as a beneficial, though time-consuming, undertaking for the authors. There can be several purposes of such a review. First, it can be viewed as a collective reference source to aid workers in the field being reviewed and in related subjects in obtaining information about current developments. Secondly, it can be used as a means of forming a coherent unified approach to a certain set of problems, thus bringing together various aspects of the subject under study. Thirdly, it can be constructed such as to introduce the field to workers in other areas and hence serve as a cross-fertilization agent. We have here attempted the impossible, namely, to achieve all three purposes.

The study of adsorption phenomena is one of the oldest branches of physical chemistry (for a comprehensive review of early literature see ref. 1), starting with the observations by Scheele (1773) and Fontana (1777) of the uptake of gases by charcoal, and the discovery by Lowitz (1785) of the discoloration of solutions by charcoal. The distinction between regimes of adsorption phenomena, namely physical and chemical adsorption, was recognized early on and was often associated with the conditions leading to one adsorption class or the other. Among the names used to describe physical adsorption are van der Waals adsorption, low-temperature adsorption (*versus* high-temperature adsorption), secondary adsorption (*versus* primary adsorption), and capillary condensation, implying that capillary forces are responsible for physical adsorption. More modern definitions [2,3] are based on the theory of chemical binding, identifying physisorption as the state of interaction between an atom or molecule and a surface, where no chemical bonds are formed by charge rearrangement or sharing of electrons. Related to the above is the distinction of classes of adsorption according to the magnitude of the interaction energy (electron volts for chemisorption *versus* fractions of electron volts for physisorption). Modern spectroscopic techniques, especially ultraviolet photoemission spectroscopy (UPS), provide new diagnostic methods for distinguishing between the above adsorption regimes, as discussed in Section 3 (p. 34).

The traditional, and until quite recently the only, methods used for the study

---

[1] S. Brunauer, 'The Adsorption of Gases and Vapors; Volume I, Physical Adsorption', Princeton University Press, Princeton, 1945.
[2] D. M. Young and A. D. Crowell, 'Physical Adsorption of Gases', Butterworths, Washington, 1962.
[3] G. C. Bond, 'Catalysis by Metals', Academic Press, London, 1962.

of adsorption phenomena were thermodynamical in nature. In this respect, theoretical efforts towards microscopic models of the interaction preceded the development and application of microscopic experimental tools. The main microscopic theoretical approaches are reviewed in Section 2. These studies are complemented by statistical mechanical treatments of the interaction of gases with surfaces, which have been reviewed extensively recently.[1-4] Section 3 covers microscopic experimental approaches for the study of physisorption. In order to satisfy our 'third review criterion', the discussion of the results obtained by each of the experimental techniques is preceded by a brief exposition of the physical principles underlying the methods and the conventions and terminology employed. We have attempted throughout to review the methodology and results of recent studies and to indicate the link between the data and microscopic theoretical models.

Section 2 contains three main parts in which we discuss the general theory of van der Waals forces and physical adsorption (p. 5), semi-empirical calculations (p. 19), and the underlying principles and results of microscopic theories of physisorption (p. 21). In the last we have concentrated mainly on a discussion of our own theoretical studies.

Section 3 is divided into four main parts. In the first (p. 33) a classification of the methods is followed by a discussion of the use of u.v. photoemission in establishing a spectroscopic criterion for the definition of a physisorption system. Experimental methods for the study of atomic arrangement (LEED, neutron scattering, molecular beam scattering) are discussed and the results of recent studies demonstrated in the second part (p. 39). In addition the use of the above techniques for investigations of dynamical properties of physisorbed atoms (p. 61) and the atom–surface interaction potential (p. 77) are discussed. In the third subsection (p. 81) experimental techniques for the study of electronic structure are discussed (UPS, XPS, and field emission techniques). The review of the results is accompanied by theoretical arguments. Finally, we discuss the application of electron spectroscopy methods (LEED and Auger) for the measurement of adsorption isotherms (p. 97).

We conclude the review, in Section 4, with a brief summary and prognosis.

2 Theoretical Approaches for the Study of Physisorption

Introduction.—The basic theoretical problem in adsorption studies is the calculation of the wavefunctions of molecules which are interacting with a solid surface as well as with each other. The procedure which offers the greatest opportunity for insight into the physics of the process is that of first calculating the potential energy of a *single* molecule interacting with the solid surface and then determining its quantum and statistical mechanical properties in this potential field. The larger problem of interaction among adsorbed molecules is much more complicated and much less progress has been made. In general, it is obvious that the potential energy of interaction is fundamental.

We are concerned with the calculation of the potential energy $W(\vec{r})$ of a neutral molecule at position $\vec{r}$ which is interacting with a solid surface through forces

---

[4] W. A. Steele, 'The Interaction of Gases With Solid Surfaces', Pergamon Press, Oxford, 1974. Many references to current theoretical studies can be found in this publication.

associated with *physical adsorption*. The definition of these forces as distinct from those involved in *chemical adsorption* (which are caused by chemical interactions) is vague.[5] It is known, however, that a neutral molecule *far* from a solid surface interacts through attractive van der Waals, or dispersion, forces. Dispersion forces may be described as *physical* in that they involve no chemical interactions such as charge transfer or rearrangement. From a theoretical viewpoint, therefore, it is natural to define the forces involved in physical adsorption as physical, producing no significant changes in electronic densities. This information provides the basis of the conventional [5] picture of physical adsorption. Indeed, physical adsorption is usually described [5] as arising mainly from van der Waals forces, even though it is not clear that the many-body forces which are responsible for attraction assume the character of dispersion forces at the relatively small distance from the surface corresponding to binding.

In analysing experiments, one must substitute operational definitions for these *a priori* ones. The most common is that binding energies associated with chemisorption are of the order of electron volts (*i.e.*, chemical binding energies) whereas those linked to physisorption are at least a factor of 10^3 smaller. Actually, there is no operational definition which clearly separates physical from chemical adsorption. For this reason, most experimental and theoretical physical adsorption studies involve rare gases, whose chemical inertness presumably precludes chemical binding. Even in this case, the distinction is not always clear. Large negative work-function variations have been observed [6,7] for noble gas adsorption on transition-metal surfaces. These indicate a strong interaction energy, which may be correlated with the empty *d* levels in these metals [7,8] (see p. 90).

Despite these cautionary words, we take the point of view here that physical adsorption is a separate phenomenon from chemical adsorption, and that the definition given above in terms of physical forces is meaningful. This point of view is in accord with that of theoretical studies [8—54] of the potential energy, most [9—52] of which, in fact, assume that van der Waals forces are responsible for binding in physisorption.

The potential energy can be Fourier-analysed in components of momentum parallel to the surface, which we assume to be planar. In the case of a single crystal surface, we can write equation (1), where $\vec{g}$ represents a reciprocal lattice

$$W(\vec{r}) = U(z) + \sum_{\vec{g}}{}' U(\vec{g},z) \exp(i\vec{g} \cdot \vec{p}) \qquad (1)$$

vector of the surface Bravais lattice, $\vec{p}$ is the component of $\vec{r}$ parallel to the surface, z is the component perpendicular to the surface (positive z is defined outward from the solid), and the primed sum is taken over all $\vec{g} \neq 0$. We have thus decomposed W into terms dependent only on z and terms dependent upon the translational symmetry of the lattice. This decomposition serves as a convenient

---

[5] J. H. de Boer, *Adv. Catalysis Related Subjects*, 1954, **6**, 67; D. M. Young and A. D. Crowell, 'Physical Adsorption of Gases', Butterworths, London, 1962; A. D. Crowell, 'The Solid–Gas Interface', ed. E. A. Flood, Marcel Dekker, New York, 1967, Vol. 1.
[6] T. Engel and R. Gomer, *J. Chem. Phys.*, 1970, **52**, 5572.
[7] J. C. P. Mignolet, *J. Chem. Phys.*, 1963, **6**, 593.

means by which to classify theoretical treatments of W. Those studies [8,39—54] which involve actual calculations of the potential energy from a more or less fundamental basis neglect all $\bar{p}$-dependence and, therefore, correspond to the first term in equation (1). In addition, these 'first-principles' studies are devoted mainly to adsorption on metal surfaces. On the other hand, treatments which involve the crystal structure of the surface [9—38] [which include the summation in equation (1)]

[8] P. M. Gundry and F. C. Tompkins, *Trans. Faraday Soc.*, 1960, **56**, 846.
[9] R. M. Barrer, *Proc. Roy. Soc.*, 1937, **A161**, 476.
[10] W. J. C. Orr, *Trans. Faraday Soc.*, 1939, **35**, 1247.
[11] A. D. Crowell and D. M. Young, *Trans. Faraday Soc.*, 1953, **49**, 1080.
[12] W. A. Steele and G. D. Halsey, jun., *J. Phys. Chem.*, 1955, **59**, 57.
[13] J. G. Aston, R. J. Tykodi, and W. A. Steele, *J. Phys. Chem.*, 1955, **59**, 1053.
[14] A. D. Crowell, *J. Chem. Phys.*, 1957, **26**, 1407.
[15] E. L. Pace, *J. Chem. Phys.*, 1957, **27**, 1341.
[16] A. V. Kiselev, 'Proceedings of the Second International Congress of Surface Activity', Butterworths, London, 1957, Vol. 2, p. 168.
[17] N. N. Avgul, A. A. Isirikyan, A. V. Kiselev, I. A. Lygina, and D. P. Poshkus, *Izvest. Akad. Nauk S.S.S.R., Otdel. Khim. Nauk.*, 1957, 1314.
[18] R. A. Pierotti and G. D. Halsey, jun., *J. Phys. Chem.*, 1959, **63**, 680.
[19] N. N. Avgul, A. V. Kiselev, I. A. Lygina, and D. P. Poshkus, *Bull. Acad. Sci. U.S.S.R., Div. Chem. Sci.* (*English Trans.*), 1959, 1155.
[20] W. A. Steele and R. Karl, jun., *J. Colloid Interface Sci.*, 1960, **28**, 397.
[21] E. L. Pace, 'Proceedings of the Fourth Conference on Carbon', Pergamon Press, New York, 1960, p. 35.
[22] J. R. Sams, jun., G. Constabaris, and G. D. Halsey, jun., *J. Phys. Chem.*, 1960, **64**, 1689.
[23] A. D. Crowell and R. B. Steele, *Phys. Rev.*, 1961, **34**, 1347.
[24] M. Ross and W. A. Steele, *J. Chem. Phys.*, 1961, **35**, 862; W. A. Steele and M. Ross, *J. Chem. Phys.*, 1961, **35**, 850.
[25] F. O. Goodman, *Phys. Rev.*, 1967, **164**, 1113.
[26] H. E. Neustadter and R. J. Bacigalupi, *Surface Sci.*, 1967, **6**, 246.
[27] F. Ricca, C. Pisani, and E. Garrone, *J. Chem. Phys.*, 1969, **51**, 4079.
[28] G. Ehrlich, H. Heyne, and C. F. Kirk, 'The Structure and Chemistry of Surfaces', John Wiley and Sons, New York, 1969, p. 49.
[29] A. Tsuchida, *Surface Sci.*, 1969, **14**, 375.
[30] A. D. Novaco and F. J. Milford, *J. Low Temp. Phys.*, 1970, **3**, 307.
[31] F. Ricca, C. Pisani, and E. Garrone, 'Adsorption-Desorption Phenomena', Academic, London, 1972, p. 111.
[32] D. E. Hagen, A. D. Novaco, and F. J. Milford, 'Adsorption-Desorption Phenomena', Academic, London, 1972, p. 99.
[33] A. D. Novaco and F. J. Milford, *Phys. Rev.*, 1972, **A5**, 783.
[34] F. O. Goodman, *J. Chem. Phys.*, 1972, **56**, 4899.
[35] W. A. Steele, *Surface Sci.*, 1973, **36**, 317.
[36] R. O. Davies and L. S. Ullermayer, *Surface Sci.*, 1973, **39**, 61.
[37] R. Dovesi, C. Pisani, and F. Ricca, *Trans. Faraday Soc.*, 1973, **69**, 79.
[38] C. Pisani, F. Ricca, and C. Roetti, *J. Phys. Chem.*, 1973, **77**, 657.
[39] J. E. Lennard-Jones, *Trans. Faraday Soc.*, 1932, **28**, 333.
[40] J. Bardeen, *Phys. Rev.*, 1940, **58**, 727.
[41] H. Margenau and W. G. Pollard, *Phys. Rev.*, 1941, **60**, 128.
[42] E. S. R. Prosen and R. G. Sachs, *Phys. Rev.*, 1942, **61**, 65.
[43] C. Mavroyannis, *Mol. Phys.*, 1963, **6**, 593.
[44] A. D. McLachlan, *Mol. Phys.*, 1964, **7**, 381.
[45] K. F. Wojciechowski, *Acta Phys. Polon.*, 1966, **29**, 119; 1968, **33**, 363.
[46] G. G. Kleiman and U. Landman, *Phys. Rev. Letters*, 1973, **31**, 707.
[47] G. G. Kleiman and U. Landman, *Phys. Rev.*, 1973, **B8**, 5484.
[48] G. G. Kleiman and U. Landman, *Phys. Rev. Letters*, 1974, **33**, 524.
[49] U. Landman and G. G. Kleiman, *J. Vacuum Sci. Technol.*, 1975, **12**, 206.
[50] G. G. Kleiman and U. Landman, *Solid State Comm.*, 1976, **18**, 819.
[51] E. Zaremba and W. Kohn, *Phys. Rev.*, 1976, **13**, 2270.
[52] W. G. Pollard, *Phys. Rev.*, 1941, **60**, 578.
[53] D. L. Freeman, *J. Chem. Phys.*, 1975, **62**, 941.
[54] J. T. Yates, jun. and T. E. Madey, *Surface Sci.*, 1971, **28**, 437.

are of a semi-empirical nature: that is, the potential energy is calculated by summing pairwise interaction energies between the adsorbate molecule and the constituent atoms of the solid. Since this type of treatment neglects the collective nature of interactions in condensed matter, these semi-empirical calculations are generally applied to interactions with insulators where the locality of electronic orbitals minimizes collective effects.

van der Waals Forces and Physical Adsorption.—The general theory of van der Waals forces has been exhaustively reviewed recently.[55,56] This discussion, therefore, is intended to provide an insight into the physical mechanisms responsible for the phenomenon and into the calculational methods most commonly employed. We start with a treatment of the van der Waals interaction between two spherically symmetric atoms originated by London,[57] which has become a standard text-book example of the use of second order perturbation theory (the early history of van der Waals studies is well reviewed in ref. 58).

Suppose that we have two atoms in their ground states, labelled by subscripts 1 and 2, which are separated by a displacement $\vec{R}$ so great that the electronic clouds of the atoms do not interpenetrate; the normal condition for dispersion interactions is that $R \gg a$, where a characterizes atomic separations (in a molecule). The dipole–dipole interaction potential, V, between the atoms is given by equations (2) and (3). The quantity $\vec{r}_{ik}$ denotes the position of the kth electron of

$$V = \frac{e^2}{R^3}\left[\vec{r}_1 \cdot \vec{r}_2 - \frac{3(\vec{r}_1 \cdot \vec{R})(\vec{r}_2 \cdot \vec{R})}{R^2}\right] \quad (2)$$

$$\vec{r}_i \equiv \sum_k \vec{r}_{ik} \quad (3)$$

the ith atom. The Hamiltonian of the isolated system, $H_0 = H_1 + H_2$, is given as the sum of the individual atom Hamiltonians, such that, in Dirac notation, equations (4)—(7) can be written. The operator $\times$ in equation (5) denotes a direct

$$H_0|1n,2m\rangle = E_{nm}|1n,2m\rangle \quad (4)$$

$$|1n,2m\rangle \equiv |1n\rangle \times |2m\rangle \quad (5)$$

$$H_1|1n\rangle = E_{1n}|1n\rangle, \; H_2|2m\rangle = E_{2m}|2m\rangle \quad (6)$$

$$E_{nm} \equiv E_{1n} + E_{2m} \quad (7)$$

product of the individual atom manifolds.

Because of reflection symmetry, the first-order energy correction, $\langle 10,20|V|10,20\rangle$ vanishes. In second order, the energy correction W_2 is given by equation (8), and

$$W_2(R) = \langle 10,20|V|\delta\psi\rangle \quad (8)$$

---

[55] D. Langbein, 'Theory of Van der Waals Attraction', Springer-Verlag, Berlin, Heidelberg, New York, 1974.
[56] Yu. S. Barash and V. L. Ginzburg, *Sov. Phys.-Usp.*, 1975, **18**, 305.
[57] F. London, *Z. Phys. Chem.*, 1930, **B11**, 222.
[58] H. Margenau, *Rev. Mod. Phys.*, 1939, **11**, 1.

the first-order eigenstate correction, $|\delta\psi\rangle$, is given by equation (9). The final

$$|\delta\psi\rangle = \sum_{n,m \neq 0} \frac{|1n,2m\rangle\langle 1n,2m|V|10,20\rangle}{E_{00} - E_{nm}} \qquad (9)$$

result, equations (10) and (11), is in terms of the oscillator strengths. The quantity M

$$W_2(R) = \frac{3}{2} \frac{e^4}{R^6} \frac{\hbar^4}{M^2} \sum_{n,m \neq 0} \frac{f_{1n} f_{2m}}{(E_{10} - E_{1n})(E_{20} - E_{2m})(E_{00} - E_{nm})} \qquad (10)$$

$$f_{in} \equiv \frac{2}{3} \frac{M}{\hbar^2} |\langle in|\vec{r}_i|i0\rangle|^2 (E_{in} - E_{i0}) \qquad (11)$$

is the free electron mass and i represents either 1 or 2 in equation (11). The appearance of oscillator strengths, f, in these equations for the interaction energy resulting from van der Waals forces manifests the intimate relation between the optical properties of matter and dispersion forces (hence the name [57,58]): classically, the oscillator strength represents the number of electrons participating in an optical transition. Furthermore, we can see from equations (8) and (9) that the characteristic R^{-6} dependence of van der Waals interaction energies in equation (10) arises from a combination of two factors: the R^{-3} dependence of the classical dipole–dipole interaction, V, and the R^{-3} dependence of the dipole induced in one atom by the other. In other words, each atom is momentarily a dipole which induces a dipole in the other atom [corresponding to the state correction $|\delta\psi\rangle$ in equation (9)] which has an R^{-3} dependence, and these two dipoles interact through V. Therefore, from this simple example, we can see that van der Waals interactions arise from *quantum fluctuations*. Later, we shall see that the phenomenon of van der Waals forces can be attributed to quantum fluctuations of the electromagnetic field:[59] this is the reason for the appearance of oscillator strengths in equation (10).

The first treatment of the van der Waals interaction between a molecule and a solid surface was that of Lennard-Jones,[39] who considered atom–metal interactions. He assumed that the metal corresponds to a perfect conductor. Consequently, he considered the interaction between the molecule and its image in the metal. The potential energy of a molecule with Z electrons whose centre is located at a distance l from the surface, resulting from this interaction is given by equations (12) and (13). The quantity $\vec{r}_i$ denotes the position of the ith electron relative to the

$$V = \frac{-e^2}{16l^3} (r^2 + z^2) \qquad (12)$$

$$\vec{r} \equiv \sum_{i=1}^{Z} \vec{r}_i ; z \equiv \sum_{i=1}^{Z} z_i \qquad (13)$$

centre of the molecule (z_i is the component of $\vec{r}_i$ in the direction perpendicular to the surface, where positive z is defined outward from the surface). In Lennard-Jones' treatment, it is assumed that all of the electrons are in spherical orbits, so that $\langle z^2 \rangle = \langle r^2 \rangle/3$, where the bracket signifies quantum mechanical averaging. The

---

[59] E. M. Lifschitz, *Sov. Phys.-JETP*, 1956, **2**, 73.

result for the van der Waals interaction energy is given in equation (14). This

$$W_{\text{L-J}}(l) = \frac{-e^2}{12l^3} \langle r^2 \rangle \qquad (14)$$

expression represents the attractive energy of interaction between a molecule and a surface. As in the other fundamental calculations discussed in this section,[40-45] all short-range repulsive interactions are neglected. Therefore, it is important to note that these theories cannot provide information about the binding of molecules without some extra assumption about the equilibrium position at which binding takes place.

The result in equation (14) is the first to show that the van der Waals interaction between a molecule and surface has an l^{-3} dependence. In later work, however, Bardeen[40] demonstrated that the image method provides a lower limit for the true interaction energy, which is negative. In deriving his result, Lennard-Jones assumed that the metal responds instantaneously to the charge density fluctuations in the molecule. Bardeen, on the other hand, pointed out that corrections arising from the finite kinetic energy of the metallic electrons reduce the magnitude of the binding energy. He considered two systems, A and B, connected by a perturbation, V. Defining W_A as the change of the energy of system A caused by V when the electrons of B are fixed (with a similar definition for W_B), he was able to show, employing second-order perturbation theory, that $|W|$, the magnitude of the exact interaction energy, is smaller than either $|W_A|$ or $|W_B|$ and is closer to the smaller of the two. He derived the approximate expression (15). In applying this

$$W \simeq W_A W_B / (W_A + W_B) \qquad (15)$$

formula to the interaction of a continuum metal (*i.e.*, system A) and a one-electron atom (*i.e.*, system B), it follows that W_A is given by equation (14). W_B was calculated to second order by assuming that the metallic electrons obey Fermi–Dirac statistics and by accounting for their mutual repulsion in an approximate way. Application of this method yields,[40] from equation (15), the expressions (16) and (17)

$$W(l) \simeq W_{\text{L-J}}(l) F_A / (1 + F_A) \qquad (16)$$

$$F_A \equiv Ce^2 / 2r_s \Delta_A \qquad (17)$$

The factor C in equation (17) results from the approximate treatment of the electronic repulsion; for the specific approximation made,[40] $C = 2.6$ and it is expected[40] that C varies slowly with r_s at ordinary densities (*i.e.*, r_s is the radius of a sphere of a volume sufficient to contain one electron). The quantity Δ_A is an average excitation energy of the atom. Bardeen was able to show that the Lennard-Jones image force result[39] applies only for $r_s \to 0$ (high densities). In other words, the quantity F_A in equation (17) represents the correction produced by the finite kinetic energy of the metallic electrons.

In a later study, Margenau and Pollard[41] also recognized the necessity for accounting for the non-instantaneous response of the metal electrons. Their procedure involved applying the London formula, equation (10), to the interaction between an atom and an infinitesimal element of the metal (which they also treated as a continuum) so that one of the oscillator strengths appearing in equation (10) is characteristic of the metal. They integrated the resulting expression over the metallic half-space in order to derive the total van der Waals interaction energy. In their final expression, they represented the response of the atom by a single transition to level 1, say, and assumed that only transitions involving small energy differences are important for the metal. Their equation is given by the expressions (18) and (19), in which $i = $ a or m, where a and m refer to the atom and metal,

$$W(l) = \frac{-1}{8l^3} \left\{ \frac{e^2 \hbar}{M} a_m(\nu_1) \frac{f_{a1}}{\nu_1} + \bar{E}_m a_a(0) \right\} \tag{18}$$

$$a_i(\nu) \equiv \frac{e^2 \hbar^2}{M} \sum_{n \neq 0} \frac{f_{in}}{(E_{i0} - E_{in})^2 - (h\nu)^2} \tag{19}$$

respectively. The quantity $\bar{E}_m$ is an average excitation energy of the metal, $\nu_1 \equiv (E_{a1} - E_{a0})/h$, the dynamic polarizability, is given by equation (19), and a_m is the metallic polarizability per unit volume.

Another calculation, by Prosen and Sachs,[42] was based on the observation that, in calculations in which a metal is considered to be composed of small elements,[41] there are two restrictions which determine the range of validity of the results. First, multipole–multipole interactions are unimportant only so long as the dimensions of the interacting bodies are much smaller than the separation between them. On the other hand, polarizabilities are well defined only within the context of a macroscopic volume (whose dimensions are much greater than a lattice constant). These two restrictions imply that the Margenau–Pollard[41] formula, equations (18) and (19), applies only for l much greater than a lattice constant. Since Margenau and Pollard show that their and Bardeen's results are related, the same criticism presumably holds for Bardeen's work.

Instead of characterizing the metal by a polarizability defined at each point, Prosen and Sachs[42] described it by a product of one-electron wavefunctions. They then calculated the interaction energy for distances where the metallic and molecular wavefunctions do not overlap appreciably so that they could use second-order perturbation theory. Their result[42] (where they neglect high-energy metallic transitions) for a non-degenerate solid (applicable to lightly doped semiconductors and metals at high temperatures, perhaps) is given by equation (20), where ρ is the

$$W(l) = \frac{-a_a(0)}{2} \frac{\pi \rho e^2}{l} \tag{20}$$

electronic density of the solid. For degenerate materials (*e.g.* metals), the corresponding result is equation (21), where $k_F \equiv (3\pi^2 \rho)^{\frac{1}{3}}$ is the Fermi momentum.

$$W(l) = \frac{-a_a(0) e^2 \pi}{(2\pi)^3} k_F^2 \frac{\ln(2k_F l)}{l^2} \tag{21}$$

The results in equations (20) and (21) break down for distances larger than a Bohr radius because electron–electron interactions (*i.e.*, effects resulting from the many-body nature of the electron gas of the metal) become important at these distances. Furthermore, the perturbation calculation[42] was designed for regions of small metallic–molecular wavefunctions overlap; at very small l, therefore, the results in equations (20) and (21) are not valid. There is, then, at best, a narrow region of l where these results hold.[42]

These early efforts [39—42] preceded the development of the modern theory of many-body effects [60] and, therefore, were performed without the benefit of the powerful and rigorous techniques of that theory.[60] Although these were pioneer studies in the field, it is important to note that they were performed prior to the evolution of modern surface science ideas (for example the importance of self-consistency of electronic wavefunctions at a surface [61]) and hence employed model descriptions of the system [39,41] which are not in accord with more recent findings. In addition, some of these studies used such approximation methods, for the evaluation of the interaction energy, that the validity and accuracy of the results are limited. Moreover, they were suitable for interactions at small distances, and as such offered no possibility for explanation of long-range forces.

That interactions at large distances have a different character was first shown experimentally when it was found [62] that neutral molecules interacting at large distances do not obey an R^{-6} interaction energy law.[57] It was suggested [62] that the London analysis [57] neglects the finite nature of the speed of light and, therefore, ignores retardation effects in the propagation of the electromagnetic field from one particle to the other. This led Casimir and Polder [63] to compute the force between two neutral polarizable particles by perturbation theory in quantum electrodynamics, and for large separations they indeed found a fall-off in the interaction energy as R^{-7}. From the simple mathematical structure of the results of perturbation theory, Casimir [64] deduced that the interaction can be understood as arising from the change in quantum mechanical zero point energy as the particles are brought together. This understanding serves as the basis of a calculational technique [65] which has been commonly used [55,56,65] and which we will discuss later.

The aforementioned difficulties with early calculations of the attractive interaction energy between a neutral, polarizable particle and a solid, as well as the contemporary development of many-body theory, led Lifschitz [59] to treat the problem within the context of the classical electrodynamics [66] of fluctuating fields.[67] Instead of considering the quantum fluctuations of electronic charge densities as in the derivation of the London [57] formula, Lifschitz observed that it is equally valid to consider only long-wavelength fluctuations of the electromagnetic field,[66,67] and describe the matter in the system by response functions [66] which are properties of the material bodies (*i.e.*, dielectric functions). Fluctuations in the position and motion of the charges in matter produce spontaneous electronic and magnetic moments. If we consider only local spatial coupling of the electric fields, then the

---

[60] A. A. Abrikosov, L. P. Gor'kov, and I. E. Dzyaloshinskii, 'Methods of Quantum Field Theory in Statistical Physics', Prentice-Hall, Englewood Cliffs, 1963.
[61] N. D. Lang and W. Kohn, *Phys. Rev.*, 1970, **B1**, 4555.
[62] E. J. W. Verwey and J. Th. G. Overbeck, 'Theory of Stability of Tycophobic Colloids', Elsevier, Amsterdam, 1948.
[63] H. B. G. Casimir and D. Polder, *Phys. Rev.*, 1948, **73**, 360.
[64] H. B. G. Casimir, *Proc. Koninkl. Ned. Akad. Wetenschap*, 1948, **51**, 793.
[65] N. G. Van Kampen, B. R. A. Nijboer, and K. Schram, *Phys. Letters*, 1968, **A26**, 307.
[66] L. D. Landau and E. M. Lifschitz, 'Electrodynamics of Continuous Media', Pergamon, Oxford, 1960, p. 361.
[67] L. D. Landau and E. M. Lifschitz, 'Statical Physics', Pergamon, Oxford, 1958, Sec. 123.

electric displacement is given by [59] equation (22), where we ignore anisotropy and

$$\vec{D} = \epsilon \vec{E} + \vec{K} \tag{22}$$

where $\vec{K}/4\pi$ corresponds to the spontaneous electric moment per unit volume; strictly speaking, all the fields should be regarded as quantum mechanical operators,[66] but the results of fluctuation theory [67] are unaffected [66] by the semi-classical treatment and we shall regard the fields as classical, for simplicity. More generally, of course, equation (22) should include both anisotropy and magnetic moments.

Lifschitz's procedure is to calculate the long-wavelength $\vec{E}$ and $\vec{H}$ fields from Maxwell's equations subject to appropriate boundary conditions so that these fields are linear functionals of $\vec{K}$.[66] He applied the technique to the calculation of the force per unit area between two planar, semi-infinite, continuous solid media separated by a third medium of thickness l. From classical electromagnetism [68] the total force on an object is given by equations (23) and (24), in which the quantity

$$\frac{d\vec{P}}{dt} = \int_S \overleftrightarrow{T} \cdot \vec{n} \, dS \tag{23}$$

$$\overleftrightarrow{T} \equiv \frac{1}{4\pi} [\vec{E}\vec{E} + \vec{B}\vec{B} - \overleftrightarrow{I}(E^2 + B^2)] \tag{24}$$

$\vec{P}$ corresponds to the total (*i.e.*, mechanical plus electromagnetic) momentum, t denotes time, $\overleftrightarrow{T}$ is the Maxwell stress tensor (double-headed arrows denote tensors), S represents a surface surrounding the object ($\vec{n}$ is the surface normal) and $\overleftrightarrow{I}$ stands for the unit tensor. In Lifschitz's problem, the surface of integration is taken just inside the front and back surfaces of one of the media. The force per unit area is then identified with $\langle T_{zz} \rangle$, where the bracket denotes averaging over a statistical ensemble.

The characteristic feature of fluctuation theory appears in the manner in which the fluctuations are treated. In a purely formal manner, it is equivalent to treat the fields $\vec{K}$ as external fields [66,67] which Lifschitz characterized by purely random statistical fluctuations. That is to say, the correlation function in a medium, of the spatial component of frequency ω of $\vec{K}$, is represented by

$$\langle K_i(\vec{r},\omega) K_j(\vec{r}',\omega) \rangle = 2\hbar \mathrm{Im}\, \epsilon(\omega)\, \delta_{ij} \delta(\vec{r} - \vec{r}') \coth\left(\frac{\hbar\omega}{2k_B T}\right) \tag{25}$$

where ϵ is the momentum-independent dielectric permittivity of the medium (Im denotes the imaginary part), k_B is Boltzmann's constant, T is the temperature, δ_{ij} is the Kronecker delta and $\delta(\vec{r})$ is the Dirac delta function.

The properties of the medium are incorporated in ϵ in equation (25). Utilizing the linear relation between $\vec{K}$ and the $\vec{E}$ and $\vec{H}$ fields resulting from Maxwell's equations, and inserting the result of equation (25) into $\langle T_{zz} \rangle$ in equation (24) one obtains, by means of this relation, the expression for the force per unit area, F.

---

[68] J. D. Jackson, 'Classical Electrodynamics', Wiley, New York, 1962.

The general expressions are complicated and have been discussed in detail.[59,60,66,69] We are interested here in applications to physisorption and, except for commenting that Lifschitz's results agree with all previous rigorous calculations (such as those of London[57] and Casimir and Polder[63]), we restrict ourselves to discussion of these applications.

It is of interest to examine the relation between the particle interaction energy, W, and the force F. Assuming that one of the media is characterized by constant particle density, N, the total interaction energy per unit area, U, of the interacting media, as well as F, are functionals of the separation, l, and N. Thus, in the limit of a rarefied medium (*i.e.*, $N \to 0$), we derive equation (26), also expressed as equation (27) where the convention that an attractive force is negative, is used.

$$U(l,N) = \int_l^\infty dz F(z,N) = \int_l^\infty NW(z)dz \tag{26}$$

$$W(l) = \lim_{N \to 0} F(l,N)/N \tag{27}$$

The limit is taken in equation (27) because F depends, in general, upon the collective nature of the media. The force at small distances between a rarefied gas of molecules (subscript M) and a solid medium (subscript s) is given by equation (28),

$$F(l,N) = \frac{-\hbar}{8\pi^2 l^3} \int_0^\infty d\xi \frac{[\epsilon_M(i\xi) - 1][\epsilon_s(i\xi) - 1]}{[\epsilon_M(i\xi) + 1][\epsilon_s(i\xi) + 1]} \tag{28}$$

which is independent of temperature.[59,69] Since the dielectric permittivity of a gas is given by $\epsilon_M = 1 + 4\pi N a_M$, where a_M is the dynamic molecular polarizability, we can derive equations (29) and (30) for the attractive interaction energy between a molecule and a solid. In these equations the l^{-3} dependence which is characteristic

$$W(l) = -C_{VW}/l^3 \tag{29}$$

$$C_{VW} \equiv \frac{\hbar}{4\pi} \int_0^\infty d\xi a_M(i\xi) \frac{[\epsilon_s(i\xi) - 1]}{[\epsilon_s(i\xi) + 1]} \tag{30}$$

of media in which electromagnetic coupling is local is displayed. Figure 1 illustrates the dependence of the van der Waals energy of an He atom interacting with a metal upon the parameters characterizing the metal and the atom:[46,47] here the metal is described by $\epsilon_s(\omega) = 1 - \omega_p^2/\omega^2$, where ω_p is the plasma frequency of the (free electron) metal [*i.e.*, $\omega_p \equiv (4\pi n e^2/m)^{\frac{1}{2}}$, where n is the metallic electron density and e and m are the free electron charge and mass, respectively]. Figure 1 clearly demonstrates that an accurate representation of the atomic properties is *essential* for an adequate calculation of the van der Waals energy. Curves (a), (b), and (c) correspond, respectively, to a_{He} calculated from equation (19) with only the first transition, the first 12 transitions,[70] and the first 12 transitions plus the continuum

---

[69] I. E. Dzyaloshinskii, E. M. Lifschitz, and L. P. Pitaevskii, *Adv. Phys.*, 1961, **10**, 165; *JETP*, 1958, **37**, 229; I. E. Dzyaloshinskii and L. P. Pitaevskii, *JETP*, 1959, **36**, 1797.

[70] W. L. Wiese, M. W. Smith, and B. M. Glenon, *Nat. Bur. Stand. Ref. Data Series*, NBS4, 1966, 1.

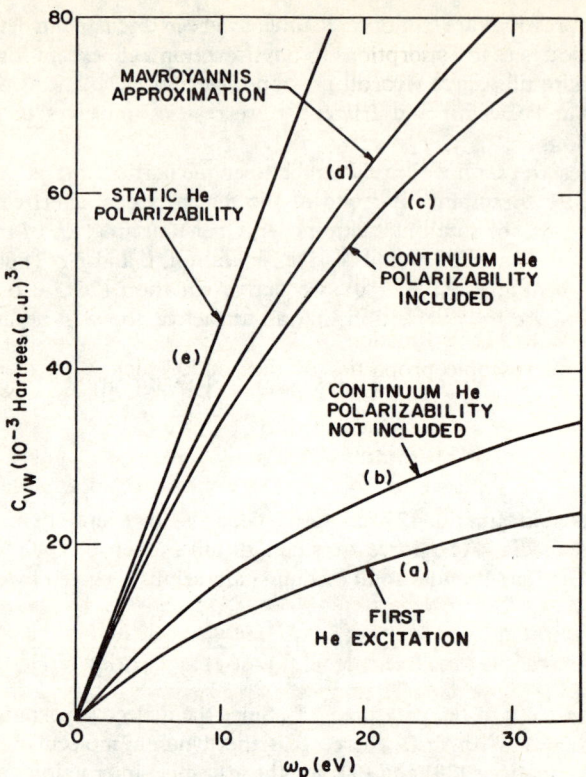

Figure 1 Helium van der Waals constant C_{VW} in equation (30) as a function of plasma frequency ω_p. C_{VW} increases monotonically with ω_p for curves (a)—(e). Five models are considered: (a) only the first excitation is included; (b) twelve discrete ground-state excitations; (c) twelve discrete ground-state excitation and transitions to the continuum; (d) one excitation with oscillator strength of 2 and energy $\sqrt{2/\alpha(0)}$; (e) static polarizability alone. The importance of an adequate description of the frequency response of the atom is illustrated
(Reproduced by permission from *Phys. Rev.*, 1973, **B8**, 5484)

contribution [47,71] of the He atom. The curve labelled (d) represents the approximation [43] in which the polarizability of He is modelled by one level with an oscillator strength of Z (*i.e.*, the total He oscillator strength) with a corresponding energy equal to $\sqrt{2/\alpha_{He}(0)}$. Finally, curve (e) results from replacing $\alpha_M(i\xi)$ by the static polarizability, $\alpha_{He}(0)$, in equation (30). This represents an upper bound upon C_{VW}. Evidently, a major contribution to C_{VW} originates in the continuum levels. Significant errors occur, therefore, when crude approximations of the atomic properties (*i.e.*, polarizability) are used in evaluating the van der Waals energy. In addition, we note a systematic increase in attraction correlated with the plasma frequency of the metal.

[71] S. S. Huang, *Astrophys. J.*, 1948, **108**, 354.

Following these results of Lifschitz,[59] calculations were performed, one of which [43] used the S-matrix method of quantum electrodynamics and in the other [44] the fluctuation–dissipation theorem was employed.[72] In these calculations the method of images was used in order to incorporate the microscopic nature of the dielectric. The results of these calculations agree with those of Lifschitz.[59]

It may seem strange that such methods as either inserting random *external* fluctuations into classical electromagnetic equations or combining the method of images with rigorous quantum statistical mechanical techniques [43,44] are capable of describing the microscopic nature of the interacting metal–dielectric system correctly. In rigorous quantum-field theoretical treatments,[60,69] however, it was shown that the photon temperature Green's function is simply connected to the ordinary (retarded) Green's function of the electromagnetic fields. The former is related to the microscopic properties of the solid, which, in principle, may be calculated to arbitrary accuracy [60,69] (although, in practice, the situation is less pleasant). This relationship is provided through the spatially dependent dielectric function, or polarization operator.[60] Determination of the photon temperature Green's function is reduced to solving Maxwell's equations with appropriate boundary conditions. In this way, the connection between quantum field theory and classical electromagnetism (*e.g.*, the connection between dispersion forces and optical properties) becomes clear.

Lifschitz's results [59] were rederived [69] within this rigorous formalism by using a somewhat more complicated Maxwell stress tensor calculation than that of Lifschitz (direct calculation of the total free energy was too complicated [69]). Besides justifying Lifschitz's theory,[59] these modern quantum field theoretical calculations [69] enabled the theory to be extended to bodies separated by a liquid layer and to the study of the properties of liquid films. An important point [59,69] is that a solid body cannot be treated as a sum of atoms because of modification of electronic properties by juxtaposing atoms in a solid.

These studies, therefore, showed how to apply the rigorous field theoretic techniques of modern many-body theory [60] to the calculation of dispersion forces. The methods of calculation,[59,69] however, do not lend themselves readily to extension to more complicated cases (*e.g.*, either more complicated geometries or momentum dependent dielectric functions). In the physical basis of these theories, however, is the germ of a technique [65] which is much simpler than the stress-tensor methods, although, perhaps, less rigorously justifiable.[56]

In these field theoretic treatments,[59,69] the interaction is viewed as occurring through the medium of the fluctuating electromagnetic field which is *always* present in any absorbing medium (because of thermodynamic fluctuations) and which extends throughout all space. Even at the absolute zero temperature, the field does not vanish – it is associated here with the zero point vibrations of the radiation field.[59,69] It is this physical reasoning which indicates the connection [73] between quantum field theoretical calculations [59,69] of dispersion forces and Casimir's idea [64] of calculating the forces from summing over the normal modes of the electromagnetic field.[65] The technique is most easily illustrated by considering [56] two planar

---

[72] H. B. Callen and T. Welton, *Phys. Rev.*, 1951, **83**, 34.
[73] T. H. Boyer, *Ann. Phys.*, 1970, **56**, 474.

transparent media (indices 1 and 3) separated by a third one (index 2) of width l. Solution of Maxwell's equations yields the dispersion equations (31) and (32) for

$$\Delta(\vec{k}_\|,\Omega_i) \equiv 1 - h(\Omega_i)e^{-2k_\| l} = 0 \tag{31}$$

$$h(\omega) \equiv \frac{[\epsilon_2(\omega) - \epsilon_3(\omega)][\epsilon_2(\omega) - \epsilon_1(\omega)]}{[\epsilon_2(\omega) + \epsilon_3(\omega)][\epsilon_2(\omega) + \epsilon_1(\omega)]} \tag{32}$$

the l-dependent eigenfrequencies, Ω, of the normal modes of the field, in the limit $c \to \infty$ (*i.e.*, neglect of retardation). We display $\vec{k}_\|$, the component of momentum parallel to the surfaces, explicitly and distinguish the different modes by subscript i. Since we are considering transparent media in this example (*i.e.*, Im$\epsilon_j = 0$ for $j = 1,2,3$) all of the Ω_i are real, from equation (31). The resulting l-dependent total free energy of the system (each mode is a harmonic oscillator in the normal co-ordinates of the field) is given by equations (33) and (34), from elementary

$$f(l) = \sum_i f_s\,[\Omega_i(l)] \tag{33}$$

$$f_s(\omega) \equiv k_B T \ln\left[2\sinh\left(\frac{\hbar\omega}{2k_B T}\right)\right] \tag{34}$$

statistical mechanics. In equation (34), f_s is the free energy of a harmonic oscillator of frequency ω, and the sum in equation (33) is over all positive Ω_i and $\vec{k}_\|$. From the theory of complex variables [65] equation (35) is derived, where g_1 is an

$$\frac{1}{2\pi i}\oint_C g_1(z)\,\frac{d}{dz}[\ln g_2(z)]\,dz = \sum_\alpha g_1(z_{\alpha 0}) - \sum_\beta g_1(z_{\beta p}) \tag{35}$$

analytic function in the complex region bounded by closed contour C and g_2 is a function with zeroes $z_{\alpha 0}$ and poles $z_{\beta p}$ in the same region. Applying relation (35) to the evaluation of f in equations (33) and (34), we identify g_1 with f_s, g_2 with Δ and the contour C with a semicircle in the right complex half-plane whose base is the imaginary axis and whose radius is infinite; in addition we note that the poles of Δ correspond to l-independent modes (*i.e.*, modes associated with the isolated media). Since the integral over the infinite semicircle vanishes, it is easy to integrate by parts along the imaginary axis [equations (36) and (37)]. The

$$f(l) = \sum_{\vec{k}_\|} \frac{1}{2\pi i}\oint_C dz\, f_s(z)\,\frac{d}{dz}\{\ln[\Delta(\vec{k}_\|,z)]\}$$

$$= \sum_{\vec{k}_\|} \frac{-1}{2\pi i}\int_{-i\infty}^{i\infty} dz\, f_s(z)\,\frac{d}{dz}\{\ln[\Delta(\vec{k}_\|,z)]\}$$

$$= \sum_{\vec{k}_\|} \frac{1}{2\pi i}\int_{-i\infty}^{i\infty} dz\,\frac{\hbar}{2}\ln[\Delta(\vec{k}_\|,z)]\coth\left(\frac{\beta\hbar z}{2}\right)$$

$$= \sum_{\vec{k}_\|} \frac{k_B T}{2} \sum_{n=-\infty}^{\infty} \ln[\Delta(\vec{k}_\|,i\xi_n)] \tag{36}$$

$$\xi_n \equiv 2\pi n\,\frac{k_B T}{\hbar} \tag{37}$$

quantities ξ_n (Matsubara indices [60]) are the poles of the coth term in equation (36) and n is an integer. The third line in equation (36) is derived from integration by parts. The force per unit area at $T = 0$ is derived by differentiating f with respect to l and letting $\sum_n \to \int_{-\infty}^{\infty} dn \to \frac{\hbar}{2\pi k_B T} \int_{-\infty}^{\infty} d\xi$ (i.e., $T \to 0$ implies $\xi_{n+1} - \xi_n = d\xi$), so that finally equation (38) emerges. This expression agrees exactly with the

$$F(l) = -\frac{\hbar}{2\pi} \sum_{k_\parallel} k_\parallel \int_{-\infty}^{\infty} d\xi \, [h^{-1}(i\xi)e^{2k_\parallel l} - 1]^{-1}$$

$$= -\frac{\hbar}{16\pi^2 l^3} \int_0^{\infty} d\xi \int_0^{\infty} x^2 dx \, [h^{-1}(i\xi)e^x - 1]^{-1} \tag{38}$$

corresponding one derived from quantum field theory.[59,69] In the general case, one must solve Maxwell's equations including retardation effects for transverse electric (TE) as well as transverse magnetic (TM) modes [i.e., equations (31) and (32)]. For this more general case, the formulas agree [74] with the exact results. These results explain the rather peculiar prescription for deriving equation (38) (i.e., we evaluate the free energy at $T \neq 0$, but calculate the force at $T = 0$). An exact calculation which includes retardation [69] shows that temperature is unimportant so long as $lk_B T/c\hbar \ll 1$, so that the small distance formula takes the form given in equation (28). But calculations without retardation (i.e., $c \to \infty$) are, therefore, independent of temperature: it is inconsistent to calculate retardationless forces at $T \neq 0$. On the other hand, we wished to illustrate the method [65] for calculating $f(l)$, which is independent of the exact form of Δ (i.e., with or without retardation). For these tutorial reasons, we chose the prescription we have used. Only at zero temperature is the *derivation* of equation (38) correct. Equation (36), for f for $T \neq 0$, is correct *only* if the exact form of Δ, including retardation, is inserted: following this insertion, of course, manipulations can be made in order to derive equation (38), for example, which is valid for small distances at all temperatures.

The derivation of equations (36) and (38) was made for the case of transparent media. In the more general case,[59,69] Im$\epsilon \neq 0$ and not all the eigenfrequencies Ω_i are real. It has been demonstrated,[56] however, that even though equations (33) and (34) lose their meaning in this case, the formula in equation (36) is *still valid* (provided, of course, that Δ includes retardation and both the TE and TM modes). This result [56] was derived from the electromagnetic internal energy and the Nyquist theorem,[66] since the fluctuating fields are in thermodynamic equilibrium with the solid media. We should note that $\epsilon = \epsilon^R$, the retarded dielectric permittivity (which is analytic in the upper half-plane) for $n > 0$, and $\epsilon = \epsilon^A$, the advanced dielectric permittivity (which is analytic in the lower half-plane) for $n < 0$ in equation (36). These functions are related to the retarded and advanced photon Green's functions [60] and $\epsilon^R(z) = [\epsilon^A(z^*)]^*$, where z is in the upper half-plane and the star denotes complex conjugation.[75]

Our last topic in this section is that of the influence of spatial dispersion [i.e., $\epsilon(\vec{r}, \vec{r}', \omega) \neq \epsilon(\vec{r}, \omega) \delta(\vec{r} - \vec{r}')$], which is equivalent to the local coupling case

---

[74] F. Gerlach, *Phys. Rev.*, 1971, **B4**, 393.
[75] Manifold references to the zero-point energy method are cited in refs. 55 and 56.

heretofore treated]. A point has been raised in the literature [56] that some calculations [48,76-78] incorporating this effect are in error. This opinion stems from a rigorous quantum field theoretic derivation [56] of the change in free energy produced by assembling an inhomogeneous system of material bodies. This derivation employs the method of integration over the coupling constant and defines the interaction as that arising from coupling with *all* the long-wavelength electromagnetic fluctuations. A result of the derivation is that the change in free energy involves the spatially non-local dielectric function, which is influenced by the interaction. It is demonstrated [56] that it is precisely this interaction-dependent ϵ which produces a deviation from the formula arising from the zero-point energy method, *i.e.* equation (36). It is thus opined that not only is it *not* generally true [78] that the zero-point energy method gives the change in free energy, but also calculations [48,76,77] applying this method to spatially dispersive media have led to unreliable results.

Two comments are in order. First, the authors employing the zero-point energy method [48,76,77] have studied *model* systems in which it is assumed that the dielectric permittivity is the same before and after assembly; this assumption is the same as the one used in earlier studies [59,69] employing local dielectric permittivities. Therefore, within the context of these models, the calculations [48,76,77] are correct (*i.e.*, by construction the dielectric permittivities are independent of the interaction). It is necessary to determine the interaction dependence of realistic dielectric permittivities before and after assembly in order to verify the reliability of such an assumption – which brings us to the second comment.

In the derivation,[56] the interaction involved *all* long-wavelength fluctuations, as we have mentioned, *even those existent before assembly*. Therefore, the interaction includes modes which characterize the isolated media. For example, in the case of rare gas crystals, the system existing before turning on the interaction would be unbound (since long-range forces bind these crystals together). This implies that the aforementioned dependence of the dielectric permittivity upon the interaction does not necessarily have very much to do with fluctuations associated with the separation of the bodies (which produce the van der Waals forces). An isolated metal crystal, for example, would have a different dielectric permittivity after turning on this interaction. In our opinion, then, it is not likely that the separation-dependent interaction responsible for van der Waals forces influences the dielectric permittivity enough to invalidate [56] the zero-point energy method [65] (*i.e.*, the corresponding fluctuations are surface dependent).

Although a number of studies have been made [48,76-78] of the influence of spatial dispersion upon van der Waals forces, only one [48] has been applied to physical adsorption. From equations (27) and (36),[47,65,79] the van der Waals contribution, W, to the interaction energy between an atom in vacuum and a metal surface is given by equations (39) and (40). The summation in equation (39) is over integer

$$W(l) = -\lim_{N \to 0} \frac{k_B T}{\pi} \sum_{n=0}^{\infty} \left(\frac{\xi_n}{c}\right)^3 \int_1^\infty \frac{p^2 dp}{N} (\Delta_1^{-1} + \Delta_2^{-1}) \qquad (39)$$

[76] B. Davies and B. W. Ninham, *J. Chem. Phys.*, 1972, **56**, 5797.
[77] J. Heinrichs, *Solid State Comm.*, 1973, **13**, 1595.
[78] Calculations incorporating spatial dispersion in van der Waals studies are cited in ref. 56.
[79] B. W. Ninham, V. A. Parsegian, and G. H. Weiss, *J. Stat. Phys.*, 1970, **2**, 323.

Microscopic Approaches to Physiorption

$$ck_{||} \equiv \xi_n(p^2 - 1)^{\frac{1}{2}} \quad (40)$$

values of n, where the $n = 0$ term is given half weight. The dispersion relations for the normal surface modes (TM and TE) of the electromagnetic field (i.e., surface plasmons) are represented by Δ_1 and Δ_2 [specified in equation (44) below]. Calculation of W, therefore, reduces to a solution of Maxwell's equations (atomic units are used exclusively here).

In the presence of metallic spatial dispersion the relationship in the metal between the Fourier components of the electric displacement, $\vec{D}$, and the electric field, $\vec{E}$, can be written as (41), in which ω is the frequency and the metal fills the

$$\vec{D}(z, \vec{k}_{||}, \omega) = \int_{-\infty}^{0} dz' \, \epsilon(z, z', \vec{k}_{||}, \omega) \, \vec{E}(z', \vec{k}_{||}, \omega), \quad z < 0 \quad (41)$$

negative-z half-space. Maxwell's equation with the insertion of equation (41) cannot be solved generally, and we utilize a model, equation (42), which corresponds to the bulk metallic hydrodynamic constant,[80] where ω_P is the plasma frequency

$$\epsilon_M(\vec{k}, \omega) = 1 - \omega_P^2 \, [\omega(\omega + i/\tau) - (\alpha k)^2]^{-1} \quad (42)$$

(i.e., $\omega_P^2 = 4\pi n_+$, where n_+ is the electron density for free-electron metals), τ is a damping lifetime, and α^2 is set equal to $0.6 V_F^2$ (V_F is the Fermi velocity) in order to recover the bulk-plasmon dispersion relation from the hydrodynamic equation.[80-83] In relating equations (41) and (42), we employ two solvable models of the surface electronic boundary conditions: specular reflection [84] and diffuse reflection.[85] Specular reflection is equivalent to assuming that the field components parallel (perpendicular) to the surface plane are even (odd) under inversion about the surface plane; the exact solution of Maxwell's equations for this case are well known.[86] The diffuse-reflection boundary condition corresponds to assuming that equation (43) holds wherein we have suppressed the ω argument. This boundary

$$\epsilon(z, z', \vec{k}_{||}) = \int_{-\infty}^{\infty} (dk_z/2\pi) \exp\,[ik_z(z - z')] \epsilon_M(\vec{k}) \equiv \epsilon_M(z - z', \vec{k}_{||}) \quad (43)$$

condition, therefore, corresponds to the same contribution from each volume element of dielectric as in the bulk constitutive relation. Exact solutions of Maxwell's equations for diffuse reflection have been found independently by several workers.[87-89]

The influence of spatial dispersion upon the van der Waals interaction can be

---

[80] R. H. Ritchie, *Prog. Theor. Phys.*, 1963, **29**, 607.
[81] J. Harris, *Phys. Rev.*, 1971, **B4**, 1022.
[82] L. Kleinman, *Phys. Rev.*, 1973, **B7**, 2288.
[83] In ref. 82 it was pointed out that α should properly be considered a function of frequency for a hydrodynamic solution of the Boltzmann equation (i.e., see ref. 79).
[84] G. E. H. Reuter and E. H. Sondheimer, *Proc. Roy. Soc.*, 1948, **A195**, 336.
[85] J. Heinrichs, *Solid State Comm.*, 1973, **12**, 167; *Phys. Rev.*, 1973, **B7**, 3487.
[86] R. Fuchs and K. L. Kliewer, *Phys. Rev.*, 1971, **B3**, 2270.
[87] G. Agarwal, D. N. Pattanyak, and E. Wolf, *Phys. Rev. Letters*, 1971, **27**, 1022; *Optics Comm.*, 1971, **4**, 255.
[88] A. A. Maradudin and D. L. Mills, *Phys. Rev.*, 1973, **B7**, 2784.
[89] G. G. Kleiman and U. Landman, in 'Proceedings of the Thirty-third Annual Conference on Physical Electronics', Berkeley, California, 1973 (unpublished).

clarified by simple physical arguments, without resort to the calculational details of the exact solutions with retardation, which are quite complicated. The general form of the dispersion relations of TM(Δ_1) and TE(Δ_2) surface modes in equation (39) are evaluated on the imaginary frequency axis $\omega = i\xi$ [equations (44)—(47)]. The quantity ϵ_A is the dynamic dielectric function of the atom.

$$\Delta_j(i\xi,p) \equiv D_j(i\xi,p) \exp(2p\xi l/c) - 1, \quad (j = 1,2) \qquad (44)$$

$$D_1(i\xi,p) = [(S_A + p\epsilon_A)/(S_A - p\epsilon_A)]C_{1M} \qquad (45)$$

$$D_2(i\xi,p) = [(S_A + p)/(S_A - p)]C_{2M} \qquad (46)$$

$$S_A = (\epsilon_A - 1 + p^2)^{\frac{1}{2}} \qquad (47)$$

The dispersion relation of the modes of the isolated metal are specified by $C_{jM}^{-1}(k_{\parallel},\omega) = 0$; the surface plasmons correspond to C_{1M}.

Because of the exponential in equation (44), the major contribution to W comes from terms for which $p\xi \simeq c/2l$. For the small l with which we are concerned the important limit is $p \to \infty$, since $\epsilon_A \to 1$ as $\xi \to \infty$. In the case of a *local* metallic dielectric function, $C_{1M}(p \to \infty, \xi)$ is a finite function of ξ alone [$C_{2M}(p \to \infty, \xi) \to \infty$]. The finiteness of C_{1M} in this limit reflects the fact that the surface–plasmon frequency $\omega_s \to \omega_p/\sqrt{2}$ as $k_{\parallel} \to \infty$ for a local metallic dielectric function [*i.e.*, see equation (39)]. A simple transformation of variables [59] in equation (39) yields the result that $E_{VW} \sim l^{-3}$. This reasoning is the justification for the conventional inverse-cube law of the van der Waals force.

In the case of the model non-local dielectric functions we are considering, however, *both* C_{1M} and $C_{2M} \to \infty$ as $p \to \infty$. This has the consequence that there is negligible contribution to the integral in equation (39) from terms with large p. The l dependence of W cannot be deduced from a transformation of variables and is more complicated than the inverse-cube law. This result follows from the fact that, for both of our models, the surface mode is admixed with volume modes;[86] the admixture increases with increasing $k_{\parallel}$, so that $\omega_s \to \infty$ as $k_{\parallel} \to \infty$.[86,88] The negligible contribution from terms with large $k_{\parallel}$ exists also in the random-phase approximation solved with specular-reflection boundary conditions.[86] Another consequence of the admixture is that the contribution to W of modes localized at the surface (*i.e.*, l-dependent modes) decreases relative to their contribution in the local-dielectric-function case (*i.e.*, $\alpha = 0$). We expect, therefore, that W will be weaker with spatial non-locality than without.

These results are illustrated in Figure 2, in which we plot $C_{VW} \equiv l^3 W$ as a function of l for He on a typical free-electron metal, Mg. The polarizability of He includes both discrete and continuum contributions.[46,47] The salient points are that W varies more slowly than the inverse-cube law in the spatial dispersion models and that the van der Waals interaction is weaker in these cases than in the case of the corresponding local dielectric function (*i.e.*, $\alpha = 0$), in accord with the previous discussion. Another important feature of Figure 2 is that the effect of spatial dispersion on W is much more significant than that of the different boundary conditions.

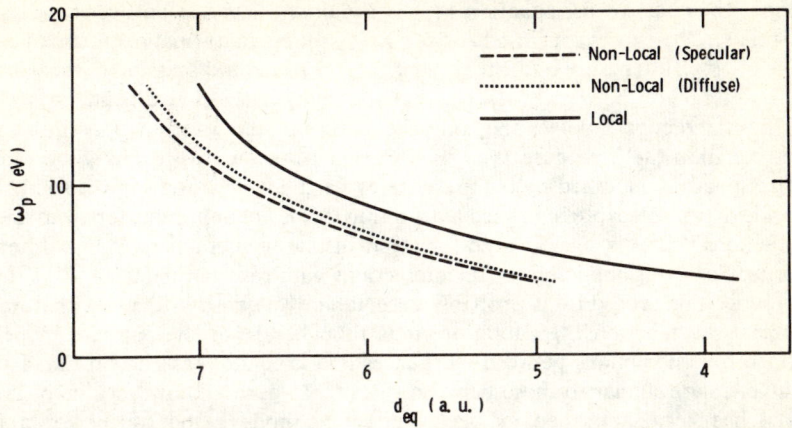

Figure 2 van der Waals constant (i.e., $C_{VW} \equiv l^3 E_{VW}$) for He adsorbed on Mg (i.e., $\alpha = 0.56$ a.u.). Results are shown for spatial dispersion with specular reflection (*dashed line*) and diffuse reflection boundary (*dotted line*) conditions and for the local dielectric function model (*solid line*). Infinite lifetime τ is assumed here
(Reproduced by permission from *Phys. Rev. Letters*, 1974, **33**, 524)

In addition to the zero-point energy method,[65] Lifschitz's[59] method has been extended to spatially dispersive media.[90]

In this subsection, we have presented the salient features of contemporary knowledge of the theory of van der Waals forces as applied to physical adsorption. In so doing, we have brushed over many technical difficulties involved in the calculations (such as divergences arising in free-energy calculations[60,73]). Our main purpose was, instead, that of elucidating the physical and mathematical bases of applicable theories and demonstrating applications to the problem of physical adorption. In this connection, it should be pointed out that the most advanced theories[59,65,69] have not been formulated to include the effects of surface periodicity (*i.e.*, the surfaces are uniform). Recent studies[51,56] point to the importance of this effect and present methods for their calculation.[56]

Semi-empirical Calculations.—Details of semi-empirical treatments have been adequately reviewed.[91,92] Our purpose is to elucidate the physical basis of these treatments[9—38] and to offer some remarks within the context of our understanding of van der Waals forces. These treatments start from the basic equation (48),

$$W(\vec{r}_i) = \sum_j U_{ij}(\vec{r}_{ij}) \qquad (48)$$

where W is the interaction energy of a physisorbed atom of type i at $\vec{r}_i$ interacting with a solid comprising atoms labelled by j. U_{ij} is the total energy of interaction between a pair of atoms separated by displacement $\vec{r}_{ij}$. Since W is associated with a physisorbed atom, the attractive part of U_{ij} corresponds to a van der Waals

---

[90] D. R. Chang, R. L. Cooper, J. E. Drummond, and A. C. Young, *J. Chem. Phys.*, 1973, **59**, 1232.
[91] J. P. Hobson, *Crit. Rev. Solid State Sci.*, 1974, **4**, 221.
[92] H. Margenau and N. R. Kestner, 'Theory of Intermolecular Forces', Pergamon, Oxford, 1969.

energy. In other words, equation (48) is a pairwise summation over single pair potentials. The expressions used for U_{ij} are assumed mathematical forms: hence, the semi-empirical nature of the treatments.[9-38] From our discussion of the van der Waals interaction between a neutral, polarizable particle and a solid, it is clear that the interaction involves the collective nature of the wavefunctions in a solid [*e.g.*, equation (38)], because the wavefunction associated with a single atom in the gas phase is modified by the presence of neighbouring atoms in the solid.[59,69] For this reason, an expression such as equation (48) is not applicable for atom–metal interactions,[26] and is suspect even for atom–insulator interactions.[59,69] Although in principle [69] this applies even for interactions with rare-gas crystals,[24,27,35,37,38] in practice this procedure is probably acceptable [25] because of the local nature of rare-gas wavefunctions. In addition, it is thought [91] that the pairwise model is justified for interactions between inert gases and graphite: gas–carbon interaction parameters are similar to those between gases.[35] In general, however, the pairwise model has been confirmed by very indirect methods [25] and except for a few cases [24,27,30-33,35,37,38] even this confirmation is lacking.

One of the salient features of equation (48) is that it incorporates the lattice periodicity of the surface, a feature unfortunately lacking in the van der Waals energies we considered previously. In many applications,[5,93,94] however, a one-dimensional model is assumed for W, or else the solid is treated as a continuum of integration,[92] so that this advantage of equation (48) is lost. The reason for this and other approximations [9-34,37,38] is that the sum in equation (48) is usually of importance as input into other calculations such as those of either statistical mechanical quantities [35] or eigenfunctions [27,30-33,37,38] or in scattering problems.[25,29,94-96] Therefore, numerically determined values of W for each r_i are harder to use. Recently, however, techniques for giving analytic representations of W have been developed.[35,36] The techniques involve evaluating the Fourier coefficients of equation (48) where W is Fourier-analysed in the surface periodicity [*i.e.*, see equation (1)]. Although the derivations are independent [35,36] the final formulas for the Fourier coefficients are identical. The value of these calculations is that only a few Fourier coefficients are needed to represent the potential rather accurately.[35,36] Indeed, since the higher Fourier components [see equation (1)] decay rapidly with z, a one-dimensional treatment may be sufficient, especially for metals. The difference between these treatments [35,36] is that in one,[36] all planes of the solid are summed and the results tabulated, while in the other,[35] only the surface plane is considered important and analytic expressions are given. In addition, in one [36] the potentials for simple cubic and alkali-halide type lattices are computed while in the other,[35] graphite and the (100) and (111) planes of fcc crystals are treated and statistical mechanical quantities are calculated for the computed potentials.

Both calculations represented $U_{ij}(r_{ij})$ by the commonly used [5] Lennard-Jones 6–12 potential [equation (49)],[23] wherein the potential energy represents that

---

[93] M. Kaminsky, 'Atomic and Ionic Import Phenomena on Metal Surfaces', Academic, New York, 1965.
[94] A. F. Devonshire, C. Strachan, and J. E. Lennard-Jones, *Proc. Roy. Soc.*, 1935, **A150**, 442, 456; 1936, **A156**, 6, 29, 37.
[95] F. O. Goodman, *Surface Sci.*, 1971, **24**, 667.
[96] R. M. Logan, 'Solid State Surface Science', Marcel Dekker, New York, **3**, 1.

$$U_{ab}(r) = \varepsilon_{ab}\left\{\left(\frac{r_{ab}}{r}\right)^{12} - 2\left(\frac{r_{ab}}{r}\right)^{6}\right\} \tag{49}$$

$$r_{ab} = (r_{aa} + r_{bb})/2 \tag{50}$$

$$\varepsilon_{ab} = (\varepsilon_{aa}\varepsilon_{bb})^{\frac{1}{2}} \tag{51}$$

between atoms of species a and b. The well depth, ε_{ab}, and equilibrium separation, r_{ab}, are calculated from the corresponding parameters for interactions between similar atoms (e.g., ε_{aa} is the well depth for two atoms of species a). Equations (50) and (51) for the determination of these parameters are empirically determined.[23]

Despite their shortcomings, in the absence of more accurate calculations of the potential energy, these studies have provided considerable insight [5,91—98] into the properties of physisorbed systems.

Fundamental Calculations of Physisorption Binding Energies (Heats of Adsorption).— We discuss here those calculations of the potential energy [43,46—53] which derive the binding energy of physisorbed particles with a minimum of phenomenological assumptions, in contrast to the situation among semi-empirical studies. An additional dividend of most of these calculations is the determination of the equilibrium distance, since these studies include both attractive and repulsive contributions to the binding energy.

A pioneering attempt to account for both the attractive and repulsive contributions to the interaction energy has been presented by Pollard.[52] In this method the van der Waals interactions is described by classical dipole–dipole interaction [41] between the atom and the solid substrate. The repulsive term is evaluated as the exchange interaction resulting from a Heitler–London coupling scheme. Implicit in this approach is the formation of a 'one-electron bond' between localized orbitals of the surface and valence orbitals of the adsorbate. Moreover, the model of the surface employed in the calculation does not allow for 'leakage' of electronic charge into the vacuum. In addition, interactions of valence electrons of the adsorbate with the substrate electrons which may be of the same order as the Heitler–London bonding energy are neglected in the calculation. Coupled with a number of approximations used in the evaluation,[52] the results of the above method do not conform with more recent knowledge of surface structure characteristics,[61,99,100] and are not in good agreement with available experimental data. The calculation of Mavroyannis,[43] on the other hand, is related to the work of Lifschitz [59] and Dzyaloskinskii et al.:[69] the attractive energy corresponds to equations (29) and (30). It employs a uniform continuous model of the surface and in calculating the interaction energy ignores the presence of repulsive forces entirely. The calculation is performed by assuming that a noble-gas atom resides at a distance from the metal which is the average of the nearest-neighbour distances of the metal atoms in the metal and rare-gas atoms in a rare-gas crystal. The physisorption energy is then set equal to the van der Waals energy at the average

[97] J. H. de Boer, Adv. Catalysis, 1956, **8**, 18.
[98] E. C. Beder, Adv. Atomic Mol. Phys., 1967, **3**, 205.
[99] N. D. Lang, Solid State Phys., 1973, **28**, 225.
[100] A. J. Bennett and C. B. Duke, Phys. Rev., 1967, **160**, 541.

distance and an approximate formula for the energy is used. The neglect of repulsive contribution to the energy of physisorption, the above assumption of equilibrium position, and the approximations introduced in the expression cause serious difficulties.[46]

Another series of studies [46–50] attempted to derive a microscopic formulation of the physical adsorption interaction consonant with modern studies of bare-metal surfaces.[61,99] The fundamental assumption [46–50] is that a physisorbed atom (or molecule) resides sufficiently far from the surface, so that there is only weak coupling between the substrate and the incident particle. This is equivalent to the assumption that there is no chemical interaction (either charge transfer or rearrangement or other mechanism) involved: the particle and solid interact through the intermediary of the long-wavelength electromagnetic field, or through van der Waals forces. This assumption agrees with the usual picture of physical adsorption. The model determined by this hypothesis implies that the mechanisms responsible for attraction at short distances from the substrate (*i.e.*, exchange, correlation, and electrostatic interactions) transform into the van der Waals attraction alone at physisorption distances. This is analogous to Bardeen's result [40] for the image force on an electron. Extending this picture further, it is concluded [46–50] that the repulsion between the physisorbed entity and the surface is provided by the remaining portion of the Hamiltonian: the electronic kinetic energy, which is a result familiar from the theory of diatomic molecules.[101]

In this work, therefore, the attractive energy is identified with the van der Waals energy. In order to calculate this realistically from equation (29), for example, it is necessary to measure the distance from the correct origin. For a stationary classical external charge particle, the image-interaction energy should be measured from the centroid of the induced charge density.[99] However, van der Waals forces originate from the electromagnetic field fluctuations in the solid, extending beyond its boundaries, which perturb spontaneous fluctuations in the atom. In this case, therefore, we are considering a high-frequency interaction to which the electron density is not expected to be able to respond instantaneously. It has been shown [102] that the density fluctuations induced by a high-frequency external charge are related to the derivatives of the static isolated bare-metal electron density. In the framework of the jellium model of a metal surface,[61] the centroid of the high frequency induced charge density [102] is located very close to the edge of the jellium background, in contrast to zero-frequency results.[99] Since van der Waals forces have been derived from an image method,[44] the distance in the expression for the van der Waals energy should be measured from the centroid of the derivative of the isolated-metal-electron density. However, in making consistent comparisons with the jellium model of a bare-metal surface, we shall measure the van der Waals distance from the edge of the jellium background. Since the energies of the transitions involved in the van der Waals interaction are high but not infinite, one expects small corrections due to the diffuseness of the surface region of the bare metal.[102]

A calculation of the origin from which van der Waals forces associated with continuum metal should be measured has recently been reported.[51] One result is

[101] J. C. Slater, 'Quantum Theory of Molecules and Solids', McGraw-Hill, New York, 1963, **1**.
[102] J. Harris and R. O. Jones, *J. Phys.* (*C*), 1974, **7**, 3751.

that the origin may not be identifiable generally with the centroid of induced charge; it is, however, a reasonable approximation to make this identification in the absence of a general proof of its applicability. Another conclusion is that van der Waals energies could be increased considerably by inclusion of the correct origin rather than the edge of the jellium background.[46—50] Calculations [51] of the binding energy of Xe on noble metals with the insertion of empirically determined equilibrium separations into Lifschitz's theory [*i.e.*, equations (29) and (30)] with these origin corrections indicate that neglect of the repulsive energy has serious effects.

Whereas the attractive portion of the energy in Kleiman–Landman theory [46—50] is derived from van der Waals energies [*i.e.*, see equations (29) and (38) and Figures 1 and 2], the repulsion is derived from the increase in electronic kinetic energy upon assembling the system, as we have remarked. This increase is a consequence of the Heisenberg uncertainty principle, since assemblage restricts the electrons to a smaller volume. A convenient method for calculating the change in kinetic energy is the density-functional formalism.[103,104] According to this theory, the ground-state energy of an interacting many-electron system in an external potential is a unique functional of the electron density $n(\vec{r})$. Formally the ground-state energy $E_v[n]$ can be written as equation (52).[103] The quantities T and E_{xc}

$$E_v[n] = T[n] + F[n] \tag{52}$$

$$F[n] \equiv E_{xc}[n] + E_{es}[n] \tag{53}$$

$$E_{es}[n] \equiv \int v(\vec{r})n(\vec{r})\mathrm{d}^3r + \frac{1}{2}\int\int \frac{n(\vec{r})n(\vec{r}')\mathrm{d}^3r\mathrm{d}^3r'}{|\vec{r}-\vec{r}'|} \tag{54}$$

represent kinetic and exchange-correlation energies, respectively, which are unique functions of n. In addition, we have extracted the electrostatic energy E_{es} as usual.[103] In equation (54), v is the external potential which we define as the potential produced by all the ion cores in the system.

In the following we derive the change in kinetic energy, ΔT, which is produced by juxtaposing the atom and metal surface. Under the assumption of weak coupling, ΔT can be expressed in terms of the isolated atomic and metallic electron densities, to first order in the coupling parameter. It is not necessary, therefore, to determine the self-consistent electron density of the combined system to this order of approximation.[46—50] The ground-state energy satisfies the following variational principle [103] set out in equations (55)—(57). The symbol μ represents

$$\mu = \frac{\delta E_v[n]}{\delta n} = \frac{\delta T[n]}{\delta n} + V(v,n,\vec{r}) + \varepsilon_{xc}[n] \tag{55}$$

$$V(v,n,\vec{r}) \equiv v(\vec{r}) + \int \frac{n(\vec{r}')}{|\vec{r}-\vec{r}'|}\mathrm{d}^3r' \tag{56}$$

$$\varepsilon_{xc} \equiv \delta E_{xc}[n]/\delta n \tag{57}$$

the chemical potential of the system. An equivalent system of self-consistent

---

[103] P. Hohenberg and W. Kohn, *Phys. Rev.*, 1964, **136**, B864.
[104] W. Kohn and L. J. Sham, *Phys. Rev.*, 1965, **140**, A1133.

one-electron equations is given, in atomic units, by equations (58)—(61).

$$\{h[n] + V(\nu,n,\vec{r},)\}\psi_i(\vec{r}) = \varepsilon_i \psi_i(\vec{r}) \tag{58}$$

$$h[n] \equiv -\tfrac{1}{2}\nabla^2 + \varepsilon_{xc}[n(\vec{r})] \tag{59}$$

$$T[n] = \sum_{i=1}^{N} \varepsilon_i - \int d^3r \{V(\nu,n,\vec{r}) + \varepsilon_{xc}[n(\vec{r})]\}n(\vec{r}) \tag{60}$$

$$n(\vec{r}) = \sum_{i=1}^{N} |\psi_i(\vec{r})|^2 \tag{61}$$

In applying the above equations to the physisorption problem, it is important to realize that equations (52) and (58) are formally exact, even though we do not, at present, know the form of E_{xc}.[46-50] Thus, these equations are correct for both atoms and condensed matter. N is the total number of electrons in the system (*i.e.*, the ψ_i are a complete set of orthonormal states labelled in order of increasing energy). In the context of physisorption, all energies are measured relative to the vacuum, in agreement with the zero energy of the van der Waals forces. According to the assumption of weak coupling the electron densities of the combined system undergo only small changes from their isolated systems values. In particular, in the case of He and other rare-gas atoms, this consequence is further supported by the chemical inertness of these elements. In other words, the metal electrons 'see' the atom as almost neutral, while the atomic electrons experience a small pertubation due to the metal electron density in the surface region. Thus, electron states in the combined system can be identified with states of the isolated systems. We label the atomic He states by $i = 1,2$ and the metal states by $i = 3, \ldots, N+2$, where N is the number of electrons in the metal. The quantity of interest is the change in kinetic energy, ΔT, of the combined system [equation (62)], where T_d and

$$\Delta T[n_c] = T_d[n_c] - T_\infty[n_A, n_M] \tag{62}$$

T_∞ are the kinetic energies of the atom and metal electrons at a distance d and infinite separation, respectively. n_c, n_A, and n_M are the electron number densities of the combined atom–surface system, isolated atom and metal, respectively, and n_c^0 corresponds to $n_A + n_M$. The change in kinetic energy, ΔT, in equation (62), is calculated[47] by combining first-order perturbation theory and the variational principle expressed in equations (55)—(57). To zeroth-order, ΔT can be identified with the change in kinetic energy of metallic electrons.[47] The result[46-50] is shown in equation (63). Thus, the zeroth-order change in kinetic energy can be

$$\Delta T[n_c] \cong \int d^3r n_M \left(\frac{\delta T[n_c^0]}{\delta n} - \frac{\delta T[n_M]}{\delta n} \right) \tag{63}$$

determined approximately without resorting to self-consistent solutions. The notation $n_A(d)$ indicates that the nucleus is located at separation d. The scheme of calculation involves choosing a functional $T[n]$, which is an adequate description of the kinetic energies of both the atomic and metallic electrons when they are separated by an infinite distance. The functional we use is derived from the extended Thomas–Fermi version of the density functional formalism.[103] For the case of slowly varying electronic densities one may perform an expansion of $T[n]$

in successive orders of the gradient operator which, to first order, yields the expression (64). In the case of the He atom, n_A is represented by the hydrogenic

$$T[n] = \int d^3r [0.3(3\pi^2)^{2/3} n^{5/3} + \tfrac{1}{72} (\vec{\nabla} n)^2 / n] \tag{64}$$

variational solution of the Schrödinger equation.[105] The resulting form is a good approximation to the exact number density and is given by equation (65), in which

$$n_A(r) = \left(\frac{2}{\lambda^3 \pi}\right) e^{-2r/\lambda} \tag{65}$$

$\lambda = \tfrac{16}{27}$ and r is measured from the He nucleus. The kinetic energy resulting from insertion of equation (65) into equation (64) is 91.8% of the variational kinetic energy.[47] Since the variational ground-state energy is within 1.9% of the experimental value,[105] it follows from the virial theorem that the variational result for the kinetic energy has the same degree of accuracy. Equations (64) and (65), therefore, constitute a good approximation for the atomic kinetic energy within the context of physisorption.* Application of equation (64) to the description of bare-metal surfaces [106] and work-function calculations [107] resulted in adequate agreement with experimental values. Rather than using the density-functional variational solutions for the metal number density, we employ in this example a parameterized form of n_M specified by (66),[46–50, 106] where n_+ is the positive jellium

$$n_M = n_+ - \tfrac{1}{2} n_+ \exp[\beta z], \quad z < 0$$
$$n_M = \tfrac{1}{2} n_+ \exp[-\beta z], \quad z > 0 \tag{66}$$

charge density and β is a variational parameter. Equation (64) is a functional form for the kinetic energy, which in conjunction with the electron densities in equations (65) and (66) is an adequate representation of both the isolated atom and metal. In accord with the density-functional formalism, we assume that equation (64) describes the combined system. Equation (67) for the repulsive energy E_R therefore

$$E_R(d) \cong \int_{V_B} d^3r \, n_M \left[0.5(3\pi^2)^{2/3} (n_A^{2/3} - n_M^{2/3}) \right.$$
$$\left. + \tfrac{1}{72} \left(\frac{(\vec{\nabla} n_A)^2}{n_A^2} - \frac{(\vec{\nabla} n_M)^2}{n_M^2} - \frac{2\nabla^2 n_A}{n_A} - \frac{2\nabla^2 n_M}{n_M} \right) \right] \tag{67}$$

results.[46–50] The approximate form in equation (67) results from noting that n_M varies much more slowly than n_A (i.e., $2/\lambda \simeq 3\beta$), and $n_A \gg n_M$ in the region of the nucleus. Consequently, from equation (63) we observe that the major contribution comes from the region around the nucleus. The region V_B is a sphere of radius b, centred at the nucleus such that $n_A \geqslant n_M$ for $r \leqslant b$. Contributions from regions with $r > b$ are neglected because the density in equation (65) possesses an exponential tail. This is important only in calculating differences in energies. The neglect corresponds to cutoffs used in conventional Thomas–Fermi treatments.[50]

In Figure 3 we display the He repulsive energy presented in equation (67) for

---

* Better approximations may be obtained by varying the value of λ in equation (65).

[105] L. I. Schiff, 'Quantum Mechanics', McGraw-Hill, New York, 1955.
[106] J. R. Smith, *Phys. Rev.*, 1969, **181**, 522.
[107] N. D. Lang and W. Kohn, *Phys. Rev.*, 1971, **B3**, 1215.

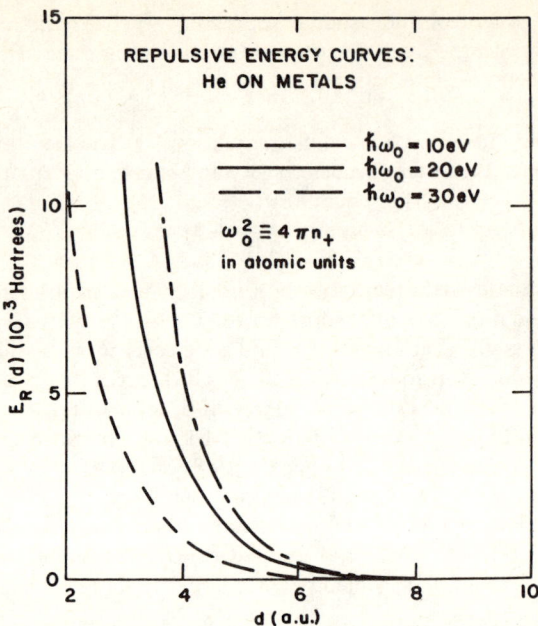

Figure 3 *The repulsive energy E_R for* He *on metals. As ω_0 increases E_R becomes stronger*
(Reproduced by permission from *Phys. Rev.*, 1973, **B8**, 5484)

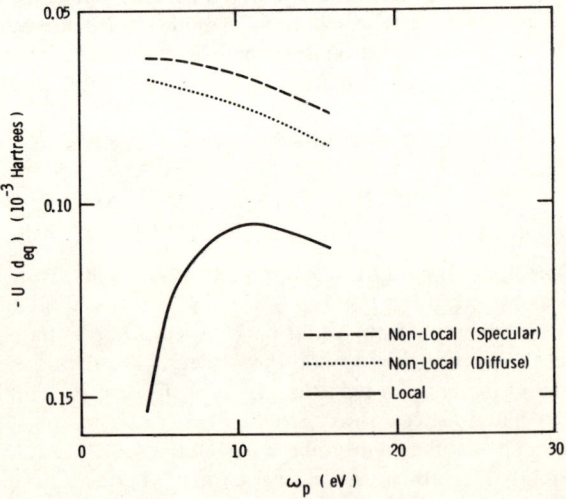

Figure 4 *Illustration of the relation between U, E_{VW}, and E_R for a typical 'ideal' metal (i.e., $\omega_p = \omega_0 = 20$ eV). The arrow indicates the equilibrium position, d_{eq}. The largest contribution to U is from E_{VW}, indicating that $U(d_{eq})$ is mostly independent of temperature at physisorption distance*
(Reproduced by permission from *Phys. Rev.*, 1973, **B8**, 5484)

three different positive jellium charge densities.[46-50] The systematic strengthening of E_R with increasing $\omega_0^2 \equiv 4\pi n_+$ (in a.u.) is evident.

The total energy of interaction $U(d)$ is derived from equation (68),[46-50] where

$$U(d) = E_{VW}(d) + E_R(d) \qquad (68)$$

E_{VW} is the negative van der Waals energy given in terms of the plasma frequency, ω_p [i.e., $\epsilon_s = 1 - (\omega_p^2/\omega^2)$] in equations (29) and (38). Illustrated in Figure 4 [46-50] are $U(d)$, $E_{VW}(d)$, and $E_R(d)$ in atomic units (with no spatial dispersion) for a 'typical' metal (i.e., $\omega_p = \omega_0 = 20$ eV). At the equilibrium position (vertical arrow), the van der Waals energy is the major component of the physisorption energy. From our

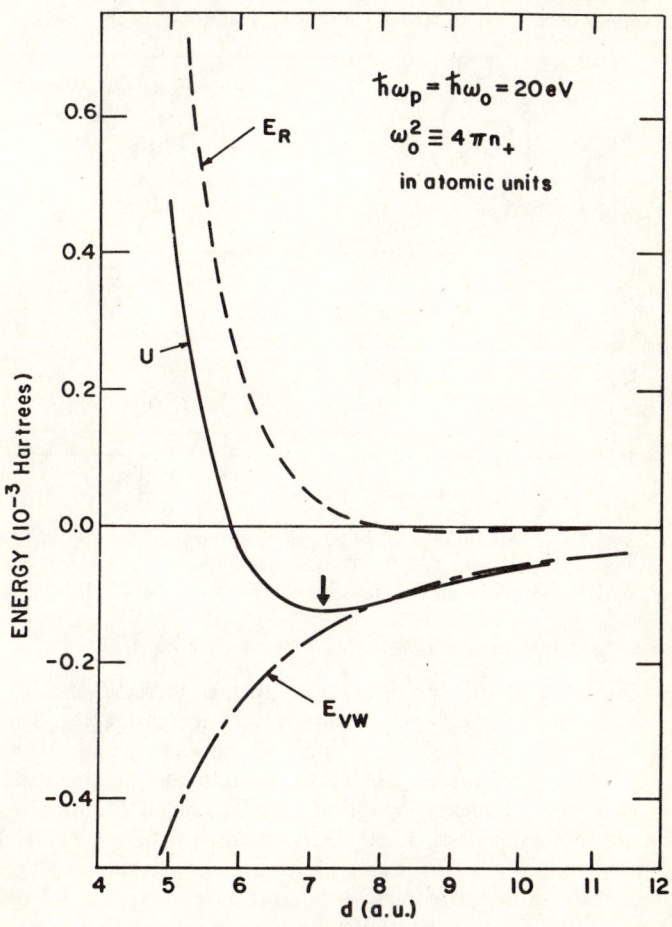

Figure 5 *Physisorption energies for He adsorbed on free-electron metals for the three models of electromagnetic coupling. The monotonic increase in* $-U(d_{eq})$ *with plasmon frequency for non-local couplings illustrates a distinction from the case of local coupling. Infinite lifetime τ is assumed here*
(Reproduced by permission from *Phys. Rev. Letters*, 1974, **33**, 524)

discussion above (pp. 2—19) we conclude that $U(d_{eq})$, where d_{eq} is the equilibrium position of the atom, is largely independent of temperature. Another interesting feature illustrated in Figure 4 is that as the particle approaches the surface, E_R first decreases because the atomic and metallic electrons overlap and then increases because the electrons are 'squeezed' into a smaller volume. This is exactly analogous to the situation for the binding in diatomic molecules.[101]

In Figures 5 and 6 we plot the physisorption energy, $-U(d_{eq})$, and equilibrium position, respectively,[46—50] for He on a range of free electron metals in atomic units, *i.e.*, ω_p in equation (29) = ω_0, for local coupling and the two models of non-local coupling [46—50] discussed earlier (*i.e.*, see the discussion associated with Figure 2). The solid line in Figure 5 corresponds to $-U(d_{eq})$ for local coupling in ideal metals, where it is apparent that $U(d_{eq})$ first increases (weaker binding)

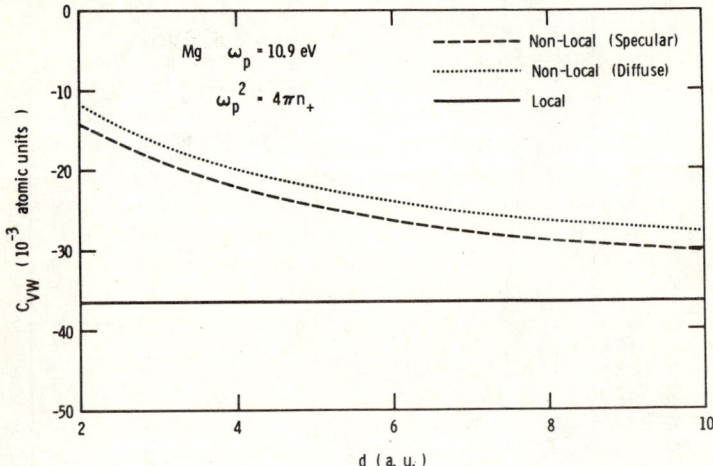

Figure 6 *Equilibrium position d_{eq} of physisorbed He atoms on free-electron metals (i.e., $\omega_p^2 = 4\pi n_+$) for our three models. The relative insensitivity of d_{eq} to the degree of dispersion and different electronic boundary conditions is illustrated. Infinite lifetime τ is assumed here*
(Reproduced by permission from *Phys. Rev. Letters*, 1974, **33**, 524)

with increasing d_{eq} and then decreases with further increase of d_{eq}. Figure 6 demonstrates that d_{eq} increases monotonically with increasing ω_0. This increase is first very rapid (*i.e.*, for $\omega_0 \lesssim 10$ eV) and then markedly slower. Note that the values of d_{eq} for the three models are within 1 a.u. of one another and that the changes in electronic boundary conditions produce small changes in d_{eq}. The equilibrium position, therefore, is relatively insensitive to the model boundary conditions used to describe the metal. In contrast to local coupling,[46—50] the monotonic increase in the physisorption interaction energy, *i.e.*, $-U(d_{eq})$, of He on metals with increasing electron density for the two non-local models is exhibited in Figure 5. This monotonic behaviour is to be contrasted with the peak in $-U(d_{eq})$ for local coupling. Comparison of the surface–plasmon solutions of the hydrodynamic and Boltzmann equations [82] indicate that $\frac{1}{3} < a^2/V_F^2 < 0.6$ for $\omega < \omega_p$; this corresponds to small ξ values which make the major contribution to E_{VW} given in equation (39) [*i.e.*, a is defined in equation (42)]. Such a reduction in a brings the magnitudes of the binding energies for the non-local models into closer agreement with those for the local model. From the results of our calculations we observe that within the errors in the currently available experimental data the degree of agreement achieved between the calculated and experimental values is not changed upon the inclusion of spatial dispersion.[108] We would like to remark here that caution should be exercised in applying model calculations to comparisons with experimental data because of substantial discrepancies in the present experimental situation. Therefore, the need for additional well-controlled experiments on physisorption systems and further theoretical studies is indicated.

In order to compare with experiments,[46—50] both the plasma frequency, ω_p, needed to calculate E_{VW} [*i.e.*, $\epsilon = 1 - (\omega_p^2/\omega^2)$ for local coupling as given in equation (42)], and the electron density n_+, necessary for E_R, must be specified. In applications,[46—50] ω_p is determined from experimentally determined values.[109—111] For free electron metals,[109] ω_0 differs only slightly from ω_p because of interband transitions.[109]

In the case of transition metals, there is a potential problem because the choice of a free-electron density appropriate for a surface calculation is not clearly indicated. Fortunately, a prescription has been formulated[112] for choosing the 'free' valence electrons in these metals. This prescription, which identifies the free electrons with the most stable oxidation state of the isolated atom, has been shown to be a rather adequate parametrization of surface energy data in the density

Table 1 *Physisorption energies, equilibrium positions, and parameters used in the calculation of helium adsorption on 'free' and transition metals (ref. 47). The column labelled E_{expt} represents values of the scattering well depth in helium scattering studies (refs. 114 and 115)*

Metal	ω_0/eV	ω_p/eV	β	d_{eq}/a.u.	$-U(d_{eq})/10^{-3}$Hartree	$E_{expt}/10^{-3}$Hartree
K	4.3	3.9	1.32	3.8	0.129	
Na	5.9	5.9	1.27	4.9	0.123	
Li	8.0	7.1	1.24	5.9	0.097	
Mg	10.9	10.6	1.22	6.5	0.103	
Al	15.8	15.3	1.24	7.0	0.109	
Be	19.0	19.0	1.26	7.1	0.121	
Ag	12.7	23.0	1.22	6.3	0.179	0.128[a]
Zn	13.5	22.9	1.22	6.4	0.169	
Cu	15.3	20.0	1.24	6.8	0.140	
Co	19.3	21.0	1.27	7.1	0.129	
Ni	19.4	22.9	1.27	7.1	0.138	
Mo	23.0	25.0	1.30	7.3	0.137	
W	23.0	23.0	1.30	7.3	0.129	0.159[b]
Pt	19.1	23.0	1.34	7.6	0.141	0.167[a]

[a] Ref. 114; [b] Ref. 115.

---

[108] See Table 1.
[109] D. Pines, 'Elementary Excitations in Solids', Benjamin, New York, 1964.
[110] D. Pines, 'Solid State Physics', Academic, New York, 1955.
[111] J. M. Burkstrand, F. M. Propst, T. L. Cooper, and D. E. Edwards, *Surface Sci.*, 1972, **29**, 663.
[112] J. Schmit and A. A. Lucas, *Solid State Comm.*, 1972, **11**, 419.

parameter, r_s, i.e., $r_s \equiv (4\pi n_+/3)^{-1/3}$. Even though there is some disagreement [113] about the validity of the theory, nevertheless the parametrization can be considered as phenomenological, providing support to the applicability of the prescription. In the following, we denote the plasma frequency used in computing E_{VW} by ω_p and that used in calculating E_R by ω_0. In Table 1 are displayed results of the calculations of He physisorption energies for various free-electron and transition metals (i.e., local coupling) along with parameters used in the calculations, and compare with available experimental values. In comparing the results [46,47] with experiments, we use values derived from He scattering experiments,[114,115] which

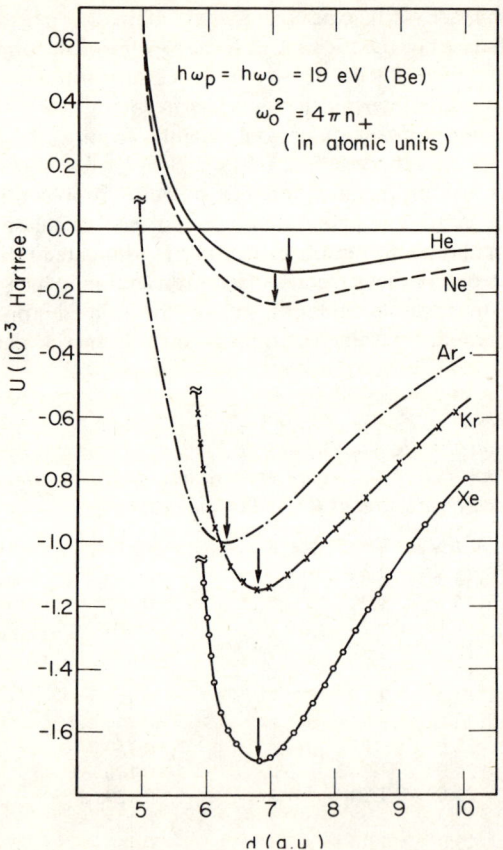

Figure 7 *Interaction potentials of rare gases with Be ($\hbar\omega_p = 19$ eV). The He results were obtained as described. Values of the static atomic polarizabilities used in the calculations: 2.6424, 10.9875, 16.6161, and 27.0306 (a_0^3) for Ne, Ar, Kr, and Xe, respectively. Note variations in characteristics of the potentials between He, Ne and Ar, Kr, Xe* (Reproduced by permission from *Solid State Comm.*, 1976, **18**, 819)

[113] M. Jonson and G. Srinivasan, *Phys. Letters*, 1973, **A43**, 427.
[114] R. Sau and R. P. Merrill, *Surface Sci.*, 1973, **34**, 268.
[115] W. H. Weinberg, Ph.D. Thesis, University of California, Berkeley, 1971 (unpublished).

were performed under controlled conditions. Other adsorption measurements have been made [116,117] on substrates whose surface conditions have not been specified; we, therefore, do not include these results.

Recently, interaction potentials of other rare gases interacting with metal surfaces have been calculated [50] with the same formalism. A local dielectric formalism and the density functional method [103] with Thomas–Fermi atomic charge densities were used in the calculation of the attractive and repulsive energies, respectively. Depth and curvature variations of the potentials suggest a classification into light (He and Ne) and heavy rare gases. In addition it is observed that contrary to the 'sum-of-radii' rule, the atom-surface equilibrium distance decreases in the series He, Ne, Ar followed by a slight increase for Kr and Xe, and is larger than the sum-of-radii of the gas and metal atoms. Figure 7 shows interaction energies calculated [50] for rare gases interacting with Be. In calculating these energies, a non-self-consistent metal density [106] was employed. For reference, we present, in Table 2, comparison between experimental energies, U_{exp}, and theoretical energies (and corresponding d_{eq}) calculated for non-self-consistent U(nsc) [106] and self-consistent U(sc) [61] metal densities. The degree of agreement with experiment and the effect of the level of description of the metal substrated (self-consistent *versus* non-self-consistent) are noted. The results shown in Figure 7 and similar ones for other metal substrates could be used in further calculations of quantities characterizing atom–metal interaction systems. Among these we mention the virial coefficients, trajectory analysis, and 'bound-state resonances'. Preliminary calculations of the eigenvalues corresponding to the above potentials indicate the possibility of performing selective adsorption experiments for rare-gas–metal systems (see p. 74). However, due to the characteristics of the interaction potentials and the increased probability for trapping and other inelastic processes in the series He to Xe, the need for monochromatic incident beams and efficient detection methods with increased signal to noise ratios is emphasized.

Finally, a calculation has been presented [53] of rare gases physisorbed on graphite in which the Gordon–Kim version [118] of the density functional technique [103] is used. This method has proven to be successful in reactions involving rare gases.[118] The surface calculation [53] involves calculation of the difference in total energy [*i.e.*, equation (52)] between rare-gas–graphite systems in the assembled and isolated states. The attractive energy involves explicit exchange and correlation density functionals so that the interaction between the atom and the first graphite layer predominates. This treatment,[53] however, neglects the long-range van der Waals forces arising from the other graphite layers. Comparison with Lennard-Jones pairwise calculations [35] provides an error estimate of 58% in this procedure [53] (the error in ref. 53 is underestimated by a factor of two). In addition, the technique of subtracting total energy functionals neglects the variational condition in equation (55), which is critical in determining properties associated with self-consistency.[99] Also, the inadequate graphite density functionals employed [53] add to the uncertainties in the results.

---

[116] H. Chon, R. A. Fischer, R. D. McCammon, and J. G. Aston, *J. Chem. Phys.*, 1961, **36**, 1378.
[117] F. Pollock, H. Logan, H. Hobgood, and J. G. Daunt, *Phys. Rev. Letters*, 1972, **28**, 346.
[118] R. G. Gordon and Y. S. Kim, *J. Chem. Phys.*, 1972, **56**, 3122.

Table 2 Binding energies and equilibrium distances for rare gases on metals (ref. 50)

System*	$-U_{exp}/10^{-3}$Hartree	Ref.	$-U(sc)/10^{-3}$Hartree	d_{eq}/a.u.	$-U(nsc)/10^{-3}$Hartree	d_{eq}/a.u.
Ne/Pt	0.526	a	0.398	6.0	0.287	6.9
Ne/Ag	0.351	b	0.462	5.4	0.392	6.0
Ar/Cu	3.332	a	1.863	4.7	1.303	5.7
Ar/Zn	2.503	a	2.984	3.0	1.710	5.2
Ar/Ni	1.103	c	1.830	5.1	1.194	6.2
Ar/W	2.869	d	1.713	5.3	1.092	6.4
Ar/Pt	2.105	a	1.857	5.2	1.223	6.1
Kr/Ag	2.232	e	2.490	4.8	1.964	5.6
Kr/Pt	3.364	a	1.960	5.7	1.378	6.6

* Polarizabilities of the rare gases as in caption to Figure 7. Values of the metal plasma frequencies (ω_p) and the variational parameters (β) as given in Table 1.

[a] H. Chon, R. A. Fisher, R. D. McCammon, and J. G. Aston, *J. Chem. Phys.*, 1962, **36**, 1378; [b] R. Sau and R. P. Merrill, *Surface Sci.*, 1973, **34**, 268 (Table 1); [c] P. M. Gundry and F. C. Tompkins, *Trans. Faraday Soc.*, 1960, **56**, 846; [d] R. Gomer, *Austral. J. Phys.*, 1960, **13**, 391; [e] R. F. Steiger, J. M. Morabito, G. A. Somorjai, and R. H. Muller, *Surface Sci.*, 1969, **14**, 299.

3 Microscopic Experimental Methods for the Study of Physisorption

Introduction.—*General Introductory Remarks.* The objective of microscopic approaches to the physical interaction of atoms and molecules with surfaces is to provide a fundamental understanding about the adsorption process and the interaction mechanism. The construction and evaluation of microscopic theoretical models of the interaction and examinations of the significance and influence of the model parameters upon the results, form the basis for understanding of the complex problem of gas–surface interaction and the interpretation of many related observations. A necessary step in the development of theoretical models for these interaction systems is to compare and correlate the results of calculations with experimental data. This can be achieved *via* direct comparison of the calculated values of certain observables with experimental results or by using theoretical values for parameters which enter into a model description of an experiment.

The traditional experiments on physisorption systems are macroscopic in nature. In these experiments thermodynamic quantities of the ensemble of atoms or molecules (gas and surface) are measured. To obtain information about the interaction parameters of an atom or a molecule with the surface, a statistical mechanics calculation of the measured thermodynamic quantity is employed in which the atomic or molecular interaction parameters are introduced and then an ensemble average is performed.[119,120] Consequently, in attempting to extract microscopic information from such data, difficulties due to the macroscopic nature of the experiments are encountered. In making comparisons with experiments and in order to assess the adequacy of *microscopic models* of the interaction process, data from probes which are sensitive to the *microscopic structure* of the surface are preferred. Such probes became available to the surface scientist only recently. The reasons for the 'late arrival' of microscopic methods to this area are mainly technological. The nature of these experiments dictates stringent requirements on the experimental conditions. In particular, to achieve well-characterized reproducible conditions, an ultra-high-vacuum (UHV) environment of the order of 10^{-10} Torr has to be maintained. In addition, methods of sample preparation and manipulation and new detection techniques had to be developed. In fact, as emphasized by Hobson,[121] the great majority of devices for creating and measuring UHV are devices in which the solid interface plays a dominant role. Thus advances in the former discipline enhance progress in the latter and *vice versa*. Current progress in UHV technology and other advances in measurement techniques greatly increase the accessibility and use of microscopic surface probes for absorption studies.

Some of the recent applications of microscopic probes to physical adsorption systems are reviewed here. In addition, we briefly indicate the microscopic information which could be gained from thermodynamic measurements (the thermodynamic and statistical mechanics approaches have been reviewed recently [119,120,122]). As an overall approach we chose to illustrate (when available) measurements on adsorp-

---

[119] W. A. Steele, 'The Interaction of Gases With Solid Surfaces', Pergamon, Oxford, 1974.
[120] J. R. Sams, 'Solid State Surface Science', ed. M. Green, Marcel Dekker, New York, 1973, vol. 3, p. 1.
[121] J. P. Hobson, *Adv. Colloid Surface Sci.*, 1974, **4**, 79, and references therein.
[122] J. G. Dash, 'Films on Solid Surfaces', Academic Press, New York, 1975.

tion systems in which metal substrates and rare gases have been used as absorbent and absorbate, respectively. Owing to the number and diversity of techniques covered, we are limited to rather short discussion of the physical principles of the methods (detailed discussions can be found in the indicated references).

A complete characterization of gas–solid interaction systems involves the analysis of geometrical, electronic, and vibronic factors. Since these aspects of the problem are interdependent it is difficult to classify the measurement techniques by those categories. However, it is convenient to perform such a classification following the common practice of the various probes, noting the multiple use of certain techniques when necessary. Thus in our discussion we classify the microscopic techniques as follows: (i) techniques which probe the geometrical and vibronic characteristics (p. 39): low energy electron diffraction (LEED), neutron scattering, low energy atomic and molecular scattering (LEMS); (ii) techniques which probe electronic characteristics (p. 82): u.v. photoemission (u.v. photoemission spectroscopy – UPS), X-ray photoemission spectroscopy (XPS), electron energy loss spectroscopy (EELS), and field emission microscopy (FEM). In addition, we also discuss the use of modern spectroscopic methods for accurate measurement of thermodynamic quantities, mainly the use of LEED and Auger emission in adsorption isotherm studies (p. 100).

On the Use of UPS in Characterization of the Physisorption State. Before turning to a detailed discussion of the above methods we return to the definition of the physisorbed state. As remarked before the customary convention for drawing a distinction between physical and chemical adsorption is on the basis of binding-energy magnitudes (of the order of $10^{-2} - 10^{-1}$ eV and several eV for the former and latter, respectively. Alternatively, somewhat related to the above, physisorption is considered as that state of interaction which does not involve major charge rearrangement and/or chemical bond formation, whereas the chemisorption state involves electron transfer or sharing of electrons between the components of the interacting atom–solid system. An experimental 'verification' of these definitions on the microscopic level requires a spectroscopic probe of the electronic structure of the system. In particular, an experiment which is sensitive to the *valence orbital energy levels* and which can detect variations of these upon adsorption is desired (although core-level shifts can also be detected, see p. 85). Ultraviolet photoemission spectroscopy (UPS) is a method which satisfies the above requirements, and its potential use in investigations of adsorption and reactions of gases with solids has recently been demonstrated.[123–125] Moreover, its utility as a diagnostic tool for differentiating physical and chemical absorption is now recognized.

In the UPS experiment [126–131] (Figure 8a) a u.v. photon of energy $\hbar\omega$ is absorbed

---

[123] J. E. Demuth and D. E. Eastman, *Phys. Rev. Letters*, 1974, **32**, 1123.
[124] W. E. Spicer, K. Y. Yu, I. Lindau, P. Pianetta, and D. M. Collins, in this series, Vol. 5, p. 103.
[125] C. R. Brundle, 'Electron Spectroscopy for the Investigation of Metallic Surfaces and Adsorbed Species' to be published in: 'Electronic Structure and Reactivity of Metal Surfaces', ed. D. Reidel.
[126] D. E. Eastman, in 'Techniques of Metal Research', ed. R. P. Bunshah, Wiley, 1972, Vol. 6, chap. 6.
[127] B. Feuerbacher and R. F. Willis, *J. Phys. (C)*, 1976, **9**, 169.
[128] W. E. Spicer, comments in *Solid State Phys.*, 1973, **5**, 105.

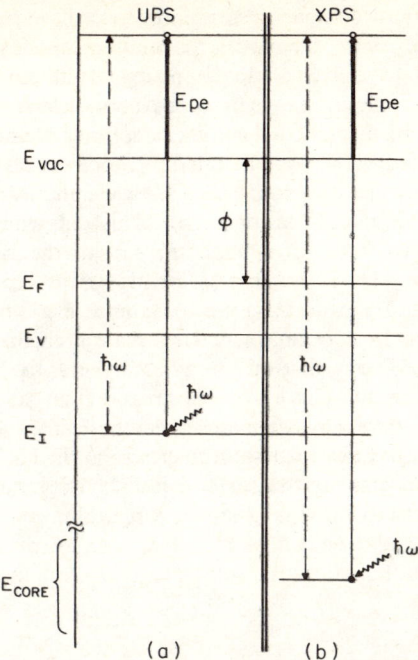

Figure 8 *Schematic one-electron energy-level diagrams for* (a) *UPS and* (b) *XPS. Levels are denoted as:* E_{vac} – *vacuum,* E_F – *Fermi,* E_{core} – *core levels;* ϕ *is the work function of the sample, and* E_{pe} *is the photoelectron energy. Dashed vertical lines indicate known energy quantities, the solid vertical lines indicate electron transitions and a heavy solid line indicates the electron energy measured in the technique*

by a valence electron, exciting it to a state above the vacuum level thus making it possible for the excited electron to leave the sample. A measurement of the final-state energy, E_F, of the photoelectron, determines the initial state energy E_I according to equation (69). Clearly, the range of initial states accessible in UPS

$$E_I = E_F - \hbar\omega \qquad (69)$$

depends on the radiation source, and extends at most $(\hbar\omega - \phi)$ below the Fermi energy, where ϕ is the work function of the sample under study (Figure 8a). In conventional studies resonance lines of noble-gas atoms and ions ($\hbar\omega \sim$ 16—40 eV) are used as photon sources. Recently, u.v. radiation from a synchrotron source has been employed. The surface sensitivity of UPS derives from the short mean free path (5—10 Å) in solids of electrons in the low-energy range of interest.

The basic idea underlying the use of UPS to draw a distinction between the physisorption and chemisorption states of an atom or a molecule interacting with

---

[129] H. D. Hagstrum and E. G. McRae, 'Surface Structure – Experimental Methods', in 'Treatise on Solid State Chemistry', ed. N. B. Hannay, Plenum, New York, Vol. 5.
[130] J. A. R. Samson, 'Techniques of Vacuum Ultraviolet Spectroscopy', Wiley, New York, 1967.
[131] T. N. Rhodin and D. L. Adams, 'Adsorption of Gases on Solids', to be published in 'Treatise on Solid State Chemistry', ed. N. B. Hannay, Plenum, New York, Vol. 5.

a solid surface is that of comparing the photoelectron energy distribution curve (EDC) from the adsorption system with the EDC of the atom or molecule in the gas phase. To facilitate such a comparison, it is useful to discuss first some of the fundamental electronic mechanisms which occur in the photoelectric process. The basic quantity to be investigated is the valence orbital energies, ε_j, of the atom or molecule under study. The photoemission process does not provide a direct measure of ε_j but measures the difference between the initial state of the *total system* before and after the ionization event. This difference is only equal to ε_j, if upon the excitation of the *j*th orbital, the rest of the system remains frozen (Koopman's theorem). However, this 'adiabatic' approximation is not valid in our case, since relaxation processes due to valence electrons responding to the creation of a positive hole occur on a time-scale compatible with that of the photoelectric process. The relaxation energy contributes to the stabilization of the final state by reducing the energy of the frozen final state by an amount E_j^R (Figure 9). The relaxation may originate from atomic and extra-atomic relaxation processes, the first associated with valence electrons of the ionized atom and the latter with other electrons in the system.[132—134] The amount of extra-atomic relaxation contribution to the final state of a photoionized adsorbed molecule is likely to be larger than for the free molecule because of the larger number of electrons available, both from the substrate and from neighbouring adsorbed molecules (Figure 9).

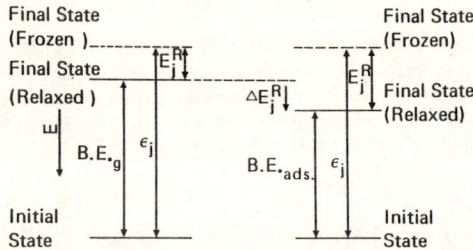

Figure 9 *Relationship between orbital energy ε_j, experimental binding energy* BE, *and relaxation energy E_j^R for a free molecule* (left) *and for an adsorbed molecule* (right) *in the limit of no bonding interactions between the molecule and the surface:* ads *and* g *denote quantities of the adsorbed and gaseous* (free) *states, respectively.* ΔE_j^R *is the relaxation-shift of orbital j*
(Reproduced by permission from 'Electronic Structure and Reactivity of Metal Surfaces')

An additional term involved in comparing electronic binding energies (BE) of gaseous and adsorbed species is the *reference level*. The gas-phase binding energies, BE_g, are experimentally determined with respect to the vacuum level and the condensed state values, BE_{ads}, are given with respect to the Fermi level of the substrate. These two reference levels differ by the work function ϕ of the clean substrate and any changes in ϕ due to electrostatic fields, $\Delta\phi$ (see p. 90). Thus

---

[132] D. A. Shirley, *Chem. Phys. Letters*, 1972, **16**, 220.
[133] P. H. Citrin and D. R. Hamann, *Chem. Phys. Letters*, 1973, **22**, 301.
[134] See also refs. 123 and 125.

we arrive at an approximate relation between the binding energies of the two phases, which in the absence of chemical bonding in the absorption system, is given as (70). In the case of a chemisorptive interaction, an additional term,

$$\Delta_{g,ads} \equiv (BE_g - BE_{ads})_j = (\phi + \Delta\phi) + \Delta E_j^R \qquad (70)$$

$$\Delta E_j^R = E_{j,ads}^R - E_{j,g}^R \qquad (71)$$

$\pm \Delta E_j^B$, is added to the right-hand side of equation (70) to account for modifications in the energy of the valence orbitals involved in bonding and in those levels which respond to the change in valence electron densities due to bond formation.

It has been observed that in many cases the values of ΔE_j^R obtained upon condensation (physisorption) are constant for all levels j and lie between 1 and 3 eV. Consequently, the following diagnostic test has evolved:[123,125] *An atom or a molecule interacting with a solid is in the physisorbed state if all the gas-phase valence energy levels are observed in its photoemission EDC, and their relative position is the same as in the gas phase, and the value of* $[BE_{ads} - BE_g + (\phi + \Delta\phi)]_j$ *is small* (< 3 eV) *and close to being constant for all valence orbitals j*. On the other hand, in the case of chemisorption the near constancy of ΔE_j^R for all j is not to be expected (especially for those orbitals which participate in chemical bond formation). In addition one cannot experimentally separate shifts due to bonding and relaxation ($\pm \Delta E_j^B + \Delta E_j^R$). In the chemisorption case, $\Delta_{g,ads}$ for the valence levels not involved in bonding is expected to be larger than for the physisorption situation. For the other energy levels, $\Delta_{g,ads}$ may be larger or smaller with reference to the condensed state, depending on the sign of ΔE_j^B.

The above principles, have been used in a number of studies of the adsorption of atoms and molecules on surfaces. The original study of Demuth and Eastman[123] of the adsorption of hydrocarbons on Ni(111) is chosen here for illustration. In Figure 10a the photoemission spectra $N(E)$ for clean Ni(111) and upon exposure to benzene at *ca*. 300 K are shown. To enhance the adsorbate-induced changes, a *difference curve*, $\Delta N(E)$, is given in Figure 10b. The dashed background base line estimates the contribution to $\Delta N(E)$, due to attenuation of the Ni *d*-band emission and increased secondary electron emission. The difference curve for several layers of benzene condensed on top of the chemisorbed layer at *ca*. 150 K is shown in Figure 10c. The spectrum for gaseous benzene is given in Figure 10d (note the ionization energy scales). Examination of Figures 10c and 10d reveals that all the relative orbital energies and intensities are essentially unchanged, except that all *ionization energies* measured relative to the vacuum level (ϕ and $\Delta\phi$ are given in the figure) are reduced by $\Delta E_j^R \simeq 1.4$ eV (for all j). Thus, this case satisfies the physisorption criterion given above. Analysis of the results shown in Figure 10b reveals that while the binding energies of all the lower-lying σ orbitals (energies $\gtrsim 6$ eV) are further reduced (owing to an increased relaxation shift, as discussed above), the binding energies of the uppermost two-fold degenerate π-orbitals is increased by *ca*. 1.2 eV. Adsorbate-induced changes in the *d*-band region ($\lesssim 3$ eV) are also noted. The results of the adsorption studies of benzene and for acetylene and ethylene are summarized in Figure 11. It is also interesting to note that saturated hydrocarbons (C_2H_6, C_3H_8, C_4H_{10}, and C_6H_{12}) lacking π-orbitals (which participate in binding on Ni, as demonstrated above), adsorbed only for $T \lesssim 150$ K in the physical interaction mode (no noticeable bonding shifts). Further discussion of the use of UPS in physisorption studies is given on p. 82.

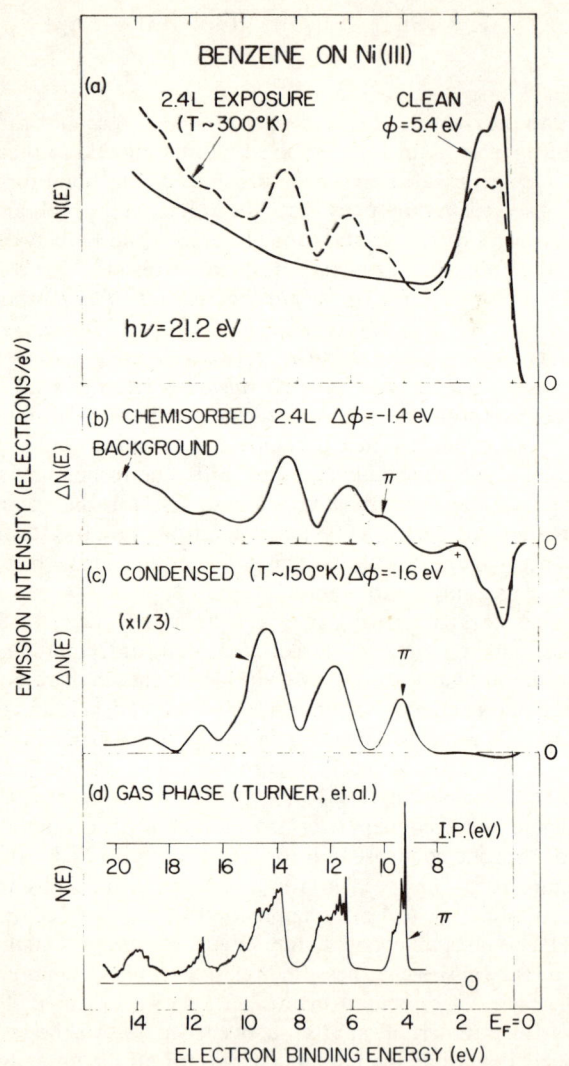

Figure 10 (a) *Photoemission spectra $N(E)$ for Ni(111) and with 2.4×10^{-6} Torr s^{-1} benzene exposure at $T \sim 300$ K;* (b) *adsorbate-induced difference in emission, $\Delta N(E)$, from the clean surface for chemisorbed benzene;* (c) $\Delta N(E)$ *for a condensed benzene layer formed at $T \sim 150$ K with a benzene pressure of 2×10^{-7} Torr;* (d) *gas-phase photoelectron spectra for benzene* (D. W. Turner *et al.* 'Molecular Photoelectron Spectroscopy', Wiley, New York, 1970). *Note that ionization energies ε_i (vacuum level reference) are given by $\varepsilon_i = E_i$ (binding energy) $+ \phi(Ni) + \Delta\phi$*
(Reproduced by permission from *Phys. Rev. Letters*, 1974, **32**, 1123)

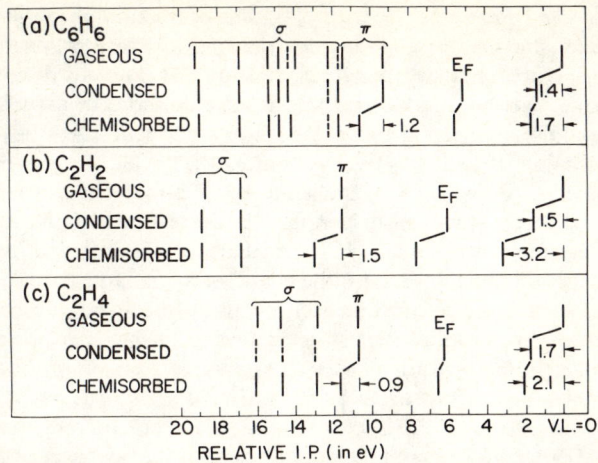

Figure 11 *Vertical ionization energies, Fermi (E_F) and vacuum levels (VL) for the gaseous, condensed and chemisorbed phases of* (a) *benzene,* (b) *acetylene, and* (c) *ethylene, all plotted relative to σ-orbital gas-phase levels. Relaxation shifts are given by the vacuum level shifts while bonding shifts are given for relevant π-orbital shifts. The dotted levels represent less certain orbital ionization energies*
(Reproduced by permission from *Phys. Rev. Letters*, 1974, **32**, 1123)

Experimental Techniques for the Study of Atomic Arrangement in Physisorption Systems.—The geometrical parameters of an interaction system are of fundamental importance for the understanding of interaction processes and related physical phenomena. Accurate measured values of interatomic distances and other structural information are needed for comparison with the results of theoretical calculations. In the particular case of physical adsorption systems, the energetics of the interaction exhibit a marked sensitivity to changes in the structural variables.[119]

The primary source for obtaining structural information about matter is *via* scattering phenomena.[135,136] Both electromagnetic radiation and particles are used, the basic requirement of the experiment being that the wavelength of the incident beam be of the same order as atomic distances in the system under study. Thus, appropriate incidence energies are chosen as dictated by the relation between wavelength and energy, which for *X*-rays, neutrons, and electrons is given as (ε in eV): $\lambda(\text{Å}) \cong 12.4 \times 10^3/\varepsilon, 0.28/\varepsilon^{\frac{1}{2}}$, and $12/\varepsilon^{\frac{1}{2}}$, respectively. Additional requirements on the experimental techniques for surface structure analysis of physisorption systems are surface sensitivity and non-destructiveness. In this section we discuss the recent applications to the study of the atomic geometrical structure of physisorption systems of three spectroscopic methods, namely LEED, neutron diffraction, and atom scattering. They can be operated in either the elastic (energy conserving) or inelastic modes. The former mode is the more useful in geometrical structure studies, the latter in the determination of electronic structure (p. 90).

---

[135] 'Advances in Structure Research by Diffraction Methods', ed. W. Hoppe and R. Mason, Pergamon, Oxford, 1974, Vol. 5.
[136] R. H. Hosemann and S. N. Bagchi, 'Direct Analysis of Diffraction by Matter', North-Holland, Amsterdam, 1974.

LEED: Background Remarks.[129,131,137—141] The diffraction phenomena exhibited in the characteristic energy and angular distributions of low-energy electrons scattered from a predominantly single crystal of Ni(111) were discovered half a century ago by Davisson and Germer [142] in a classic series of experiments which combined ingenuity and a fortunate laboratory accident. At the time of the discovery its major impact was in the context of the de Broglie particle–wave duality hypothesis, for which the Davisson and Germer observations served as confirmation. These authors also identified basic features of the interaction of low-energy electrons with solids. In particular, they concluded that electrons scattered by the topmost layers of the solid predominantly contribute to the measured backscattered electron intensity, owing to the strong absorption of the incident electrons in the solid; hence the scattering of low-energy electrons from solids provides a surface-sensitive probe. Despite the recognition of the potential of the new phenomena for surface structure studies, it was only in the early 1960s [143,144] that interest in LEED as an analytical tool revived, mainly due to advances in UHV technology and the development of the post-acceleration method which allows for the simultaneous display of diffracted beams on a fluorescent screen. To facilitate the presentation of results, we introduce first a few basic terms and concepts connected with LEED and other diffraction experiments.

As we indicated above, the propagation of an incident electron with energy in the range $10 \text{ eV} \leqslant E \leqslant 10^3 \text{ eV}$, in the direction normal to the surface plane is strongly attenuated (mean free path ~ 10 Å). Thus, the 'operational' crystal as 'seen' by the low-energy electron can be most simply described as a set of a small number of atomic layers stacked parallel to a prescribed surface plane. Assuming a periodic atomic arrangement in the planes, they serve as two-dimensional gratings. It is convenient to decompose the wave vector $\vec{k}$ of the incident electron (with wavelength $k = 2\pi/\lambda$) into components normal and parallel to the surface plane, as indicated in Figure 12a and equation (72). The translational symmetry

$$\vec{k} \equiv (k_\perp, \vec{k}_\parallel) \tag{72}$$

in the planes leads to a conservation law for the parallel momentum [equation (73)],

$$\Delta \vec{k}_\parallel \equiv \vec{k}'_\parallel - \vec{k}_\parallel = \vec{g}_{hk} \tag{73}$$

where $\vec{k}'_\parallel$ and $\vec{k}_\parallel$ are the parallel wave vectors of the exit and incident momentum, respectively, and $\vec{g}_{hk}$ is the hk reciprocal vector of the two-dimensional net (h and k are the Miller indices of the net). We can also write equations (74) and (75),

$$k_\parallel = \left(\frac{2mE}{\hbar^2}\right)^{\frac{1}{2}} \sin \theta \tag{74}$$

---

[137] J. B. Pendry, 'Low Energy Electron Diffraction', Academic Press, London, 1974.
[138] P. J. Estrup and E. G. McRae, *Surface Sci.*, 1971, **25**, 1.
[139] M. B. Webb and M. G. Lagally, in 'Solid State Physics', ed. H. Ehrenreich, F. Seitz, and D. Turnbull, Academic Press, New York, 1973, Vol. 28.
[140] T. N. Rhodin and S. Y. Tong, *Phys. Today*, 1975, October, p. 23.
[141] D. L. Adams and U. Landman, 'The Use of Direct Methods in the Analysis of LEED', to be published in 'Advances in Characterization of Metal and Polymer Surfaces', ed. L. H. Lee.
[142] C. J. Davisson and L. H. Germer, *Phys. Rev.*, 1927, **30**, 705.
[143] J. J. Lander, in 'Progress in Solid State Chemistry', Pergamon, 1965, Vol. 2, p. 26.
[144] H. E. Farnsworth, *Adv. Catalysis*, 1964, **15**, 31.

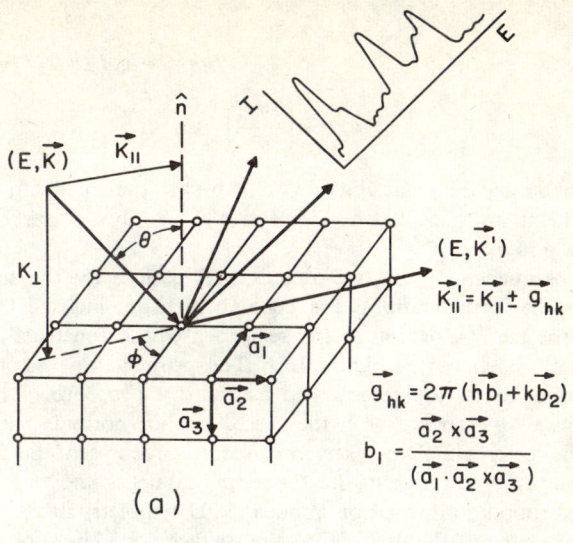

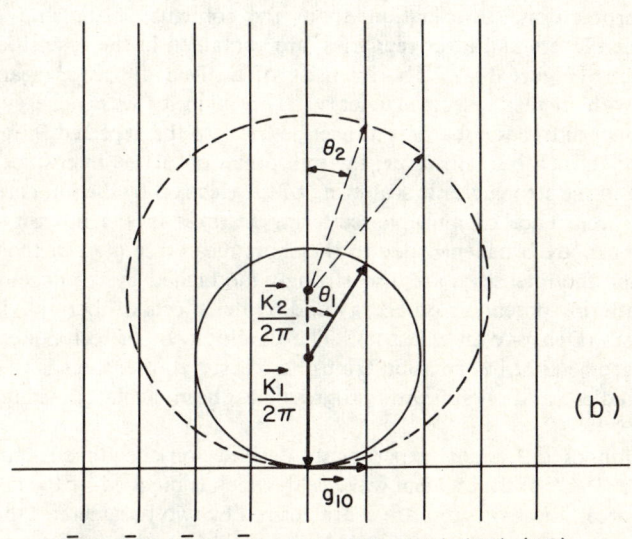

Figure 12 *Kinematical description of low-energy electron diffraction from a surface. (a) An incident beam (angles of incidence θ and ϕ) of primary energy and wave vector, E and $\vec{k}$ respectively, diffracts from a lattice characterized by basis vectors $\vec{a}_1, \vec{a}_2$, and $\vec{a}_3$. The diffracted beams (heavy lines) are characterized by the energy E (elastic) and wave vector $\vec{k}'$. Owing to translational invariance in the plane parallel to the surface the component of the momentum in the surface plane, $\vec{k}_{\parallel}$, is conserved modulus, a reciprocal net vector $\vec{g}_{hk}$, where h and k are the Miller indices of the reciprocal 2D net. A typical record of the intensity vs. primary electron energy for one of the diffracted beams is included. (b) A cross-section in the Ewald-sphere construction, along the (h0) direction, for the case of normal incidence. The constructions for two primary wave vectors ($k_2 > k_1$) are shown. The allowed diffracted beams occur at the intersections of the circle with the reciprocal-net rods. It is also seen that as the energy (wave vector) of the incident beam increases ($k_2 > k_1$) the corresponding diffracted beam extends a smaller angle ($\theta_2 < \theta_1$) with the surface normal*

$$k'_{||} = \left(\frac{2mE}{\hbar^2}\right)^{\frac{1}{2}} \sin \theta'_{hk} \tag{75}$$

where θ and θ'_{hk} are the polar angles with respect to the normal to the surface plane, of the incident and exit beams, respectively (for a specularly diffracted beam $h = k = 0$, and $\theta_{00} = \theta$).

As a consequence of the two-dimensional periodicity the scattered electrons emanate in the form of diffraction beams which are indexed by the (hk) Miller indices of the net. A section in the reciprocal space construction, known as the Ewald sphere, is shown in Figure 12b. The propagation vectors of the incident and diffracted beams are represented as radii of a sphere of radius $k/2\pi$. The intersections of the sphere with the set of surface normals passing through the points of the reciprocal 2D net (*reciprocal-net rods*) represent the allowed diffraction beams. The resulting image of the 2D reciprocal net is known as a *diffraction spot pattern*, and supplies information about the 2D symmetry in the planes, and about the size and shape of the unit cell of the surface net. The spot patterns are open to simple interpretation. Information about the contents of the unit cell, the spacings between layers and layer registries, are contained in the intensities of the individual beams (Figure 12a). The intensity of a given diffracted beam varies continuously with incident electron energy. The intensity *versus* energy records exhibit prominent diffraction features at energies near to the expected Bragg values for inter-planar diffraction. However, the interpretation of the intensities is complicated owing to the strong elastic scattering of the electron by the ion-cores which results in the occurrence of multiple-scattering features in the intensity record. In addition, peaks are broadened due to the short mean free path of the electron in the solid, and the intensities are also strongly modulated by the dependence of the atomic scattering potentials on energy and by the effects of thermal vibrations of the scatterers (Debye–Waller factor). The development of techniques for the extraction of geometrical information from the measured intensities is the subject of current studies, and significant progress has been achieved using several methods.[137,139—141]

The resolution of diffraction experiments depends on the dimensions of the coherence zone [145,146] of the incident wave field, which is dictated by the characteristics of the source. Electron guns are characterized by two parameters: the energy spread in the beam (limited temporal coherence), ΔE, which is determined by thermal spread of the emitted electrons ($\sim 3kT/2$), and the angular spread, β, of the beam due to the finite size of the source (typically, $\beta \sim 10^{-3}$—10^{-2} radians). *The coherence zone* is defined as the largest surface region such that the phase difference between any two points in it differs by less than a specified amount ($\sim \pi/4$ radians) from the phase difference for a plane wave having the same nominal energy and direction as the primary beam. For a typical LEED apparatus the coherence zone is of dimensions 100—500 Å. Consequently, LEED is suitable for the detection of ordered phases over rather small areas (containing, typically, as few as 10^3 atoms). Thus, even if the surface of the sample contains topographical

---

[145] R. D. Heidenreich, 'Fundamentals of Transmission Electron Microscopy', Wiley Interscience, New York, 1964, p. 104.
[146] A. Chutjian, *Phys. Letters*, 1967, **24A**, 516.

imperfections, the existence of small, flat, and ordered portions would give rise to a diffraction pattern.

A remark about nomenclature is appropriate here.[137,138,147,148] Conventionally, the periodicity of crystal surfaces is given with reference to the substrate net. Let us designate the substrate net basis vectors by $\vec{a}_1$ and $\vec{a}_2$ and those of the overlayer by $\vec{b}_1$ and $\vec{b}_2$. If the angles $\angle\,(\vec{b}_1,\vec{b}_2)$ and $\angle\,(\vec{a}_1,\vec{a}_2)$ are equal and the unit cell of the overlayer 2D net is rotated through an angle α relative to the substrate unit mesh, the structure would be designated as $\left(\frac{|\vec{b}_1|}{|\vec{a}_1|} \times \frac{|\vec{b}_2|}{|\vec{a}_2|}\right)\alpha$. If $\alpha = 0$, it is omitted.[149]

LEED Studies of Physisorption Systems. The first application of LEED to the study of 2D ordered structures in physisorption systems is due to Lander and Morrison,[150] who undertook their investigation 'because the existence of such structures has been the subject of much speculation,[151-154] and because it has not been clear that LEED can detect them successfully even if they exist. For example, it has been argued that due to the weakness of physisorbing forces, the electron beam of the instrument would disturb the molecules significantly. The very successful nature of the results demonstrates that LEED provides a powerful means of investigating physisorption and related phenomena.'

The substrate material in the above studies was a single crystal of graphite and the diffraction patterns for the adsorption of a variety of chemical compounds (Br_2, I_2, $C_6H_3Br_3$, GeI_4, $FeCl_3$, As_2O_3, and ZnI_2) and atoms (Cs and Xe) were observed. The temperature behaviour of some of the adsorbed phases was studied, a phase diagram for bromine was constructed, and order–disorder transformations were investigated. As noted by the authors, most of the systems studied by them, with the exception of Xe, are better characterized as weak chemisorption rather than physisorption systems. In the case of Xe adsorption, a $\sqrt{3} \times \sqrt{3}$ 30° unit mesh was observed at 90 K and 1×10^{-3} Torr xenon pressure, and was ascribed to a Xe overlayer in registry with the substrate, with the atoms centred on the substrate hexagons. Heating the specimen by a few degrees caused the extra spots due to the Xe overlayer to become fuzzy and develop into rings of equal diameter, surrounding each of the spots due to the graphite substrate. A further increase in temperature caused the rings to disappear, followed by desorption of the Xe atoms. In the ordered phase, the Xe area per atom was 15.7 Å^2 as compared with 16.8 Å^2 in the close-packed plane of solid (fcc) Xe. The change in diffraction pattern with temperature was interpreted as a sequence of transformations, from 2D ordered crystal → 2D liquid → 2D gas (mobile layer). However, the strong effects of contaminants and the inability to operate at temperatures lower than 90 K cautioned the authors from drawing conclusive deductions.

---

[147] E. A. Wood, *J. Appl. Phys.*, 1964, **35**, 1306.
[148] R. L. Park and H. H. Madden, *Surface Sci.*, 1968, **11**, 188.
[149] When $\angle\,(\vec{b}_1,\vec{b}_2) \ne \angle\,(\vec{a}_1,\vec{a}_2)$ the matrix notation proposed in ref. 148 may be used.
[150] J. J. Lander and J. Morrison, *Surface Sci.*, 1967, **6**, 1.
[151] D. M. Young and A. D. Crowell, 'Physical Adsorption of Gases', Butterworths, London, 1962.
[152] S. Ross and J. P. Oliver, 'On Physical Adsorption', Wiley, New York, 1964.
[153] W. A. Steele and E. R. Kebbekus, *J. Chem. Phys.*, 1965, **43**, 292.
[154] D. Brennan *et al.*, *Phil. Trans. Roy. Soc.*, 1965, **A258**, 325, 347, 375.

Table 3 LEED Studies of Physisorption Systems

Substrate	Gas	Measurements	Ref.
graphite	C_6H_3Br, GeI_4, GeI_2, $FeCl_3$, As_2O_3, ZnI_2, Br_2, I_2, Cs, Xe	diffraction patterns, effect of temperature on diffraction, phase changes, P–T diagram (Br_2), contact potential (Br_2,Cs) and variations with temperature	a
graphite	Xe, Kr	diffraction patterns, effect of temperature on intensities, Auger, Auger amplitude vs. coverage (isotherms), phase changes, isosteric heat	b—d
graphite	Kr	diffraction patterns (high resolution), phase changes	e
Pd(100)	Xe	diffraction patterns, work function and Auger amplitude vs. exposure, temperature effects, P–T diagram, isosteric heat vs. coverage	f
Ir(100)	Xe, Kr	mainly studies of epitaxially grown crystals, diffraction patterns, intensities, temperature effects, Auger, Auger amplitudes vs. coverage, attenuation θ_D	g—j
Nb(100)	Ne, Ar, Kr, Xe	studies of epitaxially grown crystals, diffraction patterns, intensities, electron-beam effects (desorption curves), characteristic-loss spectra	k—m
Cu(100), (110), (111), (211), (311)	Xe	diffraction patterns, variations with exposure, surface potential, Auger	n—p
Ag(111), (110), (211)	Xe	diffraction patterns, variations with exposure, surface potential, Auger	p
Cu(211), Ag(111)	Kr	diffraction patterns, variations with exposure, surface potential	q
Ag(111)	Xe	surface potential vs. pressure (isotherms), Auger amplitude vs. pressure (isotherms), isosteric heat, electron energy loss spectrum	r
Ag(111)	Xe	diffraction patterns, intensities, temperature effects, P–T diagram, isosteric heat, θ_D	s
Ag(100), (110)	Xe, Kr	diffraction patterns, temperature effects, work function, combined studies with ellipsometry	t

[a] J. J. Lander and J. Morrison, *Surface Sci.*, 1967, **6**, 1; [b] J. Suzanne, J. P. Coulomb, and M. Bienfait, *Surface Sci.*, 1973, **40**, 414; 1974, **44**, 141 (Xe); [c] J. P. Coulomb, J. Suzanne, M. Bienfait, and P. Masri, *Solid State Comm.*, 1974, **15**, 1585 (Xe); [d] H. M. Kramer and J. Suzanne, *Surface Sci.*, 1976, **54**, 659; [e] M. F. Chinn and S. C. Fain, jun., *J. Vacuum Sci. Technol.*, 1977, **14** (to be published); [f] P. W. Palmberg, *Surface Sci.*, 1971, **25**, 598; [g] A. Ignatiev, A. V. Jones, and T. N. Rhodin, *ibid.*, 1972, **30**, 573; [h] A. Ignatiev and T. N. Rhodin, *Phys. Rev.*, 1973, **B8**, 893; [i] S. Y. Tong, T. N. Rhodin, and A. Ignatiev, *ibid.*, 406; [j] A. Ignatiev, T. N. Rhodin, and S. Y. Tong, *Surface Sci.*, 1974, **42**, 37; [k] J. M. Dickey, H. H. Farrell, and M. Strongin, *ibid.*, 1970, **23**, 448; [l] H. H. Farrell, M. Strongin, and J. M. Dickey, *Phys. Rev.*, 1972, **B6**, 4763; [m] H. H. Farrell and M. Strongin, *ibid.*, p. 4711; [n] M. A. Chesters and J. Pritchard, *Surface Sci.*, 1971, **28**, 460 [Cu(100)]; [o] H. Papp and J. Pritchard, *ibid.*, 1975, **53**, 371 [Cu(311)]; [p] M. A. Chesters, M. Hussain, and J. Pritchard, *ibid.*, 1973, **35**, 161 [Cu(100)], (110), and (111); [q] R. H. Roberts and J. Pritchard, *ibid.*, 1976, **54**, 687; [r] G. McElhiney, K. Papp, and J. Pritchard, *ibid.*, 1976, **54**, 617; [s] P. I. Cohen, J. Unguris, and M. B. Webb, *ibid.*, 1976 (to be published); [t] R. F. Steiger, J. M. Morabito, jun., G. A. Somorjai, and R. H. Muller, *ibid.*, 1969, **14**, 279.

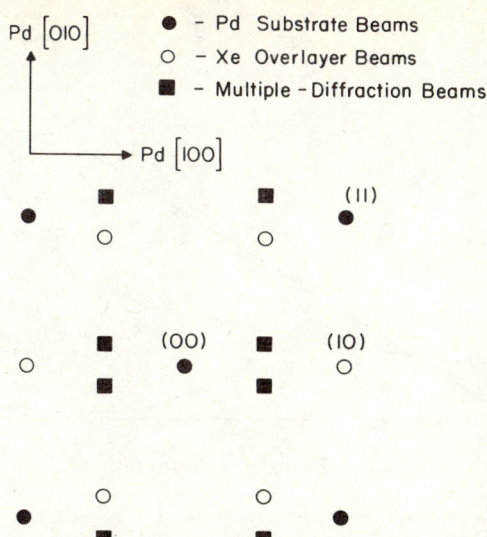

Figure 13 *Schematic representation of observed LEED pattern from hexagonal overlayer structure of* Xe *on* Pd(100) *at* 77 K
(Reproduced by permission from *Surface Sci.*, 1971, **25**, 598)

Since the original study by Lander and Morrison, a number of studies of physisorption systems using LEED have been performed. A list of the systems studied along with the type of measurements made and literature references is given in Table 3. In the following we illustrate some of the results of these investigations, and demonstrate the variety of observations which can be performed.

Apart from the studies of the adsorption of rare gases on graphite, all other LEED studies of the physisorption of rare gases employed metal substrates. The adsorption of Xe on the (100) surface of Pd has been examined by Palmberg [155] using LEED, work function and Auger electron spectroscopy (AES). The diffraction spot pattern obtained at 77 K and monolayer coverage is shown in Figure 13. It was interpreted as due to two equivalent orthogonal domains with hexagonal close packing. The Xe atoms occupy an area of 17.2 Å^2. Only at near to monolayer coverage was long-range order detected, indicating that the lateral Xe–Xe forces are dominant in determining the structure (Xe–Xe spacing in the ordered layer was determined to be 4.48 Å, compared to 4.37 Å for bulk Xe). The role of the substrate is to orient the close-packed hexagonal overlayer with respect to the substrate net. The kinetics of formation of the hexagonal monolayer was monitored by measuring the work function, the intensity of the (10) LEED beam from the Xe overlayer and the Xe Auger peak amplitude as a function of Xe exposure as shown in Figure 14. The exposure after which the Auger amplitude and LEED beam intensity remain constant was taken as corresponding to the ordered monolayer coverage (5.8×10^{14} atoms cm^{-2}). The isoteric heat of

---

[155] P. W. Palmberg, *Surface Sci.*, 1971, **25**, 598.

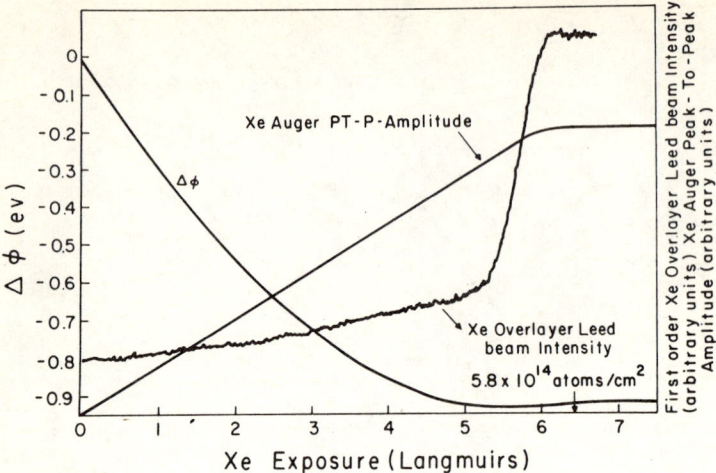

Figure 14 *Work function*, ϕ, Xe *Auger peak-to-peak amplitude and* Xe *overlayer LEED intensity vs.* Xe *exposure at* 77 K; 1 Langmuir = 10^{-6} Torr s^{-1}
(Reproduced by permission from *Surface Sci.*, 1971, **25**, 598)

adsorption, q, was obtained by the continuous work function technique.[155,156] In this method the work function ϕ is used as a measure of surface coverage θ, in combination with measurements of the variations in the intensity of an overlayer beam. By measuring these quantities as a function of temperature for a series of gas pressures, and through the use of the Clausius–Clapeyron equation (76),

$$\left|\frac{d(\ln P)}{d(1/T)}\right|_\theta = -q/k_B \tag{76}$$

where θ is the coverage, the isosteric heat can be evaluated from a family of θ *vs.* T isobars (P = constant). The isosteric heat exhibits a monotonic decrease as a function of coverage (see Figure 15), indicative of a repulsive lateral Xe–Xe interaction.

The effect of the substrate on the atomic arrangement in the physisorbed layer is a topic of interest from both the theoretical and practical aspects. The determination of surface areas *via* physical adsorption measurements involves estimation of the monolayer capacity and then conversion of this capacity into an equivalent area.[157] This conversion involves a model of atomic packing in the complete monolayer. The two main models proposed are the lattice-packing (or site adsorption) model [158,159] and the close-packing model.[155,160] The above results for the adsorption of Xe on Pd(100) favour the latter model. Additional evidence that close packing is favoured in the case of absorption of rare gases on metals

---

[156] J. C. Tracy and P. W. Palmberg, *Surface Sci.*, 1969, **14**, 274.
[157] S. J. Gregg and K. S. W. Sing, 'Adsorption, Surface Area and Porosity', Academic Press, London, 1967.
[158] J. R. Anderson and B. G. Baker, *J. Phys. Chem.*, 1962, **66**, 482.
[159] D. Brennan and M. J. Graham, *Phil. Trans. Roy. Soc.*, 1965, **A258**, 325.
[160] M. A. Chesters, M. Hussain, and J. Pritchard, *Surface Sci.*, 1973, **35**, 161.

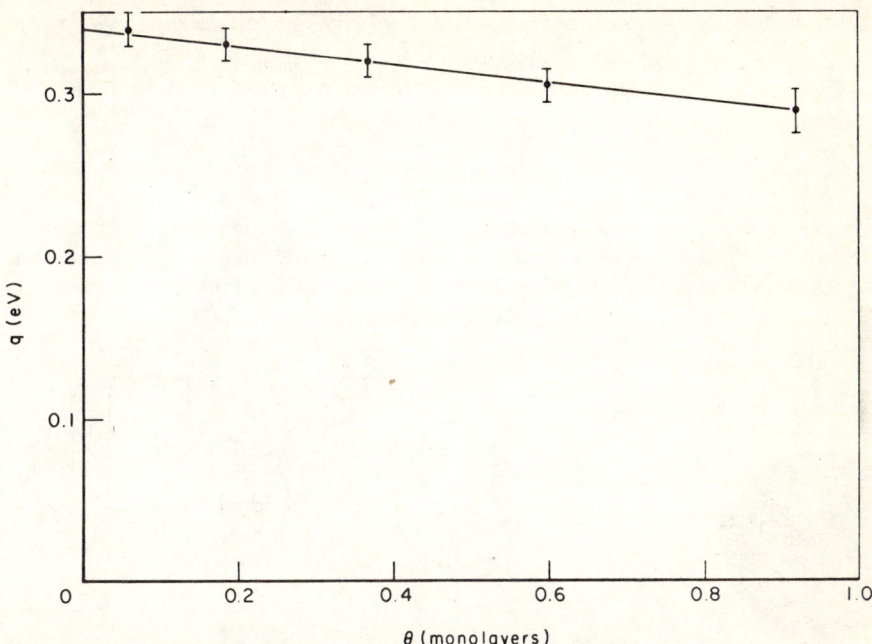

Figure 15 *Isosteric heat of adsorption versus overlayer coverage, for Xe adsorption on Pd(100) as obtained via the Clausius–Clapeyron equation. The decrease of q with coverage indicates repulsive interactions in the adsorbed layer*
(Reproduced by permission from *Surface Sci.*, 1971, **25**, 598)

is provided by the extensive studies of Pritchard and his collaborators [160—164] on the adsorption of Xe and Kr on a number of faces of fcc metals (Ag, Cu). Whereas site adsorption may take place at the initial stages of adsorption (and consequently, a marked dependence on the crystal face of the substrate would be expected), a close-packed arrangement is observed at coverages nearing completion of a monolayer (indeed, apart from adsorption on Cu(110),[160] no long-range-ordered sub-monolayer structures have been observed by these authors). A comparison between the results derived from the examination of LEED patterns for the adsorption of Kr and Xe on Ag(111) and Cu(211) (which are markedly different crystallographically) serves as a demonstration of the above argument. In Table 4 interatomic spacings and occupied areas per adatom for the above adsorption systems and for the rare-gas solids (close-packed) are given. Comparing the spacings and area ratios between the solid and adsorption systems, and between the adsorption systems themselves, it is concluded that within the accuracy of the measurements the atomic arrangement in the monolayer is close-packed hexagonal, on both Ag(111) and the 'stepped' Cu(211) surfaces.

As mentioned above, ordered structures of the overlayer in registry with the

---

[161] M. A. Chesters and J. Pritchard, *Surface Sci.*, 1971, **28**, 460.
[162] H. Papp and J. Pritchard, *Surface Sci.*, 1975, **53**, 371.
[163] G. McElhiney, H. Papp, and J. Pritchard, *Surface Sci.*, 1976, **54**, 617.
[164] R. H. Roberts and J. Pritchard, *Surface Sci.*, 1976, **54**, 687.

Table 4 *Spacing and area occupied per adatom (after Roberts and Pritchard[164])*

	Spacing/Å		Occupied area/Å^2		Spacing ratio	Area ratio
	Xe	Kr	Xe	Kr	Xe:Kr	Xe:Kr
Solid	4.38	4.06	16.6	14.3	1.08	1.16
Ag(111)	4.51	4.19	17.7	15.2	1.08	1.16
Cu(211)	4.38	4.18	16.9	15.2	1.06	1.12

substrate, which are not close-packed, can occur at submonolayer coverages. Figure 16a illustrates such a case: the diffraction pattern from ~0.2 monolayer of Xe (similar results were obtained for Kr) on (1 × 5) Ir(100)* at a temperature

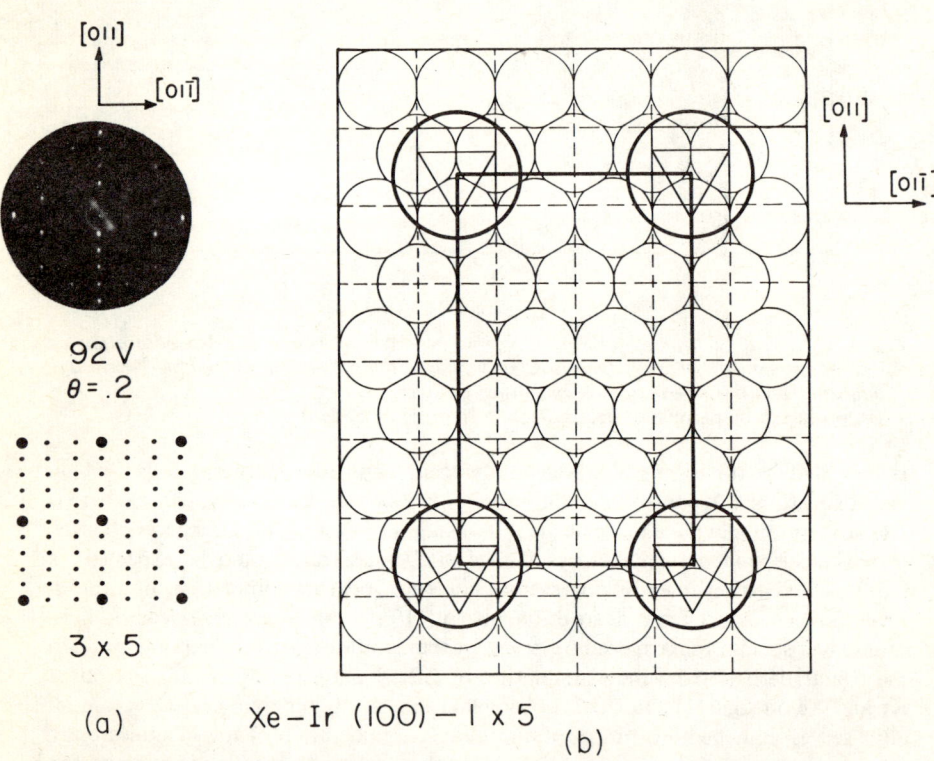

Figure 16 (a) *The 3 × 5 diffraction pattern of Xe adsorbed on 1 × 5 Ir(100) at 55 K with a coverage of 0.2 monolayer (similar results were obtained for the adsorption of Kr under the same conditions).* (b) *Hard-sphere structure analysis (real space description) of the diffraction patterns shown in (a). Substrate atoms are indicated by light circles and overlayer atoms by heavy circles with the unit cell denoted by the heavy lines. Note the adsorbed xenon at the sites of coincidence of hexagonal layer three-co-ordinate sites (triangles) and typical fcc(100) plane four-co-ordinate sites (squares)*
(Reproduced by permission from *Surface Sci.*, 1972, **30**, 573)

---

* Two surface structures of Ir(100) have been observed. The (1 × 1) structure can be transformed into a stable (1 × 5) structure upon heating to 2100 K for more than 1 h in a vacuum of 1 × 10^{-10} Torr

of 55 K, measured by Ignatiev, Jones, and Rhodin.[165] The diffraction pattern contains extra spots due to a (3 × 5) structure of the overlayer. A direct-space model of the structure is shown in Figure 16b. Hard-sphere models of Xe adsorbed on the (1 × 1) and (1 × 5) Ir(100) surfaces are shown in Figure 16c. These models were constructed on the basis of the assumption that in the preferred configuration the Xe atom would maximize its co-ordination to the substrate atoms. The two different Xe–Ir interlayer d_z spacings (Figure 16c, d_{z_1} = 3.14 Å, and d_{z_2} = 2.96 Å) should lead to extra peaks in the LEED intensity *versus* energy spectra. Because of the kinematical nature of electron scattering from Xe, these extra peaks are expected to be simple Bragg-type, and analysis of their location in energy is open to simple geometrical interpretation. Normalized intensity curves of the (00) beam are shown in Figures 17 and 18 for the (1 × 5) and (1 × 1) Ir(100) adsorption systems, respectively. Comparing the intensities for the clean and adsorbed situation in each case, the extra peaks due to the Xe overlayer are readily identified and labelled as to their order number. A simple Bragg law analysis, in which the energy of the extra peaks is plotted against the sequence of the diffraction order (E vs. n^2 plot) yields straight lines with slopes proportional to the interlayer spacing.

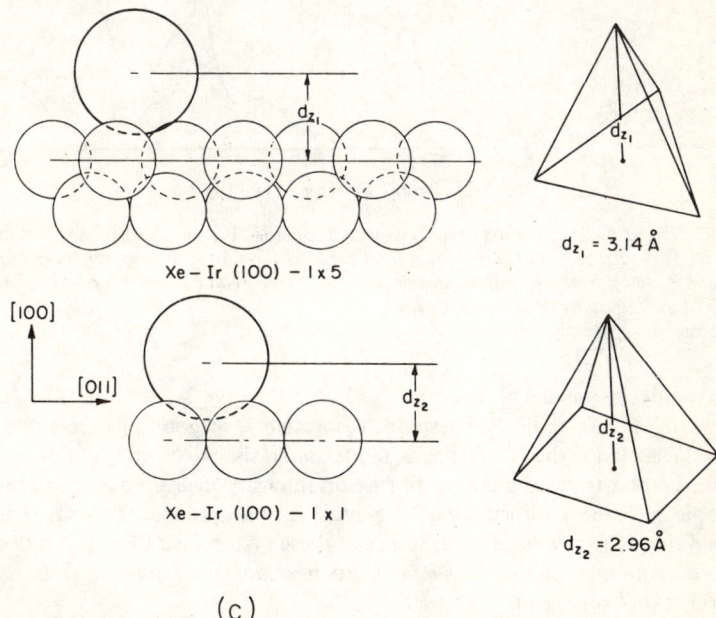

Figure 16 (c) *Hard-sphere models of xenon adsorbed on both iridium surface structures. The xenon on the 1 × 5 Ir(100) is adsorbed in three-co-ordinate sites on the hexagonal surface with the schematic picture showing the interlayer d-spacing (d_{z_1}) equal to 3.14 Å. The Xe on the 1 × 1 Ir(100) is adsorbed in four-co-ordinate sites and the schematic picture shows the interlayer spacing (d_{z_2}) is equal to 2.96 Å*
(Reproduced by permission from *Surface Sci.*, 1972, **30**, 573)

[165] A. Ignatiev, A. V. Jones, and T. N. Rhodin, *Surface Sci.*, 1972, **30**, 573.

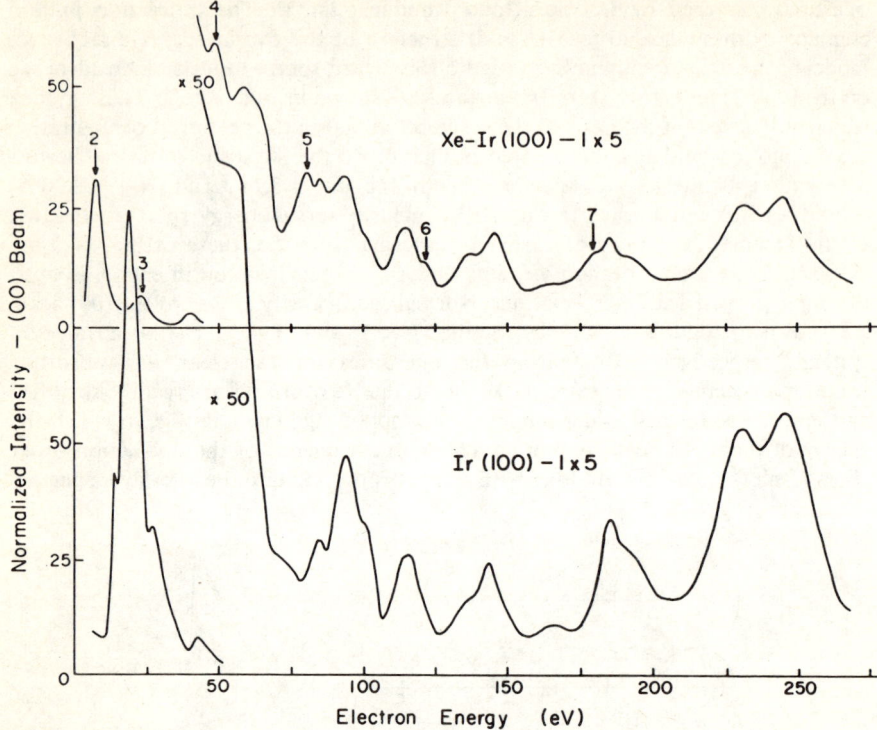

Figure 17 *The LEED intensity–energy spectra of the* 1×1 *Ir(100) surface and the* 1×1 *Ir(100) surface with 0.2 monolayer of adsorbed xenon. Both spectra were taken at 55 K and 5° away from normal incidence. All the extra peaks in the adsorbed xenon spectrum are labelled as to their order number*
(Reproduced by permission from *Surface Sci.*, 1972, **30**, 573)

Such an analysis yielded [47] the values 3.16 ± 0.03 and 2.95 ± 0.03 Å for the Xe–(1×5)Ir and Xe–(1×1)Ir systems, respectively, in good agreement with the values deduced from the hard-sphere model analysis of the diffraction patterns. The above demonstrates the use of diffraction intensity measurements in obtaining microscopic structural information of physisorption systems. Indeed, it is now recognized that in future structural studies of these systems *via* LEED, it is desirable to perform *diffraction pattern and intensity* measurements over a *wide range of temperatures and coverages*.

Most recently, a detailed LEED investigation of Xe monolayer adsorption on Ag(111) has been performed by Cohen, Unguris, and Webb.[166] To circumvent the problem of contamination of the surface by impurities from the residual gas which is accentuated at cryogenic temperatures, special precautions have been taken to obtain an especially good vacuum *via* the use of a cryopump shielding which surrounds the central dewar and which can be cooled to near

---

[166] P. I. Cohen, J. Unguris, and M. B. Webb, *Surface Sci.*, 1976, **58**, 429.

Microscopic Approaches to Physisorption

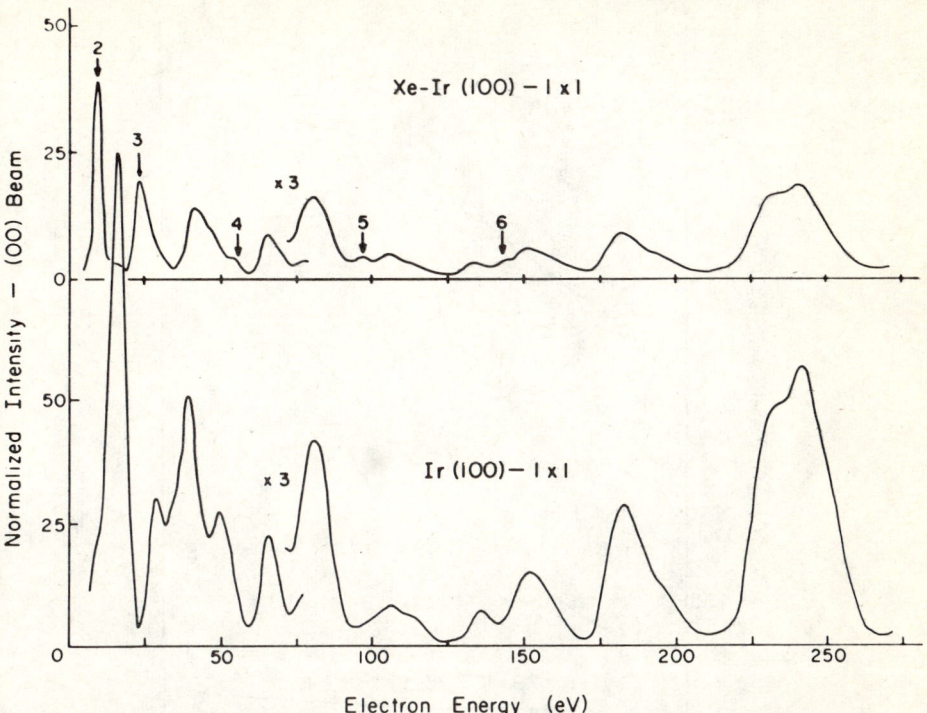

Figure 18 *The LEED intensity–energy spectra of the* 1×5 *Ir(100) surface and the* 1×5 *Ir(100) surface with 0.2 monolayer of adsorbed xenon. Both spectra are taken at 55 K and 5° away from normal incidence. All extra peaks in the xenon-adsorbed spectrum are labelled as to their order number*
(Reproduced by permission from *Surface Sci.*, 1972, **30**, 573)

liquid He temperatures, while the crystal is still at several hundred Kelvin. Measurements over a wide temperature range (25 K and up) were performed. From the observed diffraction patterns (presented as intensity *versus* angle of exit profiles), the behaviour of the Xe–Xe interatomic distance *versus* temperature was determined as shown in Figure 19. A measurement of the lateral spacing for a thick film at 25 K is also included. The spacing of bulk Xe and 1.5 times the spacing of Ag, both from X-ray data [167,168] are shown, for reference. It is observed that at 25 K the Xe–Xe spacing is 2.2% larger than the X-ray value for Xe and 1.7% larger than the thick-film value. The structure is distinctly out of registry with the substrate. The lateral thermal expansion coefficient, α, of the layer is 2×10^{-4} K^{-1}, nearly the same as for bulk Xe but an order of magnitude larger than for Ag. These observations provide additional evidence that the xenon atoms are 'quite oblivious of the structure of the silver substrate'.[166]

[167] G. L. Pollack, *Rev. Mod. Phys.*, 1964, **36**, 748.
[168] 'American Institute of Physics Handbook', ed. D. E. Gray, McGraw-Hill, New York, 1972.

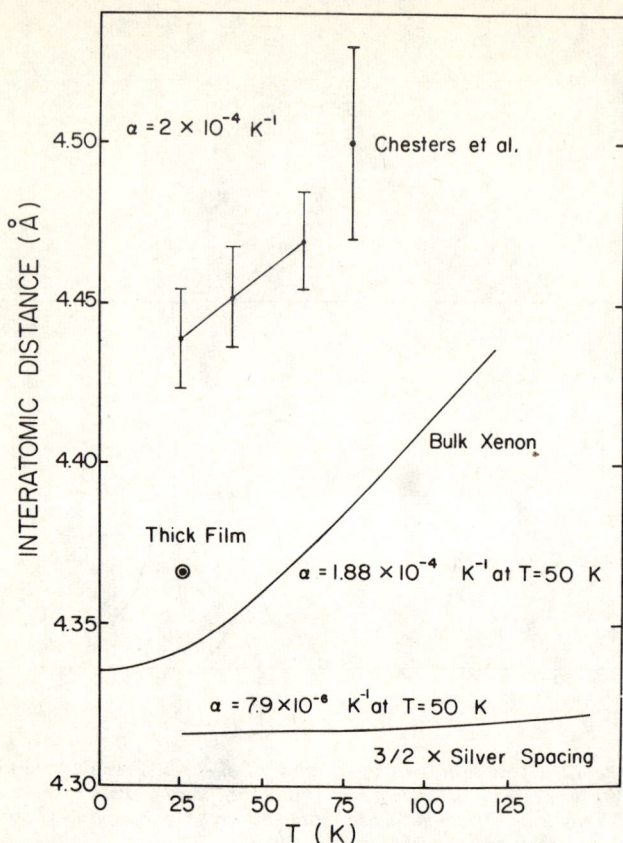

Figure 19 *Interatomic spacing at various temperatures for the full monolayer of* Xe *on* Ag(111) *(after compression) and for the thick* Xe *film, derived from LEED measurements. Bulk* Xe *and* Ag *values are shown for comparison. The thick film measurement is less precise than those for the monolayer because of more diffuse scattering and a larger angular width of the diffracted beam from the single preparation examined*
(Reproduced by permission from *Surface Sci.*, 1976, **58**, 429)

Extensive measurements of intensity *versus* incident energy profiles have been performed by the above authors.[166] For illustration we show in Figure 20 intensity records for the specular beam from both the clean Ag(111) (solid) and Xe/Ag(111) (dash) systems, measured at 25 K for polar angles of incidence 0—18°. An interesting feature of the intensities is the similarity of the energy values of peak positions between the two systems. This example emphasizes the importance of relative intensity considerations in the analysis of LEED intensities for structure determination.[166,169] From the analysis of the intensities, a value of 3.5 ± 0.1 Å was determined for the Xe–Ag layer spacing (compared to 2.36 Å for Ag(111)). In order to compare this result with hard-sphere models, two locations of the Xe

---

[169] U. Landman and D. L. Adams, *J. Vacuum Sci. Technol.*, 1974, **33**, 585.

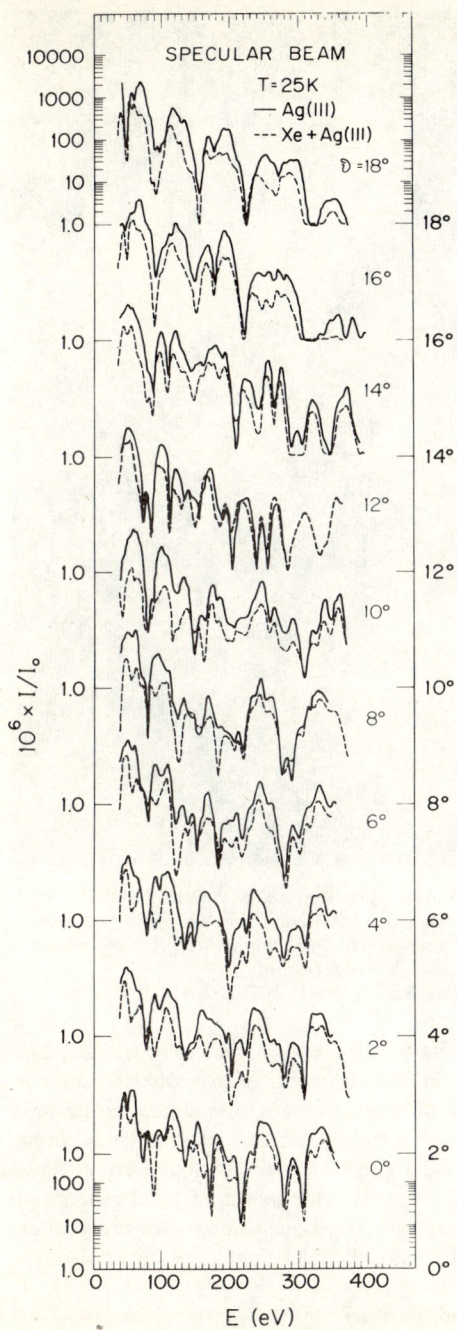

Figure 20 *Intensity vs. incidence energy records of the Xe(10) beam for polar angles of incidence $0° < \theta < 18°$. Intensities for clean Ag(111) are denoted by a solid line and for the Xe covered substrate by dashed lines. The sample temperature was 25 K. The angles given in the right-hand margin indicate the base lines for the corresponding curves. Note the similarity in peak positions but not in relative intensities between the clean and adsorption systems*
(Reproduced by permission from *Surface Sci.*, 1976, **58**, 429)

Figure 21 (a) *LEED photograph of krypton overlayer on graphite at* 67 K, 1.1 × 10⁻³ Pa. *Photo was taken with a* 144 eV, 3 nA *normally incident beam and* −92 V *suppressor grid voltage. The outer six spots result from graphite first-order plus krypton second-order beams. The inner six spots are due to krypton first-order beams*
(Reproduced by permission from *J. Vacuum Sci. Technol.*, 1977, **14**, 314)

atom with reference to the substrate have been considered. The hard-sphere distances $[\sigma_{gs} = \frac{1}{2}(\sigma_{gg} + \sigma_{ss})]$ depend on the assumed adsorption site and are given as 3.2 and 3.31 Å for the three-fold dimple and the saddle point, respectively, and 3.61 Å for adsorption directly on top of a substrate atom. Within the accuracy of the determination, the stated value of 3.5 ± 0.1 Å agrees with the latter value, while being significantly larger than the values corresponding to the first two locations. This observation is in agreement with the conclusions of our theoretical studies which indicate that in general the rare-gas metal spacing is larger than the value obtained by the 'sum-of-radii rule'.[2]

Finally, we illustrate some of the results of a recent study by Chinn and Fain of the adsorption of krypton on the basal plane of graphite.[170] In this study, a

[170] M. P. Chinn and S. C. Fain, jun., *J. Vacuum Sci. Technol.*, 1977, **14**, 314.

Figure 21 (b) *LEED photograph of krypton overlayer on graphite at 57 K, 9×10^{-4} Pa. Photo was taken with a 144 eV, 3 nA normally incident beam and -48 V suppressor grid voltage. The triple splitting of the inner spots indicates compression of the krypton layer out of registry*
(Reproduced by permission from *J. Vacuum Sci. Technol.*, 1977, **14**, 314)

LEED apparatus was designed, in which spot patterns were observed using a flat-channel electron multiplier plate with phosphor screen. This design permitted high-resolution investigations at much lower current densities (3 nA at 144 eV) than normally employed, thus greatly reducing beam-related problems such as local heating, electron-induced desorption, or electron-assisted reactions with residual gases (see p. 97). To demonstrate the quality and resolution achieved in this study, we show in Figure 21a the spot pattern of a krypton overlayer on graphite obtained at 67 K and a pressure of 1.1×10^{-3} Pa. The outer six spots result from graphite first-order plus krypton second-order beams. The inner six spots are due to krypton first-order beams. Upon exposure to 9×10^{-4} Pa at 57 K (a pressure about 10^2 larger than the monolayer condensation pressure at this temperature), a triple splitting of the inner krypton spots was resolved, as shown in Figure 21b. The above splitting indicates a compression of the krypton monolayer of $3 \pm 0.5\%$. In conventional LEED measurements, such splitting would have been seen as increase in spot size. The unique high-resolution obtained by Chinn and Fain permits very precise measurements of misregistry of the overlayer, which in turn can be used in assigning characteristic features in the phase diagram of the system under study (see p. 97).

The studies reviewed above illustrate the diversity of microscopic structural information which can be derived from LEED investigations of physisorption systems. In addition to the structural data, LEED studies provide information about vibrational and electronic characteristics (pp. 62 and 90) and can be used in mapping phase-diagrams of physisorbed films (p. 97).

Neutron Scattering: Background Remarks.[135,171-173] The scattering of slow neutrons (*i.e.* energies of the order 10^{-2} eV) is one of the routine methods in structural analysis of matter. Both the elastic and the inelastic modes are used, the former for structural studies and the latter for the investigations of excitation spectra and their dispersion (in particular phonons and magnons).

The theory of X-ray scattering by crystals [174-176] is closely related to the theory of neutron scattering with the replacement of the atomic scattering factor, f, in the former theory by the scattering length, b, in the latter.[135] The neutron scattering length does not vary in a systematic manner with atomic number as does f, and the small nuclear size means that b is independent of the collisional momentum transfer. (The quantity b represents the coherent scattering amplitude from an atom.[172] As long as $k|b| \ll 1$, the elastic scattering cross section is given as $\frac{d\sigma}{d\Omega} = b^2$.) As we indicated on pp. 40—43 the reciprocal space of a stack of 2D ordered layers consists of a 2D reciprocal net and a set of reciprocal-net rods perpendicular to the 2D net. The hk structure factor for such a system (containing n atoms per unit cell) can be written as equation (77), where b_j is the scattering length of atom j

$$F_{hk} = \sum_{j=1}^{n} b_j \exp\{-i[2\pi(hx_j + ky_j) + d_z K_\perp z_j]\} \tag{77}$$

in the unit cell, x_j, y_j, and z_j are its reduced coordinates, d_z is the inter-layer spacing, and $K_\perp$ is the component of the scattering vector normal to the 2D lattice mesh. The first two terms in the exponent are the result of the satisfaction of the two (planar) corresponding Laue equations, while the third term corresponds to the normal direction (it becomes a Bragg point, characterized by the Miller index l in the 3D periodic case). In the studies to be described below a modified version of the theory of X-ray diffraction in random layer lattices [177] was employed. In

---

[171] G. Placzek and L. van Hove, *Phys. Rev.*, 1954, **93**, 1307.
[172] L. van Hove, *Phys. Rev.*, 1954, **95**, 249.
[173] C. Kittel, 'Quantum Theory of Solids', Wiley, New York, 1963, chap. 19.
[174] B. E. Warren, '*X*-Ray Diffraction', Addison-Wesley, Reading, Mass. 1969.
[175] L. V. Azaroff, R. Kaplow, N. Kato, R. J. Weiss, A. J. C. Wilson, and R. A. Yound, '*X*-Ray Diffraction', McGraw-Hill, New York, 1974.
[176] A. Guinier, '*X*-Ray Diffraction', W. H. Freeman, San Francisco, 1963.
[177] B. E. Warren, *Phys. Rev.*, 1941, **54**, 693.

this theory [178] the expression for the intensity of the (hk) diffraction beam is given by equations (78) and (79), where 2θ is the scattering angle, m_{hk} is the multiplicity

$$I_{hk}(\theta) = Nm_{hk}|F_{hk}(k_\perp)|^2 G(\theta) \qquad (78)$$

$$G(\theta) = \frac{f^2(\theta)e^{-2M}}{(\sin\theta)^{\frac{3}{2}}} \left(\frac{L}{\pi^{\frac{1}{2}}\lambda}\right)^{\frac{1}{2}} F(a)H(\gamma) \qquad (79)$$

of the hk-th reflection, F_{hk} is the structure factor, $f(\theta)$ is the molecular form factor, e^{-2M} is the Debye–Waller factor, λ is the wavelength, N the number of scatterers, L is the array size parameter, and

$$F(a) \equiv \int_0^\infty \exp[-(x^2 - a^2)]dx \qquad (80)$$

where

$$a = (2\pi^{\frac{1}{2}}L/\lambda)(\sin\theta - \sin\theta_{hk}) \qquad (81)$$

and

$$\theta_{hk} = \sin^{-1}(\lambda/d_{hk}) \qquad (82)$$

with d_{hk} being the 2D plane spacing for the hk-th reflection. $H(\gamma)$ is a probability function, which has been introduced in order to account for deviation from complete orientational randomness in the system (orientational alignment). Clearly the structural information contained in the intensities of the diffracted beams is due to the structure factor term $|F_{hk}(k_\perp)|^2$, which is modulated by the other terms in equation (78). In analysing the neutron scattering experiments (described below), a parametrical fitting of the theoretical expression to the observed intensities has been performed allowing the extraction of structural data.[178]

Thus far we have considered the elastic scattering mode only. An incident beam of monoenergetic slow neutrons undergoes an interaction with the crystal during which its energy and momentum, E and $\hbar k$, respectively, are changed to the final values E' and $\hbar k'$. This change in energy and momentum is accomplished through the creation or annihilation of a phonon in the crystal. Since the energy of such a phonon is $\hbar\omega$ and its momentum is $\hbar\vec{q}$, we can write equations (83) and (84), where

$$E' - E = (k'^2 - k^2)\hbar^2/2m = \hbar\omega(\vec{q}) \qquad (83)$$

$$(\vec{k} - \vec{k}') = (2\pi\vec{\tau} + \vec{q}) \qquad (84)$$

$\vec{\tau}$ is an arbitrary lattice vector. When $\vec{q} = 0$ the equation reduces to the Bragg scattering formula. Since the dispersion of the frequency has the property $\omega(\vec{q}) \equiv \omega(\vec{q} + 2\pi\vec{\tau})$, equation (84) is equivalent to (85) and equation (83) to (86).

$$\omega(\vec{q}) = \omega[(\vec{k} - \vec{k}')] \qquad (85)$$

$$\hbar(k'^2 - k^2)/2m = \omega(\vec{q}) \qquad (86)$$

This expression is applicable to the experimental determination of the dispersion curve $\omega(\vec{q})$. By measuring the intensity over a large number of exit angles a number of $\vec{q}$ values are obtained, which together with the corresponding energy transfer

---

[178] J. K. Kjems, L. Passell, H. Taub, J. G. Dash, and A. D. Novaco, *Phys. Rev.*, 1976, **B13**, 1446.

$\omega(\vec{q})$ yield the dispersion curve of the elementary excitation in the material.[179,180] Such experiments for the adsorption of ^{36}Ar on graphite [181] are described below.

An important criterion for a reliable experimental probe in structure analysis is that it should be non-destructive. Some of the LEED studies mentioned above have indicated that at the experimental conditions of LEED, electron-stimulated desorption of weakly adsorbed species occurs (see also p. 97). Although diffraction experiments of low-energy electrons can be performed on almost any surface which exhibits long-range periodicity of the order of the LEED coherence length with no restrictions of large specific areas, use of the technique may be limited to the more firmly bound physisorbed systems (*e.g.* those systems which involve Xe and Kr as adsorbents as compared to Ne or He). On the other hand, when neutrons are used, a broad range of coverage and temperatures of weakly bound systems can be investigated since desorption by the neutron beam is negligible. In addition to the geometrical structure analysis, the inelastic neutron scattering methods allow the study of the excitation spectra of these systems with high precision. However, there are several severe limitations of neutron diffraction. The major problem is the background scattering from the substrate due to the strong penetration of neutrons into the material, which restricts the choice of gases and substrates. The neutron scattering and cross-sections of several gases are given in Table 5, and it can be seen that the neutron scattering from N_2 and ^{36}Ar is

Table 5 *Neutron scattering and capture cross-sections of some gases commonly used for physisorption studies (after Kjems et al.[178])*

Gas	$4\pi(\Sigma_j b_j)^2/b^a$	$l/\text{Å}^b$	$4\pi(\Sigma_j b_j)^2 j_0(\tfrac{1}{2}Ql)^2/b^c$	σ_a/b^d
^{4}He	1.13		1.13	—
Ne	2.66		2.66	0.032
^{36}Ar	74.20		74.20	0.005
Kr	6.88		6.88	25.0
Xe	2.90		2.90	24.5
H_2	7.03	0.742	6.13	0.66
D_2	22.36	0.742	19.49	0.001
N_2	44.41	1.094	32.75	3.7
O_2	16.90	1.207	11.64	0.001
Cl_2	46.13	1.988	15.47	66.0

^{a} Coherent scattering amplitudes of bound atoms from C. G. Shull (private communication to Kjems *et al.*). ^{b} Bond lengths from G. Herzberg, 'Molecular Spectra and Molecular Structure; I, Spectra of Diatomic Molecules', 2nd edn., Van Nostrand, New York, 1950. ^{c} Scattering cross-section evaluated at a wave vector, Q, equal to 1.71 Å^{-1}, corresponding to the diffraction peak for the $\sqrt{3} \times \sqrt{3}$ registered phase on graphite. ^{d} Capture cross-section at $E_n = 0.0253$ eV.

superior to that from the other gases in the list. Thus if these gases are deposited on relatively transparent substrates (small scattering and capture cross-section), the surface signal should be enhanced. As was pointed out by Kjems *et al.*,[178]

---

[179] E. W. Montroll, 'Theoretical Basis of Techniques for the Investigation of Molecular Dynamics and Structure of Solids', in 'Molecular Dynamics and Structure of Solids', National Bureau of Standards, Washington D.C., 1969, and references therein.

[180] A. A. Maradudin, E. W. Montroll, G. H. Weiss, and I. P. Ipatova, 'Theory of Lattice Dynamics in the Harmonic Approximation', Academic Press, New York, 1971.

[181] H. Taub, L. Passell, J. K. Kjems, K. Carneiro, J. P. McTague, and J. G. Dash, *Phys. Rev. Letters*, 1975, **34**, 654.

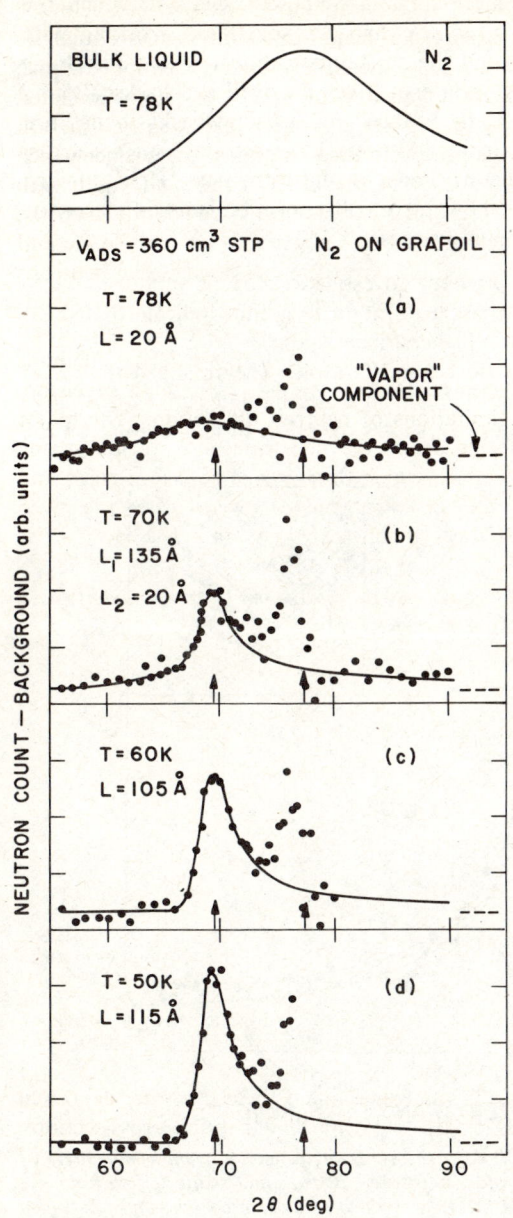

Figure 22 *Temperature dependence of the diffraction observed with an adsorbed volume of 360 cm³ STP (0.87 layers). The solid curves are calculated lineshapes. A constant 'vapour' contribution to the scattering (indicated by the dashed line) was assumed for each scan. The scan at 70 K was fitted by assuming it to be a composite of the scans at 78 and 50 K, i.e., assuming it to be scattering from a mixture of disordered and $\sqrt{3} \times \sqrt{3}$ registered phases. The arrows indicate the peak position corresponding to a nearest-neighbour distance $a_{nn} = 4.25$ Å and the position of the 002 graphite reflection ($2\theta = 77.3°$). The solid curve at the top represents scattering from bulk liquid nitrogen at 78 K observed under identical experimental conditions*
(Reproduced by permission from *Phys. Rev.*, 1976, **B13**, 1446)

an improvement in substrate transparency can be achieved by a proper choice of the neutron wavelength. This method applies to those systems in which the adsorbed layer has a more open structure than the substrate. Working with neutrons of wavelength near or beyond the crystalline substrate cutoff for Bragg scattering, but not too large for Bragg scattering from the overlayer, backscattering from the substrate can be substantially reduced. Other drawbacks in neutron scattering experiments are the requirement for very large substrate surface areas, and the present inadequacy of the methods of data analysis. However, the application of neutron scattering to the study of physisorption systems is a recent development, and its importance should increase.

The Adsorption of N_2 and ^{36}Ar on Graphite. Investigations of the vapour-pressure isotherms and heat capacities of physisorbed monolayer films indicate that these systems exhibit unique two-dimension-like characteristics.[122]

Kjems *et al.*[178] investigated N_2 adsorbed on Grafoil. They mapped the phase diagram of the system by examining the neutron scattering intensities for a range of temperatures and coverages. Figure 22 shows the effect of temperature on the

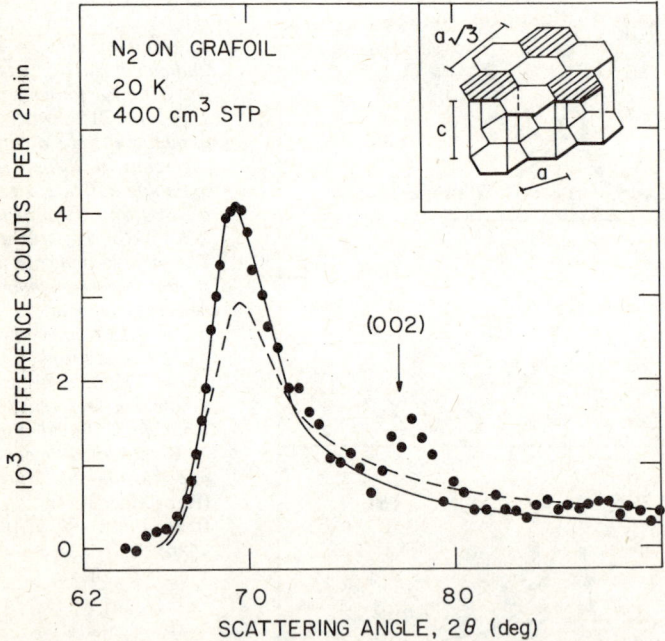

Figure 23 *Diffraction profile at $T = 20$ K of the 10 Bragg reflection from a monolayer of N_2 in $\sqrt{3} \times \sqrt{3}$ registry on the Grafoil substrate. Background scattering with no N_2 in the cell has been subtracted. The solid curve denotes calculated lineshapes in which allowance was made for non-randomness. The dashed curve indicates the lineshape expected with random orientation of the substrate crystallites. The peak near $2\theta = 77°$ is related to the background from the (002) graphite reflection as discussed in the text. Illustrated in the inset is the $\sqrt{3} \times \sqrt{3}$ lattice in registry on the graphite basal plane*
(Reproduced by permission from *Phys. Rev.*, 1976, **B13**, 1446)

scattering intensity for submonolayer coverage. A scattering peak develops at $2\theta = 69.5°$ on lowering the temperature. This peak grows rapidly at the expense of scattering from the disordered phase. The peak at $2\theta = 77.3°$ is due to the (002) graphite reflection. The effective dimensions, L, of the 2D crystallites responsible for the scattering, obtained by curve fitting using equation (78) to the data (solid curve) are given in the figure. These dimensions show a similar increase to that of the diffraction peak. Similar behaviour is characteristic of coverages between 350 and 400 cm³ STP (~0.87 layer and monolayer, respectively). The diffraction peak at $2\theta = 69.5°$ is characteristic of a N_2 layer in $\sqrt{3} \times \sqrt{3}$ registry on the Grafoil substrate as was determined via the analysis of the diffracted intensity from a monolayer at 20 K shown in Figure 23. No additional diffraction peaks from other ordered 2D structures have been observed.

A summary of the variation of the intensity of scattering from the registered phase with temperature is shown in Figure 24. The discontinuous change at 84.5 K corresponding to the registered phase–disordered phase transition, and a transition between a dense and registered phase at < 70 K, are noted.

Another system which has been studied is ^{36}Ar monolayer films on Grafoil.[181] The neutron diffraction results for this system at 5 K are shown in Figure 25. The structure of the monolayer determined from this data is a two-dimensional $\sqrt{3} \times \sqrt{3}$ triangular lattice with nearest neighbour spacing of 3.88 Å. This net

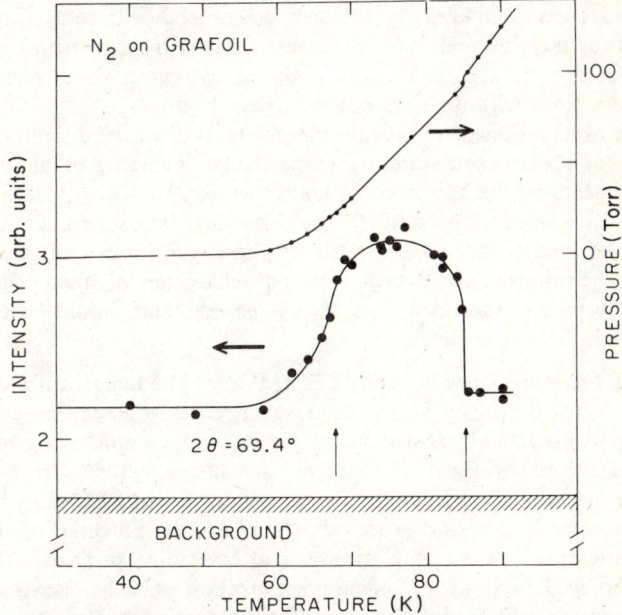

Figure 24 *The temperature dependence of the neutron intensity at the $\sqrt{3} \times \sqrt{3}$ registered phase peak ($2\theta = 69.4°$) and the corresponding vapour pressure in the sample cell. The adsorbed volume on the substrate varied from 460 cm³ STP at $T = 40$ K to 415 cm³ STP at $T = 85$ K. The neutron intensity is proportional to the number of molecules in the registered phase*
(Reproduced by permission from *Phys. Rev.*, 1976, **B13**, 1446)

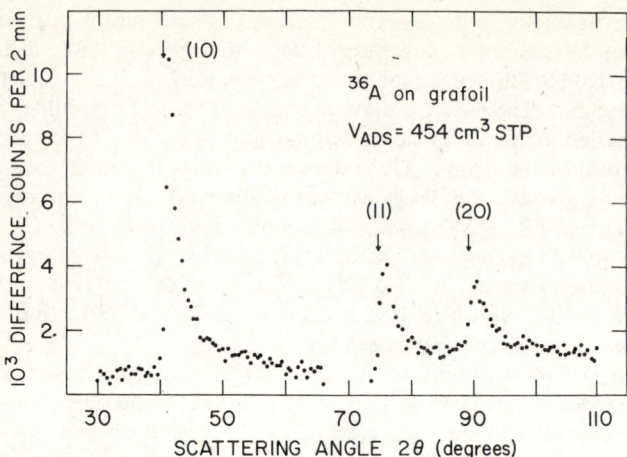

Figure 25 *Neutron diffraction from monolayer films of* ^{36}Ar *adsorbed on graphite at* $T = 5$ K. *Background scattering from the substrate has been subtracted. The apparent negative swing between 66° and 73° is due to screening of the substrate scattering by* ^{36}Ar (Reproduced by permission from *Phys. Rev. Letters*, 1975, **34**, 654)

is incommensurate with the substrate and confirms the view that the geometrical structure of rare-gas monolayers is primarily determined by intralayer interactions. The above study includes preliminary measurements of inelastic neutron scattering which compared quite adequately with a simple two-dimensional phonon model in which interactions with the substrate were neglected.

The above observations demonstrate the potential of neutron scattering for the investigation of physisorbed systems. In particular, mapping of phase diagrams and measurements of the spectrum of collective excitations provide microscopic data which in conjunction with thermodynamical measurements permit the quantitative testing of theoretical models of phase transitions and physical interactions. The demonstrated two-dimensional character of these films makes comparisons between experiment and theory feasible, and should lead to further developments in both disciplines.

The Study of Thermal Vibrations using LEED.[137,139] The temperature dependence of the intensity of diffracted beams in low-energy electron scattering from solid has been already observed by Davisson and Germer in their pioneering investigation and was interpreted by them as the Debye–Waller effect.[142] The observation that temperature effects are generally more prominent in LEED than in X-ray or high-energy electron scattering experiments suggests the potential of the method for the investigation of the lattice dynamics of crystalline surfaces. The method has been employed in studies of a number of clean and chemisorption surface systems [182] and recently in investigations of physisorption systems.

The quantities which characterize the vibrational dynamics of surfaces are the mean-square displacement amplitudes of surface atoms and the dispersion curves of the surface normal, modes which include: Rayleigh waves, transverse modes

---

[182] G. A. Somorjai and H. H. Farrell, *Adv. Chem. Phys.*, 1972, **20**, 215, and references therein; see also ref. 21.

with polarization vectors parallel to the surface, mixed modes which consist of a superposition of longitudinal surface waves and transverse bulk waves, and combination modes in which longitudinal and transverse bulk waves superimpose.[183] As in X-ray diffraction, the energy and angular resolution in typical LEED experiments are insufficient to observe losses or gains due to phonon annihilation or creation (neutron scattering is a more obvious tool for such studies). Investigations of this type require high resolution and special instrumentation [184-186] and are performed in the inelastic mode of low energy electron diffraction (ILEED). A methodology for the analysis of ILEED data was developed recently [187,188] for the determination of the dispersion relation and damping of surface collective electronic excitations (surface plasmons), and a similar approach could prove useful for the investigation of surface phonon modes *via* ILEED measurements of loss and angular intensity profiles.

In typical LEED studies of surface atom dynamics, temperature effects on the intensity of elastically diffracted beams are measured. The interpretation of these experiments proceeds *via* a formalism similar to that used in X-ray studies.[174,176] According to this model, a thermal average of the expression for the diffracted intensity from a set of vibrating scatterers is performed (since the motion of the atoms is rapid compared to the time of measurement). The zero-phonon part of the resulting expression, in the kinematical (single-scattering) approximation is given as equation (87),[139] where $\vec{k}$ and $\vec{k}'$ are the wave vectors of the incident and

$$\langle I(\vec{k}' - \vec{k}) \rangle = I_0(\vec{k}' - \vec{k}) e^{-\langle [(\vec{k}' - \vec{k}) \cdot \vec{u}]^2 \rangle} \equiv I_0(\vec{k}' - \vec{k}) e^{-2M} \quad (87)$$

diffracted beams, respectively, $\vec{u}$ is the displacement of the scattering atom from its mean position, and $I_0(\vec{k}' - \vec{k})$ is the intensity scattered from a rigid lattice. The angular brackets indicate temperature averaging, and e^{-2M} is the Debye–Waller factor. For an hk diffracted beam the exponent takes the form of equation (88),[189]

$$2M \equiv \langle [(\vec{k}' - \vec{k}) \cdot \vec{u}]^2 \rangle = \tfrac{1}{2}[g_{hk,x}^2 \langle u_x^2 \rangle + g_{hk,y}^2 \langle u_y^2 \rangle] + \tfrac{1}{2}[k_\perp + k'_\perp]^2 \langle u_z^2 \rangle \quad (88)$$

where $g_{hk,\alpha}$ ($\alpha = x,y$) is the hk reciprocal vector of the 2D net in the α-th cartesian direction, $\langle u_\alpha^2 \rangle$ is the cartesian component of the mean square displacement and $k_\perp, k'_\perp$ are the perpendicular components of the electron momentum inside the solid for the incident and diffracted beams, respectively (in the above it is assumed that motions in different cartesian directions are uncorrelated, *i.e.* $\langle u_\alpha u_\beta \rangle = 0$ for $\alpha \neq \beta$). The assumption of a Debye model for the phonon spectrum leads to equations (89) and (90) for the mean-square vibrational amplitude, the former applying in the low-temperature limit and the latter in the high-temperature

$$\langle u_\alpha^2 \rangle = \frac{3\hbar^2}{m k_B \theta_D(\alpha)} \left[1.642 \frac{T^2}{\theta_D^2(\alpha)} + \tfrac{1}{4} \right] \quad (89)$$

---

[183] For a survey of theoretical work on surface lattice dynamics see T. Wolfram, R. E. DeWames, and E. A. Kraut, *J. Vacuum Sci. Technol.*, 1972, **9**, 685, and references therein.
[184] F. M. Propst and T. C. Piper, *J. Vacuum Sci. Technol.*, 1966, **4**, 53.
[185] H. Ibach, *Phys. Rev. Letters*, 1970, **24**, 1416; 1971, **27**, 253.
[186] H. Ibach, *J. Vacuum Sci. Technol.*, 1972, **9**, 713.
[187] C. B. Duke, U. Landman, and J. O. Porteus, *J. Vacuum Sci. Technol.*, 1973, **10**, 183.
[188] C. B. Duke and U. Landman, *Phys. Rev.*, 1973, **B7**, 1368.
[189] S. Y. Tong, T. N. Rhodin, and A. Ignatiev, *Phys. Rev.*, 1973, **B8**, 906.

limit ($T \geqslant \theta_D/2\pi$), where T is the absolute temperature, m the mass of the atom,

$$\langle u_\alpha^2 \rangle = \frac{3\hbar^2 T}{mk_B \theta_D^2(\alpha)} \tag{90}$$

k_B Boltzmann's constant, and $\theta_D(\alpha)$ the Debye temperature for vibration in the α-th cartesian direction (for an isotropic model of the vibrating system

$$\langle u_\alpha^2 \rangle = \tfrac{1}{3} \langle u^2 \rangle \tag{91}$$

and

$$\theta_D(\alpha) = \theta_D \tag{92}$$

for $\alpha = x,y,z$). It is thus observed from the above, that the diffracted intensity is attenuated by the temperature-dependent Debye–Waller factor. The procedure for determining the mean-square vibrational amplitudes (and thus the effective Debye temperature) of surface atoms, is dictated by the above relationships. Combining equations (87) and (88), it follows that $d\ln\langle I \rangle/dT$ is linearly proportional to T (in the high-temperature limit). The procedure consists of [182] first, recording the intensity of prominent peaks in the intensity versus incident energy profiles for several elastically diffracted beams. Secondly, a semi-log plot of the intensity of each individual peak as a function of temperature is constructed. According to this model the above plot should result in a straight line. From the slope of that line an effective Debye temperature can be determined. It is of interest to note that in the case of anisotropic vibrations the mean-square vibrational amplitude in the normal direction, $\langle u_z^2 \rangle$, can be determined by performing the above measurements for the specular diffraction beam, i.e. $h = k = 0$, since in this case only the term involving $\langle u_z^2 \rangle$ in equation (88) is non-vanishing. In addition, from a measurement of the intensities of a non-specular diffraction beam, the mean-square amplitudes in the direction normal to the surface plane ($\langle u_z^2 \rangle$) and the in-plane amplitude ($\langle u_x^2 \rangle + \langle u_y^2 \rangle \equiv \langle u_\parallel^2 \rangle$) can be determined as follows. First, the procedure described above is followed from which M' (defined as $M' = M/T$) is determined. It is noted that equation (88) can be written as

$$2M' = \frac{\Delta k_\parallel^2 \langle u_\parallel^2 \rangle}{2T} + \frac{\Delta k_\perp^2 \langle u_z^2 \rangle}{2T} \tag{93}$$

where $\Delta k_\parallel$ and $\Delta k_\perp$ are the momentum transfer parallel and normal to the surface plane, respectively. Consequently, from a plot of $2M'$ versus $\Delta k_\perp^2$, $\langle u_\perp^2 \rangle/2T$ and $\Delta k_\parallel^2 \langle u_\parallel^2 \rangle/2T$ can be obtained as the slope and intercept, respectively.

The main assumption of the model discussed above is that the intensity of the elastically diffracted beams is dominated by the contribution of single-scattering processes. Incorporation of temperature effects into the dynamical (multiple-scattering) theories of LEED, and model calculations based on these theories reveal that upon inclusion of multiple scattering the intensity still decreases exponentially with temperature but the slope is not simply related to $\langle u^2 \rangle$. However, these calculations reveal that the dynamic effects are relatively small, yielding values of the Debye temperature which differ from the kinematical ones by less than 15%.[139] Other effects which enter the estimation of the accuracy of the method are due to contributions to the intensity from thermal diffuse scattering (in addition to the Bragg or zero-phonon contribution) and thermal-expansion shifts in the energies of peaks in the I vs. E records.[166] The effects of thermal diffuse scattering can in principle be accounted for by subtracting from the measured intensities a theoretical background curve, but this involves a rather complicated calculation.[139] Errors due to thermal expansion can be minimized by following the location of the diffraction beam whose intensity is measured, as the temperature is changed.[166] The above effects, coupled with uncertainties due to multiple scattering and an incomplete knowledge of the variation of the electron attenuation coefficient with energy yield an error bound of $\sim 20\%$ on the values obtained *via* LEED measurements of mean-square vibrational amplitudes of surface atoms.[166]

The above degree of accuracy may be sufficient in order to discriminate between various models of the force-field and methods of calculation. Such a comparison has been performed, quite successfully, for Xe adsorption on the (0001) face of graphite,[190] using LEED data and a dynamical model based on Lennard-Jones and Mie potentials. Two additional interesting observations were made in the above study. First, the ratio $\langle u^2 \rangle_{Xe}/\langle u^2 \rangle$ graphite was found to be about 10, which means that xenon lies on a practically frozen graphite substrate. Secondly, a comparison between $\langle u^2 \rangle$ for adsorbed Xe and that of the (111) surface of crystalline Xe as determined by Tong et al.[189] and bulk crystalline Xe,[191,192] shows that in the physisorbed state, Xe vibrates less than in the crystalline state. This is probably due to the attractive force exerted by the graphite substrate on the adsorbed Xe layer. An investigation of the thermal vibrations of Xe adsorbed on Ag(111) has been recently performed by Cohen et al.[166] Their results for the Xe (10) beam intensity plotted *versus* temperature at fixed energies and normalized at 25 K, are shown in Figure 26. From these data and a plot of $2M'$ vs. $\Delta k_\perp^2$ (as described above) the authors determined the value $\langle u_\perp^2 \rangle/T = (1.8 \pm 0.4) \times 10^{-4}$ Å^2 K^{-1}. At 25 K the r.m.s. displacement for Xe atom in the monolayer physisorbed state is 0.07 ± 0.1 Å compared to 0.15 Å for the surface of bulk xenon [165,193] and 0.04 Å for the surface of silver.[194] It is thus observed that the metal substrate, even at low temperatures, is less static with respect to the vibrations of the physisorbed layer than a graphite surface.

Another related property which can be studied using temperature effects on LEED measurements is the effective thermal expansion coefficient of the adsorbed layer,[139,166,193,195] defined by equation (94), where ΔE_B is the shift in peak energy

$$\alpha = -(\Delta E_B/2E_B)\Delta T \tag{94}$$

in the intensity *versus* incidence energy profile. A value for α can be obtained as the slope of a plot of $\Delta E_B/2E_B$ vs. $(\Delta T)^{-1}$ for a given Bragg maximum E_B. Unfortunately, values of α determined for different peaks in the I vs. E profile exhibit scatter, beyond experimental uncertainties,[139] which may be due to changes of dynamic

---

[190] J. P. Coulomb, J. Suzanne, M. Bienfait, and P. Masri, *Solid State Comm.*, 1974, **15**, 1585.
[191] R. E. Allen, F. W. de Wette, and A. Rahman, *Phys. Rev.*, 1969, **179**, 887.
[192] R. E. Allen and F. W. de Wette, *Phys. Rev.*, 1969, **179**, 873; 1969, **188**, 1320.
[193] A. Ignatiev and T. N. Rhodin, *Phys. Rev.*, 1973, **B8**, 893.
[194] J. T. McKinney, E. R. Jones, and M. B. Webb, *Phys. Rev.*, 1967, **160**, 523.
[195] L. W. Bruch, P. I. Cohen, and M. B. Webb, *Surface Sci.*, 1976, **59**, 1.

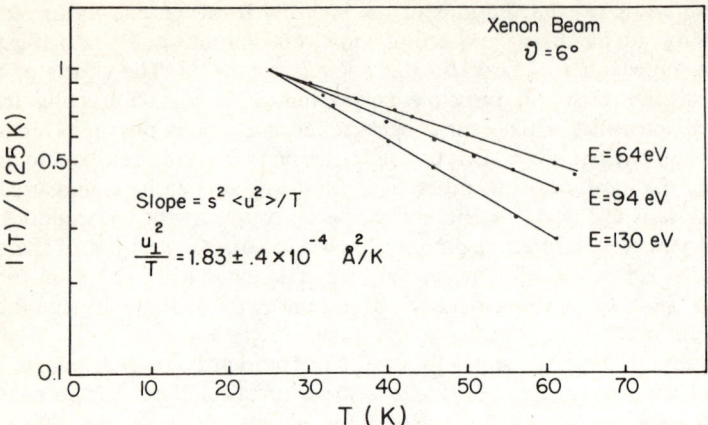

Figure 26 *The diffracted intensity at* 64, 94, *and* 130 eV *of the* Xe(10) *beam vs. temperature, normalized to the intensity at* 25 K. *The variable* s^2 *in the figure denotes the normal momentum transfer* ($\Delta k_\perp^2$ *in the text*)
(Reproduced by permission from *Surface Sci.*, 1976, **59**, 1)

diffraction conditions in the surface region since thermal expansion there is neither isotropic nor uniform, and due to different penetration depths for electrons with different primary energies. Improved understanding of the energy dependence of the attenuation of the propagation of low-energy electrons at the surface region of solids, and of LEED data over extended temperature ranges would allow better estimates of this quantity.

The above discussion demonstrates the potential of LEED studies in providing microscopic information about atom dynamics in physisorption systems. Such data would improve greatly our theoretical models of the gas–surface force laws and of interatomic interactions in the physisorbed layer, and would allow quantitative assessments of theoretical studies of surface dynamics.[191,192,196,197]

Low Energy Molecular (Atomic) Beam Scattering (LEMS). The method of molecular and atomic beam scattering from surface shares a common origin with low-energy electron scattering, namely the de Broglie hypothesis. Indeed, as in the case of the Davisson and Germer's discovery of the diffraction phenomena in the scattering of electrons from crystals, the experiments of Esterman, Frisch, and Stern[198—200] in which diffraction of He atoms from the (100) face of a LiF crystal was observed, served as experimental proof of the particle–wave duality hypothesis. Advances in UHV technology, sample preparation, beam sources, and detection techniques mean that LEMS is of great importance in the study of atom–surface interactions. The experimental and theoretical aspects of molecular beam scattering from surfaces

[196] R. F. Wallis, B. C. Clark, and R. Herman, in 'The Structure and Chemistry of Solid Surfaces', ed. G. Somorjai, Wiley, New York, 1969, chap. 17.
[197] A. A. Maradudin, D. L. Mills, and S. Y. Tong, in 'The Structure and Chemistry of Solid Surfaces', ed. G. Somorjai, Wiley, New York, 1969, chap. 16.
[198] I. Esterman and O. Stern, *Z. Physik*, 1930, **61**, 95.
[199] I. Esterman, R. Frisch, and O. Stern, *Z. Physik*, 1931, **73**, 348.
[200] R. Frisch and O. Stern, *Z. Physik*, 1933, **84**, 430.

Microscopic Approaches to Physisorption

have been reviewed extensively recently.[119,121,201—210] The following discussion emphasizes the microscopic information which could be derived from the data.

Though the scattering of atoms and molecules from surfaces is often considered as a technique for geometrical structure determination (atomic arrangement at the surface), in our view the main use and importance of the technique is in investigations of the interaction potential between individual atoms and molecules and solid surfaces. The interaction potential governs the dynamics and trajectory of the scattered particles. Consequently, given a model of the scattering process, analysis of atomic scattering data could provide a way for evaluation and parameterization of theoretically constructed interaction potentials (see below). Apart from the above unique merit, LEMS can provide direct information of the following surface properties:[202] (i) surface structure from diffraction experiments; (ii) dynamics of adsorbed layers; (iii) extent of surface coverage by adsorbed gases; (iv) surface roughness; (v) cross-sections for energy transfer between the gas and the surface-sticking coefficient and accommodation factors; (vi) surface phonon spectrum; and (vii) kinetics of adsorbed gas reactions on surfaces (catalysis) and gas–surface reactions.

The kinematic description of atom-scattering from the surface of single crystals is a consequence of the momentum and energy conservation laws, and is similar to that of the scattering of low-energy electrons. The projectile in LEMS experiments can be atomic, molecular, or a free radical, which permits the study of the interaction mechanism of different chemical species and in addition allows the performance of experiments for a large range of de Broglie wavelengths. For example, beams of slow H atoms with $\lambda \simeq 20$ Å (velocity = 200 ms^{-1}) have recently been produced[202] (this corresponds to electrons of energy 0.5 eV). In early work, the beam source consisted of a Knudsen effusion cell with variable temperature control. This source yields well-collimated, velocity-selected beams of low intensity and poor signal-to-noise ratio in the scattered beam.[119] In modern investigations, high-pressure nozzle expansion sources are used which yield high-intensity beams with typical Mach numbers ($\Delta v/v$) of the order 5×10^{-2}—10^{-1}. In modern LEMS studies the temperatures of both the beam and the substrate can be varied. Velocity selection of the scattered beams can be done by mechanical velocity selectors ($\Delta v/v \sim 10^{-2}$), by the time-of-flight ($\Delta v/v \sim 2.5 \times 10^{-2}$—$6 \times 10^{-2}$) and phase-sensitive methods ($\Delta v/v \lesssim 10^{-1}$), and by other methods.[202,208] The scattering of low-energy atoms or molecules is highly surface-sensitive, and the interaction is limited almost exclusively to the very outermost layer of the target, thus avoiding the question of penetration depth. This sensitivity brings about

---

[201] F. Goodman, *Surface Sci.*, 1971, **26**, 327.
[202] J. P. Toennies, *Appl. Phys.*, 1974, **3**, 91.
[203] R. M. Logan, in 'Solid State Surface Science', ed. M. Green, Marcel Dekker, New York, 1973, Vol. 3, p. 1.
[204] G. A. Somorjai and S. B. Brumbach, *CRC Crit. Rev. Solid State Sci.*, 1974, **4**, 429.
[205] J. P. Hobson, *CRC Crit. Rev. Solid State Sci.*, 1974, **4**, 221.
[206] H. Saltsburg, *Ann. Rev. Phys. Chem.*, 1973, **24**, 493.
[207] A. G. Stoll, jun., R. E. White, J. J. Erhardt, R. I. Masel, and R. P. Merrill, *J. Vacuum Sci. Technol.*, 1975, **12**, 192.
[208] W. H. Weinberg, *Adv. Colloid Interface Sci.*, 1975, **4**, 301.
[209] R. E. Stickney, in 'The Structure and Chemistry of Solid Surfaces', ed. G. A. Somorjai, Wiley, New York, 1969.
[210] J. N. Smith, jun., *Surface Sci.*, 1973, **34**, 613.

other difficulties in the analysis of the experimental data owing to the increased dependence of the scattering on the detailed topography of the surface (kinks, steps, dislocations, impurities, and other imperfections; this can be turned into an advantage if an estimate of surface roughness or cleanliness is required).

Experimental Observations: Scattering Regimes. Atomic and molecular scattering from solid surfaces can be conveniently classified into reactive and non-reactive scattering. Only the latter will be considered in the following (for a review of reactive scattering see ref. 204), particularly those experiments in which rare gases were used as projectiles and metallic single crystals as targets, since these systems serve as prototypes in our physisorption studies.

Non-reactive scattering events can be further classified according to the changes in momentum and energy which accompany the collision process. Denoting the wave vectors of the incident and scattered beams by $\vec{k}$ and $\vec{k}'$, respectively, and their components parallel to the surface by $\vec{k}_{\parallel}$ and $\vec{k}'_{\parallel}$ (see Figure 10), and letting ω_n, $\vec{q}_n$ be the frequency and tangential component of the lattice phonon and $\vec{g}_{hk}$ a reciprocal vector of the two-dimensional surface net, the conservation laws are given as equations (95) and (96).

$$k'^2 = k^2 - (2m/\hbar^2) \sum_{n=1}^{N} (\pm \hbar \omega_n) \qquad (95)$$

$$\vec{k}'_{\parallel} = \vec{k}_{\parallel} + \vec{g}_{hk} - \sum_{n=1}^{N} (\pm \vec{q}_n) \qquad (96)$$

A classification of the collision processes with reference to these conservation laws is given in Table 6. From the analysis of thermal scattering experiments of rare gases from several metal surfaces [W(110), Pt(100), and the (111) surface of Au, Ag, Ni, and Pt] a classification of the scattering events into the following

Table 6 *Classification of surface collision processes by energy and momentum conservation (after Toennies* [202])

Collision process	Final surface state	Final projectile state
Specular reflection (elastic)	$\omega_n = 0$ (all n) $\vec{q}_n = 0$ (all n)	$k'^2 = k^2$ ($E' = E$) $\vec{k}'_{\parallel} = \vec{k}_{\parallel}$
Non-specular diffraction (elastic)	$\omega_n = 0$ (all n) $\vec{q}_n = 0$ (all n)	$k'^2 = k^2$ ($E' = E$) $\vec{k}'_{\parallel} = \vec{k}_{\parallel} + \vec{g}_{hk}$
Inelastic scattering, one-phonon annihilation	$-\omega_{n'} \neq 0, \omega_n = 0$ ($n \neq n'$) $\pm \vec{q}_{n'} \neq 0, \vec{q}_n = 0$ ($n \neq n'$)	$k'^2 = k^2 + \dfrac{2M_g}{\hbar^2} \hbar \omega_{n'}$ $\vec{k}'_{\parallel} = \vec{k}_{\parallel} \pm \vec{q}_{n'} + \vec{g}_{hk}$
Inelastic scattering, one-phonon creation	$\omega_{n'} \neq 0, \omega_n = 0$ ($n \neq n'$) $\pm \vec{q}_{n'} \neq 0, \vec{q}_n = 0$ ($n \neq n'$)	$k'^2 = k^2 - \dfrac{2M_g}{\hbar^2} \hbar \omega_{n'}$ $\vec{k}'_{\parallel} = \vec{k}'_{\parallel} \pm \vec{q}_{n'} + \vec{g}_{hk}$
Selective adsorption	$\omega_n = 0$ (all n) $\vec{q}_n = 0$ (all n)	$\vec{k}'_{\parallel} = \vec{k}_{\parallel} + \vec{g}_{hk}$ $E'_z < 0$
Trapping	$\Sigma \hbar \omega_n = E^2$ $\mp \Sigma \vec{q}_n = \vec{k}_{\parallel}$	$k' = 0$ $k'_{\parallel} = 0$

three regimes was suggested.[211,212] (i) The quasi-elastic regime, which is typified be helium scattering from clean metal surfaces. The angular distribution of the

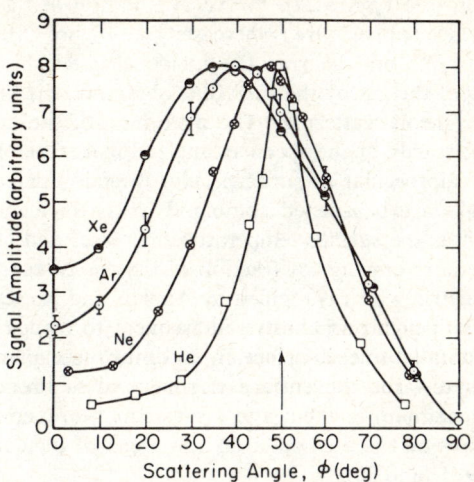

Figure 27 *Experimental scattering angular distributions for various rare gases scattered from* Ag(111). *The temperatures of the surface and gas were 560 and 300 K, respectively, and the angle of incidence 50°. Note the shift from specularity as the gas becomes heavier* (Reproduced by permission from *J. Chem. Phys.*, 1971, **54**, 163)

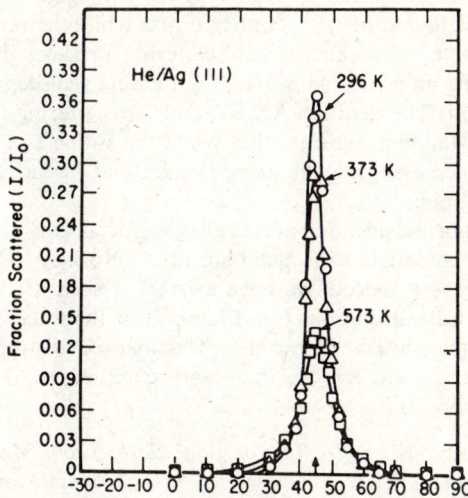

Figure 28 *Angular distribution of* He *scattered from* Ag(111) *single crystal as a function of surface temperature. Note the specularity of the scattered intensity and the attenuation of the peak of the intensity as the surface temperature increases* (Reproduced by permission from *Surface Sci.*, 1973, **94**, 268)

[211] W. H. Weinberg and R. P. Merrill, *J. Chem. Phys.*, 1971, **54**, 163.
[212] R. Sau and R. P. Merrill, *Surface Sci.*, 1973, **34**, 268.

scattered beam is peaked at the specular direction independent of gas or surface temperature, or polar angle of incidence, and is only slightly broader than the incident distribution (see Figure 27).[213] An attenuation of the scattered beam is observed as the surface temperature is increased (see Figure 28) owing to a Debye–Waller effect (see p. 62), and also as the incidence angle becomes less grazing. Negligible trapping of the gas by the surface is observed. (ii) The inelastic regime, which is typified by neon scattering. The maximum of the angular distribution does not occur necessarily at the specular angle (Figure 27). The maximum can occur in either the subspecular or supraspecular direction depending on the ratio of T_s/T_g. The peaks are broadened compared to the incident distribution, and they attenuate both as the surface temperature increases and also as the angle of incidence becomes more grazing. A fraction of the gas is trapped on the surface. (ii) The trapping regime, which is typified by Ar, Kr, and Xe scattering. Trapping of the gas occurs with high probability. Subsequent to trapping, the gas atom is emitted with a random (cosine) distribution and thus the maximum in the angular distribution moves towards the surface normal and is broadened significantly (Figure 27). The maximum of the scattering distribution increases as the surface temperature increases and is attenuated as the angle of incidence becomes more grazing (Figures 29 [214] and 30).

Several qualitative features (curvature and depth) of the gas–surface interaction potentials can be derived from the above data. (i) The scattering data (quasi-elastic) suggest that He interaction with metals occurs almost exclusively through the repulsive part of the gas–surface potential. The binding minimum of the potential is shallow as evidenced by the negligible trapping. (ii) The data for Ne scattering (quasi-elastic and inelastic) indicate that while the repulsive part of the interaction potential is dominant in the scattering process, the attractive part of potential has some influence. However, the binding well-depth of the potential is rather shallow. (iii) The data for Ar, Kr, and Xe scattering (Ar, Kr – inelastic and trapping, Xe – trapping) indicate that while the stiffness of the repulsive part of the potential is increased, the curvature is such that the attractive well is deep enough to allow trapping.

The above suggests a classification of the rare gases according to the characteristics of their interaction potentials with metal surfaces, into light (He, Ne) and heavy (Ar, Kr, Xe) rare gases. Indeed, we have arrived at such a classification on the basis of our theoretical calculations (see Figure 7) of the interaction potentials of the rare gases with a variety of metals. The marked distinction between the potentials of He, Ne and Ar, Kr, Xe, in curvature and depth, is evident from the results shown in Figure 7.

Experimental Observations. Diffraction of Rare Gases From Metals. The original experiments of Esterman, Frisch, and Stern [198–210] have inspired numerous attempts to observe diffraction effects in the scattering of atomic and molecular beams from crystalline surfaces. However, most of these studies used the (100) face of alkali halides, LiF in particular (for reviews see refs. 202, 210, and 211). Attempts to observe diffractive scattering from clean metal surfaces were generally unsuccessful,

---

[213] H. Saltsburg and J. N. Smith, jun., *J. Chem. Phys.*, 1966, **45**, 2175.
[214] A. G. Stoll, jun., D. L. Smith, and R. P. Merrill, *J. Chem. Phys.*, 1971, **54**, 163.

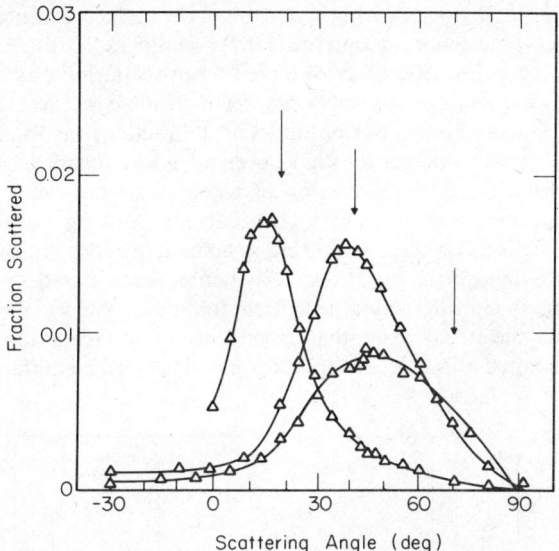

Figure 29 *The effect of changes in the polar angle of incidence on the scattered intensity of a 295 K Ar beam scattered from a Pt(111) surface*
(Reproduced by permission from *J. Chem. Phys.*, 1971, **54**, 163)

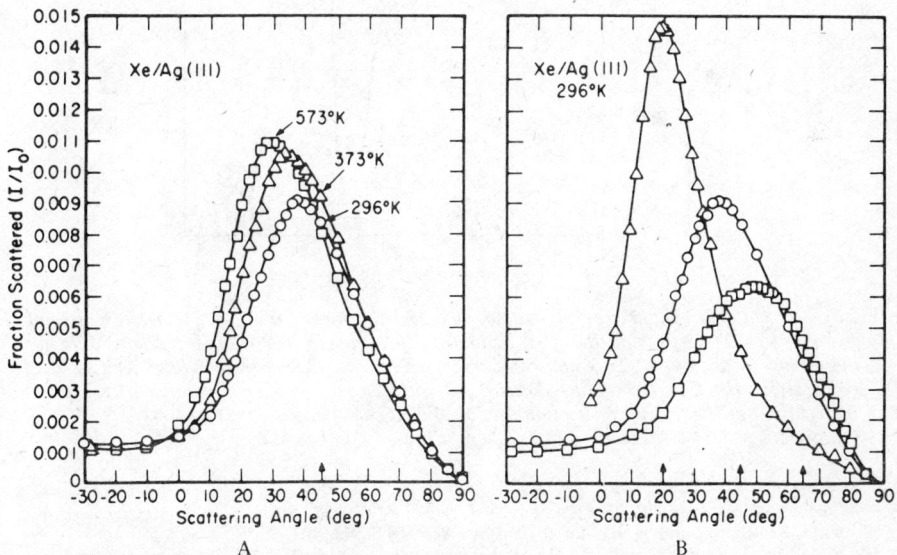

Figure 30 *Angular distribution of Xe scattered from a Ag(111) surface as a function of surface temperature (A) and incidence angle (B)*
(Reproduced by permission from *Surface Sci.*, 1973, **34**, 268)

with the exception of He scattering from the (112) surface of tungsten when the azimuthal plane of incidence is along the [1$\bar{1}$0] crystallographic direction [207,215—219] (see Figure 31). No diffraction was observed when scattering was along the [11$\bar{1}$] azimuthal direction. In the case of a Ne beam incident on W(112) in the [1$\bar{1}$0] azimuthal direction rainbow scattering (and weak diffraction) was observed [207,218—220] (see Figure 32). Various suggestions have been advanced for the lack of diffraction (dominant specularity) in the scattering of rare gases from clean metal surfaces (with the exception noted above).[221] The current thinking is that the general absence of diffraction for these scattering systems is a reflection of the weakness of the structure-dependent variations (the components parallel to the surface plane) in the gas–solid interaction potentials for these systems. This explanation is in accord with the strong azimuthal dependence of the scattering characteristics from W(112) as noted above, since this surface is close-packed in the [11$\bar{1}$] direction

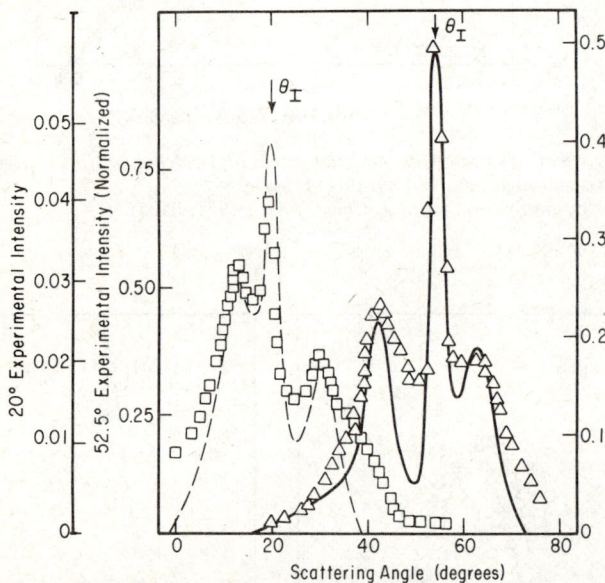

Figure 31 *Calculated and experimental angular distributions of a* 300 K He *beam scattering from a* W(112) *surface in the* [1$\bar{1}$0] *azimuth. The squares and triangles denote data for angles of incidence of* 21.5 *and* 50°, *respectively.*[218] *The corresponding semiclassical calculations are denoted by the dashed and solid curves. For a discussion of the parametrization of the potential function used in the calculations, see text*
(Reproduced by permission from *J. Chem. Phys.*, 1976, **64**, 45)

[215] D. V. Tendulkar and R. E. Stickney, *Surface Sci.*, 1971, **27**, 516.
[216] A. G. Stoll, jun., and R. P. Merrill, *Surface Sci.*, 1973, **40**, 405.
[217] R. I. Masel, R. P. Merrill, and W. H. Miller, *Surface Sci.*, 1974, **46**, 681.
[218] A. G. Stoll, jun., J. J. Erhardt, and R. P. Merrill, *J. Chem. Phys.*, 1976, **64**, 34.
[219] R. I. Masel, R. P. Merrill, and W. H. Miller, *J. Chem. Phys.*, 1976, **64**, 45.
[220] R. E. White, J. J. Erhardt, and R. P. Merrill, *J. Chem. Phys.*, 1976, **64**, 41.
[221] See discussion and cited references in ref. 119, pp. 300—307.

Microscopic Approaches to Physisorption

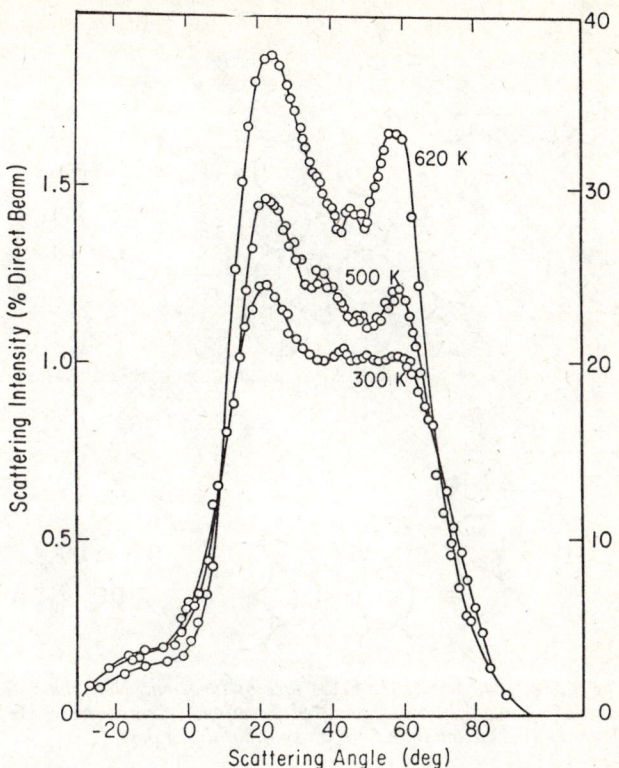

Figure 32 *Angular scattering distributions of Ne scattering from W(112) at the [1$\bar{1}$0] azimuth, with a polar angle of incidence of 45°. The surface temperature was 173 K, and the beam temperatures 300 K, 500 K, and 620 K. The scattered distribution exhibits rainbow scattering characteristics, and some of the fine structure between the peaks appears to be correlated with the positions of expected diffraction peaks, but a definite statement could not be made*
(Reproduced by permission from *J. Chem. Phys.*, 1976, **64**, 41)

but consists of pronounced atomic rows of spacing 4.47 Å in the [1$\bar{1}$0] direction (see Figure 33). As a consequence of the geometrical atomic arrangement the (112) face of tungsten as viewed along the [1$\bar{1}$0] azimuthal direction presents a 'rough', long-range ordered surface with the periodic variations in the potential parallel to the surface plane, large enough to give rise to diffraction. Since the dominant contribution to the structure-dependent component of the gas–surface potential is expected to be due to repulsion between the incident atom and the scattering surface, the above reasoning is consistent with the conclusions derived from our theoretical calculations of the interaction of rare gases with metals, in which we have shown that in equilibrium the incident atom resides at a distance greater than is expected on the basis of hard-sphere models ('sum of radii rule'). This leads to a reduction in the strength of the periodic components in a two-dimensional Fourier representation of the potential. These results are also in

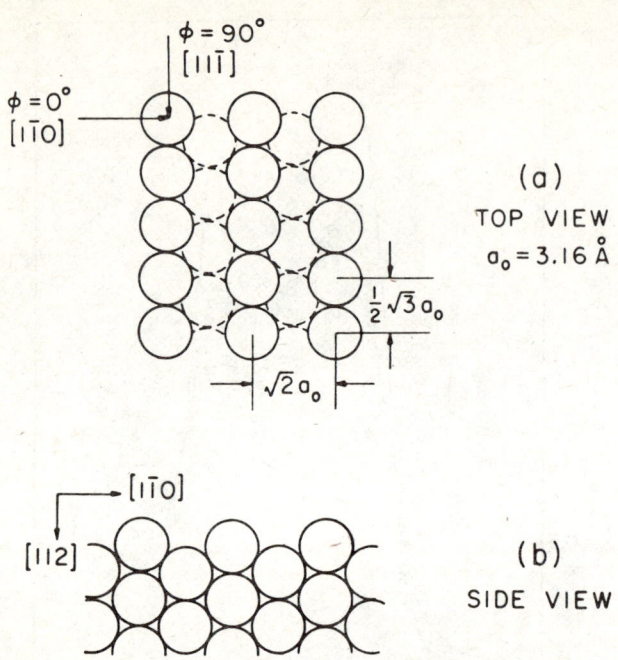

Figure 33 *The atomic structure of the* (112) *face of tungsten. Note the anisotropy of the atomic arrangement along the two azimuthal directions. Consequently,* He *diffraction is observed along the* [1$\bar{1}$1] *azimuth but not along other directions*

agreement with a recent analysis of He diffractive scattering from W(112) [217,219] (see also p. 77).

Experimental Observations. Selective Adsorption. The phenomenon of selective adsorption was first observed by Frisch and Stern,[200] and its study consists of the measurement of small intensity losses in the angular intensity distributions of the diffracted beam. The first explanation was given by Lennard-Jones and Devonshire (LJD).[222,223] Selective adsorption has been the subject of numerous experimental [224,225] and theoretical [226—229] studies. Although the method has been applied exclusively to scattering of light gases from alkali halide surfaces, the underlying physical principles of the phenomena will be briefly discussed in the following, since these experiments provide data of major relevance for the assessment, testing, and parametrization of atom–surface interaction potentials.

A description of selective adsorption, on a most elementary level, can be given in terms of conservation laws, and first-order perturbation treatment of the elastic

[222] J. E. Lennard-Jones and A. F. Devonshire, *Nature*, 1936, **137**, 1069.
[223] J. E. Lennard-Jones and A. F. Devonshire, *Proc. Roy. Soc.*, 1937, **A158**, 242; *ibid.*, p. 253.
[224] See ref. 202 and references therein.
[225] J. A. Mayers and D. R. Frankel, *Surface Sci.*, 1975, **51**, 61.
[226] A. Tsuchida, *Surface Sci.*, 1969, **14**, 375; 1974, **46**, 611.
[227] N. Cabrera, V. Celli, F. O. Goodman, and R. Manson, *Surface Sci.*, 1970, **19**, 67.
[228] G. Wolken, jun., *J. Chem. Phys.*, 1973, **58**, 3047.
[229] H. Chow and E. D. Thompson, *Surface Sci.*, 1976, **54**, 269.

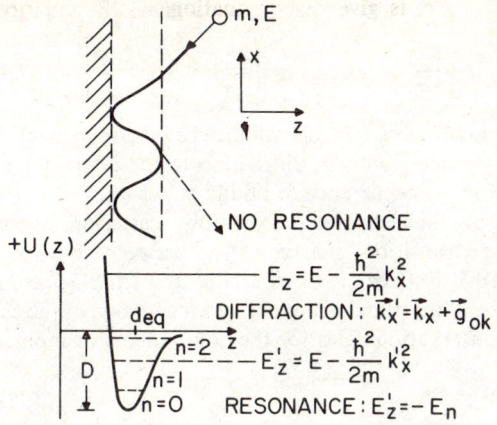

Figure 34 *Schematic illustration of the mechanism leading to selective adsorption. The upper part shows the trajectory of a selectively adsorbed particle (solid) and that of a normally scattered one (dashed). The lower panel is a section of the potential surface in the plane perpendicular to the surface plane. $U(z)$ is the interaction potential in the normal (z) direction, D is the depth of the potential well and d_{eq} is the equilibrium distance of the incident particle from the surface. The dashed horizontal lines ($n = 0,1,2$) denote bound states of the potential. Prime quantities associated with the scattered beam and unprimed quantities are associated with the incident particle. $\vec{g}_{h0}$ is the ($h0$) reciprocal vector of the 2D surface net*
(Reproduced by permission from *Appl. Phys.*, 1974, **3**, 91)

scattering. Consider a scattering geometry as shown in Figure 34. For a general direction of incidence the translational energy of the incident particle, E, can be written in terms of its cartesian components as equation (97), where E_z, the

$$E = \frac{\hbar^2}{2m}(\vec{k}_z^2 + \vec{k}_x^2 + \vec{k}_y^2) \equiv E_z + E_x + E_y \equiv E_z + E_{\parallel} \qquad (97)$$

'perpendicular energy' of the particle, is positive and depends on the incidence angle.

The interaction potential is assumed to be periodic in the surface plane (xy) and have the general form of equation (98), where g_{hk} is the hk reciprocal vector

$$V(\vec{\rho}, z) = \sum_{h,k} B_{hk}(z) \exp[i\vec{g}_{hk} \cdot \vec{\rho}] \qquad (98)$$

of the 2D surface net and $\vec{\rho}$ is a 2D-translational vector of the net. The asymptotic form ($z \to \infty$) of the wavefunction corresponding to a solution of the Schrödinger equation with this potential, for elastic scattering is given by equation (99), where

$$\psi = e^{i\vec{k}\cdot\vec{r}} + \sum_{hk} A_{hk} \exp[(\vec{k}_{\parallel} + \vec{g}_{hk})\cdot\vec{\rho} + k'_z z] \qquad (99)$$

k'_z is the perpendicular component of the momentum after scattering.

For simplicity we specialize to a simplified form of the above potential which, for a square lattice with lattice spacing a, is given by equation (100), and treat

$$V(x,y,z) = V(z) + 2U_1(z) \cos\left[\frac{2\pi x}{a} + \cos\frac{2\pi y}{a}\right] \equiv V(z) + H' \quad (100)$$

the periodic term H' as a perturbation. The translational invariance of the potential in the xy plane results in diffraction in the plane, while the non-periodic part, $V(z)$ allows for the occurrence of bound states ($E'_z < 0$).

A description of the selective adsorption scattering process can be given in terms of the transition matrix M_{fi}, between free and bound states, $M_{fi} = \langle\psi_f(\text{bound})|H'|\psi_i(\text{free})\rangle$. Considering a scattering geometry as shown in Figure 34 where the plane of incidence coincides with one of the cartesian directions, the momentum conservation rules for the components of momentum in the plane yield

$$k'_y = k_y \text{ and } k'_x = k_x \pm 2\pi/a \quad (101)$$

or

$$k'_x = k_x \text{ and } k'_y = k_y \pm 2\pi/a \quad (102)$$

Together with the conservation of energy ($E = E'$), allowing for the occurrence of bound states

$$(k_x^2 + k_y^2 + k_z^2) = (k_x'^2 + k_y'^2 - |k_z'|^2) \quad (103)$$

the above yields the following relations

$$E'_z \equiv E - E'_{||} = -\frac{\hbar^2|k'_z|^2}{2m} \quad (104)$$

where

$$E'_{||} = \frac{\hbar^2}{2m}(k_x'^2 + k_y'^2) \quad (105)$$

The above relation can also be expressed as a parabolic relation between k_x (or k_y) and k_z, given as

$$k_z^2 = \pm\frac{4\pi}{a}k_\alpha + \left(\frac{2\pi}{a}\right)^2 - k_z'^2 \quad (\alpha = x,y) \quad (106)$$

Inspection of equation (104) reveals the following three cases: (i) $E'_z > 0$, corresponding to ordinary diffraction; (ii) $E'_z = 0$, corresponding to the diffraction threshold, also called 'surface resonance' where the diffracted beam just grazes the surface; and (iii) $E'_z < 0$, corresponding to the condition for selective adsorption. Ordinarily the matrix element M_{fi} coupling the incident wave with an outgoing wave for which $E'_z < 0$ will be small. When $|E'_z| = |E_n|$, where E_n is a bound-state energy eigenvalue of the non-periodic component of the potential ($V(z)$) in equation (100), a 'resonance' occurs and the matrix element is large. When the above condition is satisfied (dictated by a combination of angle of incidence and primary beam energy) the particle is 'caught' by the substrate. The energy released by the binding is converted into translational energy of motion in a direction parallel to the surface plane. Eventually the particle will be either trapped or diffusely scattered by some imperfection. In either case a change (usually a decrease) in the intensity of the specularily diffracted beam is observed From a measurement of the scattered specular intensity as a function of scattering angle at a constant angle of incidence, the locations of the above-noted changes in the intensity can be assigned and the corresponding bound-state energies, E_n, evaluated [see equation (104)]. Such data provide an experimentally measured spectrum of eigenvalues (bound states) of the interaction potential. Consequently, it can be used in deducing values of parameters which occur in assumed or theoretically evaluated potential functions (see next section).

Selective adsorption has not been observed yet in scattering experiments from metals. However, in some of the cases for which diffraction occurs, improved resolution should allow the measurement of selective adsorption effects in the intensities. Considering the scattering of rare-gas atoms from metallic surfaces, the balance between two conflicting effects has to be optimized in order for the above experiments to be successful. The first effect is that significant diffractive scattering has been measured to date only with He as the projectile, and since the potential well for He is rather shallow, perhaps exceedingly high resolution is needed to observe the effect of bound-state resonances. The second observation is that whereas the potential of interaction for heavier gases is deep enough to allow for a well-resolved eigenstate spectra, their scattering is dominated by trapping and inelastic processes which greatly reduce the intensity of the specular beam. We believe that improvements in detection techniques and increased resolution would allow the application of selective adsorption methods to some rare-gas–metal scattering systems. Such measurements would assist theoretical investigation of the microscopic origins of the gas–surface interaction process.

On Interaction Potentials used in the Theoretical Analysis of LEMS. A complete discussion of the theories of atomic and molecular scattering from surfaces is beyond the scope of this article. Hence, we review briefly the characteristics of potential functions which are used in several formulations of the problem.

Traditionally, theoretical analysis of data about gas–surface interactions centred around quantities termed 'accommodation coefficients', for example, the energy accommodation coefficient, which is defined by equation (107), where $\bar{E}$ is the

$$a = (\bar{E}_r - \bar{E})/(\bar{E}_s - \bar{E}) \qquad (107)$$

mean energy per atom incident on the surface, $\bar{E}_r$ is the actual mean energy per atom leaving the surface, and $\bar{E}_s$ is the mean energy per atom leaving the surface in the limiting case when the gas has come into thermal equilibrium with the surface. The quantity a varies between 0 (when $\bar{E}_r = \bar{E}$, *i.e.* elastic scattering) and 1 (when $\bar{E}_r = \bar{E}_s$, *i.e.* the incident particle is thermalized by the interaction with the surface-trapping). Thus a measures the efficiency with which the gas comes into equilibrium with the surface. However, such accommodation coefficients do not describe the details and dynamics of the gas–surface collision process. The objectives of modern theories in this field are to describe and explain the velocity and angular distributions of the scattered particles (which are observable in modern molecular beam experiments). Such a description involves a formulation of details of the scattering process and a statement of the interaction potential involved. Thus besides the dependence of the results of the calculations on the model assumptions concerning the collision process, they are influenced by the choice of the potential function being employed in the theory.

In order to illustrate this, we resort first to a classical description of the problem. The scattering system which we deal with is composed of a gas atom impinging on a lattice. Thus, in calculating the trajectory of the scattered gas atom the reaction (response) of the lattice has to be included in the formulation. We start by writing the classical equation of motion (Newton's law) for the gas atom

$$m_g \ddot{\vec{r}} = -\vec{F}(t) \tag{108}$$

where m_g and $\vec{r}$ are the mass and position of the gas atom and $\vec{F}(t)$ is the force exerted on it by the solid. The force, $\vec{F}(t)$, is derived from the interaction potential $V(\vec{r},\vec{R})$, where $\vec{R}(t)$ is the momentary position vector of atoms in the solid. The displacements of the atoms in the solid are given by a set of equations which describe the motion of atoms in the semi-infinite solid under the influence of mutual central and non-central interactions and the gas–solid force. This model formulation reduces to a formidable computational problem which can be tackled only if certain approximation schemes are invoked.[230] In the simplest approximation a rigid lattice is considered, which yields equation (109). Substitution in equation (108)

$$F(\vec{r}) = -\frac{\partial}{\partial \vec{r}} V(\vec{r}) \tag{109}$$

gives three simultaneous differential equations (in the three cartesian co-ordinates) for the trajectory of the scattered particle.[231] The solution involves the integration of the equations of motions, which brings us to our main topic as to the form of gas–solid interaction potential to be used in these calculations. Before discussing some of the potential functions used in the theory or gas–surface scattering, we should note that a statement of the potential function is an essential step in *any* theory of the process. Thus, in quantum mechanical treatments of the problem the potential enters the Schrödinger equation of the scattering system. Two schemes have been used in the development of quantum mechanical theories of gas–surface scattering.

(a) *A perturbation theory approach*, in which the hamiltonian is partitioned into a zeroth-order and perturbation components, and the zero-order set of states serves as a basis for the evaluation of the perturbative correction to the solution. An example of this approach is the LJD theory of scattering[222,223] which was mentioned in the previous subsection. In their original study, Lennard-Jones and Devonshire used a Morse potential function, equation (110), where D is the well

$$V = D \{\exp[-2\kappa(z_g - z_s)] - 2\exp[-\kappa(z_g - z_s)]\} \tag{110}$$

depth, κ is the inverse length parameter describing the 'stiffness' of the potential and z_g, z_s are the normal distances from the surface plane of the gas and surface atom, respectively. Denoting the equilibrium position of the surface atom by z_{so},

---

[230] For a review see ref. 203, sections 3.2—3.5, and references cited therein. See also ref. 208, section B.
[231] For a review see ref. 98.

Microscopic Approaches to Physisorption

the Schrödinger equation for the scattering of the gas atom from a rigid lattice, equation (111), is solved subject to certain boundary conditions.[232] The perturba-

$$-\frac{\hbar^2}{2m_g}\frac{d^2\psi_2}{dz_g} + V(z_g, z_{so})\psi = E\psi \qquad (111)$$

tion potential is defined as

$$H' \equiv V(z_g, z_s) - V(z_g, z_{so}) \sim (z_s - z_{so})V'(z_g)$$

$$+ \tfrac{1}{2}(z_s - z_{so})^2 V''(z_g) + \ldots \qquad (112)$$

where

$$V'(z_g) = [dV(z_g, z_s)/dz_s]_{z_s = z_{so}} \qquad (113)$$

$$V''(z_g) = [d^2 V(z_g, z_s)/dz_s^2]_{z_s = z_{so}} \qquad (114)$$

[Lennard-Jones and Devonshire considered only the first term in equation (112), which restricts the theory to the consideration of transition involving single-phonon excitations. The theory was extended to higher order terms of the potential by several authors.[233,234]] The perturbation theory of LJD is non-unitary, and a modification to it was obtained by Goodman.[234] Comparison of the accommodation coefficients, calculated according to this approach, with experimental values yielded unsatisfactory agreements, which is attributed, among other reasons, to the one dimensionality [234–236] of the model and the potential function employed.

(b) *'Exact' quantum mechanical scattering theories.* An exact solution of the quantum mechanical equation is possible only for simplified, and in most cases unrealistic, forms of the potential (for example, a square-well interaction potential). Assuming 2D translational invariance in the surface plane, a Fourier decomposition of the interaction potential can be performed yielding equation (115), where $\vec{g}_{hk}$

$$V_{g_{hk}}(z) = A \int_{\text{surface}} V(\vec{r}) \exp(-i\vec{g}_{hk} \cdot \vec{r}) d\vec{r} \qquad (115)$$

is a reciprocal vector of the 2D net. Expansion of the wavefunction in a similar Fourier series and substitution in the wave equation yields a series of coupled equations ('beam equation') which can be solved using matrix techniques.[237] In this theory both diffractive and nondiffractive scattering is considered and it has been successful in predicting bound-state resonances and selective adsorption characteristics in the scattered intensities. Analysis of ^{3}He and ^{4}He scattering from LiF using a pairwise Morse potential [equation (110)] resulted in adequate agreement with experiment.[238] A recent study [228] of the elastic scattering of an atom from a surface in the close-coupling formulation, in which a LJD model potential [equation (100)] with a Morse potential for $V(z)$ and an exponential

---

[232] A. F. Devonshire, *Proc. Roy. Soc.*, 1937, **A158**, 269.
[233] R. T. Allen and P. Fener, in 'Rarefied Gas Dynamics' (Proc. 7th Internat. Symp., Pisa, 1970); see also ref. 235.
[234] F. O. Goodman, *Surface Sci.*, 1971, **24**, 667, and references therein.
[235] F. O. Goodman and J. D. Gillerlain, *J. Chem. Phys.*, 1971, **54**, 3077.
[236] D. M. Gilbey, *J. Phys. Chem. Solids*, 1962, **23**, 1453.
[237] E. C. Beder, *Surface Sci.*, 1964, **1**, 242.
[238] F. O. Goodman, *Surface Sci.*, 1970, **19**, 93.

potential $U_1(z)$ [equation (116)] were used, emphasizes the influence of the chosen form of the potential on the results.

$$U_1(z) = -2\beta D \exp[-2\kappa(z_g - z_s)] \tag{116}$$

A modified Morse potential has been suggested in the context of a classical continuum theory of scattering [239] by Trilling.[240] This potential is characterized by a range parameter b and a well depth D [which occurs at a distance given by $\exp(-r/b) = \frac{1}{2}$, where r is the gas-surface separation], and is given by equation (117), where $C = 4$, $a = 1/b$. Comparison between experimental data

$$V(r) = CD[\exp(-2ar) - \exp(-ar)] \tag{117}$$

and calculated values of the energy accommodation coefficient using this model yielded satisfactory agreement for values of the potential parameters, comparable to those obtained by other methods. Examination of the values of the parameters for the scattering of rare gases from metals indicates a marked insensitivity of the range parameter to the components of the scattering system while the well depth increases for heavier gases as expected. However, the value for the depth for a given gas did not vary for different metal substrates. These results indicate, in our opinion, a serious problem which is common to many methods which use parametric potentials. Different metal substrates are expected, by physical reasoning, to exhibit different interaction potential characteristics towards impinging rare gases, and indeed our detailed calculations (see Section 2) show this dependence. Thus, while agreement with experiment (in particular accommodation coefficients) may be achieved, the derived potential function should be examined critically.

An additional remark about Morse-type potentials is worth noting. The eigenvalues of the Morse potential [equation (117)] are given as

$$E_n = -CD\left[1 - (n + \tfrac{1}{2})\frac{\hbar a}{\sqrt{2mCD}}\right]^2 \qquad (n = 0, 1, 2 \ldots) \tag{118}$$

This closed form expression has been used [225] (with $C = 1$) for the analysis of recent selective adsorption experiments of ^{4}He scattering from LiF (001) surfaces yielding very poor agreement, which adds to the doubts concerning the validity of this potential for the description of this system.

We have already mentioned (p. 68) some of the main forms of the interaction potentials used in gas–surface scattering theories. In the following we briefly discuss some others.

Among these, the *hard and soft cube models* have enjoyed much interest. The first was introduced [241,242] to explain qualitative features of scattering data, and was followed and to a large extent superceded by the soft-cube model.[243] The main attractive features of this model are the relative ease of analysis involved, and the

---

[239] L. D. Landau, *Phys. Z. Sowjetunion*, 1935, **8**, 489.
[240] L. Trilling, *Surface Sci.*, 1970, **21**, 337.
[241] R. M. Logan and R. E. Stickney, *J. Chem. Phys.*, 1966, **44**, 195.
[242] R. M. Logan, J. C. Keck, and R. E. Stickney, in 'Rarefied Gas Dynamics' (Proc. 5th Internat. Symp., Oxford, 1966), ed. C. L. Brunolin, Academic Press, New York, 1967, Vol. 1, p. 49.
[243] R. M. Logan and J. C. Keck, *J. Chem. Phys.*, 1968, **49**, 860.

good agreement with experimental observation which has been achieved in some cases (with the reservations expressed above). The main assumptions of the hard-cube model are: (i) The gas and surface atoms are regarded as rigid elastic particles and thus the gas–solid repulsion potential is impulsive. There is no attractive component to the potential. (ii) The scattering potential is uniform in the surface plane and hence there is no change in the tangential component of the incident particle velocity. In addition the surface atoms are represented as an ensemble of independent hard cubes confined by a square-well potential and they possess a Maxwellian velocity distribution.

In the soft-cube model some of the failings of the above model have been eliminated. The assumptions of the model are: (i) The interaction potential has a stationary attractive component (square well) and an exponential repulsive part [of a form given in equation (116)]. (ii) The gas–solid interaction potential is uniform in the plane of the surface. In addition the surface atom involved is connected by a one-dimensional 'spring force' to a remainder fixed lattice, and the ensemble of oscillators has an equilibrium energy distribution at the temperature of the solid. This model allows the derivation of a closed-form expression of the scattered intensity for a mono-energetic incident beam (otherwise numerical integration is required). Unlike the hard-cube model, which is not parametric, the soft-cube one contains three adjustable parameters: the depth of the attractive well, the range of the repulsive potential, and the natural frequency of the surface oscillator. The limitations of the model can be deduced from the above listed assumptions. In view of these and the analysis of recent experimental data,[211,212,214] it appears that certain apparent agreements between experiments and calculated results using the cube models have been fortuitous.

Among the other potentials we mention first the pairwise Lennard-Jones (LJ) potentials, commonly used in the theory of intermolecular forces.[244] In particular we mention the superimposed LJ (n,m) potentials used by McClure[245] ($n = 12$, $m = 6$) in classical trajectory calculations. Equation (119) is the general form of

$$U = CD[(b/r_j)^n - (b/r_j)^6] \qquad n = 7, 8, \ldots, 12 \qquad (119)$$

the LJ potential (see also Section 2), where D is the binary binding energy, b is a range parameter, and r_j is the separation of the gas atom from the jth surface atom. In the above calculations the long-range van der Waals' potential is taken to be one-dimensional, and its strength taken as an adjustable parameter. Trajectory analysis of Ne scattering from LiF (001) resulted in satisfactory agreement with experimentally measured angular distributions. Owing to the assumed pairwise additivity of the potentials, this model is not expected to apply well to scattering from metal substrates. In this context we should mention the trajectory calculations performed by Oman and his associates [246–248] of the scattering of rare gases from Ni and Ag(111) surfaces. In these calculations three potential functions have been postulated, corresponding to three specified distance (r)

[244] I. M. Torrens, 'Intermolecular Potentials', Academic Press, New York, 1972.
[245] J. D. McClure, *J. Chem. Phys.*, 1972, **57**, 2810, 2823.
[246] R. A. Oman, A. Bogan, and C. H. Li, in 'Rarefied Gas Dynamics' (Proc. 4th Internat. Symp.), ed. J. H. De Leeuw, Academic Press, New York, 1966, Vol. 2, p. 396.
[247] R. A. Oman, *J. Chem. Phys.*, 1968, **48**, 3919.
[248] V. S. Calia and R. A. Oman, *J. Chem. Phys.*, 1970, **52**, 6184.

regimes between the gas atom and the surface; (i) $r > 5a$ (where a is the lattice spacing) – an LJ (6,12) potential between the incident particle and a semi-infinite solid is assumed, which when integrated over the surface yields a van der Waals (VDW) attraction proportional to r^{-3}; (ii) $5 \geqslant r \geqslant 2.5a$ – the above VDW potential combined with a discrete LJ (6,12) potential is used; and (iii) $r < 2.5a$ – an LJ (6,12) potential is used.

Despite the authors' claim of good agreement between their calculations and available experimental data, there can be little statistical confidence in their calculation because of the exceedingly small number (50) of trajectories used, and as such, these calculations do not support the validity of the potential model employed in them.

Finally, we mention the semiclassical calculations of He and Ne scattering from W(112) by Masel, Merrill, and Miller,[217,219] in which a modified LJ (3,9) potential was used. This potential is given as equation (120), where x is the distance along

$$V(x,z) = D \left\{ \frac{2}{15} \frac{(\sigma_{gs}[1 + B\cos(2\pi x/a_x)])^{12}}{\sigma_{gs} z^9} - \frac{\sigma_{gs}^6}{\sigma_{ss}^3 z^3} \right\} \tag{120}$$

the surface in the (110) direction, z is the distance normal to the surface, D is the well depth, σ_{ss} is the interatomic distance in the solid, σ_{gs} is the hard-sphere collision parameter between the incident gas and the surface atoms, and a_x is the lattice dimension in the x direction. The parameter B is the amplitude of the surface periodicity, which is estimated from a geometrical construction of the surface. The well depth is estimated from the heat of adsorption of the gas on the metal substrate and σ_{gs} is an adjustable parameter. It was observed that the most significant parameter of the model is the gas–surface distance σ_{gs}. As σ_{gs} is reduced, independent of B, the potential gets stiffer. This increases the back-scattered peaks at the expense of the forward scattered intensity. The 'best fit' value for σ_{gs}, yielding good agreement with experiment (a sample of the results is shown in Figure 31) was 3.2 Å, for He scattering from W. This value is larger than the value of 2.65 Å which is obtained on the basis of the hard sphere 'sum of radii' rule, indicating that the repulsive potential of the surface extended considerably further than just the crystal atomic radius. This result is in good agreement with our microscopic calculations of the interaction potential for this system in which an equilibrium distance of 3.3 Å was determined (see Table 1, p. 29). Although this agreement is not taken as 'proof', it is an encouraging indication that progress in the development of theoretical models of atom–surface interaction and scattering processes can provide detailed information, on the microscopic level, about the interaction potentials. Although an analytical form of the potential simplifies the calculations, numerical integration methods could be easily implemented allowing the use of tabulated potentials derived *via* microscopic theories.

Experimental Techniques for the Study of the Electronic Structure of Physisorbed Systems.—*Ultraviolet Photoemission Spectroscopy* (*UPS*). The principles of the use of UPS in distinguishing between chemical binding and physical states of adsorption on the basis of valence level shifts are outlined on p. 34. Similar ideas have been employed in a number of studies of the adsorption of organic and inorganic compounds on metal surfaces, in the monolayer and multilayer regimes. (For reviews of this subject see refs. 124—126, 128, 131, 249, and 250. References 251—269 are a representative though incomplete, list of original studies.) As emphasized in some of these investigations, UPS studies of adsorption systems over a range of temperature and pressure may provide information about gas–surface adsorption and reaction mechanisms. Apart from the diagnostic value, UPS studies of physisorption systems and comparisons with gas-phase spectra enable the isolation of certain effects connected with UPS measurements of adsorption systems, where the complications due to chemical binding are minimized.

Yu, McMenamin, and Spicer [254] studied the physical adsorption (or condensation) of several gases on the cleavage face of MoS_2. The main conclusions of their studies are as follows. (i) The photoemission from the substrate appears to be unaffected by the condensed gas. This is to be contrasted with the strong effects of chemisorption on metal and semiconductors (Si, Fe, Ni, Cu) on the emission from the substrate. (ii) The condensation of gas molecules on MoS_2 causes a lowering of the photoemission threshold (reduction in work function). (iii) The importance of an adequate subtraction of substrate and background emission from the spectrum of the adsorption system, prior to comparison with gas-phase spectrum, is emphasized. (iv) The main contribution to the relaxation shift is attributed to the surrounding molecules, since the shift was found to be independent of coverage.

In a recent study of Xe physisorption on W(100) by Waclawski and Herbst,[266] evidence for surface-crystal-field effects has been reported. The delineation of the effect is made possible by the physical rather than chemical nature of the gas–solid interaction. The photoelectron energy distribution for clean W(100) at 80 K and after exposure to 5 Langmuirs of Xe are shown in the bottom panel of Figure 35. The corresponding difference curve is shown in the top panel of the figure. First,

[249] D. E. Eastman and J. E. Demuth, 'Proc. Internat. Conf. on Solid Surfaces', 1974 [*Jap. J. Appl. Phys.* Supp. 2(2), 827].
[250] C. R. Brundle, *J. Vacuum Sci. Technol.*, 1976, **13**, 301.
[251] J. E. Demuth and D. E. Eastman, *J. Vacuum Sci. Technol.*, 1976, **13**, 283.
[252] J. E. Demuth and D. E. Eastman, *Phys. Rev.*, 1976, **B13**, 1523.
[253] J. E. Demuth and D. E. Eastman, *Solid State Comm.*, to be published.
[254] K. Y. Yu, J. C. McMenamin, and W. E. Spicer, *Surface Sci.*, 1975, **50**, 149.
[255] K. Y. Yu, W. E. Spicer, I. Lindau, P. Pianetta, and S. F. Lin, *J. Vacuum Sci. Technol.*, 1976, **13**, 277.
[256] S. J. Atkinson, C. R. Brundle, and M. W. Roberts, *Discuss. Faraday Soc.*, 1974, **58**, 62.
[257] C. R. Brundle and A. F. Carley, *Discuss. Faraday Soc.*, 1975, **60**, 51.
[258] P. J. Page and P. M. Williams, *Discuss. Faraday Soc.*, 1974, **58**, 80.
[259] H. D. Hagstrum, *Science*, 1972, **178**, 275.
[260] E. W. Plummer, in 'Topics in Applied Physics', ed. R. Gomer, Springer, Berlin, 1975.
[261] W. F. Egelhoff, J. W. Linnett, and D. L. Perry, in *Discuss. Faraday Soc.*, 1974, **58**.
[262] W. F. Egelhoff and D. L. Perry, *Phys. Rev. Letters*, 1975, **34**, 93.
[263] B. Feuerbacher and B. Fitton, *Phys. Rev. Letters*, 1972, **29**, 786; *Phys. Rev.*, 1973, **B8**, 4890.
[264] B. Feuerbacher and M. R. Adriaens, *Surface Sci.*, 1974, **45**, 553.
[265] G. Brodén and T. N. Rhodin, *Discuss. Faraday Soc.*, 1975, **60**, 112.
[266] B. J. Waclawski and J. F. Herbst, *Phys. Rev. Letters*, 1975, **35**, 1594.
[267] T. V. Vorburger, D. R. Sandstrom, and B. J. Waclawski, *J. Vacuum Sci. Technol.*, 1976, **13**, 287.
[268] J. M. Burkstrand, G. G. Kleiman, G. G. Tibbetts, and J. C. Tracy, *J. Vacuum Sci. Technol.*, 1976, **13**, 291.
[269] B. J. Waclawski and E. W. Plummer, *Phys. Rev. Letters*, 1972, **29**, 783.

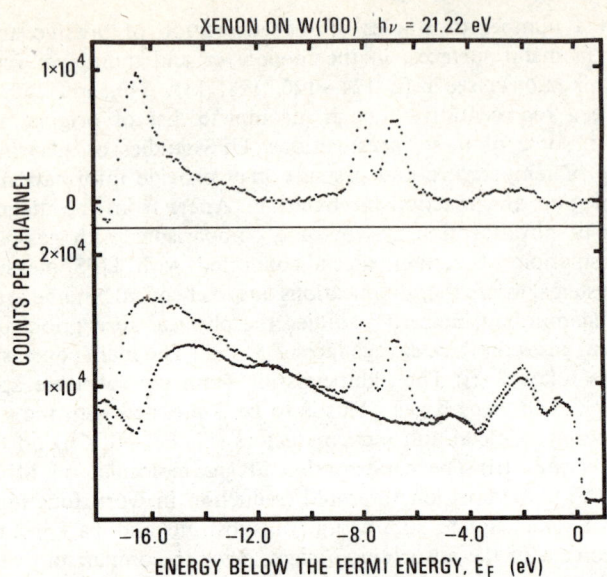

Figure 35 *Photoelectron energy distributions. The lowermost curve is for* W(100) *at* ~80 K, *and directly above it is the distribution after exposure to* 5 Langmuirs *of xenon. The difference curve of these two measurements is given as the top panel. Note that the structure at* −0.5 eV, *assigned as a surface state of tungsten, persists after exposure to xenon, indicating the physical nature of the interaction*
(Reproduced by permission from *Phys. Rev. Letters*, 1975, **35**, 1594)

it is noted that the peak at −0.5 eV, which has been identified previously [263,269] as a surface state, persists even after exposure to the rare gas, whereas exposure to H_2, O_2, N_2, or CO suppresses the peak. This demonstrates the non-chemical nature of the interaction. Secondly, the peaks at −6.8 and −8.1 eV are identified with emission due to the two spin–orbit split Xe 5P state ($5P_{\frac{3}{2}}$ and $5P_{\frac{1}{2}}$, respectively). A comparison between the gas-phase spectrum of Xe and the above portion of the adsorbed Xe spectra is given in Figure 36. The spin–orbit split in both cases is 1.3 eV and the areas under the peaks yield essentially the same ratio. However, it is seen that the $5P_{\frac{3}{2}}$ peak is significantly broadened with respect to the $5P_{\frac{1}{2}}$ peak, and the former may be considered as consisting of two peaks having the same width as the $5P_{\frac{1}{2}}$ peak but separated by ~0.3 eV. A first-order perturbation calculation for a configuration in which the Xe atom is partially embedded in the dipole layer of the tungsten atoms and is located at the four-fold site (centre) of the square (100) unit cell of the substrate, has been performed. Since, to first order, a $P_{\frac{1}{2}}$ hole state which has a spherically symmetric charge density will not split in any crystal field, this model provides a mechanism by which the substrate crystal field may preferentially affect the $5P_{\frac{3}{2}}$ peak. This work demonstrates that studies of weak-coupling physisorption systems in which complications due to charge rearrangement and bond formation do not occur, allow the identification and investigation of fundamental gas–solid interaction mechanisms.

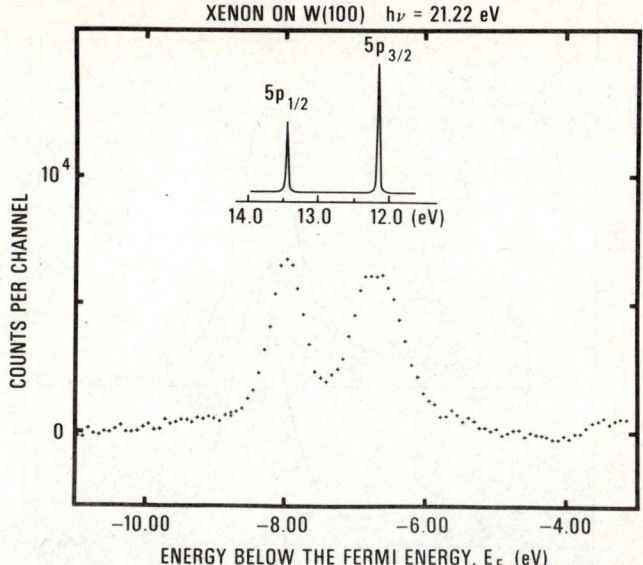

Figure 36 *Expanded view of the two Xe 5P peaks, from the difference spectrum shown in Figure 35. Note the broadening of the $5P_{\frac{3}{2}}$ peak (right) with respect to the $5P_{\frac{1}{2}}$ peak (left). The insert is the photoelectron distribution of gaseous Xe*
(Reproduced by permission from *Phys. Rev. Letters*, 1975, **35**, 1594)

X-Ray Photoemission Spectroscopy (*XPS*).[125,129,131,270—275] In XPS, electron excitation from a core level to a final state above the vacuum level is caused by an incident X-ray photon (see Figure 8b). XPS experiments usually employ unmonochronized $K_{\alpha 1,2}$ radiation from Mg (1254.6 eV) or Al (1486.6 eV) having a linewidth of about 1 eV. The energy balance of the X-ray photoemission process is the same as that of the u.v. photoemission process (p. 34). The surface sensitivity of the technique derives from the limited mean free path (5—15 Å) of electrons in the energy range of interest. Core-level binding energies, measured by XPS, are characteristic of individual atoms even if they are in some aggregate state (molecular or other interaction), since the effects of chemical shifts on core levels are usually small: hence the routine application of XPS in elemental analysis and the common name for it, electron spectroscopy for chemical analysis (ESCA).[270] Nevertheless, various chemical dependent effects in XPS can be detected. Among those we mention: chemical shifts, spin–orbit splitting, multiple structure, and

---

[270] K. Siegbahn, C. Nordling, A. Fahlman, R. Nordberg, K. Hamrin, J. Hedman, G. Johansson, T. Bergmark, S. E. Karlsson, I. Lindgren, and B. Lindberg, 'ESCA: Atomic, Molecular and Solid State Structure Studied by Means of Electron Spectroscopy', Almqvist and Wiksells, Uppsala, 1967.
[271] C. R. Brundle, in this series, Vol. 1, p. 171.
[272] D. T. Clark, in 'Electron Emission Spectroscopy', ed. W. DeKeyser, L. Fiermans, G. Vanderkelen, and J. Vennik, Reidel, 1973.
[273] W. N. Delgass, T. R. Hughes, and C. S. Fodley, *Catalysis Rev.*, 1971, **4**, 179.
[274] 'Electron Spectroscopy', ed. D. A. Shirley, North-Holland, Amsterdam, 1972.
[275] C. S. Fadley, in 'Progress in Solid State Chemistry', ed. G. Somorjai and J. McCaldin, Pergamon, Oxford.

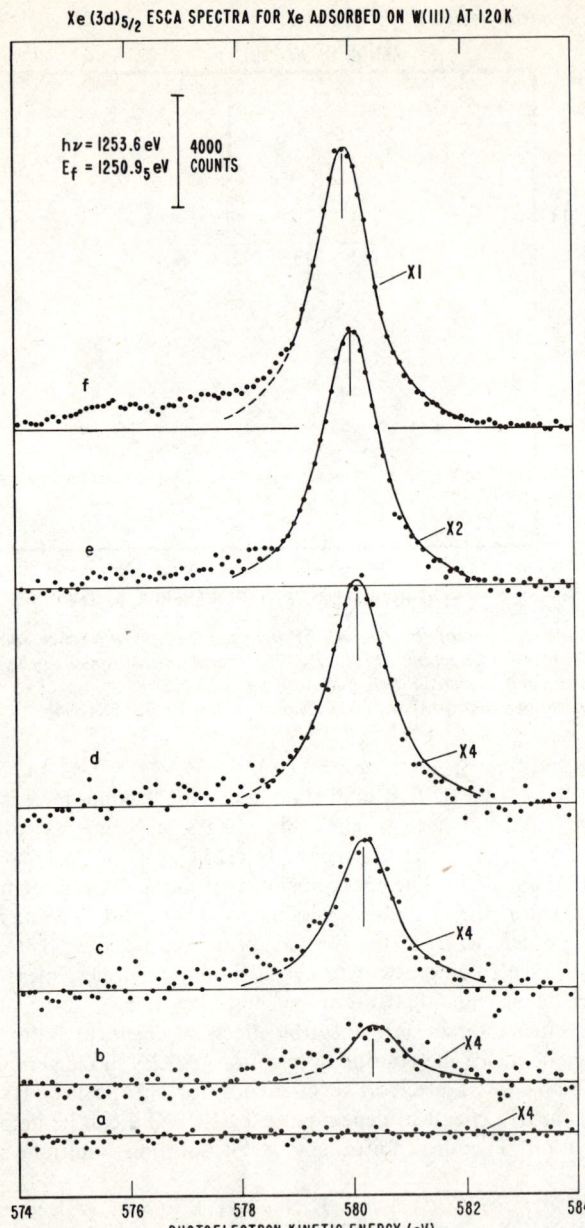

Figure 37 Xe($3d_{\frac{5}{2}}$) *ESCA spectra for* Xe *adsorbed on* W(111) *at* 120 K. *Following* 2500 K *cleaning, the crystal is cooled in* 4×10^{-10} Torr *vacuum for* 7 min *prior to Xe adsorption. All spectra at* 1 s *per channel,* 0.1 V *per channel,* 4 *scans. Background count level* = 6×10^3 *counts per channel. Lorentzian curves have fwhm* = 1.24 eV. *Kinetic energies can be converted into binding energies (relative to the Fermi level) by subtracting abscissa from* 1250.95 eV. *A small satellite peak near* 576 eV *is due to the* Xe ($3d_{\frac{3}{2}}$) *peak excited by* Mg $K_{\alpha 3,4}$ *radiation.* (a) *Clean* W(111); (b) Xe *exposure* = 8.5×10^{-8} Torr s^{-1}; (c) Xe *exposure* = 4.8×10^{-7} Torr s^{-1}; (d) Xe *exposure* = 7.0×10^{-7} Torr s^{-1}; (e) Xe *exposure* = 1.3×10^{-6} Torr s^{-1}; (f) Xe *exposure* = 5.0×10^{-6} Torr s^{-1} (Reproduced by permission from *Surface Sci.*, 1974, **44**, 489)

satellite structure. XPS spectra are usually less convolved than UPS energy distribution curves. Coupled with the quantitative elemental analysis power of XPS, the combination of the two techniques brings distinct advantages to the study of adsorption and reaction mechanisms on surfaces.

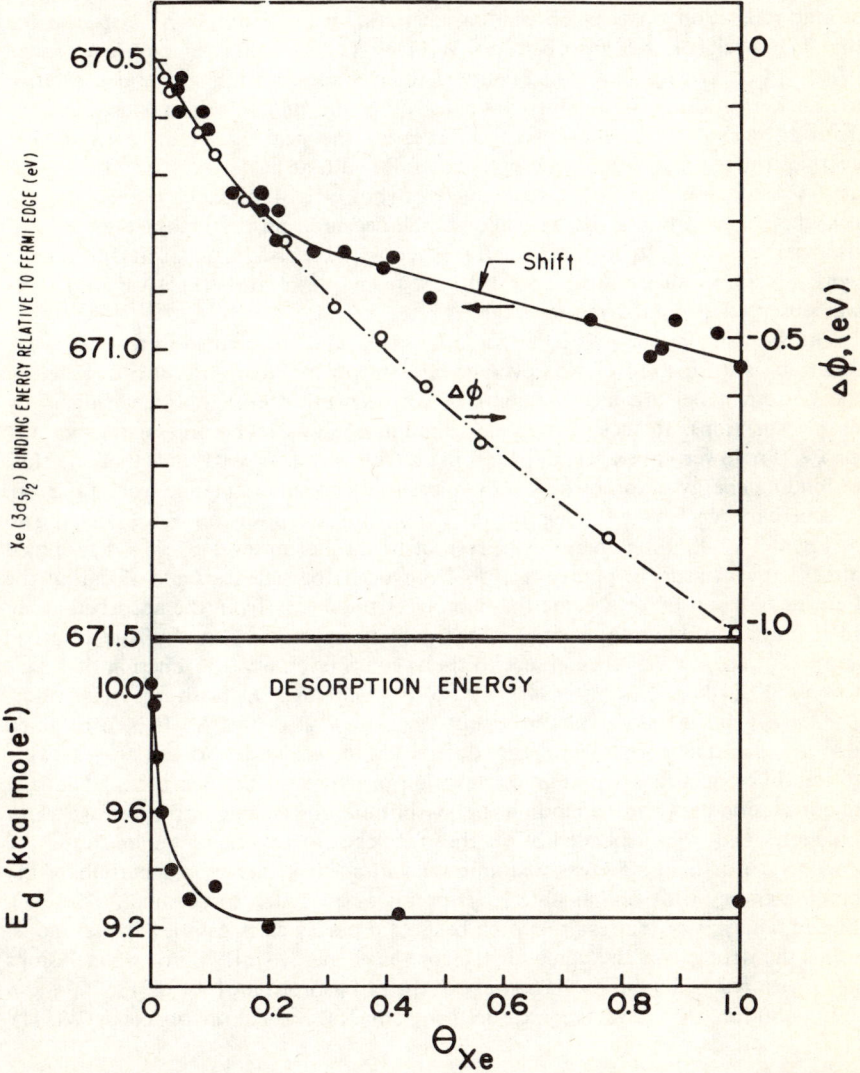

Figure 38 *Shift of* $Xe(3d_{\frac{5}{2}})$ *energy with increasing Xe coverage on* W(111) *at* 120 K, *work function behaviour, and the desorption energy, as determined separately (see ref.* 277). *The rapid changes below* $\theta_{Xe} \sim 0.2$ *are attributed to xenon adsorption on extraneous sites or defects on the* W(111) *crystal*
(Reproduced by permission from *Surface Sci.*, 1974, **44**, 489)

In order to study the effect of physisorption on core-level binding energies, and to provide information about processes associated with X-ray photoionization, free from complications due to chemisorptive binding effects, we discuss in the following the XPS study carried by Yates and Erickson [276] on the adsorption of Xe on the (111) face of tungsten.[277] In particular, a comparison of XPS spectra for gaseous and adsorbed Xe should permit an evaluation of the effect of extra-atomic relaxation on electron binding energies. Representative XPS spectra for the $3d_{\frac{3}{2}}$ peak of Xe adsorbed on W(111) at 120 K for a coverage range $0.045 \leqslant \theta_{Xe} \leqslant 1$ are shown in Figure 37 (conversion to binding energies, relative to the Fermi Level, is obtained by substrating the photoelectron kinetic energy from 1250.95 eV). As the coverage increases, the peak shifts to lower kinetic energies (increasing binding energy). The full width at half maximum of the peak at $\theta_{Xe} > 0.1$ was constant (spectrometer fwhm = 1.24 ± 0.02 eV), and thus the peak heights relative to the base-line points measured on the high-energy edge of the peak can be used to measure Xe coverage (see also p. 97). In addition, a work function reduction of -1.1 ± 0.1 eV from the clean W(111) value (4.4 eV) was observed at monolayer coverage.

The shift in the energy of the Xe ($3d_{\frac{3}{2}}$) peak, and the change in average work function as a function of Xe coverage, are shown in Figure 38 (also included in the bottom panel are measurements of Xe desorption energy as determined by flash desorption). It is seen that the core binding energy shift becomes monotonically greater for lower coverages, approaching 2.6 eV at $\theta_{Xe} = 0.05$, and that the shift in binding energy does not follow the measured change in average work function. A schematic energy-level diagram for xenon physisorption on W(111) is shown in Figure 39. Binding energy with respect to the Fermi level E_B^f was determined directly by measuring photoemission from both the valence band (locating the Fermi edge) and by measuring the photoelectron energy from the adsorbed atom. In relating core binding energies of the gaseous atom ($E_{B,g}^o$) and of the adsorbed atom ($E_{B,ads}^o$), both with reference to the vacuum level, a work function of 3.3 eV was used yielding $E_{B,ads}^o = 674.3 \pm 0.1$ eV, which together with the gas value, $E_{B,g}^o = 676.4 \pm 0.1$ eV, results in a shift $\Delta_{ads,g} = -2.1 \pm 0.2$ eV. It is to be noted that the final state core-hole ions are different in the gas and adsorbed phases owing to the differences in extra-atomic relaxation processes in the two cases. The lack of correlation between the binding energy shift and the change of the average work function, and the dependence of the former on coverage, demonstrate the importance of final state extra-atomic relaxation mechanisms in determining the binding energy shift (initial-state contribution is estimated to be small, ~ 0.1 eV). Indeed, analysis of the results on the basis of a purely electrostatic (dipole) model yields the wrong sign and value for the core-level energy shift. Similar conclusions have been reached from an XPS study of the physisorption of Xe on Pt.[278]

In addition to the above measurements of Xe adsorption on clean W(111),

---

[276] J. T. Yates, jun. and N. E. Erickson, *Surface Sci.*, 1974, **44**, 489.
[277] M. J. Dresser, T. E. Madey, and J. T. Yates, jun., *Surface Sci.*, 1974, **42**, 533. In this work Xe adsorption on W(111) was investigated using flash desorption and work function methods. In addition, effects due to Xe interaction with preadsorbed oxygen were studied.
[278] N. E. Erickson, *J. Vacuum Sci. Technol.*, 1974, **11**, 226.

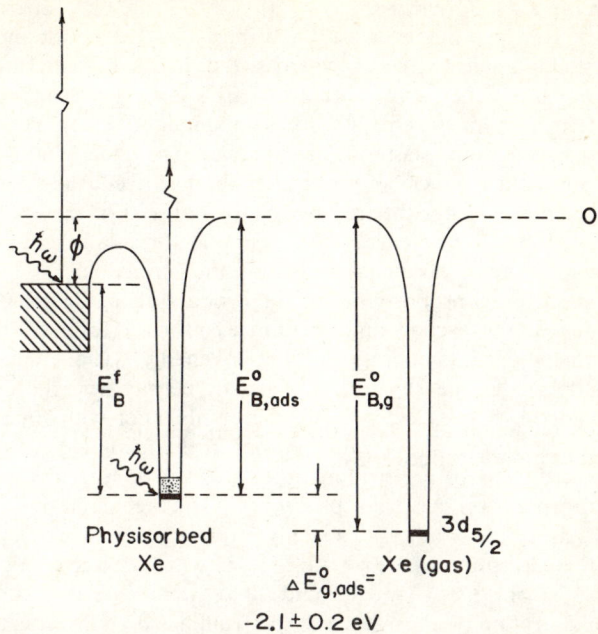

Figure 39 *Schematic diagram of energy levels for xenon physisorption on W(111). The binding energies of gaseous and adsorbed atoms, with reference to the vacuum level (0) are denoted by $E^o_{B,ads}$, respectively. The shift between the two is denoted by $\Delta E^o_{ads,g}$, and ϕ is the work function of the sample*
(Reproduced by permission from *Surface Sci.*, 1974, **44**, 489)

effects due to oxygen coadsorption were studied, and a downward shift in Xe($3d_{\frac{5}{2}}$) binding energy with increasing oxygen coverage was measured.

The ESCA studies discussed above demonstrate that XPS could be used for the study of the electronic structure of physisorbed systems (in the above work a sensitivity to 0.05 monolayer and up was estimated). The current sensitivity of the method, compared to Auger spectroscopy, is poorer by a factor of 10 to 100. However, since electron beam damage (electron-induced desorption – EID, see p. 100) which affects Auger measurements does not occur in XPS work, the latter may be used in studies of *weak adsorption systems*. Coupled with the demonstrated sensitivity of the method, the main value of XPS studies of physisorption systems may be due to the ability to examine effects associated with core ionization of adsorbed atoms and molecules, *in the absence of chemical binding effects*. Such studies would improve our understanding of core-hole relaxation mechanisms. In addition, such data for atoms and molecules, which under different conditions (*e.g.* temperature) participate in either a physical or chemical interaction with the solid, would provide a reference for the study of effects due to chemical binding.[125,250]

Electron Energy Loss Spectroscopy (EELS).[129,142,163,279—285] Electron scattering can be operated in two modes, elastic and inelastic. The elastic mode is used in surface crystallographical studies whereas the inelastic scattering probes the electronic structure. The basic process of EELS is two-electron in character:[142] a primary electron of kinetic energy E loses an amount of energy $\hbar\omega$ to an electron (or a collective excitation, plasmon) of the solid. The underlying mechanism is a two-step process, since either before or after the loss event the electron is turned around by being elastically scattered by the ion-cores. For the excitation of core electrons the probability of energy loss near the threshold for the excitation is proportional to the density of final states above the Fermi level.[129,286] The intensity of the loss spectrum is proportional to the product of the final-state densities of the transitions of the excited and exciting electrons. However, the final state density of the latter is smoothly varying in energy, yielding an approximate proportionality of the intensity to the final-state density near the Fermi Level of the transition of the excited electron. EELS in combination with other spectroscopic techniques (like UPS) can be a valuable tool in disentangling the energetics and electronic mechanisms of adsorption processes, and in the investigation of the effects of adsorption on substrate properties. The technique has not been used much in physisorption studies. The only study known to us is by McElhiney *et al.*[163] of the adsorption of Xe on Ag(111), in which characteristic losses due to a Xe monolayer at 13.5 eV and 68 and 78 eV were observed. In addition the effect of Xe adsorption on the silver loss spectrum has been investigated,[163] providing further information about the origin of certain transitions.

Field Emission Techniques. Experiments on rare-gas–metal physisorption systems with field emission techniques have provided valuable information about the nature of the physical adsorption interaction, the dependence of the interaction on the substrate crystal structure, and the mobility of physisorbed atoms, and have allowed the extraction of physical quantities such as heats of adsorption, diffusion coefficients, and pre-exponential frequency factors. Since the work function of the system under study is basic for the description of the field-emission process, we discuss first this quantity, followed by a schematic description of the experiment and a discussion of some of the results.

The work function ϕ of a solid is a thermodynamic quantity defined[287,288] as the work done against the chemical potential, μ, of an electron in the bulk and the difference, $\Delta\Phi$, in the electrostatic potential across the surface of the solid, *i.e.*

$$e\phi = e\Delta\Phi - \mu \tag{121}$$

where e is the charge on the electron. The surface sensitivity of the work function

---

[279] H. Raether, 'Springer Tracts in Modern Physics', ed. G. Holer, Springer Verlag, Berlin, 1965, Vol. 38, p. 84.
[280] G. E. Laramore, *J. Vacuum Sci. Technol.*, 1972, **9**, 525.
[281] C. B. Duke and U. Landman, *Phys. Rev.*, 1973, **B8**, 505, and references cited therein.
[282] H. Ibach and J. E. Rowe, *Phys. Rev.*, 1974, **B9**, 1951.
[283] J. E. Rowe and H. Ibach, *Phys. Rev. Letters*, 1973, **31**, 102.
[284] J. Kuppers, *Surface Sci.*, 1973, **36**, 53.
[285] F. Steinrisser and E. N. Sickafus, *Phys. Rev. Letters*, 1971, **27**, 992.
[286] R. L. Park and J. E. Houston, *J. Vacuum Sci. Technol.*, 1973, **10**, 176.
[287] C. Herring and M. H. Nichols, *Rev. Mod. Phys.*, 1949, **21**, 185.
[288] N. D. Lang and W. Kohn, *Phys. Rev.*, 1971, **B3**, 1215.

has been amply demonstrated by observing variations in ϕ with crystal plane of the same material and by the large changes in the work function which commonly occur upon adsorption.

The surface potential, $\Delta\Phi$, is usually associated with a dipole layer caused by the separation of positive and negative charges at the surface. Electrons at the surface region lower their kinetic energy by tunnelling through the surface potential barrier into the vacuum.[289,290] This produces a negative outward layer. Furthermore, a dipole moment of opposite sign is produced by lateral flow of negative charge to smooth the variation in the potential along the surface.[291] This later process accounts for the lower work function of the less densely packed surfaces. Modifications in the surface potential due to adsorption are caused by a spatial rearrangement of the charge distributions, and by charge transfer between adsorbate and adsorbent. Formally, the change in work function is given in terms of an effective dipole moment p (per adsorbed particle) as equation (122), where N is

$$e\Delta\phi = 4\pi pN \qquad (122)$$

the adsorbate concentration per unit area and p is equal to qd, where q and d are the dipole charge and length, respectively. The above, discrete dipole description, is clearly an oversimplification of the problem due to the neglect of dynamical effects, the difficulties in defining the microscopic distance and the complexity of the charge distribution (and redistribution). Nevertheless, the model, and modifications to it, have been used extensively to account for various effects, like mutual dipole depolarization,[292] and semi-empirical correlations between the sign of observed dipole moments and electronegativity differences between the adsorbate and adsorbent were suggested.[293,294]

In accord with the above electrostatic dipole model for the description of the origin of changes in the work function upon adsorption, work function measurements which were carried out on chemisorption systems were not performed on physical adsorption systems, since owing to the nature of the interaction, the latter were not expected to exhibit variations in the work function (surface potential). The discovery by Mignolet[295] that the adsorption of Xe on nickel films (polycrystalline) resulted in a large decrease in the work function (large positive surface potential, as high as 0.85 V) radically changed the above trend and led to numerous studies in which variations in the work function of many physisorbed systems have been measured, by several methods. Many of the studies on the adsorption of rare gases on metals were conducted on polycrystalline films.[295-305] Experiments

[289] E. Wigner and J. Bardeen, *Phys. Rev.*, 1935, **48**, 84.
[290] J. Bardeen, *Phys. Rev.*, 1936, **49**, 653.
[291] R. Smoluchowski, *Phys. Rev.*, 1941, **60**, 661.
[292] J. Topping, *Proc. Roy. Soc.*, 1927, **A114**, 67.
[293] E. P. Gyftopoulos and J. D. Levine, *J. Appl. Phys.*, 1962, **33**, 67.
[294] A. J. Sargood, C. W. Jowett, and B. J. Hopkins, *Surface Sci.*, 1970, **22**, 343.
[295] J. C. P. Mignolet, *Discuss. Faraday Soc.*, 1950, **8**, 105.
[296] J. C. P. Mignolet, *J. Chem. Phys.*, 1953, **21**, 1298.
[297] J. C. P. Mignolet, *Rec. Trav. chim.*, 1955, **74**, 685, 701.
[298] J. C. P. Mignolet, in 'Chemisorption', ed. W. E. Garner, Butterworths, London, 1957, p. 118.
[299] R. Shurmann, E. A. Dierk, B. Engelke, H. Hermann, and K. Schulz, *Naturwiss.*, 1956, **43**, 127; *J. Chem. Phys.*, 1957, **54**, 15.
[300] J. Pritchard, *Trans. Faraday Soc.*, 1963, **59**, 437.
[301] R. Bouwman and W. M. H. Sachtler, *Ber Bunsengesellschaft phys. Chem.*, 1970, **74**, 1273.

on rare-gas adsorption on single-crystal metal substrates,[306] and measurements in which adsorption on different crystal planes could be discerned,[307—315] revealed that contrary to the traditional view, physical adsorption, in certain coverage regimes, exhibits specificity to the crystal face exposed.

Field Emission and the Probe-hole Techniques.[316—320] Among the various techniques for the measurement of work function changes in adsorption systems,[321] the field emission techniques have played an important role. In addition to work function measurements, these techniques allow the determination of heats of adsorption and rates of diffusion on metal surfaces.

Electron emission from a metal or semiconductor surface is greatly enhanced by the application of a large electric field (of 10^7—10^8 V cm^{-1}) normal to the surface. The field reduces the potential barrier, and allows electrons to 'tunnel' through and be emitted. In the field-emission microscope (FEM) the metal tip is imaged by the impingement of the field-emitted electrons onto a fluorescent screen.

The work function can be derived from measurements of the current, I, emitted from area S as a function of applied voltage V, *via* the Fowler–Nordheim equation (123), where $A = CS/(\beta^2 \phi)$, where C is a constant and β relates V to

$$I/V^2 = A \exp(-B\phi^{\frac{3}{2}}/V) \qquad (123)$$

the field strength ($\beta = V/E$). B is equal to $g(\phi,E)\beta$ and $g(\phi,E)$ is a slowly varying tabulated[318] function. In general, $\Delta\phi$ values obtained from field-emission measurements are in agreement with results obtained by other methods on macrosopic crystal planes. It should be noted that there is an intrinsic limitation of the method in attempts to measure the work function changes over individual planes, due to the averaging process involved in work function measurements.[308,312] The total current measured is a weighted sum of currents from individual crystal

[302] D. F. Klemperer and J. C. Snaith, *Surface Sci.*, 1971, **28**, 209.
[303] A. G. Knapp and M. H. B. Stiddard, *J.C.S. Faraday I*, 1972, **68**, 2139.
[304] Th. G. J. Van Oirschot and W. M. H. Sachtler, *Ned. Tijdschrift Vacuumtechniek*, 1970, **8**, 96.
[305] P. M. Gundry and F. C. Tompkins, *Trans. Faraday Soc.*, 1960, **56**, 846.
[306] These experiments were done mainly in conjunction with LEED measurements (see Table 3).
[307] G. Erlich, T. W. Hickmott, and F. G. Hudda, *J. Chem. Phys.*, 1958, **28**, 977.
[308] G. Erlich and F. G. Hudda, *J. Chem. Phys.*, 1959, **30**, 493.
[309] G. Erlich, H. Heyne, and C. F. Kirk, 'The Structure and Chemistry of Surfaces', Wiley, New York, 1969, p. 49.
[310] J. Nikliborc and Z. Dworecki, *Acta Phys. Polon.*, 1967, **32**, 1023.
[311] R. Gomer, *Austral. J. Phys.*, 1960, **13**, 391.
[312] T. Engel and R. Gomer, *J. Chem. Phys.*, 1970, **52**, 5572.
[313] W. J. M. Rootsaert, L. L. Van Reijen, and W. M. H. Sachtler, *J. Catalysis*, 1962, **1**, 416.
[314] B. E. Nieuwenhuys and W. M. H. Sachtler, *Surface Sci.*, 1974, **45**, 513.
[315] B. E. Nieuwenhuys, R. Bouwman, and W. M. H. Sachtler, *Thin Sol. Films*, 1974, **21**, 51.
[316] R. Gomer, 'Field Emission and Field Ionization', Harvard Univ. Press, Cambridge, Mass., 1961.
[317] E. W. Muller and T. T. Tsong, 'Field Ion Microscopy', American Elsevier, New York, 1969.
[318] R. H. Good and E. W. Muller, in 'Handbuch der Physik', ed. S. Flugge, Springer-Verlag, Berlin, 1956.
[319] E. W. Muller, *J. Appl. Phys.*, 1955, **26**, 732; 1957, **28**, 1.
[320] L. W. Swanson and A. E. Bell, 'Adv. Electronics and Electron Phys.', 1973, **32**, 194.
[321] For reviews see: (*a*) L. W. Swanson, A. E. Bell, C. H. Hinrichs, L. C. Crouser, and B. E. Evans, 'Literature Review of Adsorption on Metal Surfaces', Vols. 1 and 2 (NASA CR-72402, 1967); (*b*) J. C. Riviere, in 'Solid State Surface Science', ed. M. Green, Dekker, New York, 1969, Vol. 1: (*c*) P. M. Gundry and F. C. Tompkins, in 'Experimental Methods in Catalytic Research', ed. R. B. Anderson, Academic, New York, 1968.

planes. In this averaging process, the highly emitting areas (low work function) are favoured.

The probe-hole technique overcomes the above problem in admitting only those electrons emitted in a small solid angle to the detection system through a probe-hole in the fluorescent screen. The electron image can be moved across the probe-hole to bring different crystal planes into view by electrostatic deflection or by rotation of the field-emitter itself.

From equation (123) the work function changes of individual planes can be determined from a plot of $\ln(I_{hkl}/V^2)$ vs. $1/V$. i.e. from equation (124), where

$$\ln(I_{hkl}/V^2) = \ln A_{hkl} - m_{hkl}/V \tag{124}$$

$m_{hkl} = B\phi_{hkl}^{3/2}$. Comparison of the slopes m_{hkl} in these plots for a clean tip and after adsorption yields [314] the expression (125).

$$\Delta\phi_{hkl} = \frac{\phi(\text{clean}) - \phi(\text{ads})}{e} = \frac{\phi(\text{clean})}{e}[1 - m_{hkl}(\text{ads})/m_{hkl}(\text{clean})] \tag{125}$$

In early measurements of the adsorption of argon, krypton, and xenon on tungsten, the FEM was used.[308] Changes in field emission current due to adsorption were detected and variations of the patterns with temperature and coverage were investigated. A very pronounced dependence of the interaction upon the structural features of the surface was observed. Similar conclusions were deduced from the results of measurements of Xe and Kr adsorption on Rh.[309] In these latter measurements a combination of field ion microscopy and electron field emission measurements allowed a better characterization of the crystal planes under study than in the previous measurements. The specificity of the interaction was attributed to differences in the binding of gas atoms at various lattice positions and not merely due to differences in the work function increment per adatom on different planes.

In addition to the above measurements, Erlich and Hudda [308] also recorded the rate of boundary movement and disappearance on the field-emission patterns, and estimated the diffusion coefficients for adatom migration as a function of temperature and coverage. Engel and Gomer [312] repeated the measurements of Ar, Kr and Xe adsorption on a tungsten field emitter, using the field emission probe-hole technique. Although their results confirm the pronounced crystal face anisotropy of the interaction, the ordering of crystal planes according to the relative strength of binding differs from that of Erlich and Hudda. Other interesting observations made in the above study are: (i) a monotonic decrease of $\Delta\phi$ on all crystal planes with increasing coverage for all three adsorbates; (ii) for a given gas the values of $\Delta\phi$ at full coverage ($\theta = 1$) show the same trend; (iii) $|\Delta\phi|$ values for a given plane increase monotonically with increasing atomic number of the gas; and (iv) monolayer coverages for a given adsorbate are nearly constant from plane to plane (with some uncertainty for Ar). This last result indicates close-packing arrangement in the monolayer, in agreement with LEED studies (p. 47). This work emphasizes the advantages of the probe-hole technique in achieving a better definition of the system. Another advantage of the probe-hole technique is in measurements of the rate of desorption, for a determination of heats of adsorption.

The free activation energy of desorption ΔF^*_{des} can be estimated by recording the desorption at a temperature T, using equation (126), where τ_{des} is defined according

$$\Delta F^*_{des} = RT \ln \left(\frac{k_B T}{h} \tau_{des} \ln 2 \right) \qquad (126)$$

to absolute rate theory as the time in which (the coverage) changes from $\theta = 1$ to $\theta = \frac{1}{2}$ under conditions where readsorption can be neglected.[314,322] Experimentally, τ_{des} is approximated by the time during which the field-emitted electron current changes by $\frac{1}{2}|I_{\theta=1} - I_{\theta=0}|$ at constant T and V. Assuming that the activation entropy of desorption is zero, ΔF^*_{des} is equal to the activation energy of desorption E_{des}. If the activation energy for adsorption is negligibly small (as in most physisorption cases), the activation energy of desorption is equal to the heat of adsorption, q,

$$\Delta F^*_{des} \sim E_{des} = -\Delta H_{ads} = q \qquad (127)$$

Nieuwenhuys and Sachtler[314] used the above method to estimate adsorption energies (in the low and high coverage regimes) of Xe on well-defined crystal faces of iridium (using the probe-hole technique). They observed crystal face specificity and a decrease of the heat of adsorption with increasing coverage (similarly observed in LEED measurements).

Field Emission. Theoretical Considerations. The observation of large work function changes upon the physical adsorption of rare gases on transition metals is one of the challenging problems confronting theorists in this field. It is important to note at the outset that theories which attempt to calculate physisorption binding energies do not necessarily, and indeed do not, offer an explanation of the above-noted work function changes. At best, they provide indirect semi-empirical estimates of quantities related to model descriptions of adsorption mechanisms which could cause work function changes, and as such these results should be viewed with caution.

The theoretical treatments of the phenomena can be conveniently divided into electrostatic and quantum-mechanical models.[205,305,323]

In the electrostatic model,[295,305,323,324] a surface electric field F is postulated which polarizes the adsorbed molecule to yield a dipole moment $p = \alpha F$, where α is the polarizability of the adsorbed particle (corrections to the dipole moment due to the self-interaction with the polarization field produced by the dipole and depolarization effects have been presented[323]). Such a dipole, which adds to the dispersion energy of interaction a term equal to $\alpha F^2/2$, yields a change in the work function, $\Delta \phi$, as given by equation (122). The magnitude of the surface electric field, F, necessary to account for the experimental observations is of the order of 10^8 V cm^{-1}. The method of analysis used[305] consists of subtracting from the experimental heats of adsorption a calculated value for the surface dispersion

---

[322] B. E. Nieuwenhuys and W. M. H. Sachtler, *Surface Sci.*, 1973, **34**, 317.
[323] J. Patigny, Y. Barbaux, and J.-P. A. Beaufils, in 'Adsorption–Desorption Phenomena', ed. F. Ricca, Academic, London, 1972, p. 49.
[324] A. S. Schram, in 'Adsorption–Desorption Phenomena', ed. F. Ricca, Academic, London, 1972, p. 57.

energy ($aF^2/2$), and identifying the result as arising from the polarization of the adatom in the electric field [equation (128)]. This relation allows the determination

$$q_{exp} - q_{dis} \equiv Q_c = -\tfrac{1}{2}aF^2 \qquad (128)$$

of F, assuming that the polarization, a, of the adatom is known (the use of free atom polarizabilities is most probably incorrect [155]), the dipole can then be evaluated from $p = aF$, and substitution in equation (122) yields a value for the change in work function, $\Delta\phi$. There are several sources of serious uncertainties in the method: (i) the dependence of the dipole moments on coverage; (ii) the model assumed for the calculation of the dispersion-interaction contribution; (iii) the value of the adatom polarizability; (iv) the source and/or plausibility of the surface electric field as an effective polarizing field. Since the amplitude of such surface fields is expected to decay over a very short distance, it may not be sufficient to produce a significant polarization of the adatom which resides at an equilibrium distance away from the surface plane (see p. 21 and Figure 7). Indeed, the results obtained *via* this method of analysis were demonstrated to be sensitive to the above model assumptions. In particular, the uncertainty due to assumed values of the adatom equilibrium distance were noted.[306]

In the quantum-mechanical model, the surface dipole is considered to be of quantum-mechanical origin. Following Mulliken's theory of the formation of charge-transfer complexes,[325] Mignolet suggested[297] that the dipole moments responsible for the large work function changes upon the adsorption of rare gases on transition metals can be explained in terms of a charge-transfer-no-bond (CTNB) resonance. According to this model the wavefunction, ψ, of the 'surface complex', composed by the metal substrate and adatom, is written as equation (129),

$$\psi = a\psi_{MA} + b\psi_{M^+A^-} \qquad (129)$$

$$\psi_{MA} = \psi_M \psi_A \qquad (130)$$

$$\psi_{M^+A^-} = \psi_{M^+}\psi_{A^-} \qquad (131)$$

where the wavefunctions ψ_{MA} and $\psi_{M^+A^-}$ correspond to the no-bond (neutral) and ionic (dative) states of the system, respectively, where ψ_M and ψ_A are the wavefunctions (properly antisymmetrized) of the metal and atom in the unperturbed state; ψ_{M^+} and ψ_{A^-} are the wavefunctions of the corresponding ionized (perturbed) states. We define λ as the ratio b/a of the statistical weights of the two states, and denote the energies of the no-bond (MA) and dative (M^+A^-) states by W_0 and W_1 respectively. Defining the overlap integral as

$$S = \int \psi_{MA}\psi_{M^+A^-}\,d\tau \qquad (132)$$

and the resonance integral as

$$\beta = \int \psi_{MA} H \psi_{M^+A^-}\,d\tau \qquad (133)$$

where H is the hamiltonian of the system, the following results are obtained for λ and q, the heat of physical adsorption from standard perturbation theory for the case of physisorption of rare gases on metals ($W_1 > W_0$):

---

[325] R. Mulliken, *J. Amer. Chem. Soc.*, 1952, **74**, 811.

$$q \simeq (\beta - SW_0)^2/(W_1 - W_0) \qquad (134)$$

and

$$\lambda \simeq -(\beta - SW_0)/(W_1 - W_0) \qquad (135)$$

Since λ is a measure of the relative contribution of the ionic form to the state of the system it can be used to estimate the dipole moment

$$p \approx ed \frac{\lambda^2 + 2\lambda S}{1 + \lambda^2 + 4\lambda S/(2 + 2S^2)^{\frac{1}{2}}} \qquad (136)$$

where d is the dipole length, and e is the electronic charge, and thus obtain an estimate of the work-function change. Furthermore,

$$W_1 - W_0 = I - \phi - \frac{e^2}{4d} \qquad (137)$$

where I is the ionization energy of the adatom and the third term on the right is the image potential.

It is thus seen that the basic quantities of the theory are the overlap and resonance integrals. It is noted from equation (134) that since the term $W_1 - W_0$ is large (typically, of the order of several eV), the overlap integral (negative) must be large, in order for the charge-transfer state to contribute significantly. This has been used [304] to rationalize the adsorption energies of rare-gas atoms on alkali, noble, and transition metals in relation to the band structure of these materials. In addition, modifications to the theory which include the effect of broadening and shifting of the atomic level (due to the resonance) and the consideration of excited states of the adatom (with no charge-transfer) have been presented.[312] The numerical values obtained from the above model for the physical quantities of interest suffer from a high degree of uncertainty since they depend on small differences between large quantities.[312] In addition, the calculation of certain matrix elements occurring in the formalism is not straightforward and in evaluating the dipole moment [equation (136)] a value for the unknown distance, d, has to be assumed. Since the results are sensitive to this parameter, it adds to the uncertainty of the results.

The above discussion emphasizes the role of microscopic theoretical approaches in the study of physisorption systems. Characteristic microscopic parameters of the interaction, like the equilibrium distance adatom polarizability and matrix elements involving surface electronic wavefunction, occur even at the level of the approximate theoretical models discussed herein.

The above discussion may serve as a guide for a programme of theoretical studies of physisorption phenomena. First, the microscopic mechanism of the physical interaction of a single atom with a surface should be studied, and the characteristic interaction parameters evaluated. In view of the above-demonstrated specificity of the interaction, techniques which allow for crystal face anisotropy of the interaction should be developed. These studies should comparatively analyse the dependence of the results on the model of the system under study (like the influence of self-consistency of the surface electronic charge density, Table 4), in order to discern and identify the sensitivity of the results of the theory to the model assumptions and the level of description of the system under study. Finally, the mutual interaction (direct and indirect surface mediated) between adatoms should be investigated. Such studies should be complemented by adsorption experiments on well-defined single-crystal substrates (probe-hole), in various coverage regimes. In particular, to facilitate comparison with single-atom calculations, accurate measurements in the low-coverage regime are of special importance.

Measurement of Adsorption Isotherms *via* **Electron Spectroscopy.**—A powerful traditional method employed in surface studies is the measurement of adsorption isotherms,[119,151,152,326] *i.e.*, the relationship between coverage of a substrate surface by an adsorbant and the total pressure of the ambient, for a fixed temperature. The study of adsorption isotherms provides information about the kinetics, mechanism, and energetics of the adsorption process. It is of importance in surface area measurements, in distinguishing single layer from multilayer adsorption, in investigating the dynamics of adsorbed species (mobile and static adsorption and transitions between them), and in characterizing the heterogeneity of surfaces with respect to adsorption [152,327] (determination of the energy distribution of adsorption sites). In addition, these studies enable the determination of heats of adsorption and activation energies for certain processes in the adsorbed phase.

A distinction is made between volumetric,[328] gravimetric,[329–331] and desorption [277,332] methods in which surface coverage is obtained from measurement of the amount of adsorbate removed from or released into the gas phase, and methods in which surface coverage is derived from measurement of some physical property of the adsorbate whilst on the solid surface. In general, any property which is proportional to the number of atoms in the adsorbed phase can be used, subject to the requirement that an independent calibration is available for absolute coverage measurements.

We have already mentioned the use of LEED intensities (and in conjunction with work function measurements) and of Auger amplitude measurements in the construction of isotherms for the adsorption of Xe of Pd(100),[255] and the use of the Clausius–Clapeyron equation for a determination of the isosteric heat of adsorption and its coverage dependence.[255,256] Another example of such measurements, from the work of Cohen *et al.*[166] on the adsorption of Xe on Ag(111) is shown in Figure 40. In these experiments the variation with pressure of the intensity of the Xe(10) overlayer diffraction beam was recorded for fixed temperature (starting at a pressure greater than that for the formation of a monolayer). The abrupt fall in the intensity [at a pressure designated by $P_s(T)$] indicates the disappearance of the 2D ordered Xe layer. In fact, repeating the measurements for several temperatures yields a *P–T* diagram for the coexistence of the three-dimensional gas and the ordered and disordered surface phases. A plot of $P_s(T)$ *vs.* reciprocal temperature is shown in Figure 41. Since the coexistence line occurs

---

[326] S. Brunauer, 'The Adsorption of Gases and Vapors', Princeton University Press, 1945, Vol. 1.
[327] U. Landman and E. W. Montroll, *J. Chem. Phys.*, 1976, **64**, 1762, and references cited therein.
[328] G. Erlich, *Adv. Catalysis*, 1963, **14**, 256.
[329] T. N. Rhodin, *Adv. Catalysis*, 1953, **5**, 39.
[330] 'Vacuum Microbalance Techniques', ed. A. W. Czanderna, Plenum, 1971, Vol. 8.
[331] 'Ultra Micro Weight Determination', ed. S. P. Wolsky and E. J. Zdanuk, Wiley, New York, 1969.
[332] T. E. Madey, *Surface Sci.*, 1972, **33**, 355.

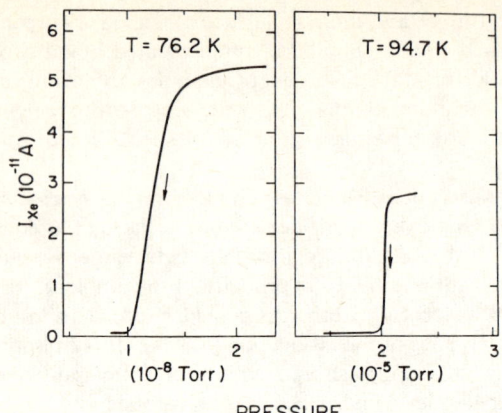

Figure 40 *The* Xe(10) *beam intensity vs. pressure [for Xe adsorption on* Ag(111)] *at two fixed temperatures, in steady-state experiments. The decrease in beam intensity as gas is slowly removed from the chamber is associated with a disordering of the* Xe *monolayer*
(Reproduced by permission from *Surface Sci.*, 1976, **58**, 429)

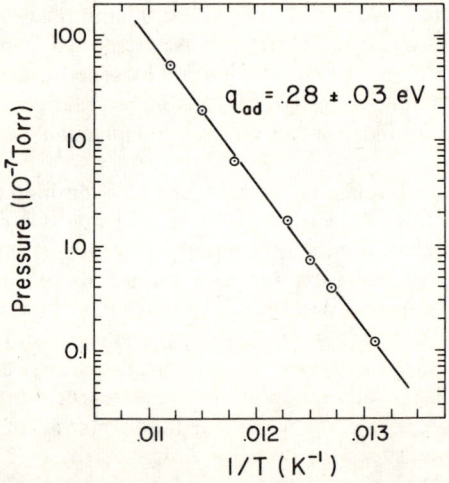

Figure 41 *The pressure-temperature dependence of the line along which the three-dimensional gas and the two-dimensional ordered and disordered phases coexist.* q_{ad} *is the isosteric heat of adsorption as derived from the Clausius–Clapeyron equation*
(Reproduced by permission from *Surface Sci.*, 1976, **58**, 429)

at essentially constant coverage, the slope of the curve in Figure 41 yields the isosteric heat of adsorption (*via* the Clausius–Clapeyron equation). The value determined in this study was $q = 0.28 \pm 0.03$ eV.

Apart from the determination of heat of adsorption, high-resolution LEED

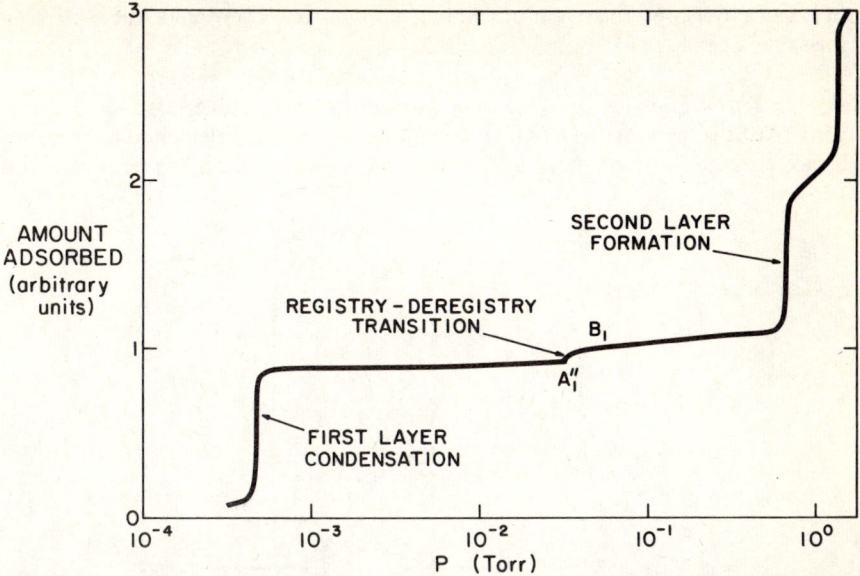

Figure 42 *Schematic representation of a typical krypton adsorption isotherm for temperatures less than the two-dimensional triple temperature (84.6 K according to ref. 337). The $A_1''B_1$ event was first published by Thomy in ref. 333, as occurring at 90.1 K. Isotherms including the $A_1''B_1$ event were reported in ref. 338 at six temperatures*
(Reproduced by permission from *J. Vacuum Sci. Technol.*, 1977, **14**, 314, and a private communication from M. D. Chinn and S. C. Fain, jun.)

measurements can provide information about the nature of certain features of the phase diagram of physisorbed systems. As an illustration Figure 42 shows a typical krypton adsorption isotherm [333—336] for temperatures less than the 2D triple temperature (84.6 K according to ref. 337). The event denoted by $A_1''B_1$ in the Figure was attributed by Thomy [338] to the first layer being forced out of registry into a more dense layer. When the temperature and pressure of this transition are extrapolated to the experimental conditions of the measurements by Chinn and Fain [170] (see Figure 21), the transition $A_1''B_1$ is predicted to occur at the same temperature and pressure as the splitting in Figure 21b occurs. Thus, it indicates that the small compression of the krypton layer out of registry could bring about a registry–deregistry transition in the phase behaviour of the krypton overlayer.

Several other non-volumetric methods have been used in adsorption isotherm

---

[333] A. Thomy and X. Duval, *J. Chim. phys.*, 1969, **66**, 1966; 1970, **67**, 1101.
[334] A. Thomy, J. Regnier, and X. Duval, in 'Thermochimie' (Colloques Internationaux du CNRS), CNRS, Paris, 1972, pp. 201, 511.
[335] X. Duval and A. Thomy, *Carbon*, 1975, **13**, 242.
[336] F. A Putnam and T. Fort, jun., *J. Phys. Chem.*, 1975, **79**, 459.
[337] Y. Larher, *J.C.S. Faraday I*, 1974, **70**, 320.
[338] A. Thomy, These, L'Universite de Nancy (1968, unpublished), p. 71.

studies of physisorption systems, including ellipsometry [339–342] and Auger electron emission. The latter will now be discussed.

Auger Electron Emission in Adsorption Isotherm Studies. Auger electron spectroscopy (AES) is the most widely used technique for chemical elemental analysis for surface characterization and in adsorption studies, and as such has been reviewed

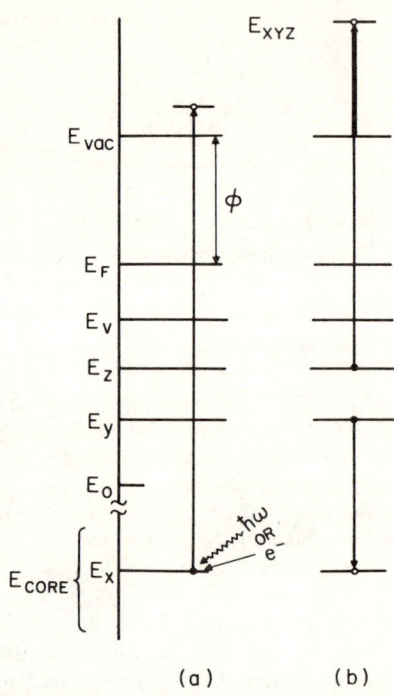

Figure 43 *A schematic description of the Auger process. (a) The process of core creation via photon or electron impact ionization. A hole is created in level E_X. (b) The Auger process. A transition of an electron from level Y fills the hole, and another electron, from level Z is emitted. The measured energy of the emitted electron is denoted by a heavy line while solid vertical lines denote electron transitions. After emission of the Auger electron, the system is left in a doubly-ionized state. Levels are denoted as: E_{vac} – vacuum, E_F – Fermi, E_v – top of valence band, E_0 – bottom of valence band, E_{core} – core levels, and ϕ the work function of the sample. The emitted Auger electron is characterized by the levels XYZ participating in the process*

---

[339] For general reviews of ellipsometry see: (*a*) 'Ellipsometry in the Measurement of Surface and Thin Films', ed. E. Passaglia, R. R. Stromberg, and J. Kruger, National Bureau of Standards, 1969; (*b*) 'Proceedings of the Symposium on Recent Developments in Ellipsometry' (Lincoln, Nebraska, 1968), ed. N. M. Bashra, A. B. Buckman, and A. C. Hall, *Surface Sci.*, 1969, **16**.
[340] R. F. Steiger, J. M. Morabito, jun., G. A. Somorjai, and R. H. Muller, *Surface Sci.*, 1969, **14**, 279.
[341] G. A. Bootsma and F. Meyer, *Surface Sci.*, 1969, **14**, 52.
[342] G. Quentel, J. M. Rickard, and R. Kern, *Surface Sci.*, 1975, **50**, 343.

extensively.[343] We limit this discussion to a schematic description of the Auger process and to illustrations of some of its applications to isotherm studies of physisorption systems.

The Auger process is a two-electron process, as shown in Figure 43. The first step is the excitation (ionization) of an electron in a core level by a photon or by electron impact (Figure 43a). Subsequently the core-hole is filled by a transition from a high-lying occupied level, and an electron from another level is emitted, leaving behind a doubly ionized atom (Figure 43b). The energy distribution of the emitted electrons is then measured. Invoking adiabatic response of the system, the energy balance of the process is given as

$$E_{XYZ} \equiv E_k = E_X - (E_Y + E_Z) - \phi \qquad (138)$$

where E_k is the kinetic energy of the emitted Auger electron and ϕ is the work function of the solid. Auger transitions are designated according to the levels participating in the transitions (positive binding energies relative to Fermi level), and their energies are a characteristic elemental property. Given an experimental chart of Auger emission cross-sections, a measurement of the Auger amplitude *versus* exposure can determine absolute concentrations of adsorbed species.[344-346] Two problems with such a procedure are (i) the production of Auger electrons by back-scattered electrons [347,348] and (ii) electron beam effects, *i.e.* electron-induced desorption (EID).[349,350] In addition there are instrumental uncertainties, due to a limited acceptance angle and transmission of the spectrometer.[344,351] Nevertheless, it has been found *via* comparative studies with independent coverage measurements, that in many cases a linear dependence of Auger intensity on adsorbate coverage holds at least in the submonolayer regime.[344,346]

Desorption effects due to incident electron beam have been studied for rare-gas crystals by Farrell *et al.*[350] who observed the rapid desorption rates for neon and argon and less for krypton and xenon, with the latter being essentially resistant to desorption. It was found that neither heating nor 'ballistic' impact could provide an effective mechanism of desorption. The dominant mechanism has been identified to be electron-induced desorption (EID).[349] The desorption proceeds through an excitation of the rare gas which, by an analogue of a vertical Frank–Condon mechanism, is transferred to a configuration of high vibrational potential energy, allowing the escape of the atom.

In the context of surface coverage measurements *via* AES, the effects of heating

---

[343] For general reviews of Auger electron spectroscopy see: (*a*) C. C. Chang, *Surface Sci.*, 1971, **25**, 53; (*b*) N. J. Taylor, in 'Techniques of Metal Research', ed. R. F. Bunshah, Interscience, 1972, vol. 7; (*c*) J. C. Tracy, in 'Electron Emission Spectroscopy', ed. W. DeKeyser, L. Fiermans, G. Vanderkehen, and J. Vennik, Reidel, 1973; (*d*) E. N. Sickafus, *J. Vacuum Sci. Technol.*, 1974, **11**, 299; (*e*) A. Joshi, L. E. Davis, and P. W. Palmberg, in 'Methods of Surface Analysis', ed. A. W. Czanderna, Elsevier, Amsterdam, 1975, Vol. 1.
[344] F. Meyer and J. I. Vrakking, *Surface Sci.*, 1972, **33**, 271.
[345] M. Perdereau, *Surface Sci.*, 1971, **24**, 239.
[346] See Table 1 in ref. 158*d*.
[347] H. E. Bishop and J. C. Riviere, *J. Appl. Phys.*, 1969, **40**, 1740.
[348] T. E. Gallon, *J. Phys.*, 1972, **D5**, 822.
[349] For reviews on electron-induced desorption, see: (*a*) D. Menzel, *Surface Sci.*, 1975, **47**, 370; (*b*) T. E. Madey and J. T. Yates, *J. Vacuum Sci. Technol.*, 1971, **8**, 525.
[350] H. H. Farrell, M. Strongin, and J. M. Dickey, *Phys. Rev.*, 1972, **B6**, 4673.
[351] J. E. Houston, *Surface Sci.*, 1973, **38**, 283.

and EID have been investigated by Baker and Sexton for the adsorption of Xe on Ni [352] and Pt [353] substrates. Comparing Auger and volumetrically measured isotherms, they demonstrated the importance of EID in coverage determination from Auger amplitude measurements and estimated the cross-section for the process to be $1-2 \times 10^{-17}$ cm^2, a value smaller than the ionization cross-section (5×10^{-16} cm^2), since excitation rather than ionization is required to promote the atom to a repulsive potential curve. These authors also suggested use of the initial substrate surface as initial standard, so that the amount of gas adsorbed can be represented as the ratio of Auger intensity peak heights Xe_{37/M_α^o}, where 37 stands for the $N_{4,5}O_{2,3}O_{2,3}$ xenon line used in the measurement and M_α^o is the line of the clean metal ($\alpha = 60$ eV for the $M_{2,3}VV$ Auger line of Ni and 58 eV for the $N_{6,7}VV$ line of Pt).

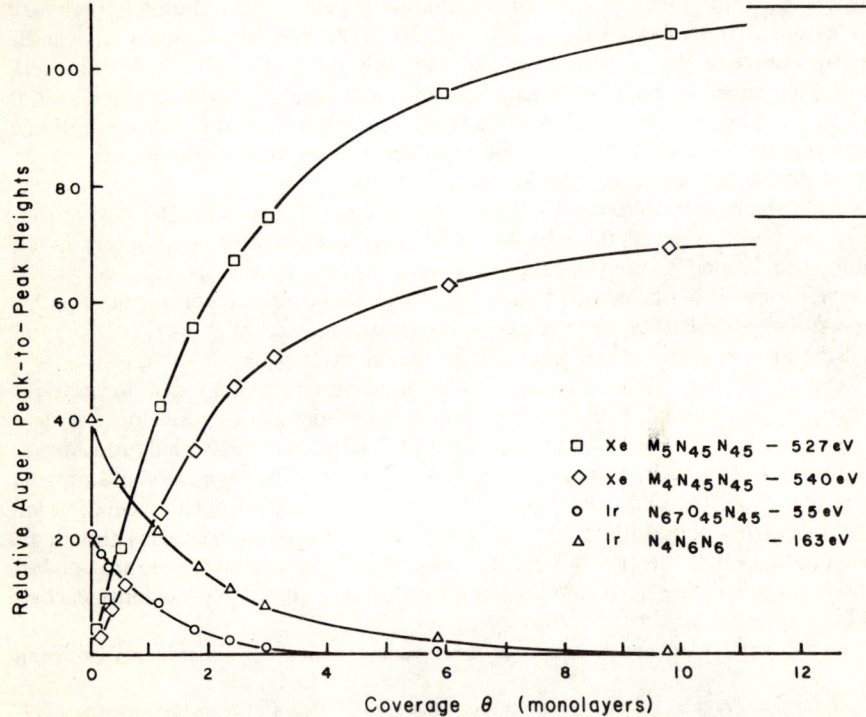

Figure 44 *Peak-to-peak heights of two xenon ($\square$ and $\diamondsuit$) and two iridium ($\bigcirc$, $\triangle$) Auger peaks in the $dN(E)/dE$ vs. E spectrum as a function of coverage θ. The limiting values at very large coverage are indicated by the horizontal lines. The correlated decrease in the Ir Auger signal with the increase in the Auger emission from Xe is evident. Calibration of exposure to coverage was performed on the basis of LEED diffraction patterns, as described in the text*
(Reproduced by permission from *Phys. Rev.*, 1973, **B8**, 893)

---

[352] B. G. Baker and B. A. Sexton, 'Proc. 2nd Internat. Conf. on Solid Surfaces' [*Jap. J. Appl. Phys.*, 1974, Suppl. 2(2), p. 275].
[353] B. G. Baker and B. A. Sexton, *Surface Sci.*, 1975, **52**, 353.

An example of a measured Auger isotherm is shown in Figure 44 for the adsorption of Xe on Ir(100).[193] The calibration of exposure to coverage in these experiments was determined from the observed 3×5 LEED diffraction pattern during the initial stages of Xe adsorption of Ir(100) 1×5 surface (see p. 43). The formation of the 3×5 pattern was very sensitive to Xe exposure and was optimized as 0.8 L. The structural interpretation of the diffraction pattern corresponds approximately to 0.2 monolayer of Xe (consequently a Xe monolayer is equivalent to ~ 4L or 6.9×10^4 atoms cm^{-2}). Beyond the linear region of the AES intensity *versus* exposure, corrections due to a reduced sticking probability have to be made. Moreover, with a build-up of Xe concentration on the surface, the incident beam excites more Xe atoms, with fewer Auger electrons being emitted due to their constant escape length. Included in the Figure are the attenuated intensities of Auger peaks of the Ir substrate *versus* Xe coverage. Auger isotherms have been also measured for the adsorption Xe on Ag(111) and compared with those measured *via* surface potential changes.

Adsorption systems for which the measurement of Auger isotherms have been studied extensively in connection with 2D phase transformations are Xe and Kr on (0001) graphite.[354,355] The authors observed good correspondence with volumetrically measured isotherms,[333—336] and in conjunction with LEED measurements have used their results to derive several physical parameters for the above adsorption systems, *e.g.* isosteric heats, adsorption entropies, and electron mean free paths in the adsorbed layers.

The main shortcomings of the use of LEED/Auger methods for the study of isotherms are: (i) the requirement of low pressure ($\gtrsim 10^{-5}$ Torr); (ii) the inability to achieve true isothermal conditions due to the contact of the gas both with a cold substrate and with some parts which are at room temperature; and (iii) electron beam desorption effects. Despite the above problems, it has been demonstrated that in some cases the measurement of isotherms using electron spectroscopy methods allows quantitative determination of the absolute number of adsorbed atoms per unit area for a wide range of coverages. In particular, measurements can be made at low pressures not accessible to volumetric methods. The great surface sensitivity of AES allows the investigation of adsorption isotherms on single-crystal substrates rather than the large area adsorbents required by most volumetric and gravimetric methods. This allows better chemical and structural characterization of the adsorption systems under study.

4 Conclusion

As stated in the Introduction the major guidelines for this review were to provide an up-to-date report about the development of microscopic approaches for the study of the physical interaction of atoms and molecules with surfaces [356] and to

---

[354] J. Suzanne, J. P. Coulomb, and M. Bienfait, *Surface Sci.*, 1973, **40**, 414; 1974, **44**, 141.

[355] H. M. Kramer and J. Suzanne, *Surface Sci.*, 1976, **54**, 659.

[356] In our discussion of microscopic experimental approaches to physisorption (Section III) we have attempted to provide a comprehensive (and to the best of our knowledge, exhaustive) state-of-the-art review of modern methods employed in the study of physisorption. A method which we have not mentioned above, and which is currently in a state of development is the application of nuclear magnetic resonance techniques to the study of physically adsorbed layers (for further information see ref. 357).

emphasize the cross-linking between theory and experiment in this field. In doing so we have attempted to supply, in a self-contained manner, sufficient background material to make the review accessible to workers in other branches of physics and chemistry. In our opinion, the exposure of a wide audience to the subject reviewed herein could bring about an influx of new ideas and techniques which would increase our knowledge and understanding of the subject. From our discussion in the preceding chapters it is clear that the application of modern theoretical techniques coupled with experimental data about the structure and interaction mechanisms in physisorption systems, on the microscopic level, enhances our understanding of the nature and properties of these systems. Furthermore, owing to the unique characteristics of the physisorption state (namely, those associated with the weak-coupling, or non-chemical, nature of the interaction), investigations of elementary processes in these systems shed light on fundamental issues in the study of atom–surface interaction, such as: hole relaxation mechanisms in adsorbed atoms, beam effects (electron-induced desorption), the mobility and dynamics of adsorbed layers, critical phenomena, and phase behaviour.

In conclusion, certain topics are reiterated which, in our opinion, deserve further study in order to achieve better model descriptions and characterization of physisorption systems and related phenomena.

1. (*a*) Calculations of the interaction potential between a physisorbed atom or molecule and a surface in which the crystal structure of the substrate is included (non-jellium models). Studies of crystal specificity of the interaction, binding energies, and work function changes. (*b*) Examination of the effects of non-locality of the response of the substrate on the interaction characteristics (see refs. 48 and 49).

2. Interactions between physically adsorbed molecules; direct and substrate-mediated interactions.[358,359] These investigations should be coupled with the study of surface concentration (coverage) dependent observations like work function changes, atomic arrangement, and critical behaviour.

3. Analysis of atomic and molecular beam scattering experiments. Such studies provide means for testing and parametrization of calculated interaction potentials. In addition, calculations of the eigenvalue spectra corresponding to the interaction potentials are of importance in the analysis of selective adsorption data as well as in statistical mechanics models.

4. The incorporation of the results of microscopic theories in statistical mechanics calculations of thermodynamical quantities (*e.g.* isosteric heat, compressibility, virial coefficients and quantum corrections, phase behaviour).

5. Dynamics and mobility in physisorption systems. (*a*) Studies of vibrational properties, directional anisotropy of the vibrations, and elementary excitations in

---

[357] See articles and references cited therein, in 'Monolayer and Submonolayer Helium Films', ed. J. G. Daunt and E. Lerner, Plenum Press, New York, 1973: (*a*) D. F. Brewer, D. J. Creswell, Y. Goto, M. G. Richards, J. Rolt, and A. L. Thomson, p. 101; (*b*) B. J. Rollefson, p. 115; (*c*) D. P. Grimmer and K. Luszczynski, p. 123; (*d*) D. L. Husa, D. C. Hickernell, and J. E. Piott, p. 133.

[358] For a recent review see: T. Takaishi, 'Progress in Surface Science', ed. S. G. Davisson, Pergamon, Oxford, 1975, Vol. 6, p. 43.

[359] See also: D. L. Freeman, *J. Chem. Phys.*, 1975, **62**, 4300.

physisorbed layers. (b) Investigation of the motion (diffusion) of physisorbed atoms, molecules, and clusters on surfaces.

6. The effects of intrinsic and induced heterogeneities. This subject is of great importance in the analysis of experimental data. Furthermore, such studies could provide a method for the characterization of surfaces in terms of energy distribution functions [327] (energy distribution of sites or equivalently the distribution of homotatic regions).

7. The role of physisorption in reaction mechanisms. Temperature dependence of reaction rates. An interesting area of experimental and theoretical research is that of physical coadsorption. Since the condensation (physisorption) of reactants on a surface decreases their motional degrees of freedom certain reactions which are difficult to produce in the gas phase may become feasible under the above conditions. For example, preliminary calculations indicate [360] that by condensing organic molecules (like C_2H_4, C_2H_2, and C_6H_6) on a surface the centrifugal barrier for their reaction with a rare-gas atom is greatly reduced. Since the polarizable π-electron system of these molecules is not perturbed by the physisorption interaction (as evidenced by UPS studies) they can form van der Waals complexes with rare-gas atoms (Kr and Xe in particular). It appears that in order to minimize the rotational motion of the organic molecule, a rather low temperature might be necessary. The degree of quenching of the rotational motion could be monitored by performing angle-dependent photoemission measurements. These experiments could provide a method for preparing a new kind of van der Waals molecule, namely organic-rare gas compound, and their investigation could provide information about the role of the physisorption state in catalysing chemical reactions.

We would like to thank the many friends and colleagues who so graciously and promptly assisted us by providing us with preprints and reproductions of results of their studies and allowed us to include their information in our review: C. R. Brundle, M. D. Chinn, J. G. Dash, J. E. Demuth, N. E. Erickson, S. C. Fain, jun., J. F. Herbst, A. Ignatiev, T. E. Madey, R. P. Merrill, P. W. Palmberg, L. Passell, T. N. Rhodin, H. Saltsburg, B. J. Waclawski, W. H. Weinberg, J. T. Yates and their colleagues.

We gratefully acknowledge the encouragement and interest of Professor E. W. Montroll and our colleagues in the Institute for Fundamental Studies.

We would like to extend our gratitude to Mrs. Shirley Brignall, and special thanks to Ms. Diana Granitto of the University of Rochester, for their excellent assistance and demonstration of asymptotically infinite patience in the course of the preparations of the manuscript.

Last but not least our thanks to our wives who bore with us, in love and care.

2
Iron-57 Conversion Electron Mössbauer Spectroscopy

BY M. J. TRICKER

1 Introduction

Few events in nuclear physics have had such significant and far-reaching consequences in other scientific disciplines as the discovery of the recoil-free emission and absorption of γ-photons in 1958 by Rudolph Mössbauer.[1] In the years since the first experiments Mössbauer spectroscopy has been applied in diverse areas of study ranging from physics to metallurgy through chemistry to the biological sciences. The power of the method lies in the extremely monochromatic nature of the recoil-free γ-rays which have linewidths of the order of 10^{-8} eV. This makes possible the measurement of small perturbations of nuclear energy levels caused by the atomic environment of nuclei in solid materials. The method is applicable in principle to many isotopes, which have suitable nuclear properties, but in practice the number is reduced by experimental considerations. Despite the fact that it is only 2% naturally abundant, the most suitable and consequently the most widely exploited isotope is ^{57}Fe. The first excited nuclear state of iron decays to the ground state with the emission of a 14.4 keV γ-photon. The excited state can be populated by the radioactive decay of ^{57}Co thereby providing a convenient source of γ-radiation. Most ^{57}Fe Mössbauer experiments have been carried out by monitoring the intensity of the γ-rays transmitted through a thin iron-containing sample as the incident beam is swept through resonance. In this transmission geometry a wealth of information relating to the environment of atoms residing in the bulk of solid materials is obtained. In principle, the atomic valence state, the symmetry and magnitude of the electric field gradient, the magnitude of the internal hyperfine magnetic field, together with information on lattice dynamics, diffusion, and particle size can all be derived from Mössbauer spectroscopy. However only under certain special and often difficult experimental conditions can this type of data be gleaned, by conventional transmission methods, about atoms at or in the surface regions of low area solids. This review describes developments in the area of back-scatter Mössbauer methods, and in particular ^{57}Fe conversion electron Mössbauer spectroscopy (CEMS), which have facilitated the study of iron-containing surfaces and have thus increased the potential of ^{57}Fe Mössbauer spectroscopy as an investigative tool in surface science. In essence ^{57}Fe CEMS involves the detection of back-scattered iron (mainly 7.3 keV K-shell) internal conversion electrons which are emitted with a 90% probability during the decay of ^{57}Fe subsequent to absorption of resonant γ-photons. Because these electrons are attenuated within the solid

[1] R. L. Mössbauer, *Z. Physik.*, 1958, **151**, 124.

material the probing depth of CEMS is limited to the outermost 300 nm or so of the surface.

In this Report the basic instrumention is described, the surface sensitivity and probing depth delineated and applications of ^{57}Fe CEMS discussed. However, first the basic principles and the parameters measured by Mössbauer spectroscopy will be outlined. Many authors [2] have dealt with these topics in detail and consequently only a brief treatment is given here.

2 Principles of the Mössbauer Effect

The energy profile of a γ-ray emitted or absorbed by a nucleus is described by a Lorentzian function

$$y(E) = y_o\left[1 + \left(\frac{E - E_o}{\frac{1}{2}\Gamma}\right)^2\right]^{-1}$$

where y_o is the maximum of the distribution, E_o the energy at the maximum and Γ the width of the peak at half height. The width of the line is determined by the uncertainty principle

$$\Gamma = \frac{0.693\,\hbar}{t_{\frac{1}{2}}}$$

where $t_{\frac{1}{2}}$ is the half-life of the state.

For the emission of 14.4 keV γ-rays from the $I = \pm\frac{3}{2}$ excited nuclear spin state of ^{57}Fe, $t_{\frac{1}{2}} \approx 10^{-7}$ s and $\Gamma \approx 5 \times 10^{-9}$ eV.

Classically it is expected that the energy of the γ-photon is degraded upon emission as a fraction of the energy is consumed by recoil processes. As a consequence the profile would broaden and the maxima of the emission and absorption profiles would be very many linewidths apart. Even if the lines could be brought back into coincidence little useful information could be gained due to the loss of resolution caused by line-broadening. Mössbauer demonstrated, however, that in a quantized solid, recoil-free emission and absorption processes were possible because recoil energy can only be taken up in quantized amounts nhw, where hw is the smallest allowed quantum of vibrational energy of the lattice with $n = 0, 1, 2, 3$ etc.

For $n = 0$ no recoil energy is transferred. The fraction f of recoil free events is given by

$$f = \exp\left(\frac{-\langle x^2\rangle 4\pi^2}{\lambda^2}\right)$$

where $\langle x^2 \rangle$ is the component of the mean square displacement of the atom in the direction of the γ-ray of wavelength λ. Large f factors are favoured for tightly bound atoms, low γ-ray energies and low temperatures. For recoil free emission and absorption processes overlap of the lines now occurs and nuclear resonance

---

[2] N. N. Greenwood and T. C. Gibb, 'Mössbauer Spectroscopy', Chapman Hall, London, 1973; V. I. Goldanski and R. H. Herber (ed.), 'Chemical Applications of Mössbauer Spectroscopy', Academic Press, New York, 1968; G. K. Wertheim, 'Mössbauer Effect: Principles and Applications', Academic Press, New York, 1964.

experiments in solid materials are possible. In order to study the resonance the γ-rays from a suitable source (^{57}Co in the case of ^{57}Fe) incident on the absorber are swept through the resonant energy by applying a small movement to the source relative to the absorber in the γ-ray direction. The velocities are of the order of a few mm s^{-1} and produce Doppler shifts in energy of the γ-ray of the order of a few linewidths. If the emitter or absorber lines do not overlap due to perturbations of the nuclear energy levels they can be brought back into resonance and the magnitudes of the energy shifts measured. A typical transmission Mössbauer spectrum consists of a plot of transmitted intensity against energy. For a chemically identical source and absorber a dip centred at zero energy modulation is observed. The background is made up of non-resonant radiation and the dip, the so-called percentage effect, is usually about 10% of the background. The full linewidth at half peak height is 2Γ because of the uncertainty in the lifetime of the states in the source and absorber.

3 Interaction of the Nucleus with its Environment

There are a number of ways in which the nucleus can interact with its environment, and as a consequence shifts and splittings of the emission or absorption lines occur. It is generally convenient to use an unsplit source and study transitions between

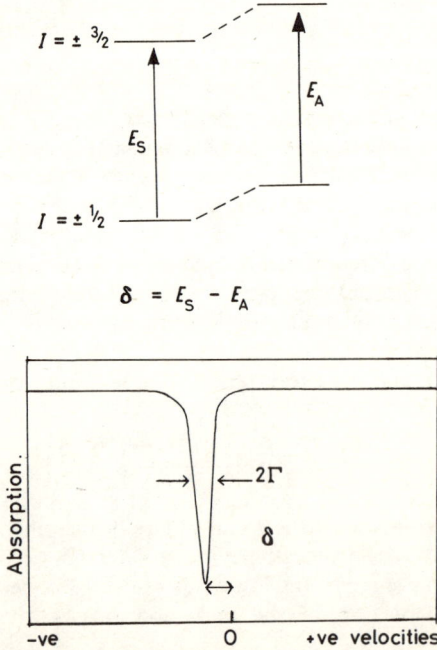

Figure 1 *The origin of the chemical isomer shift (δ). The resonant line is displaced from zero energy modulation because of the small difference in the transition energies E_s and E_a in the source and absorber respectively*

nuclear energy levels within an absorber thus obtaining a spectrum characteristic of that absorber.

Chemical Isomer Shift.—The difference in energy between the maxima of the source and absorber is known as the chemical isomer shift. It arises from the electrostatic interaction of the positively charged nucleus with electrons which have finite charge density at the nucleus. The energy shift between the source (E_s) and absorber (E_a) is given by

$$\delta = E_a - E_s = \tfrac{2}{5}\pi Z e^2 [|\Psi_o|_a^2 - |\Psi_o|_s^2] \Delta R/R$$

where $\Delta R/R$ is the fractional change in radii of the ground and excited state and $|\Psi_o|^2$ the electron density at the nucleus. The shift is shown schematically in Figure 1. The electrons which have a finite density at the nucleus are s-electrons. Isomer shifts are caused by direct changes in valence s-levels or indirectly by changes in p- or d-levels, which shield or deshield the s-electrons from the nucleus.

Quadrupole Interaction.—If the site symmetry of the Mossbauer nucleus is less than cubic there will be an electric field gradient (e.f.g.) at that site. If the nucleus has a spin (I) greater than or equal to one the degeneracy of the nuclear energy levels are partly lifted and the nuclear quadrupole moment eQ will have $2I - 1$

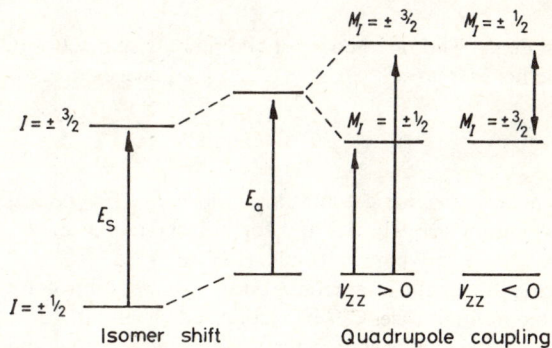

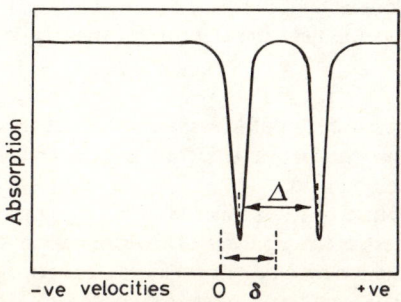

Figure 2 *The origin of quadrupole splitting in ^{57}Fe Mössbauer spectroscopy*

orientations in the e.f.g. The e.f.g. is described by a third rank tensor but with the correct choice of axes termed the principle axes, the tensor is diagonal and the e.f.g. can be defined by two parameters V_{zz}, the principle component of the e.f.g. tensor, and η the asymmetry parameter given by

$$\eta = \frac{V_{xx} - V_{yy}}{V_{zz}}$$

subject to $V_{zz} > V_{xx} > V_{yy}$. For axial symmetry of or greater than C_3 η vanishes and the e.f.g. is completely characterized by V_{zz} alone.

The ground state of ^{57}Fe does not possess a quadrupole moment as $I = \pm\frac{1}{2}$, but the first excited state has $I = \pm\frac{3}{2}$. The allowed transitions ($\Delta m = 0, \pm 1$) result in a doublet (Figure 2) separated by

$$\Delta = \tfrac{1}{2} eQV_{zz} \left(1 + \frac{\eta^2}{3}\right)^{\frac{1}{2}}$$

where Δ is the quadrupole splitting. For polycrystalline compounds the doublet intensity is 1:1. It should be noted that it is not possible to extract the sign of V_{zz} from the Mössbauer spectrum of polycrystalline iron materials from quadrupole split spectra alone. This can, however, be achieved by use of single crystals or the application of an external magnetic field.

Magnetic Interaction.—The interaction of the nuclear magnetic moment (μ) with a static magnetic field (H) results in $2I + 1$ levels of energy,

$$E_m = -\frac{\mu H m_1}{I}$$

where m_1 is the nuclear magnetic quantum number. For magnetically ordered iron materials six transitions are allowed and for polycrystalline materials a spectrum with a 3:2:1:1:2:3 intensity pattern results (Figure 3).

If quadrupole and magnetic interactions occur together quite complex spectra can result. Two simple limiting cases can be recognized. The first case occurs when the magnetic splitting is much larger than the quadrupole splitting and small shifts in the six-line pattern result (Figure 3). The second case occurs when the magnetic splitting is small compared to the quadrupole splitting. The $|\pm\frac{1}{2}, \pm\frac{1}{2}\rangle \rightarrow |\pm\frac{3}{2}, \pm\frac{3}{2}\rangle$ transition splits into a doublet and the $|\pm\frac{1}{2}, \pm\frac{1}{2}\rangle \rightarrow |\pm\frac{1}{2}, \pm\frac{3}{2}\rangle$ into a triplet. Thus by use of an external magnetic field the sign of V_{zz} may be determined for polycrystalline materials.

Spectral Areas.—From measurement of Mössbauer spectral areas and thus f factors as a function of temperature the Debye temperature and other information relating to lattice dynamics can be obtained.

A second important feature of Mössbauer spectral area measurements is that it makes possible the non-destructive analysis of multi-phase materials because each Mössbauer spectrum is a characteristic fingerprint of a particular compound. The areas of the individual spectra can be related to the concentrations of the component phases if the ratio of the f factors is known. For related materials or for different

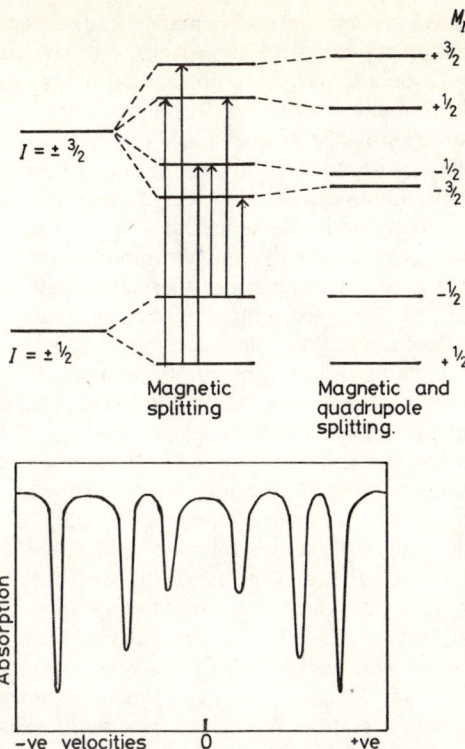

Figure 3 *The origin of magnetic splitting and magnetic and quadrupole splitting for the case where the magnetic interactions are much larger than the quadrupole interactions*

iron sites within one compound, perhaps a mineral, the assumption that the f factors are all equal can often be reasonably made.

4 Surface Studies

There are essentially three ways in which surfaces may be studied by ^{57}Fe Mössbauer spectroscopy: (i) conventional transmission methods; (ii) source studies; (iii) backscatter methods. The experimental techniques for the first two of these techniques are well established and the applications have been reviewed.[3] Consequently only a short review with selected references of these methods is given here.

Studies of ultra-fine particles some tens of ångströms in diameter by transmission methods have been particularly rewarding.[4] Materials such as iron, iron oxides,

---

[3] M. C. Hobson, in 'Characterisation of Solid Surfaces', ed. P. F. Kane and G. R. Larrabee, Plenum Press, New York and London, 1974, p. 379; H. M. Gager and M. C. Hobson, *Catalysis Rev.*, 1975, **11**, 117; M. C. Hobson, *Adv. Colloid Interface Sci.*, 1971, **3**, 1.
[4] W. Kundig, H. Bommel, G. Constabaris, and R. H. Lindquist, *Phys. Rev.*, 1966, **142**, 327; W. Kundig, K. J. Ando, R. H. Lindquist, and G. Constabaris, *Czech. J. Phys.*, 1967, **B17**, 467; K. J. Ando, W. Kundig, G. Constabaris, and R. H. Lindquist, *J. Phys. and Chem. Solids*, 1967, **28**, 2291; J. M. D. Coey, *Phys. Rev. Letters*, 1971, **27**, 1140.

or oxyhydroxides, which are normally magnetically ordered, display superparamagnetism when finally divided. The magnetic behaviour of a small particle the size of a single domain is analogous to that of a paramagnet except that the moment is a collective moment of some 10^6 electronic spins. Very often at finite temperatures the direction of the magnetic field fluctuates between easy directions in the microcrystallite. At low temperatures the time of the relaxation process is long compared to the lifetime of the nuclear excited state and the Mössbauer nucleus 'sees' an essentially static magnetic field. Consequently a magnetically split hyperfine pattern is observed. As the temperature is raised the frequency of the re-orientation processes increases as more thermal energy becomes available to overcome the energy barrier separating the various directions. Eventually the Zeeman splitting collapses when the relaxation time is short enough to average the magnetic field to zero during the time of the Mössbauer measurement. Such studies over a range of temperature when combined with a knowledge of the anisotropy constant (a measure of the height of the energy barrier) for a given material enable particle size and distributions to be determined.[4]

In the area of catalysis Mössbauer studies have been made on iron species dispersed on high-area 'inert' substrates.[5] Such investigations allow the size and chemical nature of catalytically active microcrystallites to be probed. Potential interactions between the catalyst and the 'inert' support can be elucidated [6] and the effects of, for example, dehydration,[7] oxidation and reduction cycles,[8] and chemisorbed gases [9] monitored. Transmission methods have also been applied to rather special materials with high internal surface areas such as zeolites or sheet silicates in the characterization of the environment of ion-exchanged and constitutional iron moieties and their interactions with sorbed or intercalated species.[10,11]

In order to obtain surface information pertaining to low area materials by transmission methods such as single crystals or thin films a few tens of ångströms thick it is necessary to resort to absorbers made up of layers of the required material sandwiched between inert substrates. Such methods have proved to be useful in the study of the magnetic properties of thin films.[12]

The second approach to enhance the surface sensitivity of Mössbauer spectroscopy involves the introduction of ^{57}Co at or into the surface under investigation perhaps by evaporation or electrodeposition. This source is then used in conjunction with a single-line absorber. In this way the expected anisotropy of the mean-square displacement of ^{57}Co atoms on a tungsten single crystal surface has been

---

[5] I. J. Gruverman, C. W. Siedal, and D. K. Dieterly (ed.) 'Mössbauer Effect Methodology', Plenum Press, New York, 1976, Vol. 10 and refs. cited therein.
[6] Yu. V. Maksimov, I. P. Suzdalev, and Yu. P. Yampolskii, *Doklady Akad. Nauk. S.S.S.R.*, 1972, **206**, 799.
[7] H. M. Gager, M. C. Hobson, and J. F. Lefelhocz, *Chem. Phys. Letters*, 1972, **15**, 124.
[8] W. R. Cares and J. W. Hightowers, *J. Catalysis*, 1975, **39**, 36.
[9] Yu. V. Maksimov, I. P. Suzdalev, V. I. Goldanski, O. V. Krylov, L. Ya. Margolis, and A. E. Nechitialo, *Chem. Phys. Letters*, 1975, **34**, 172; N. Malathi and S. P. Puri, *J. Phys. Soc. Japan*, 1971, **31**, 1418; H. M. Gager, J. F. Lefelhocz, and M. C. Hobson, *Chem. Phys. Letters*, 1973, **23**, 386.
[10] W. N. Delgass, R. L. Garten, and M. Boudart, *J. Phys. Chem.*, 1969, **73**, 2970.
[11] T. B. Tennakoon, J. M. Thomas, and M. J. Tricker, *J.C.S. Dalton*, 1974, 2211.
[12] T. Shinjo, *I.L.E.E. Transactions on Magnetics*, 1976, **12**, 86.

observed.[13] The source technique is also useful in the study of magnetic,[14] corrosion[15] and passivation[16] phenomena. Although, as judged from published spectra, the method is very surface sensitive in that a few monolayer equivalents of ^{57}Co can be detected on a low area solid the method suffers from the obvious disadvantage that it is not always possible to introduce the activity in a controlled and well defined fashion into or on to the desired surface.

In summary, transmission and source studies have clearly demonstrated the usefulness of Mössbauer spectroscopy in surface science but such methods are limited either by the needs of high surface area or elaborate sample preparation. An alternative approach is to use a back-scatter geometry and, as emphasized by Spijkerman,[17,18] by detecting the back-scattered conversion electrons surface studies of low-area solids should be possible. Over the past few years a number of groups have developed and explored this area and the advantages and limitations of the technique are now fairly well understood.

5 Internal Conversion

Following resonant absorption of 14.4 keV Mössbauer γ-radiation by ^{57}Fe only 10% of all the subsequent events populate the ground state by emission of a γ-photon. The remaining events are internal conversion processes which occur mainly in the K-shell resulting in the ejection of 7.3 keV internal conversion

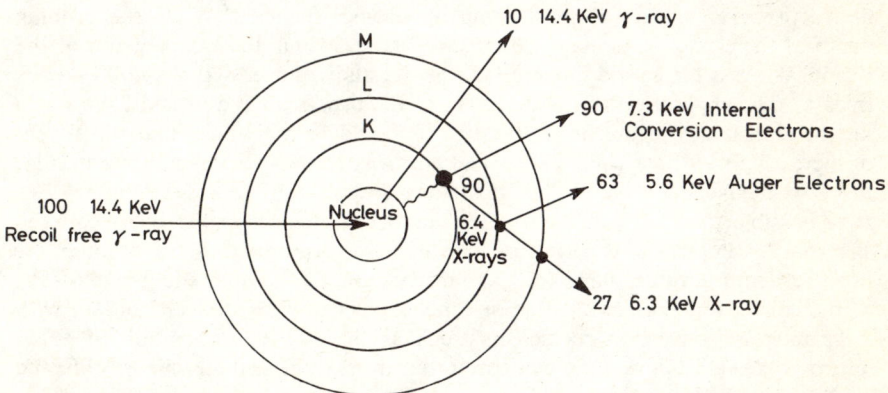

Figure 4 *Decay events following the excitation of the ^{57}Fe nucleus, indicating the relative numbers of photons and electrons produced*
(Reproduced by permission from 'Mössbauer Effect Methodology', ed. I. J. Gruverman, Plenum Press, New York, 1971, Vol. 8)

---

[13] J. W. Burton and R. P. Goodwin, *Phys. Rev.*, 1967, **158**, 218.
[14] T. Shinjo, T. Matsuzawa, T. Takada, S. Nasa, and Y. Murakani, *J. Phys. Soc. Japan*, 1973, **35**, 1032.
[15] A. M. Pritchard and C. M. Dobson, *Chem. Phys. Letters*, 1973, **23**, 514.
[16] H. Leidheiser, G. W. Simmons, and E. Kellerman, *J. Electrochem. Soc.*, 1973, **120**, 1516.
[17] J. J. Spijkerman, 'Mössbauer Effect Methodology', ed. I. J. Gruverman, Plenum Press, New York, 1971, Vol. 8, p. 85.
[18] K. R. Swanson and J. J. Spijkerman, *J. Appl. Physics*, 1970, **41**, 3155.

Table 1 *Summary of events which occur during the decay of the* $I = \pm\frac{3}{2}$ *spin state of* ^{57}Fe *for each γ-photon resonantly absorbed*

	Energy/keV	Transition probability
γ-Ray	14.4	0.09
K-Shell conversion electron	7.3	0.81
L-Shell conversion electron	13.6	0.09
M-Shell conversion electron	14.3	0.01
$K\alpha$ X-Ray	6.3	0.24
KLL Auger electron	5.4	0.57

electrons (Figure 4 and Table 1). Internal conversion also occurs in the L- and M-shell but with smaller probability as the internal conversion coefficient of a particular shell is proportional to the s-electron density of that shell at the nucleus. Following internal conversion the hole in the K-shell is filled by an electron from the L-shell thereby producing 5.4 keV Auger electrons and 6.3 keV X-rays. Auger processes in the L- and M-shells produce electrons and photons with energies of less than 1 keV.

^{57}Fe Mössbauer spectra can be accumulated in a back-scatter geometry by monitoring the γ-ray, X-ray, or electrons produced following resonant absorption. Maxima in count rates, rather than dips, are now observed at resonant energies. In all cases the requirement of a thin absorber is removed and the truly non-destructive examination of thick samples becomes possible. While the probing depth of the electrons method is limited to about 300 nm the escape depth of the γ- and X-rays is larger and therefore allows the study of overlayers several microns thick.[19] The application of the detection of the back-scattered γ- and X-rays will not be discussed in detail here, although it should be mentioned that the advent of increasingly efficient counters has caused many technologically important areas to become amenable to study by Mössbauer spectroscopy (*e.g.* see refs. 20—23). It is possible to study isotopes other than iron by detecting the back-scattered electrons. A number of experiments have been performed with ^{119}Sn although the probing depth is rather large (*ca.* 2.5 μm) because of the high energy, 19.6 keV, of the conversion electrons.[24,25] For other high-energy Mössbauer isotopes with large internal conversion coefficients, such as the 100 keV transition of ^{182}W, improved signal to noise ratio compared to transmission methods can be obtained by detecting the back-scattered electrons.[26] In this case because of the high energy of the internal conversion electron essentially bulk information is obtained, and such isotopes will not be discussed here.

---

[19] R. L. Collins, 'Mössbauer Effect Methodology', ed. I. J. Gruverman, Plenum Press, New York, 1968, Vol. 4; J. H. Terrel and J. J. Spijkerman, *Appl. Physics Letters*, 1968, **13**, 11.
[20] R. A. Levy, P. A. Flinn, and R. A. Hartzell, *Nucl. Instr. Methods*, 1972, **104**, 237.
[21] W. Meisal, *Werkstoffe Korrosion*, 1970, **21**, 249.
[22] C. J. Renken, K. J. Reimenn, and H. Berger, *Mater. Eval.*, 1974, **32**, 109.
[23] L. J. Swartzendruber, L. H. Bennett, E. A. Schoeffer, W. T. Delong, and H. C. Cambell, *Welding J.*, 1974, **53**, 51.
[24] Z. Bonchev, A. Jordanov, and A. Minkova, *Nucl. Instr. Methods*, 1969, **70**, 36.
[25] C. M. Yagnik, R. A. Mazak, and R. L. Collins, *Nucl. Instr. Methods*, 1974, **114**, 1.
[26] H. Bokemeyer, K. Wohlfahrt, E. Kankeleit, and D. Eckardt, *Z. Physik. A*, 1975, **274**, 305.

6 Theory of the Attenuation of ^{57}Fe Conversion Electrons

Krakowski and Miller [27] have performed a detailed analysis, in a general scattering geometry, of the resonant line-shape, the area under a spectrum and estimated the depth probed by the ^{57}Fe CEMS. In the calculation only the 7.3 keV conversion electrons are considered and an exponential attenuation law is assumed for both the γ-rays and electrons. The attenuation of electrons by thin films involves a series of processes, single, plural, multiple, inelastic, and elastic scatterings together with diffusion processes. None of these processes is perfectly described by an exponential law covering the whole range over which the electron slows down. However the experimentally observed transmission curves for electrons undergoing diffusion can be described by an exponential law.[28] The adoption of an exponential law therefore assumes that the conversion electrons enter immediately into a diffusion process after their production.[27] This is not unreasonable as the onset of an exponential attenuation law for an initially mono-energetic collimated beam of electrons usually corresponds to the complete randomization of the beam in the target. Consequently in CEMS the regime of single, plural, and multiple scattering can be neglected because of the isotropic, *i.e.* spatially random, nature of the emission of the conversion electrons.[27]

The area under the resonance peak produced by the back-scattered K-shell conversion electron is given by

$$\text{Area} = \pi b_1 \int_0^b E_2(\xi) I_0\left(\frac{b_1 \xi}{2}\right) \exp - \left(\frac{b_1}{2} + b_2\right) \xi d\xi$$

where $E_2(\xi)$ is a second-order Placzek exponential function, I_0 is a modified zero-order Bessel function, $b = \mu_k t$ is the reduced resonator thickness, $b_1 = \mu_R/\mu_k$ and $b_2 = \mu_E/\mu_k$. μ_k, μ_R, and μ_E are the absorption coefficients for the electrons and resonant and non-resonant γ-rays, respectively.

Numerical evaluation allows the dependence of relative count rate on reduced resonator thickness to be calculated for enriched and unenriched samples. These calculations demonstrate that the signal approaches a limiting value for depths or the order of the electron escape depth and that the signal from ^{57}Fe enriched surfaces is a factor of 10^2 larger than that from unenriched surfaces. Moreover 66% of the electrons arise from within 25 nm and 45 nm of a totally ^{57}Fe enriched and a natural iron surface respectively if a μ_k value of 2×10^5 cm is assumed.

Krakowski and Miller also calculated the electron energy distribution at the surface of the resonator. Figure 5 shows the relative probability that an electron of a given energy (E_k) has of arising from a particular depth within the surface assuming a 2% energy resolution. For conversion electrons with final energy greater than half the initial energy of 7.3 keV, spatial energy resolution exists of sufficient magnitude to permit Mössbauer spectra to be obtained from selected regions near the resonator surface down to depths of about 800 Å. However for electron energies below 3 keV the spatial resolution becomes poor.

The theory of Krakowski and Miller has more recently been extended to include

---

[27] R. A. Krakowski and R. B. Miller, *Nucl. Instr. Methods*, 1972, **100**, 93.
[28] R. L. Graham, F. Brown, J. A. Davies, and J. P. S. Pringle, *Canad. J. Phys.*, 1963, **41**, 1686.

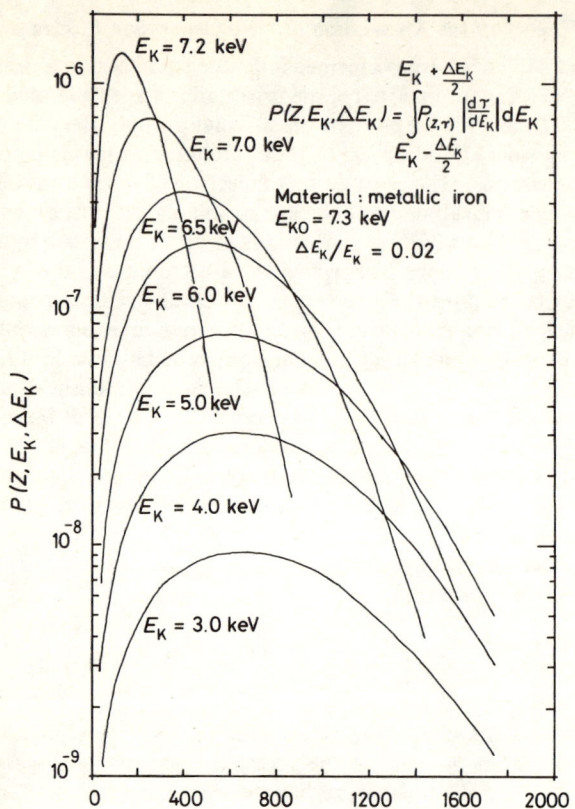

Figure 5 *Probability that an electron of energy E_k originated at a depth (z, dz) for 2% energy resolution*
(Reproduced by permission from *Nucl. Instr. Methods*, 1972, **100**, 93)

multi-layer samples and a relationship between the areas of the component spectra and an individual layer thickness derived.[29]

In essence it can be seen that two types of CEMS experiments are possible: (i) those which detect the total flux of back-scattered electrons; and (ii) depth resolved experiments which involve the accumulation of CEMS spectra with selected bands of electron energies. Since most applications of ^{57}Fe CEMS have been made by detecting the total flux of back-scattered electrons this area is treated first and discussion of the energy resolution techniques postponed till Section 10.

7 ^{57}Fe CEMS Experimental Considerations

The most widespread method of obtaining ^{57}Fe CEMS spectra is by use of the easily constructed He–CH$_4$ flow proportional counters. A number of essentially

---

[29] J. Bainbridge, *Nucl. Instr. Methods*, 1975, **128**, 531.

Iron-57 Conversion Electron Mössbauer Spectroscopy

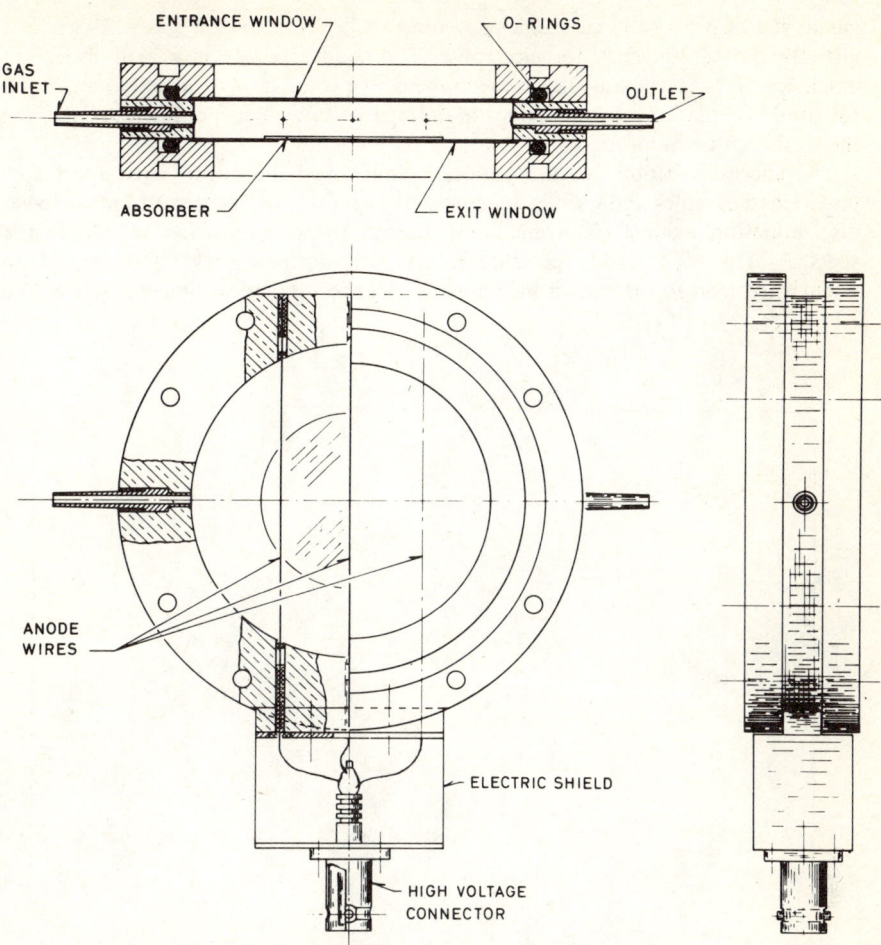

Figure 6 He–CH₄ *flow proportional counter for use in CEMS designed by Fenger.*[30] *The diameter of the counter is approximately* 10 cm
(Reproduced by permission from *A.E.K., Risø–M–1695*, 1974)

similar designs have appeared in the literature.[17,18,25,30—34] A successful design due to Fenger[30] is illustrated in Figure 6. The sample, usually *ca.* 1 cm² in area, is mounted inside a thin cell a few millimetres away from a number (usually three) of thin wire anodes held at a potential of *ca.* 1 keV and He–5% CH₄ counter gas is passed through the device at about 1 cm³ s⁻¹. Helium is chosen as a counter

---

[30] J. Fenger, *Nucl. Instr. Methods*, 1973, **106**, 203; J. Fenger, *Nucl. Instr. Methods*, 1968, **69**, 268.
[31] Y. Isozumi, D. I. Lee, and I. Kader, *Nucl. Instr. Methods*, 1974, **120**, 23.
[32] Y. Isozumi and M. Takafuchi, *Bull. Inst. Chem. Res. Kyoto Univ.*, 1975, **53**, 63.
[33] M. Takufuchi, Y. Isozumi, and R. Katano, *Bull. Inst. Chem. Res., Kyoto Univ.*, 1973, **51**, 13.
[34] M. J. Tricker, A. G. Freeman, A. P. Winterbottom, and J. M. Thomas, *Nucl. Instr. Methods*, 1976, **135**, 117; M. J. Tricker and A. G. Freeman, *Surface Sci.*, 1975, **52**, 549.

gas as it has practically zero efficiency for 14.4 keV γ-rays or 6.3 keV X-rays but virtually 100% efficiency for electrons. The counter also has a high collection efficiency of 2π steradians. By keeping the active column of the counter small and by good collimation of the γ-rays the number of unwanted photo-electrons from the walls can be minimized.

The energy resolution of the devices is small, and the pulse height spectra of unenriched samples show little structure. It is only really possible to set a lower discrimination against electronic noise during the accumulation of Mössbauer spectra. The ^{57}Fe CEM spectrum taken with normal γ-ray incidence, of an unenriched iron foil is shown in Figure 7. The accumulation time was 4 h with a

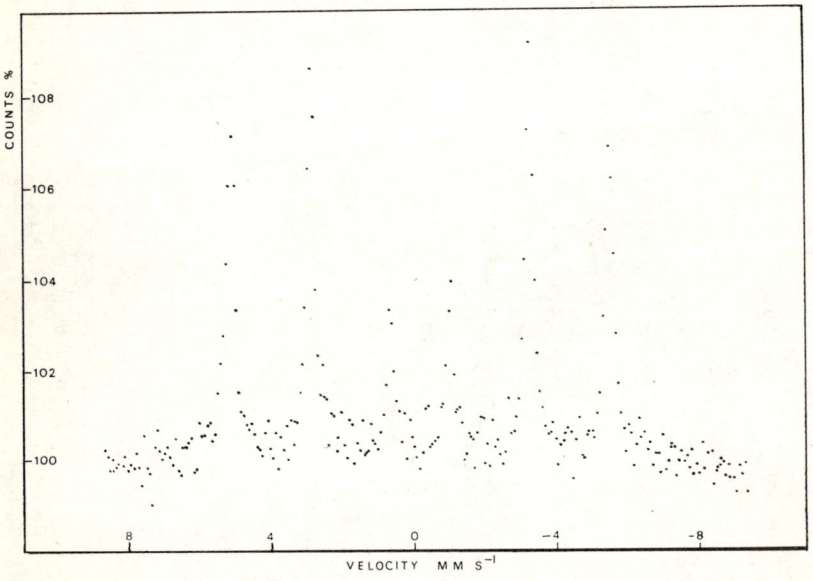

Figure 7 ^{57}Fe *conversion electron Mössbauer spectrum of a natural iron foil*

10 mCi ^{57}CoPd source. Although the percentage effect is comparable with that expected in a transmission spectrum the count rate is reduced. The background arises from photo-electrons and to a lesser extent Compton electrons produced by irradiation of the sample, and in part the walls of the counter, with 136, 122, 14.4 γ-rays and 6.3 keV X-rays from the source. The flux of photo-electrons can be reduced by filtering out the 6.3 keV X-rays. The number of background electrons from the sample will also be dependent on the cross-sections for electron production of the other elements in the sample. For a given iron concentration the best quality spectra are expected for absorbers containing elements of low atomic number.

The important feature of the iron spectrum in Figure 7 is the relative enhancement of lines 2 and 5 compared to the expected 3:2:1:1:2:3 intensity pattern. Clearly novel surface information is being obtained and the enhancement indicates that the magnetic field is polarized in the plane of the surface.[18]

CEM spectra of natural iron foils over a range of angles indicate that up to 70% increase in the count rate can be obtained at glancing γ-ray incident angles (α) of 15° compared to normal incidence.[34] Moreover up to this angle thickness effects are absent as the depth probed is small compared to the path length of the γ-photons.[34]

The expected marked improvements in the quality of CEMS spectra are obtained by the use of ^{57}Fe enriched samples and counting times can be reduced to minutes. Pulse height spectra now show a clear maxima corresponding mainly to 7.3 keV

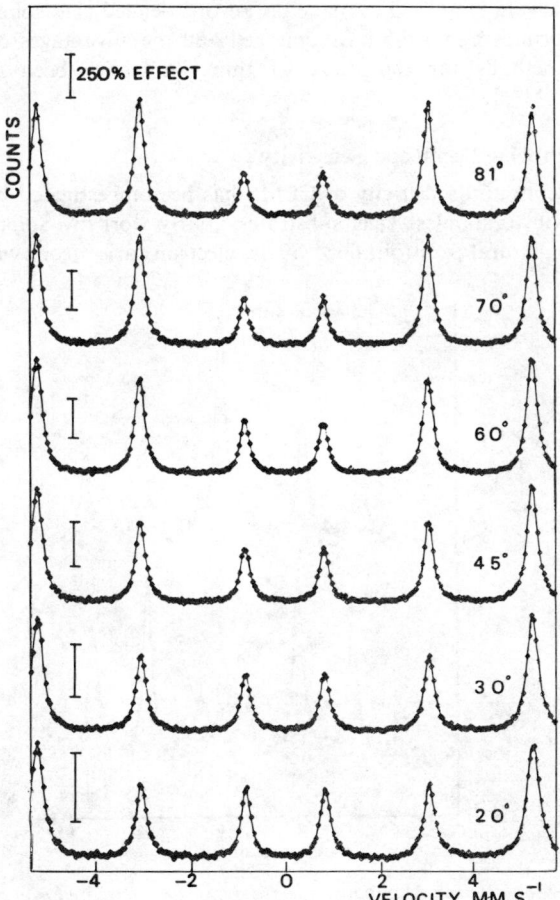

Figure 8 *Conversion electron Mössbauer spectra of a 90% ^{57}Fe enriched foil as a function of angle. Note large percentage effects and variation of the relative intensities with angle. A typical background count is 500*
(Reproduced by permission from *Nucl. Instr. Methods*, 1976, **135**, 117)

conversion electrons. Total percentage effects (*i.e.* sum of all six lines) of up to 3500% have been obtained for 90% enriched iron foils.[34] The percentage effects show a marked angular variation, with a maximum at $\alpha = 75°$ (Figure 8). This is attributable to saturation effects within the enriched absorber which are also manifest for $\alpha < 75°$ in the form of line-broadening and the deviation of the angularly independent (in the thin absorber limit) area ratios of lines 1 to 3 from 3:1.

The main disadvantages of He–CH$_4$ counters are that it is not possible to cool the sample to very low temperature or clean and treat a surface with gas *in vacuo*. Isozumi[32] has circumvented the first problem by the simple expedient of omitting the quench gas, and although the counter becomes a little unstable, satisfactory spectra can be obtained at 80 K. In order to record ^{57}Fe CEM spectra *in vacuo* Davis[35] and Oswald *et al.*[36] have made use of open-ended channel electron multipliers. Some promising results have emerged and the advantages of CEMS over transmission methods for the study of thin films have been discussed and demonstrated.[35,36]

8 Probing Depth and Sensitivity

The probing depth and sensitivity of CEMS has been investigated by evaporation of iron metal on to stainless steel substrates. Early work by Spijkerman[18] concluded that for natural iron foil, 66% of the electrons arise from within 60 nm of

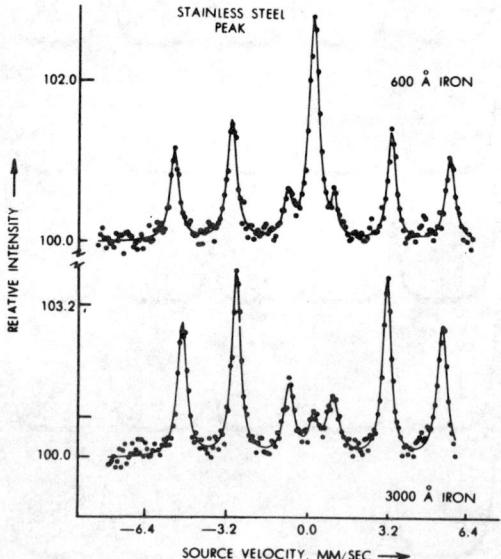

Figure 9 *Conversion electron Mössbauer spectra of vacuum-deposited iron on stainless steel foil*
(Reproduced by permission from 'Mossbauer Effect Methodology', Plenum Press, New York, 1971, Vol. 8)

[35] B. R. Davies, Ph.D. thesis, Stevens Institute of Technology, 1972.
[36] R. Oswald and M. Ohring, *J. Vac. Sci. Technol.*, 1976, **13**, 40.

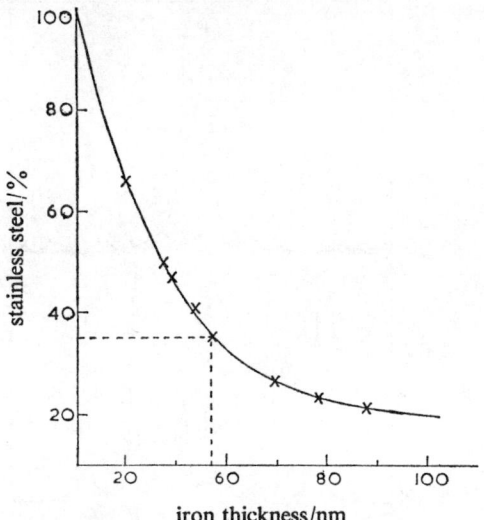

Figure 10 *Experimentally determined ratio of stainless steel to total signal plotted against thickness of iron coating*
(Reproduced from *J.C.S., Faraday II*, 1975, **71**, 1708)

the surface and 95% within 300 nm (Figure 9). This work was later extended by Thomas et al.[37] From a plot (Figure 10) of the fraction of stainless steel to total signal against iron film thickness it was determined that 66% of the electrons originating in the stainless steel are attenuated by 54 nm of iron, in good agreement with the previous experimental and theoretical estimates. From the quality of the data it was estimated that *ca.* 8 nm of natural iron or a monolayer of ^{57}Fe can be detected.[18,37] Recently Gonser and co-workers[38] have made measurements of very thin layers of ^{57}Fe a few nanometres thick. In Figure 11 spectra of vacuum deposited iron films on copper, α-Fe, and stainless steel substrates are shown. Thin ^{57}Fe films a few monolayers thick (Figure 11a, b, c), oxidized by preparation, exhibit typical superparamagnetic quadrupole split Fe^{3+} spectra at 295 K. The spectrum (Figure 12a) of a film about 3 ± 1 nm thick exhibits a doublet with a 5% effect. On this basis a single resonance line with a natural linewidth a monolayer of iron should have a 1% effect. Thus the ultimate sensitivity of CEMS is at present about 2×10^{15} ^{57}Fe atoms cm^{-2} *i.e.* about a monolayer.

It proved possible to fit the experimentally determined electron attenuation data of Figure 10 to an expression appropriate to normal γ-incidence and 2π electron collection of the form

$$\frac{I_{ss}}{I_{tot}} = \left\{ 1 + \frac{K_{Fe}}{K_{ss}} [\xi^{-1}(d/\lambda_{Fe}) - 1] \right\}^{-1}$$

where $K_{Fe} = n_{Fe}\lambda_{Fe}f_{Fe}$ and $K_{ss} = n_{ss}\lambda_{ss}f_{ss}$, and n = Fe atoms cm^{-2} f = recoil free fraction, λ = mean attenuation path length and the subscripts Fe and SS refer

[37] J. M. Thomas, M. J. Tricker, and A. P. Winterbottom, *J.C.S. Faraday II*, 1975, **71**, 1708.
[38] U. Gonser, personal communication.

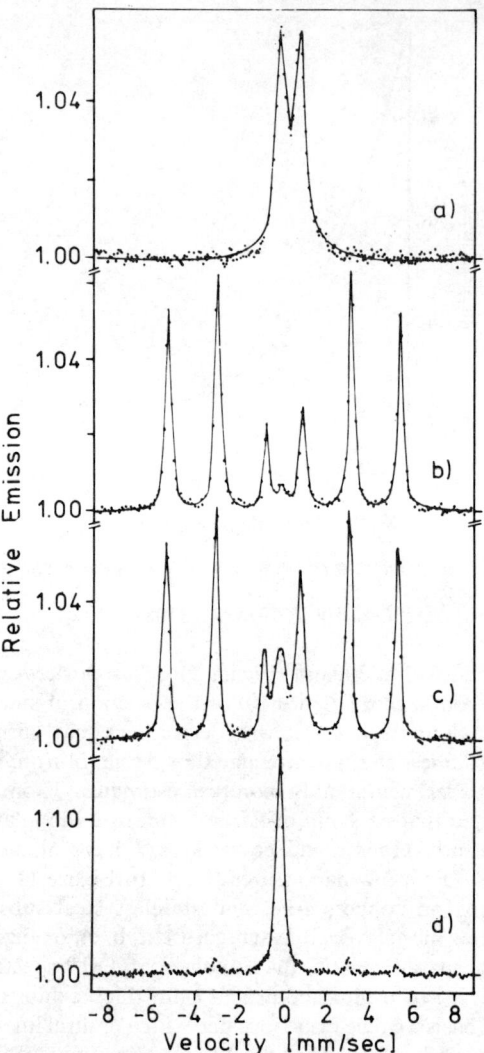

Figure 11 *CEMS spectra of iron films deposited on various substrates:* (*a*) 30 Å Fe^{57} *on* Cu; (*b*) 5 Å Fe^{57} *on natural* α-Fe; (*c*) 20 Å Fe^{57} *on natural* α-Fe; (*d*) 250 Å *natural* Fe *on stainless steel*

to iron and stainless steel, respectively. The assumption that the attenuation of the incident γ-ray in the depth probed is negligible was made. A good fit (Figure 12) to the data was obtained with a λ value of 107 nm which is approximately twice the value expected for electrons, with an initial energy of 7.3 keV, using the relationship empirically determined by Coslett and Thomas [39] *viz.*

$$\mu(p) = 1.4 \times 10^{10} E^{-\frac{3}{2}}$$

[39] V. E. Coslett and R. N. Thomas, *Brit. J. Appl. Phys.*, 1964, **15**, 883.

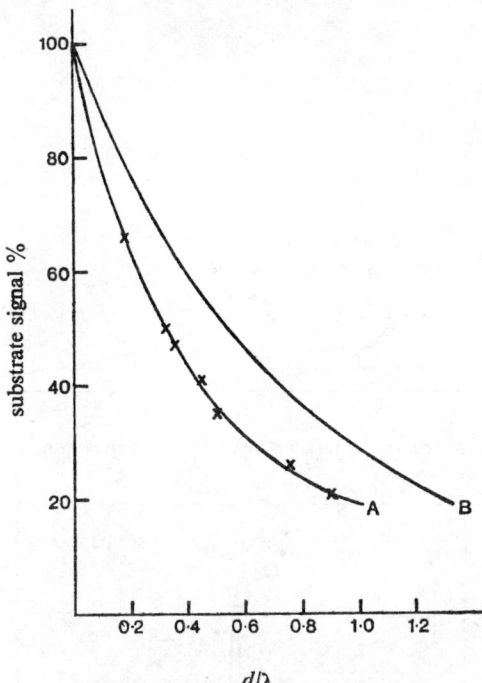

Figure 12 Graph showing agreement between calculated intensity ratio (A, lower line) and experimental points (×) for iron on stainless steel as a function of d/λ. The upper line (B) shows the calculated reduction in the steel signal, as a function of thickness for iron oxides on plain low-carbon steel. The thickness of the overlayer is d and the λ values for iron and iron oxides are 107 aad 162 nm respectively
(Reproduced from J.C.S. Faraday II, 1975, **71**, 1708)

where $\mu(p)$ is a mass absorption coefficient and E the initial electron energy. This discrepancy is perhaps not unexpected as no explicit allowance for the contribution to the CEM spectrum of other Auger and internal conversion electrons was made.

Calculation of the attenuation curves for 7.3 keV electrons assuming a simple exponential attenuation law and a $\mu(p)$ value from Coslett's data [39] (Figure 13) suggest that the 95% cut-off should occur at ca. 160 nm in iron rather than the value of ca. 300 nm experimentally observed. At first sight it might seem that the long tail is due to detection of 13.6 keV L-shell conversion electrons. However a weighted attenuation curve (assuming an exponential attenuation law and narrow slit collection) which includes 5.4 keV Auger electrons as well as 7.3 keV and 13.6 keV conversion electrons almost coincides with the curve for the 7.3 keV electron alone (Figure 13).

The origin of the tail is due to the production of photo-electrons produced in surface regions by the Mössbauer spectrum of γ- and X-rays [40] back-scattered

[40] M. J. Tricker, T. E. Cranshaw, and L. Ash, *Nucl. Instr. Methods* in the press.

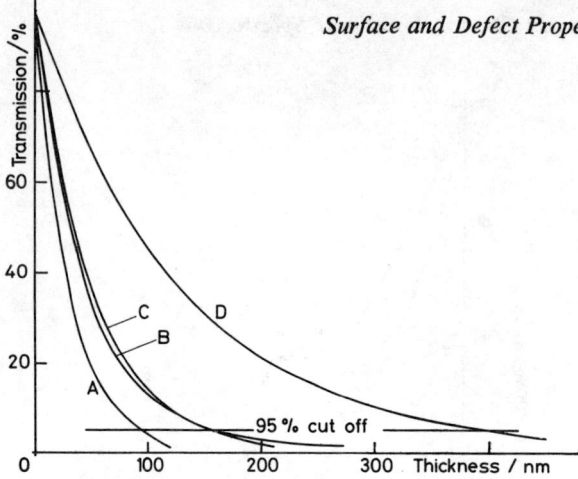

Figure 13 *Transmission curves of (A) 5.4 keV; (B) weighted sum; (C) 7.3 keV; and (D) 13.6 keV electrons in iron*

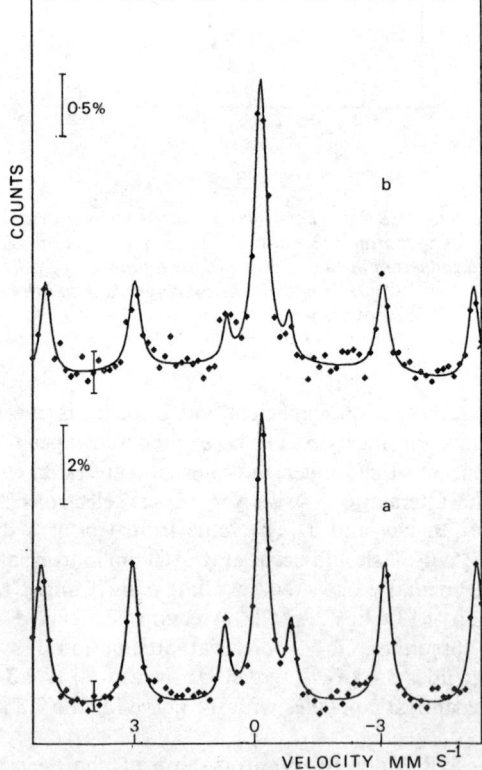

Figure 14 *^{57}Fe CEM spectrum of (a) 70 nm of iron coating on a stainless steel substrate and (b) the same sample coated with 57 nm of gold. Note the enhancement of the stainless steel signal relative to the gold in (b)*

from deep within the sample. The back-scattered γ or X-ray spectrum would not normally be detected in a He–CH$_4$ counter but the photo-electrons produced by these photons are. In effect the surface regions of the sample convert X- and γ-rays into electrons and make their detection possible. The reality of this effect is demonstrated in Figure 14 where spectra of an iron film on a stainless steel substrate with and without a gold overlayer is shown. The marked enhancement of the stainless steel signal in the overlayer case is due to the production of photo-electrons in the gold layer by X- and γ-rays back-scattered from the stainless steel. By comparison the fraction of X- and γ-rays back-scattered from the thin iron film is negligible.

In the CEMS detection methods described above the total fluxes of electrons were detected and the Mössbauer information is not spatially resolved. Although the greatest resolution of this kind comes from the use of energy dispersive β-ray

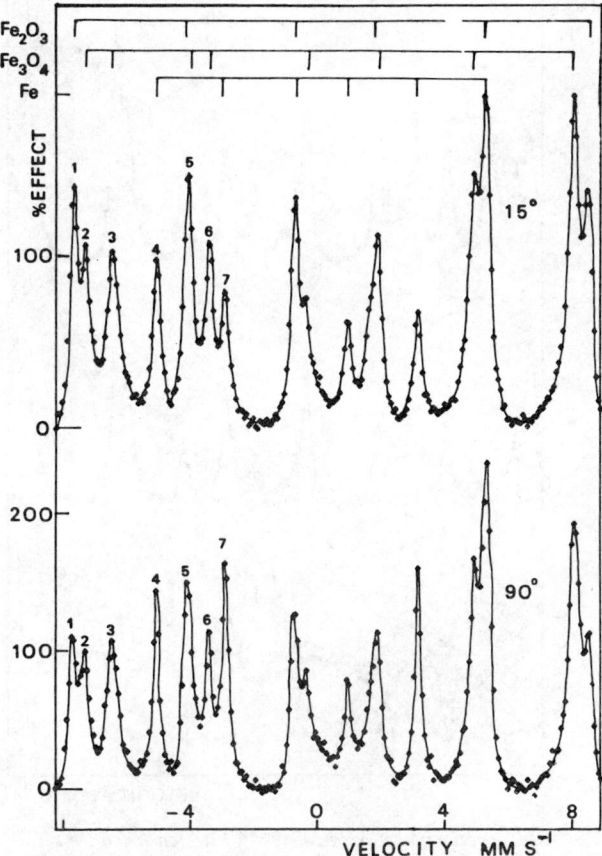

Figure 15 *CEM ^{57}Fe spectrum of an oxidized 90% ^{57}Fe enriched iron foil at $\alpha = 90°$ and $\alpha = 15°$. The oxide signal is increased relative to the substrate signal at glancing incidence*
(Reproduced by permission from *Nucl. Instr. Methods*, 1976, **135**, 117)

spectrometers (see Section 10) some cruder depth resolution has come from the use of He–CH$_4$ counters. The first method has only been applied to enriched samples and involves collecting data as a function of incident γ-ray angle.[34] Figure 15 shows spectra taken at normal and glancing γ-ray incident angles of a 90% ^{57}Fe enriched iron foil oxidized so as to produce a duplex oxide film of Fe$_2$O$_3$ (outermost) and Fe$_3$O$_4$ about 100 nm thick. The spectrum taken at glancing incidences clearly shows enhancement of the Fe$_2$O$_3$ signal compared to the substrate. This effect is due to the significant absorption of the γ-ray in the surface regions of the enriched foil at glancing incidence. For unenriched samples such surface enhancement effects would only be observed with incident γ-beams almost parallel to the surface.

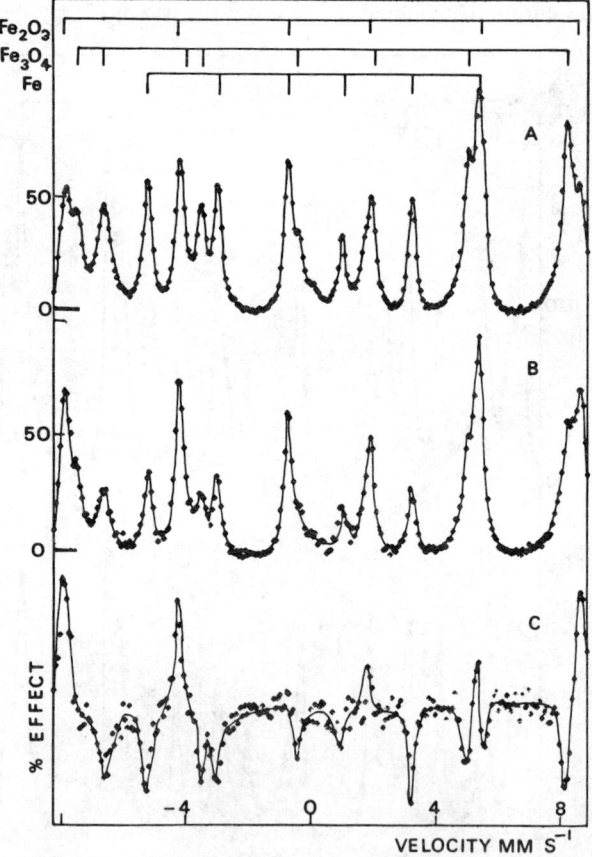

Figure 16 ^{57}Fe *CEM spectra of an oxidized* 90% ^{57}Fe *enriched iron foil obtained with (A) low energy electrons (B) high energy electrons (C) difference spectrum. This is marked enhancement of the* Fe$_2$O$_3$ *contribution in spectrum B. This is clearly seen by noting the increase in intensity of the* Fe$_2$O$_3$ *line at high positive velocities and in the difference spectrum*
(Reproduced by permission from *Nuclear Instr. Methods*, 1976, **135**, 117)

A second approach [34] makes use of the inherent, albeit poor, resolution of the He–CH_4 counters. In Figure 16 is shown the simultaneously accumulated spectra of the same oxidized foil accumulated using the high- and low-energy regions of the pulse height spectrum. Again it can clearly be seen that some degree of depth resolution has been achieved. Similar results have been obtained by Isozumi by varying the anode simple distance.[31] Thus a degree of depth resolution can be obtained by the use of He–CH_4 counters. However for unenriched samples, if only the high-energy electrons are detected, counting times become very long.

9 Applications

Although the application of ^{57}Fe CEMS is still perhaps in its infancy sufficient work has now been carried out to indicate the areas in which the method is capable of producing novel information.

Corrosion and Oxidation Studies.—The usefulness of transmission Mössbauer spectroscopy for the investigation of the advanced stages of corrosion of iron and its alloys has been well established.[41] Most studies have involved rather thick layers and often corrosion products that have been removed from the surface. *In situ* studies of thick overlayers can be made by back-scattered X-ray or γ-techniques but CEMS offers the possibility of probing the early stages of formation and properties of the films formed for example during the initial stages of low-temperature oxidation or anodic treatment.

A number of papers have appeared dealing with the feasibility of the *in situ* examination of lightly corroded iron surfaces. Fenger [30] recorded the CEM spectrum of a rusted iron foil and attributed the observed doublet ($\delta = 0.40$ mm s^{-1}, $\Delta = 0.55$ mm s^{-1}) to the presence of γ-FeOOH. A doublet with parameters corresponding to high-spin ferrous ion species has been observed by Tricker *et al.*[42] on an iron foil following brief exposure of the foil to moist HCl gas. After further exposure to the atmosphere a doublet ($\delta = 0.38$ mm s^{-1} and $\Delta = 0.70$ mm s^{-1}) was observed suggesting the formation of β-FeOOH. Interesting, simultaneously recorded transmission spectra revealed no trace of the surface species. Onodera *et al.*[43] studied the corrosion products formed on an iron foil in $NaNO_2$ solutions. The main product of this treatment identified by ^{57}Fe CEMS was α-FeOOH but some γ-Fe_2O_3 was also observed. These assignments must be treated with some caution as the measured Mössbauer parameter of the surface species differ somewhat from the bulk values. It is supposed by Onodera *et al.* that the 'oxide' in the extremely thin surface layer is in an amorphous state which results in the modification of the bulk parameters. This work in particular emphasizes the fact that although the presence of new surface phases can be readily detected by ^{57}Fe CEMS

---

[41] D. A. Channing and M. J. Graham, *Corrosion Sci.*, 1972, **12**, 271; D. A. Channing and M. J. Graham, *J. Electrochem. Soc.*, 1970, **117**, 390; D. A. Channing, S. M. Dickerson, and M. J. Graham, *Corrosion Sci.*, 1973, **13**, 933; G. M. Bancroft, J. E. O. Mayne, and P. Ridgeway, *Brit. Corrosion J.*, 1971, **6**, 119; D. D. Joyce and R. C. Axelmann, *Analyt. Chem.*, 1968, **40**, 876.
[42] M. J. Tricker, J. M. Thomas, and A. P. Winterbottom, *Surface Sci.*, 1974, **45**, 601.
[43] H. Onodera, H. Yamamoto, H. Watanabe, and H. Ebiko, *Japan. J. Appl. Phys.*, 1972, **11**, 1380.

and the presence of a particular species often eliminated it is not always possible to identify unambiguously the surface phases on the basis of one spectrum alone.

A number of studies have been made of the oxidation of iron in dry atmospheres. Simmons et al.[44] studied the ^{57}Fe enriched surfaces (by electrodeposition) of iron foils oxidized at 225, 350, and 450 °C in dry oxygen or synthetic air for periods of up to 2 h. Good quality spectra were obtained and the technique was found to

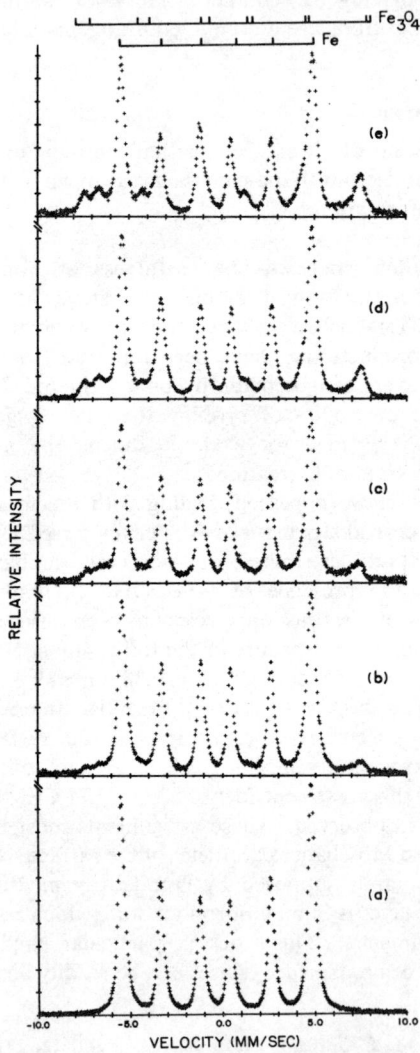

Figure 17 *Mössbauer spectra of iron oxidized at* 225 °C *for specific times* (a) *before oxidation* (b) 5 min (c) 15 min (d) 120 min *and* (e) 1000 min
(Reproduced by permission from *Corrosion* 1973, **29**, 227)

[44] G. W. Simmons, E. Kellerman, and H. Leidheiser, jun., *Corrosion*, 1973, **29**, 227.

be useful in the *in situ* study of oxide films five to several tens of nanometres thick. Non-stocheiometric magnetite corresponding to $Fe_{2.9}O_4$ or a mixture of $\alpha\text{-}Fe_2O_3$ and Fe_3O_4 was observed after oxidation at 450 °C. Oxidation at 350 °C for short times produced a duplex film consisting of Fe_3O_4 and $\alpha\text{-}Fe_2O_3$. Oxide growth was studied at 225 °C as a function of oxidation time (Figure 17). Oxide thicknesses were estimated by measuring the area of the resolved Fe_3O_4 lines at -7.5 and -6.7 mm s^{-1} of a spectrum of a sample oxidized at 225 °C for 120 min. The oxide film formed under these conditions has a nominal thickness of 10 nm and this was used as a normalization point on a curve of relative count rate against reduced resonator signal for Fe_3O_4 using the method of Krakowski and Miller.[27] It was further assumed that the electron attenuation coefficients for different materials are inversely proportional to density. The increase in oxide thickness at 225 °C was found to have a logarithmic time dependence in agreement with other studies.

Similar studies have been made over a wider temperature range by Thomas *et al.*[37] and Sette-Camara *et al.*[45] on unenriched iron samples. Similar information to that obtained by Simmons[44] can be obtained, although of course the detection limit is larger at *ca.* 10 nm, but oxide thicknesses of the order of 20 nm are needed before information can be readily extracted from the spectra. It is clear that for the complete characterization of oxide films, especially when no substrate is evident in the CEM spectrum, the use of back-scattered γ- or X-ray Mössbauer techniques provides valuable complimentary data. For example a CEM spectrum of an iron sample oxidized at 800 °C for 60 s reveals the presence of $\alpha\text{-}Fe_2O_3$ at the outermost regions of the oxide film. However, a back-scattered γ-ray spectrum detected only the presence of wustite. Taken together, the Mössbauer spectra indicate that the

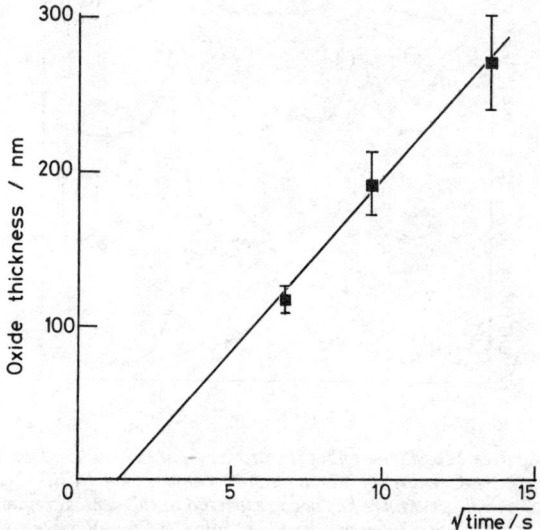

Figure 18 *Oxide thickness formed on an iron foil as a function of the square root of time*

[45] A. Sette-Camara and W. Keune, *Corrosion Sci.*, 1975, **15**, 441.

oxide film consists of a few hundreds of nanometres of α-Fe_2O_3 at the surface of a wustite layer of the order of microns thick.

In order to measure oxide film thicknesses Thomas et al.[37] adopted a similar approach to Simmons et al.[44] but measured the substrate to oxide area ratio rather than the intensity of oxide peaks. Using the data derived from the iron films on stainless steel calibration experiments it is possible to construct a curve relating the total signal area ratio to oxide thickness (Figure 12). In this way overlayer thicknesses 10—500 nm thick can be determined. Sette-Camara et al.[45] employed a similar approach but were only able to use the two calibration points reported by Spijkerman.[17,18] No allowance was made for density effect and consequently the oxide thicknesses reported for iron foils oxidized at 500 °C for periods of up to 5 min. are in all probability under-estimated by a factor of about 50%. If the oxide thicknesses are re-estimated using the calibration curve shown in Figure 12 a plot of thickness against square-root of oxidation time yields a straight line (Figure 18). The line passes closer to the origin compared to the original plot and the parabolic rate constant of $5(\pm2) \times 10^{-12}$ cm^2 s^{-1} is in better agreement with values measured by other methods.

Surface Phase Analyses of Steels.—One area of study of particular technological importance which has proved amenable to study by ^{57}Fe CEMS is the characterization of metal surfaces and particularly in the quantitative estimation of austenitic

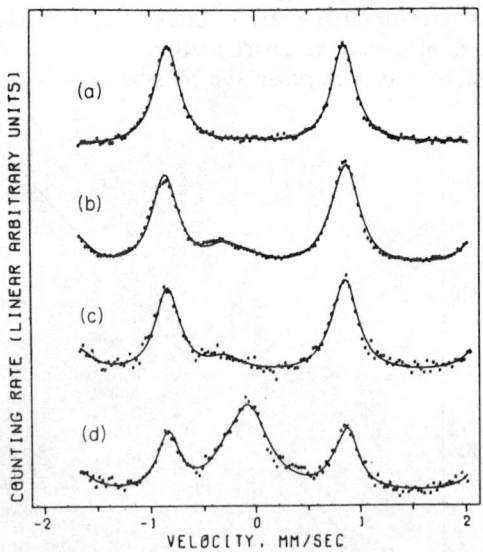

Figure 19 Back-scatter Mössbauer effect spectra on a pure Fe foil and on the ground surface of a spheroidized Fe–C alloy of the eutectoid composition (NBS Standard Reference Material 493). The velocity range has been restricted to the central region of the spectrum: (a) Fe foil; 14.4 keV γ-rays counted; (b) Fe–C alloy; 14.4 keV γ-rays counted; (c) Fe–C alloy; 6.3 keV X-rays counted; (d) Fe–C alloy; conversion electrons counted. The large central peak due to retained austenite on the ground surface is clearly evident in (d) (Reproduced by permission from Scripta Met., 1972, **6**, 737)

phases in the surface regions of the steels. Austenite is a paramagnetic phase of iron with a face centred cubic structure which appears as a single peak in a Mössbauer spectrum. It is therefore readily distinguishable from the magnetically split patterns of ferritic or martensitic phases.

Although austenite is the intermediate high-temperature form of iron, appreciable quantities of it may be retained on quenching and this has an extremely important bearing on the mechanical properties of steel. Appreciable amounts of austenite can also be formed by coarse surface grinding of carbon steels. This has been detected by conventional techniques for relatively thick (*ca*. microns) surface layers. However, Swartzendruber et al.[46] have shown by ^{57}Fe CEMS that a hitherto undetected non-magnetic layer of austenite is formed in the surface region of spheroidized Fe–C alloy of eutectoid composition even after light cutting in the presence of copious cutting oil (Figure 19). The cutting process raises the temperature of the very thin surface layer into the austenitic range. When the surface is quenched by rapid cooling provided by the substrate and cutting fluid, austenite is retained in this layer. The addition of up to 12 atom% cobalt to the steel has a considerable influence on the surface layer in that the cobalt raises the martensitic transformation temperature and consequently lowers the amount of austenite retained.[46]

Surface Stress Measurements.—Collins et al.[47] have outlined how surface stress measurements may be carried out by ^{57}Fe CEMS. Experiments indicate that the change in isomeric shift in ferrite is very small when compared with the natural linewidth of 0.25 mm s^{-1}. In order accurately to record such small shifts, Collins et al.[47] have developed a method which makes use of the small shifts in γ-ray energy brought about by the second order Doppler effect as the temperature of the source is varied. Preliminary calibration experiments have been performed and it is hoped that the method will find applications in the measurement of residual surface stress in ferrous metals generated by cutting, shaping, and finishing operations. Such examination followed by appropriate remedial treatment should lengthen component lifetimes.

Ion Implantation Studies.—Ion implanted materials have been extensively studied by Sawicki and co-workers [48-52] using CEMS. The depth probed by the technique is of the same order of magnitude as the implantation depth for Fe$^+$ ion beam energies in the range 10 to 300 keV. Stanek et al.[48] have listed some of the advantages of the method: (i) thick samples can be studied; (ii) the dangers of detecting by trans-

---

[46] L. J. Swartzendruber and L. H. Bennett, *Scripta Met.*, 1972, **6**, 737; L. J. Swartzendruber and E. Siegel, 'AIP Conference Proc. No. 18, Magnetism and Magnetic Materials', 1974, 735.
[47] R. L. Collins, R. A. Mazak, and C. M. Yagnik, 'Mössbauer Effect Methodology', ed. I. J. Gruverman and C. W. Weidel, Plenum Press, New York, 1973, Vol. 8, p. 191; *Amer. Laboratory*, 1973, **5**, 39.
[48] J. Stanek, J. A. Sawicki, and B. D. Sawicka, *Nucl. Instr. Methods*, 1975, **130**, 613.
[49] B. D. Sawicka, J. A. Sawicki, and J. Stanek, *Nukleonika*, 1976, **21**, 949.
[50] B. D. Sawicka, J. A. Sawicki, and J. Stanek, *Phys. Letters*, in the press.
[51] J. Stanek, J. A. Sawicki, B. D. Sawicka, and M. Drwiega, 'Proc. Ist. Conf. Mössbauer Spectroscopy', ed. A. Hrynkiewicz and J. A. Sawicki, Krakow, 1975, Vol. 2.
[52] J. A. Sawicki, B. D. Sawicka, J. Stanek, and J. Kowalski, *Phys. Stat. Solidi* (*b*), 1976, **77**, K1.

mission methods the impurities in the non-ion implanted regions of the samples are removed; (iii) strong sources do not overload the electronics; (iv) the method is very sensitive. Amounts of iron comparable to the doses of ^{57}Co in source experiments can be studied without resort to the implantation of radioactive species.

The method is more sensitive by a factor of 10 than transmission methods.[53] In Figure 20 is shown the CEM and transmission spectrum of 3×10^{16} ^{57}Fe cm^{-2} implanted in beryllium foil. The CEM spectrum shows a 30% effect. It should be noted however that this is a particularly favourable case because of the low probability for the production of photo-electrons in the beryllium target. The detection limits are *ca.* 10^{14} ions cm^{-2} for light targets and *ca.* 10^{15} for heavy targets. The main disadvantages in such studies are the difficulties encountered in measuring the absolute recoil free fraction of the implanted ion.

Sawicki and co-workers have studied ^{57}Fe implanted into a wide range of targets. A great variety of CEM spectra were obtained showing that the state of the ^{57}Fe is strongly host dependent.[49] In magnetic hosts (*e.g.* Fe, Co, and Ni) magnetic hyperfine patterns are observed but in para- or diamagnets single lines (V, Cr, Mo, Pd, Re, Rh, Ir, Pt) or asymmetric broad structured lines (Mn, Ti,

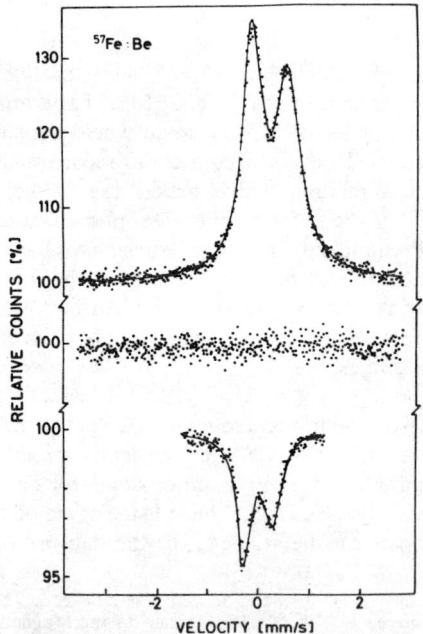

Figure 20 *Spectra of ^{57}Fe in Be. Top: spectrum of implanted side 3×10^{16} cm^2 ^{57}Fe implants. Middle: spectrum of non-implanted side. Bottom: transmission spectrum o impurities contained in* 1 mm *thick target plate*
(Reproduced by permission from *Nuclear Instr. Methods*, 1975, **130**, 613)

[53] J. A. Sawicki, B. D. Sawicka, A. Lazarski and E. Maydell, Ondrusz, *Phys. Stat. Solidi* (*b*), 1973, **57**, 143; *Phys. Stat. Solidi*, 1973, **18**, 85; J. A. Sawicki and B. D. Sawicka, *Nukleonika*, 1974, **19**, 811.

Nb, W, Mo, Cu, Au) or doublets (Zn, Ge, Si, Ag, In, Sr, Pb) resulted. The systematics of the isomer shift of diluted iron in the various transition metal targets are very similar to the behaviour observed for dilute alloys. Lattice damage therefore has little effect on the isomer shifts of an implanted iron atom but can give rise to broadening and quadrupole splitting of the resonance line. For example the CEM spectra of iron implanted into cubic host matrices of copper and gold using a 70 keV beam at 10^{16} ions cm^{-2} are broad, structured and similar to those of rapidly quenched alloys. The quadrupole splitting and structure of the Fe–Au system can be accounted for in terms of a model which attributes the origin of the e.f.g. at the iron site to random defects within the lattice. However to explain the additional structure observed in the spectrum of the Fe–Cu system clustering of the atoms has to be invoked. This is compatible with the low solubility of iron in copper compared to gold.[50] The dose dependence of ^{57}Fe$^+$ implanted into aluminium in the dopant range 10^{14} to 10^{16} ions cm^{-2} with a 70 keV beam has been investigated.[51] At 10^{14} ions cm^{-2} a single line with a 0.5% effect was observed. As the implant concentration increased an eventual 20% effect was obtained, the area under the spectra being proportional to dose. Upon annealing the concentrated samples at 400 °C the spectra sharpen, the quadrupole splittings decrease and the percentage effects increase markedly (up to 60% in some cases). This behaviour is consistent with annealing out of lattice damage. Annealing above 600 °C causes loss of the signal due to diffusion of iron into the bulk.

^{57}Fe implanted (*ca.* 10^{15} ions cm^{-2}, 70 keV beam) into silicon targets reproducibly yields symmetric doublet spectra consistent with a one iron site model if sufficient care is taken in the sample preparation procedure.[52] Subsequent annealing of a sample doped at 10^{15} ions cm^{-2} (a dopant level which greatly exceeds the solid solubility of iron in silicon) again demonstrates the recovery of the lattice at temperatures up to 600 °C. Above this temperature atom migration occurs and a marked asymmetry in the spectrum is observed. This is attributed to the precipitation of α-FeSi$_2$ in the sample surface. At higher temperatures (*ca.* 1000 °C) β-FeSi$_2$ is also precipitated, and at 1200 °C the signal disappears owing to diffusion of iron into the bulk.

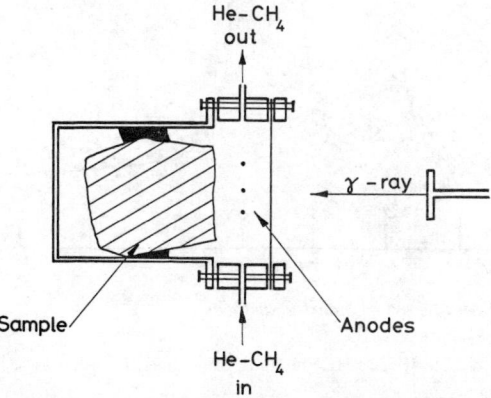

Figure 21 He–CH$_4$ *counter modified so as to allow the study of massive specimens*

The ^{57}Fe spectrum of 10^{16} ^{57}Fe ions cm^{-2} implanted to a depth of 100 nm in graphite consists of a broad doublet which sharpens in a similar manner to the iron in aluminium described above at 500 °C.[54] The data suggest that one main type of iron site is occupied and that the electronic configuration of the iron corresponds to $3d^6\,4s^{0.35}$.

Applications in Geochemistry.—The use of transmission ^{57}Fe Mossbauer spectroscopy in the study of minerals is a very fruitful area of research.[55] By simple

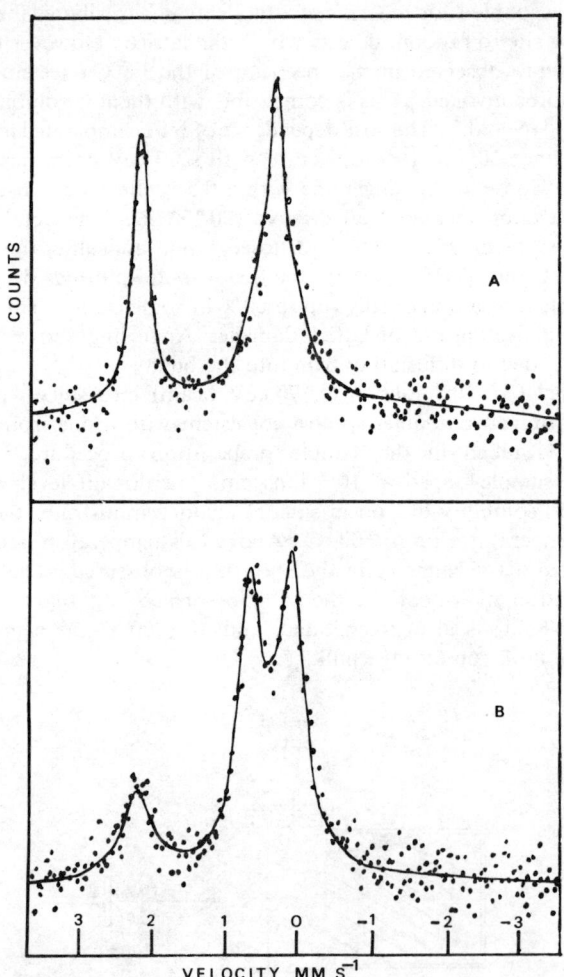

Figure 22 ^{57}Fe *CEM spectra of siderite surfaces*

[54] M. J. Tricker, R. K. Thorpe, J. H. Freeman, and G. A. Good, *Phys. Stat. Solidi (a)*, 1976, 33, K97.
[55] G. M. Bancroft, 'Mössbauer Spectroscopy, An introduction for Inorganic Chemists and Geochemists', McGraw-Hill, London, 1973 and refs. cited therein.

modifications to the usual He–CH₄ counters the surface regions of massive mineral samples may be readily probed in a non-destructive fashion (Figure 21). The method is particularly suited for the study of mineral weathering processes. For example examination of different regions of the surfaces of siderite crystals revealed in some places the predominant presence of the expected wide quadrupole doublet characteristic of high-spin Fe^{2+} but in others a doublet with parameters corresponding to Fe^{3+} iron (Figure 22). This Fe^{3+} doublet in all probability indicates the presence of superparamagnetic goethite at the mineral surface produced by weathering.

Forester [56] has examined samples of Lunar fines by CEMS in order to explore the hypothesis that the surface regions of Lunar fines contain iron particles produced by exposure to the solar wind. The CEM spectra of Lunar fines having diameters greater than the escape depth of the conversion electrons were compared

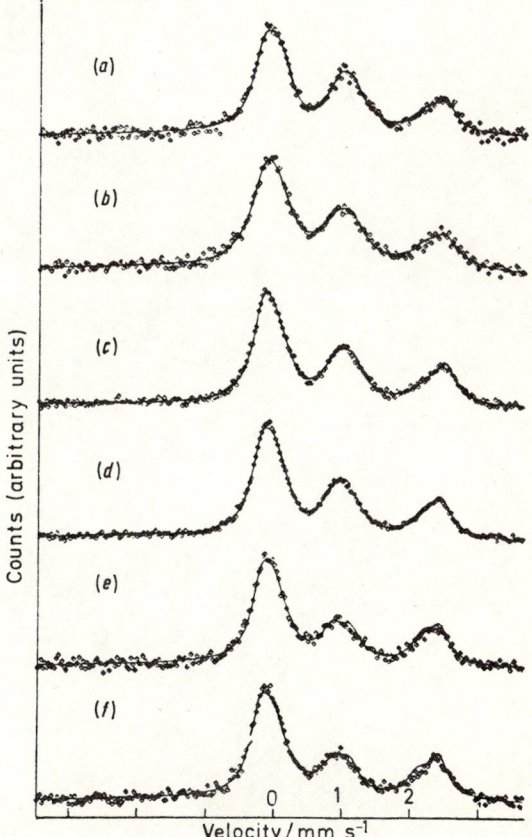

Figure 23 *CEM spectra of biotite heated at 620 K. The total time of heat treatment was* (a) 114, (b) 90, (c) 78, (d) 60, (e) 43 *and* (f) 24 h
(Reproduced from *J.C.S. Dalton*, 1976, 1289)

[56] D. W. Forester, 'Proc. 4th Lunar Sci. Conf.', Suppl. 4, *Geochem. Cosmo. Chim. Acta*, 1973, **3**, 2697.

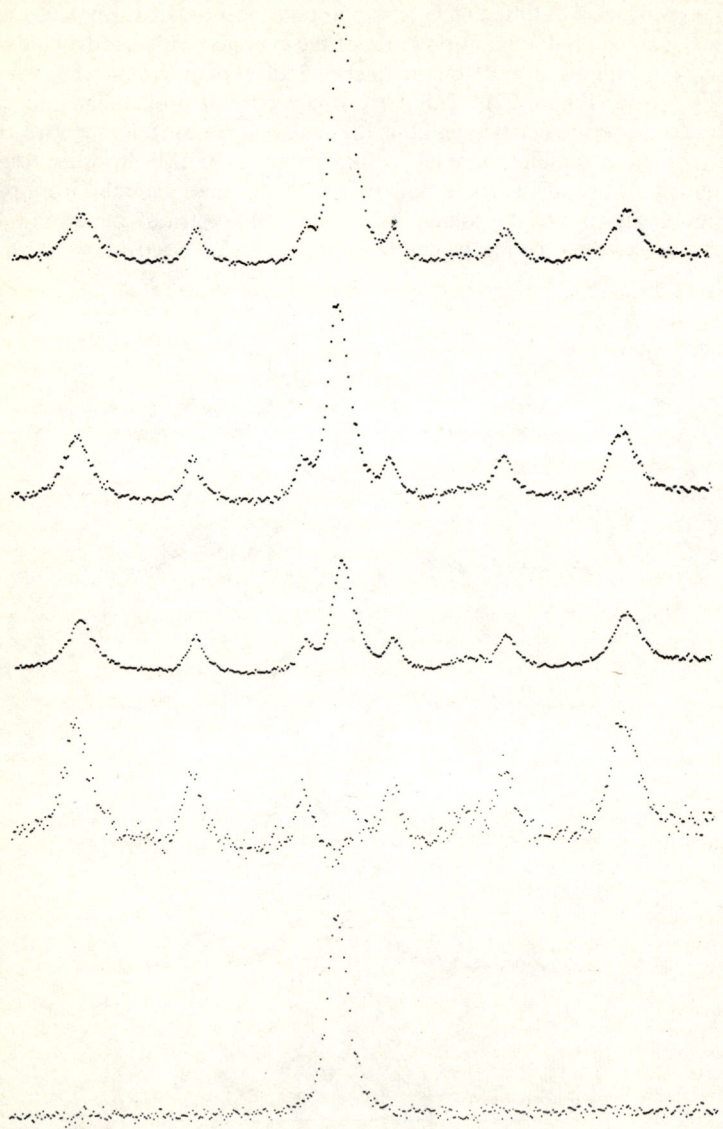

Figure 24 *Spectra for a 360 Å layer of iron on stainless steel. From top to bottom: measured spectra for various spectrometer settings, separated spectrum for the layer 0—375 Å, separated spectrum from 375 Å and inward*
(Reproduced by permission from 'Mossbauer Effect Methodology', Plenum Press, New York, 1974, Volume 9, p. 259)

with CEM spectra of unoxidized simulated fines and to transmission spectra of the Lunar and simulated samples. An excess signal close to zero velocities was observed for the spectrum of the Lunar fines compatible with the presence of superparamagnetic iron in their outermost regions.

^{57}Fe CEMS is also a viable method for following the course of the early stages of solid state reactions which initiate in the surface regions of solids. Such a study has been carried out on the air oxidation of biotite.[57] Increases in the Fe^{3+} content of the surface were clearly observed (Figure 23) which would have gone undetected if transmission methods had been used alone.

10 Depth Selective ^{57}Fe CEMS

As early as 1969 Bonchev et al.[58] experimentally demonstrated the feasibility of obtaining depth-resolved Mössbauer information in ^{119}Sn CEMS by use of a β-ray spectrometer. Following the calculations of Krakowski and Miller described in Section 6, this type of experiment has been extended to ^{57}Fe CEMS.[59-63] Much of this elegant work has been carried out by Bäverstam and co-workers at Stockholm using a magnetic β-ray spectrometer with an energy resolution of 8%.[59] Bäverstam et al. have developed a numerical method based on experimentally determined[64] electron loss curves which enables spectra of specific regions of the surface to be extracted from CEM spectra accumulated at different spectrometer settings. For example in Figure 24 is shown three spectra of 36 nm of iron on a stainless steel substrate taken at differing spectrometer settings.

The top spectrum was recorded with low-energy electrons and consequently has the largest signal from the stainless steel. The bottom two spectra are the separated spectra for the layer 0—37 nm and 37 nm inwards. It can be seen that the method is remarkably effective and it is thought that the ultimate resolution is of the order of 5 nm. Similar data have been obtained on oxidized iron foils.[65]

The application of the depth resolution technique has been limited to date, but careful study of the hyperfine field in magnetic iron indicates a 5% reduction at the surface compared to the bulk value.[66] Minkova et al.[67] have examined a 1 nm thick film of γ-FeOOH and determined that the quadrupole splitting is twice the bulk value.

[57] M. J. Tricker, A. P. Winterbottom, and A. G. Freeman, *J.C.S. Dalton*, 1976, 1289.
[58] Z. W. Bonchev, A. Jordanov, and A. Minkova, *Nucl. Instr. Methods*, 1969, **70**, 36.
[59] U. Bäverstam, C. Bohm, T. Ekdahl, D. Liljequist, and B. Ringström, 'Mössbauer Effect Methodology', ed. I. J. Gruverman, C. W. Seidel, and D. K. Dieterly, Plenum Press, New York, 1974, **9**, 259.
[60] U. Bäverstam, T. Ekdahl, C. Bohm, B. Ringström, V. Stefansson, and D. Liljequist, *Nucl. Ins'r. Methods*, 1974, **115**, 373.
[61] U. Bäverstam, T. Ekdahl, C. Bohm, D. Liljequist, and B. Ringström, *Nucl. Instr. Methods*, 1974, **118**, 313.
[62] J. P. Schunk, J. M. Freidt, and Y. Llabador, *Rev. Phys. Appliqué*, 1975, **10**, 121.
[63] T. Torijana, M. Kigaura, M. Fujioke, and K. Hisatake, *Jap. J. Appl. Physics*, Suppl. 2, 1974, 733.
[64] U. Bäverstam, C. Bohm, B. Ringström, and T. Ekdahl, *Nucl. Instr. Methods*, 1973, **108**, 439.
[65] T. Ekdahl, B. Ringström, and U. Bäverstam, *Univ. Stockholm Inst. Physics Report*, No. 74, 1974, 14.
[66] U. Bäverstam, T. Ekdahl, and B. Ringström, *J. Phys. Radium*, Suppl. C6, 1974, **35**, C6–685.
[67] A. Minkova and J. P. Schunck, *Compt. rend. Acad. bulg. Sci.*, 1975, **28**, 1171.

11 Conclusions

The surface sensitivity of ^{57}Fe CEMS depends on the energies of the internal conversion and Auger electrons and the tailing effects described in Section 8. In practice the outermost few hundreds of nanometres of a natural iron surface can be routinely probed, in a non-destructive fashion, with two-thirds of the signal arising from within *ca.* 50 nm of the surface. If the outermost surface regions are enriched in ^{57}Fe new surface films a few nanometres thick may be studied. In this way, the development of ^{57}Fe CEMS has extended the usefulness of Mössbauer spectroscopy as a method for the study of surfaces. On the other hand, although it is in principle possible to detect a monolayer equivalent of ^{57}Fe, the *in situ* examination of the atoms within the outermost monolayer of a sample is hampered by electrons originating in sub-surface regions as these may obscure the desired signal. Such studies of monolayers are therefore best carried out on inert substrates thus posing sample preparation problems of similar magnitude to those encountered in source or transmission surface studies.

The usefulness and potential of the method has been amply demonstrated by many applications in diverse areas. In many cases novel information, which would have been difficult if not impossible to obtain by other techniques, has been gained. Although the development of the instrumentation and methodology associated with CEMS has reached a stage where many problems may be routinely tackled, further work is needed before the full potential of the method is realized. Studies, conducted over a wide range of temperature, perhaps combined with depth resolution techniques enhanced in sensitivity by use of modern electrostatic electron analysers, would make possible the more unambiguous identification of surface species and allow a fuller Mössbauer spectroscopic examination of the environment of iron atoms in the surface region of solids.

Note added in proof

In the context of low temperature CEMS it is interesting to note that Sawicki *et al.*[68] have recently described a sealed He CH$_4$ proportional counter which operates in a satisfactory manner down to 77 K.

[68] J. A. Sawicki, B. D Sawicka, and J. Stanek, *Nucl. Instr. Methods*, 1976, **138**, 565.

3
The Interplay of Theory and Experiment in the Field of Surface Phenomena on Metals

BY Z. KNOR

1 Introduction

The interaction between theory and experiment in any research field occurs in: (i) the construction of the theoretical model (representing the real system); (ii) the evaluation and the interpretation of the experimental data (particularly when modern sophisticated experimental techniques are being used) and (iii) the final confrontation of the results of both approaches. In this article the first and the last point will be discussed in connection with surface phenomena.

Due to the limited possibilities of contemporary theoretical methods, one is usually forced to work with oversimplified models for a theoretical description of surface processes. On the other hand it is reasonable to include in the model as many characteristic features of the investigated system as possible. These two requirements contradict each other and one has to look for a compromise. Consequently the model becomes less universal and it can only represent the behaviour of a real system in a limited number of processes. This brings additional difficulty, namely the problem of an adequate testing of the model, without going beyond its limit of applicability.

The major problem of comparison of theory and experiment concerns the physical meaning of the quantities, resulting from theoretical calculation and from experiment. The physical meaning of these quantities, which have to be compared with each other, may differ considerably, as will be shown in this article.

2 Bare Metal Surfaces

Theoretical Models of Metal Surfaces.—Model representation of metals is primarily based on *the physical experience, i.e.* on the study of the bulk metal properties, mainly transport and magnetic phenomena. From this point of view the metals are well characterized by the high density of the nearly free valence electrons and the ion core electrons, localized at (and partly screening out) the positive nuclei, arranged in the metal lattice. Two basic models are mainly used: (i) the 'jellium' model, where the positive charge of the ion cores is smeared out into the spatially fixed, homogeneous, positively charged background – jellium – transparent for the nearly free valence electrons; (ii) the ion lattice model, where the localized character of the ion cores is preserved in the form of the periodic potential (or pseudopotential), acting on the nearly free electrons. The term 'nearly free electrons' indicates the delocalized character of the valence electrons inside the metal crystal and does not concern the partial localization of *e.g. d*-electrons.

The applicability of the above mentioned models has been tested also in the field of surface phenomena.[1,2] This extension of the applicability followed the 'classical' route of the solid state physicist, viz. when the one electron description proved to be inadequate, an improvement was looked for particularly in the mutual interactions of the electrons (exchange and correlation effects). Together with this, various theoretical methods have been exploited, leaving, however, the original over-simplified model almost untouched.[3-11] The exchange and correlation terms were introduced in exactly the same form as in the bulk problems, fitting the unknown parameters to the bulk properties.[4,6,7,9]

The jellium model will be discussed first, because it is often used for the theoretical description of surface phenomena.[8-12]

In the infinite jellium (Figure 1) each valence electron moves in the potential V,

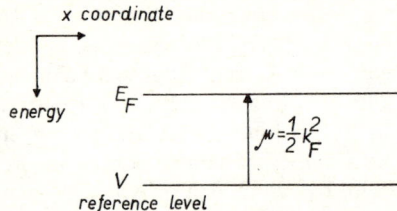

Figure 1 *Energy levels of the electrons inside the infinite jellium (E_F = Fermi energy; μ = chemical potential of the electrons; k_F = bulk Fermi wave number; V = the potential of the electrons in jellium, which is usually for the bulk problems taken as $V = 0$)*

corresponding to the sum of: (i) the electrostatic potential V_{es}, due to the field of the ion core's charge density and the compensating field of the other valence electrons; (ii) the potential V_{xc} of the positively charged holes, surrounding each particular valence electron – *i.e.* the correlation hole due to the Coulomb repulsion among the electrons and the exchange hole due to the Pauli exclusion principle.

$$V = V_{es} + V_{xc} \qquad (1)$$

Since both these potentials (due to their definition) are constant throughout the whole jellium, it represents a completely isotropic medium.

---

[1] C. Herring and M. H. Nichols, *Rev. Modern Phys.*, 1949, **21**, 185.
[2] H. J. Juretschke, in 'The Surface Chemistry of Metals and Semiconductors', ed. H. C. Gatos, Wiley, New York, 1960, p. 37.
[3] J. Bardeen, *Surface Sci.*, 1964, **2**, 381.
[4] J. R. Smith, *Phys. Rev.*, 1969, **181**, 522.
[5] C. B. Duke, *J. Vacuum Sci. Technol.*, 1969, **6**, 152.
[6] N. D. Lang and W. Kohn, *Phys. Rev.*, 1970, **B1**, 4555.
[7] N. D. Lang and W. Kohn, *Phys. Rev.*, 1971, **B3**, 1215.
[8] N. D. Lang, in 'Solid State Physics', ed. F. Seitz, D. Turnbull, and H. Ehrenreich, Academic Press, New York, 1973, **28**, 225.
[9] A. J. Bennett, in 'Critical Revs. in Solid State Sciences', ed. D. E. Schuele and R. W. Hoffman, CRC Press Inc., Chemical Rubber Co., Cleveland, 1974, p. 261.
[10] J. R. Smith, in 'Interactions on Metal Surfaces', ed. R. Gomer, Springer Verlag, Berlin, 1975.
[11] W. Peuckert, *J. Phys.* (C), 1974, **7**, 2221.
[12] F. K. Schulte, *Surface Sci.*, 1976, **55**, 427.

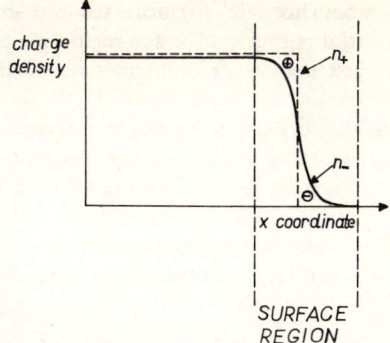

Figure 2 *Electron charge density spreading in the surface of jellium* (n_+ = *positive charge density*, n_- = *charge density of the nearly free electrons*)

When the infinite jellium is cut by a surface plane, where the positive background ends abruptly, the situation changes considerably. Some of the nearly free electrons at the surface of a semi-infinite jellium can leave the surface (because of their large kinetic energy) to some short distance, thus depleting the surface region of the metal and forming an effective dipole layer at the surface (Figure 2).[1,2] The overall potential V_{eff} acting on the nearly free electrons in this semi-infinite jellium then becomes

$$V_{eff} = V_{es} + V_{xc} + V_{dip} \qquad (2)$$

where V_{dip} represents the potential due to the effective dipole layer.

We have no direct experimental information about the potential acting on an electron passing through the potential, or acting on an electron passing through the surface region of a metal (with the exception of the field electron emission, where the high intensity of the electric field deforms the potential in the nearest

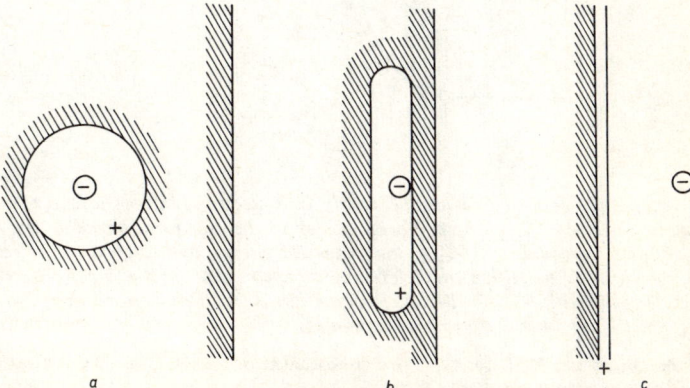

Figure 3 *Exchange and correlation holes* (a) *far from the surface*, (b) *near the surface, and* (c) *when the electron has left the metal*

neighbourhood of the surface – see later Figure 12). Several approximations of the potential curve in the surface region can be found in the literature.[13—16]

When the many-body interactions among the valence electrons are taken into account, the microscopic picture shown in Figure 3 follows:[2] if the electron moves from the volume towards the surface, its exchange and correlation holes, originally spherical, gradually flatten along the plane, parallel to the surface. When the electron finally leaves the surface, its exchange and correlation holes spread over all the surface so that there is effectively a negative point charge in front of a practically 'infinite' plane, with homogeneously spread positive charge. This is exactly the situation which can be described by the so-called image potential V_{im}:

$$V_{\text{im}} = \frac{1}{4\pi\varepsilon_0} \frac{e^2}{4x} \qquad (3)$$

where x is the distance measured from the surface plane and the other symbols have the usual meaning. The theoretical potential curves in the surface region [13—16] have to match the potential V_{eff} inside the metal and the image potential V_{im} outside the metal.

Semi-infinite jellium can be treated as a one-dimensional problem and thus easily reproduced graphically (Figure 4). The one-dimensional character of a jellium model is one of the features which makes it attractive to the theoretician.

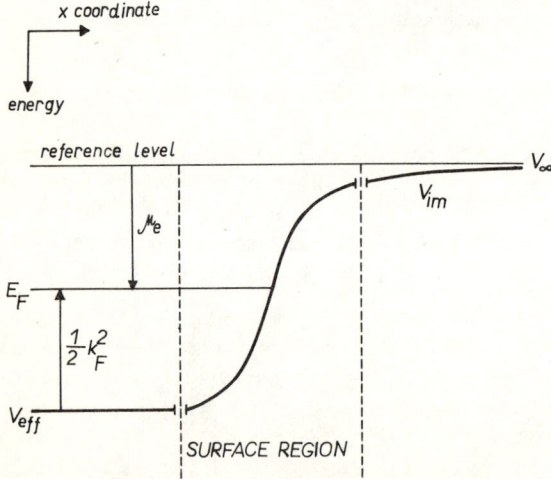

Figure 4 *Potential energy curve for the nearly free electrons in the surface region of the semi-infinite jellium (V_{eff} = the potential energy of the electrons, including the exchange and correlation contribution; V_∞ = the potential energy of the electron at rest in the infinity, which is by definition the reference level, common for the electrons both inside and outside the metal; V_{im} = the image potential; μ_{e} = the electrochemical potential of the electrons; k_{F} = bulk Fermi wave number)*

[13] E. Ya. Zandberg and N. I. Ionov, 'Poverchnostnaya ionizatsiya' (Surface Ionization), Publishing House NAUKA, Moscow, 1969.
[14] L. N. Dobrezow, 'Elektronen und Ionen-Emission', VEB Verlag Te. '-nik, Berlin, 1954.
[15] P. H. Cutler and J. C. Davis, *Surface Sci.*, 1964, **1**, 194.
[16] D. M. Newns, *J. Chem. Phys.*, 1969, **50**, 4572.

The work function ϕ of a semi-infinite jellium at ground potential is defined as

$$\phi = -\mu_e \quad (4)$$

where μ_e is the electrochemical potential of the valence electrons in the metal. *The convention that the investigated metal is at ground potential will be used throughout the text.*

The work function is an important quantity, characterizing the metal surface. It plays a decisive role in almost all surface phenomena (electron emission, surface chemical interactions, *etc.*), because it is the measure of the binding energy of the outermost electrons. The work function will, therefore, be used in this article for a detailed discussion of various models for bare metal surfaces and subsequently for comparison of the theoretical and experimental results.

The jellium model proved to be an adequate description for the alkali metals, where the free electron behaviour has been well known.[6,7] However, it does not seem to be an appropriate model for metals with a higher density of 'nearly free' electrons,[6,7] although the general tendency of the work function values is acceptable.[7,10] This fact probably does not prove the applicability of a particular theoretical approach, because it might be the consequence of the correlation between the work function and the surface density of ion cores (together with the atomic ionization potentials), empirically found by Sachtler.[17] In the case of surface energy the inapplicability of the jellium model for the high electron density metals is already indisputable, since the theoretical values of the surface energy are negative in certain cases.[6]

In the preceding paragraphs the isotropic jellium has been discussed; however, experimentally the anisotropy of various surface properties has been well established. This effect can be shown in the work function values for individual crystallographic planes of various metals (Table 1). The experimental data in Table 1 were prepared in such a way that the field emission results have been separated and the results of all other methods have been averaged. The procedure of averaging is justified for the single plane values, estimated either by one method (*e.g.* field emission, photoemission) or by several methods (with the exception of the field emission technique). The work function values for polycrystalline surfaces cannot be correctly interpreted without a detailed knowledge, both of the surface structure and of the different types of experimental technique used to measure work functions.[19] Consequently they have an illustrative value only, and in Table 1 the arithmetic mean values from the published data are given. The work function values, ϕ_{hkl}, of the alkali metals were estimated on samples, deposited on substrates (W, Ta) having the orientation (110), (100), (112), and (111). The real surface plane of the alkali-metal layers has been seldom checked.

Further experimental evidence for the strong influence of surface atomic structure on the work function values can be found in the literature. Plummer and Rhodin[67] have observed that the deposition of additional tungsten atoms onto a given crystallographic plane of tungsten can: (i) considerably decrease the work function of the densely packed plane (110); (ii) negligibly influence or even increase the work function of the less densely populated planes (112) and (111). These results have

[17] W. M. H. Sachtler, *Z. Electrochem.*, 1955, **59**, 119.
[67] E. W. Plummer and T. N. Rhodin, *Appl. Phys. Letters*, 1967, **11**, 194.

Table 1 Experimental values of the work function[a]

Metal	Lattice	ϕ_{110}/eV	ϕ_{100}/eV	ϕ_{112}/eV	ϕ_{111}/eV	$\phi_{polycryst}$/eV	Method[b]	Note[c]	Reference
Li	bcc	3.06	3.06	2.93	2.89		RP	deposited on (110) (100) (112) (111) planes of W, SV	18
						2.33	PE, SI, X	M_3	19—21
Na	bcc		2.38			2.46	PE, SI, X	M_4	20—23
							X	M (taken from 27)	27
		2.45	2.3	2.3	2.26		FE	deposited on (110) (100) (112) (111) planes of W, SV	24
		2.46					CPD	deposited on (110) plane of Ta, SV	25
		2.8					RP	deposited on (110) plane of W, SV	26
K	bcc	2.55	2.12		2.08		FE	deposited on (110) (100) (111) planes of W	28
						2.21 $s = 0.06$	PE, SI, X	M_6	19—21, 23
							X	M (taken from 27)	27
Rb	bcc		2.11			2.15	X	SV	21
							X	M (taken from 27)	27
Cs	bcc	2.06	1.82			1.89	PE, CPD, X	M_3	19, 21, 29
							CPD	deposited on (110) (100) planes of W, SV	29

Nb	bcc	2.1	1.65	2.0	1.95		FE	deposited on (110)(100)(112)(111) planes of W, SV	30
		2.18	1.76				FE	deposited on (110)(100) planes of W, SV	31
		4.66					PE	SV	32
			4.99				PE	SV	19
			4.66				FERP, T	M_2	33
			3.87				FE	SV	33
						4.30	CPD, PE, T	M_4	19, 20
Ta	bcc	4.76	4.15	4.0			PE, RP, T	M_3	19, 34, 35
							T	SV	35
						4.29	CPD, T, SI	M_8	19
						$s = 0.26$	PE		

[a] s is the standard deviation. [b] Experimental methods: RP – retarding potential technique, SI – surface ionization technique, PE – photoemission, T – thermionic technique, FERP – field emission retarding potential technique, X is used where the method has not been stated explicitly. [c] M_i is the arithmetic mean value, estimated from i-values (together with s); SV denotes single experimental value.

[18] V. K. Medvedev and T. P. Smereka, *Fiz. Tverd. Tela*, 1974, **16**, 1599.
[19] J. C. Rivière, in 'Solid State Surface Science', ed. M. Green, M. Dekker, New York, 1969, Vol. 1, p. 179.
[20] D. E. Eastman, *Phys. Rev.*, 1970, **B2**, 1.
[21] K. F. Wojciechowski, *Surface Sci*, 1976, **55**, 246.
[22] R. I. Whitefield and J. J. Brady, *Phys. Rev. Letters*, 1971, **26**, 380.
[23] Th. G. J. van Oirschot, Dissertation, Leiden, 1943.
[24] E. V. Klimenko and V. K. Medvedev, *Fiz. Tverd. Tela*, 1968, **10**, 1986.
[25] D. L. Fehrs and R. E. Stickney, *Surface Sci.*, 1971, **24**, 309.
[26] V. K. Medvedev, A. G. Naumovets, and A. G. Fedorus, *Fiz. Tverd. Tela*, 1970, **12**, 375.
[27] H. E. Albrecht, *Phys. Stat. Sol.*, 1971, **A6**, 135.
[28] A. P. Ovtshinnikov, *Fiz. Tverd. Tela*, 1967, **9**, 628.
[29] T. I. Lee, B. H. Blott, and B. J. Hopkins, *Appl. Phys. Letters*, 1967, **11**, 361.
[30] V. M. Gavriliuk, A. G. Naumovets, and A. G. Fedorus, *Zhur. Exp. i Teor. Fiz.*, 1966, **51**, 1332.
[31] L. W. Swanson and R. W. Strayer, *J. Chem. Phys.*, 1968, **48**, 2421.
[32] R. M. Oman and J. A. Dillon, *Surface Sci.*, 1964, **2**, 227.
[33] R. W. Strayer, W. Mackie, and L. W. Swanson, *Surface Sci.*, 1953, **34**, 225.
[34] D. L. Fehrs and R. E. Stickney, *Surface Sci.*, 1967, **8**, 267.
[35] O. D. Protopopov, E. V. Mikheieva, B. N. Sheinberg, and G. N. Shuppe, *Fiz. Tverd. Tela*. 1966, **8**, 1140.

Table 1—*Contd.*

Metal	Lattice	ϕ_{110}/eV	ϕ_{100}/eV	ϕ_{112}/eV	ϕ_{111}/eV	$\phi_{polycryst}$/eV	Method[b]	Note[c]	Reference
Mo	bcc	5.12		4.47	4.32		PE, T	M_3	35—37
			4.53				PE, T	M_4	35, 37, 40
		5.1	4.30		4.0		FE	SV	41
			4.80		4.67		FE	SV	39
						4.30	CPD, PE, T	M_6	19, 20
W	bcc	5.21 $s = 0.15$					CPD, T, SI, FERP	M_{13}	19, 29, 33, 37, 42—44, 47, 51—55
			4.61 $s = 0.05$				CPD, T, FERP	M_{10}	19, 29, 33, 35, 37, 41—44, 47, 49
				4.80			CPD, T	M_3	40, 43, 57
					4.45 $s = 0.08$		CPD, T, SI	M_{11}	19, 33, 35, 37, 41—44, 47, 51—52
		5.95 $s = 0.82$					FERP	M_{19}	19, 33, 38, 40, 41, 44—46, 50, 56
			4.83 $s = 0.21$				FE	M_8	33, 38, 44—46, 48—50
				4.92 $s = 0.16$			FE	M_{10}	38, 44—46, 48, 50, 56
					4.49 $s = 0.16$		FE	M_{11}	19, 33, 38, 40, 44, 49
						4.57 $s = 0.03$	T, RP, PE SI	M_9	19
Ni	fcc	5.04	5.22				PE	SV	58
			5.27		5.35		FERP, PE	M_4	19, 33, 58
						5.06	PE, T, CPD	M_7	22, 23
Ir	fcc	5.42	5.67		5.76		FERP	SV	33
					5.79		FE	SV	59
						4.57	PE	SV	22

Pt	fcc		5.84		5.93	5.50	FE	SV	60
							T, PE	M$_2$	22
							CPD, PE	M$_2$	61, 62
							CPD, PE	M$_6$	33, 61, 62
							FERP, T		
Cu	fcc	4.7	4.84	4.53	5.15	4.57	PE	SV	62
							PE, CPD	M$_3$	61, 62
							RP, CPD	M$_7$	22, 23
							PE, T		

[36] S. Berge, P. O. Gartland, and B. J. Slagsvold, *Surface Sci.*, 1974, **43**, 275.
[37] F. M. Garner, F. E. Giraud, W. L. Bloeck, and E. A. Coomes, *Surface Sci.*, 1971, **26**, 605.
[38] E. Chrzanowski, *Acta Phys. Polon.*, 1973, **A44**, 717.
[39] T. Oguri and I. Kanomata, *J. Phys. Soc. Japan*, 1964, **19**, 1310.
[40] S. Usami, *J. Phys. Soc. Japan*, 1971, **30**, 1076.
[41] B. J. Hopkins, S. Usami, in "The Structure and Chemistry of Solid Surfaces", ed. G. Somorjai, Wiley, New York, 1969.
[42] E. P. Sytaya, M. I. Smorodinova, and N. I. Imangulova, *Fiz. Tverd. Tela*, 1962, **4**, 1016.
[43] U. V. Azizov and G. N. Shuppe, *Fiz. Tverd. Tela*, 1965, **7**, 1970.
[44] C. J. Todd and T. N. Rhodin, *Surface Sci.*, 1973, **36**, 353.
[45] S. Hellwig and J. H. Block, *Z. phys. Chem. (Frankfurt)*, 1973, **83**, 269.
[46] T. Engel and R. Gomer, *J. Chem. Phys.*, 1969, **50**, 2428.
[47] L. van Someren, *Surface Sci.*, 1970, **20**, 221.
[48] T. V. Vorburger, D. Penn, and E. W. Plummer, *Surface Sci.*, 1975, **48**, 417.
[49] A. A. Holscher, *J. Chem. Phys.*, 1964, **41**, 579.
[50] B. G. Smirnov and G. N. Shuppe, *Zhur. Tech. Fiz.*, 1952, **12**, 973.
[51] V. M. Sultanov, *Radiotekh. i Elektronika*, 1964, **9**, 317.
[52] F. L. Reynolds, *J. Chem. Phys.*, 1963, **39**, 1107.
[53] E. Ya. Zandberg, *Zhur. Tekh. Fiz.*, 1960, **30**, 1215.
[54] V. K. Burkhanova and N. A. Gorbatyi, *Izvest. Akad. Nauk S.S.S.R., Ser. fiz.*, 1971, **35**, 291.
[55] G. N. Shuppe, E. P. Sytaya, and P. M. Kadyrov, *Izvest. Akad. Nauk S.S.S.R., Ser. fiz.*, 1956, **20**, 1142.
[56] C. J. Workowski, *Acta Phys. Polon.*, 1972, **A42**, 9.
[57] M. J. Dresser and D. E. Hudson, *Phys. Rev.*, 1965, **137A**, 673.
[58] B. G. Baker, B. B. Johnson, and G. L. C. Maire, *Surface Sci.*, 1971, **24**, 572.
[59] B. E. Nieuwenhuys, D. Th. Meijer, and W. M. H. Sachtler, *Surface Sci.*, 1973, **40**, 125.
[60] B. E. Nieuwenhuys and W. M. H. Sachtler, *Surface Sci.*, 1973, **34**, 317.
[61] T. A. Delchar, *Surface Sci.*, 1971, **27**, 11.
[62] P. O. Gartland, S. Berge, and B. J. Slagsvold, *Phys. Rev. Letters*, 1972, **28**, 738.

Table 1—Contd.

Metal	Lattice	ϕ_{110}/eV	ϕ_{100}/eV	ϕ_{112}/eV	ϕ_{111}/eV	$\phi_{\text{polycryst}}$/eV	Method[b]	Note[c]	Reference
Ag	fcc	4.52	4.64		4.74		PE	SV	63
							PE	SV	64
						4.34 $s = 0.19$	RP, CPD, PE	M_{12}	22, 23
Al	fcc	4.06	4.20		4.26		PE	SV	65
						4.17	RP, CPD	M_3	22
Re	hcp	ϕ_{1000} 5.53	ϕ_{1100} 5.15	ϕ_{2113} 4.34			T	SV	66
						5.01	RP, T, SI	M_6	22

[63] A. W. Dweydari and Ch. B. Mee, *Phys. Stat. Sol.*, 1975, **A27**, 223.
[64] A. W. Dweydari and Ch. B. Mee, *Phys. Stat. Sol.*, 1973, **A17**, 247.
[65] R. M. Eastment and Ch. B. Mee, *J. Phys. (F)*, 1973, **3**, 1738.
[66] O. D. Protopopov, E. V. Mikheieva, B. N. Sheinberg, and G. N. Shuppe, *Doklady Akad. Nauk Uzbek. S.S.R.*, 1966, 21.

been confirmed by Besocke and Wagner [68,76] on the macroscopic (110) plane of tungsten, using AES and LEED techniques.

All these experiments show that the higher the surface density of atoms (ion cores) the higher the work function [69] (see also Table 1).

Interesting information about the work function can be obtained from a comparison of FEM (field emission microscope) and FIM (field ion microscope) images of tungsten [70–72] and rhodium [73] surfaces. The central plane [(110) plane of tungsten, (100) plane of rhodium] as seen in FIM has a much smaller diameter than the relevant high work function area on the FEM image.[74] One can conclude that an electron leaving the terraces on the vicinals of these planes (approximating the 'hemispherical' shape of the tip in the FIM or FEM, respectively) 'feels' as high-index planes ($h'k'l'$) only those planes where the atomic steps are less than about three atoms wide. When they are wider, this region behaves as a 'patchy' surface, composed of the low index planes (hkl) (Figure 5). Therefore the general correlation between the work function and the surface density of atoms,[69] based on the theory of Smoluchowski,[75] can be applied to the low index planes only.[36,74]

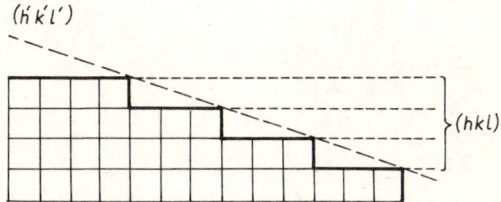

Figure 5 *The surrounding of the central plane on the apex of the field emission tip*

The characteristic features of bare metal surfaces, which a theoretical model for the work function should represent, at least qualitatively, can thus be formulated: (i) high density of 'nearly free' electrons, which changes rapidly in the surface region from the bulk to the zero value; (ii) anisotropy of different crystallographic planes. We have seen in the preceding paragraphs how important the latter feature is, although it is completely neglected in the theories based on the semi-infinite jellium model. The difference between the work function values of different crystallographic planes of one metal is approximately the same as the range of work function values of all transition and noble metals (Table 1). It seems therefore reasonable to change the original model rather than to introduce the anisotropy as only a small perturbation.[6,7]

The potential V_{eff} acting on the nearly free electrons inside the metal [equation (2)] has two isotropic terms (V_{es}, V_{xc}) and the term corresponding to the effective

[68] K. Besocke and H. Wagner, *Phys. Rev.*, 1953, **B8**, 4597.
[69] J. Müller, *Surface Sci.*, 1974, **42**, 525.
[70] G. Ehrlich and F. G. Hudda, *J. Chem. Phys.*, 1962, **36**, 3233.
[71] T. H. George and P. M. Stier, *J. Chem. Phys.*, 1962, **37**, 1935.
[72] V. Feldman and R. Gomer, *J. Appl. Phys.*, 1966, **37**, 2380.
[73] G. Ehrlich, H. Heyne, and C. F. Kirk, in 'The Structure and Chemistry of Solid Surfaces', ed. G. A. Somorjai, Wiley, New York, 1969.
[74] A. van Oostrom, *J. Chem. Phys.*, 1967, **47**, 761.
[75] R. Smoluchowski, *Phys. Rev.*, 1941, **60**, 661.

double layer at the surface, V_{dip}, which is certainly sensitive to the surface conditions and thus responsible for the experimentally observed surface anisotropy.

The anisotropy of the work function can be 'built' into the model in several ways: (i) the surface of jellium can be modified: instead of a plane boundary, the corrugated model can be used,[2,8,75] where the density of 'pyramids' simulates the density of the surface ion cores and the 'nearly free' electrons smooth the 'stepped' character of the surface (smoothing effect [75]) (Figure 6); (ii) the Wigner–Seitz polyhedral cells can be modified in the surface so that a similar smoothing effect takes place [75,77] [the smearing out of the ion core charge over each cell results in

Table 2 *Theoretical values of the work function of bulk crystals*

Metal	ϕ_{110}/eV	ϕ_{100}/eV	ϕ_{111}/eV	ϕ_{jellium}/eV	Reference
Li				2.19	79
				2.38	4
	2.40	2.40	2.30	3.37	7
	2.479	2.310			78
Na				2.15	79
				2.93	4
	3.10	2.75	2.65	3.06	7
	2.464	2.355			78
		2.38			27
K				2.20	79
				2.76	4
	2.75	2.4	2.35	2.74	7
	2.278	2.214			78
		2.03			27
Rb				2.20	79
				2.71	4
	2.65	2.35	2.30	2.63	7
	2.230	2.177			78
		2.01			27
Cs				2.15	79
				2.64	4
	2.60	2.30	2.20	2.49	7
	1.929	1.881			78
		1.90			27
Nb				3.81	4
Ta				4.12	4
Mo				4.3	4
W				4.5	4
	6.70				77
Ir				5.3	4
Pt			4.54		27
Cu				4.4	4
Ag				4.3	4
Al	3.65	4.20	4.05	4.19	7
				4.25	4

[76] K. Besocke and H. Wagner, *Surface Sci.*, 1975, **53**, 351.
[77] J. R. Smith, *Phys. Rev. Letters*, 1970, **25**, 1023.
[78] P. K. Rawlings and H. Reis, *Surface Sci.*, 1973, **36**, 580.
[79] E. Wigner and J. Bardeen, *Phys. Rev.*, 1938, **48**, 84.

Figure 6 *The corrugated model of the jellium surface*

the corrugated uniform background model, as in (i)]; (iii) the crystal lattice is simulated by 'lattice jellium' (the potential acting on the nearly free electrons is constructed in such a way that these electrons can move freely only along the lines, connecting the nearest neighbours, leaving 'dangling' bonds on the surface, which contribute to the dipole layer);[78] (iv) the ion lattice model is constructed, where the lattice consists of spheres with appropriately chosen potential.[6,7,27]

Comparison of Theoretical and Experimental Results.—Some of the theoretical values of the work functions are collected in Table 2. When comparing the theoretical (Table 2) and experimental (Table 1) results one has to bear in mind that the authors fitted their theoretical results to the old experimental values, quoted in their papers. Consequently only the general tendencies in the series of metals or crystallographic planes can be seriously judged. The results of this comparison (Figures 7—9) are not very encouraging, reflecting probably oversimplification of

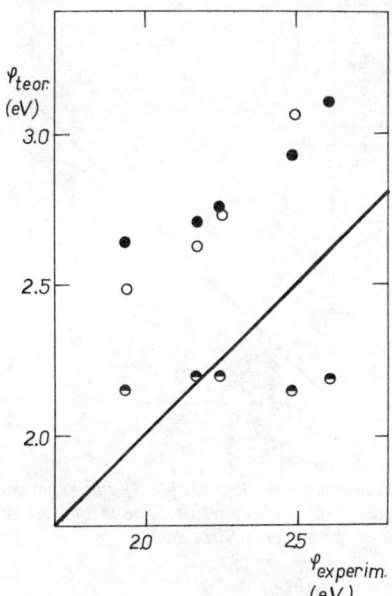

Figure 7 *The correlation between theoretical and experimental values of the work function of alkali metals. The theoretical values were calculated for jellium (● [4], ○ [7], ◉ [79]) (Table 2), the experimental values were obtained on polycrystalline samples (Table 1). The prime line corresponds to the correct theoretical representation of the experimental data*

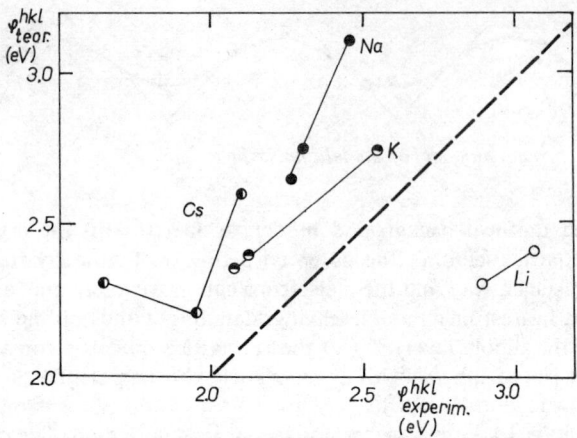

Figure 8 *The correlation between theoretical[7] and experimental values of the work function (Table 1) for individual crystallographic planes of alkali metals. The dashed prime line corresponds to the correct theoretical representation of the experimental data*

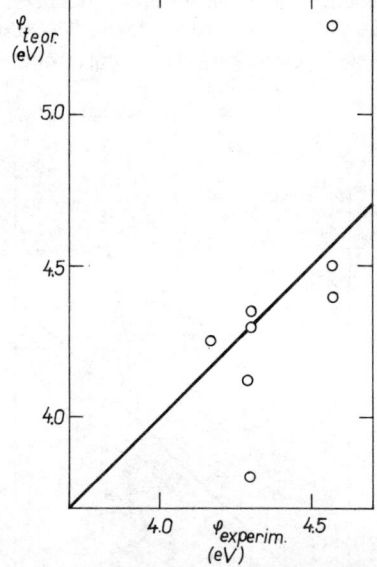

Figure 9 *The correlation between theoretical (Table 2) and experimental (Table 1) values of the work function of transition and noble metals. The prime line corresponds to the correct theoretical representation of the experimental data*

the models, rough approximations of the theoretical methods, and the influence of some uncontrolled experimental factors.

Most of the published work function values have been obtained on polycrystalline samples. The theoretical values, particularly those ones obtained for the isotropic jellium model, were therefore often fitted to these values. However, it is well known that $\phi_{polycryst}$ is strongly influenced by the surface topography, which might considerably differ in a series of samples of the same metal, depending on the experimental conditions of the surface preparation. Without an exact knowledge of the surface atomic structure, the various experimental techniques (used for the work function measurements on macroscopic samples) will supply us with some unknown 'mean' values over the whole investigated surface (*e.g.* in the case of all the emission techniques the average is strongly weighted in favour of the low work function areas [19] – see also Figure 10). One could assume that there is a limited number of the most stable (low index) planes, exposed to the gas phase in the surface of a polycrystalline sample. In this case one would expect some correlation between the values ϕ_{hkl} for the most densely populated planes (having lowest surface energy per atom and highest work function) and the polycrystalline values $\phi_{polycryst}$ (Figure 10). However, the real situation is not so simple and thus one can conclude that it is inappropriate to use polycrystalline values for any

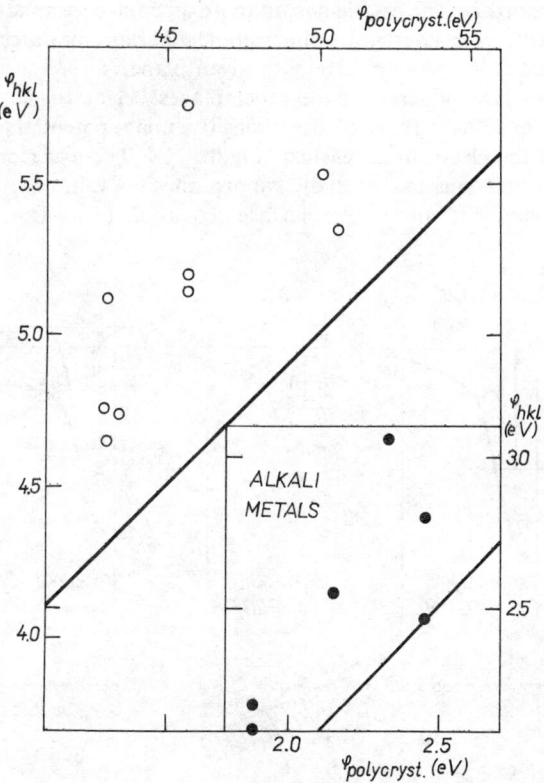

Figure 10 *The correlation between the experimental work function values of the most densely populated planes (having highest probability of occurrence in the surface, because of their lowest surface energy per atom) and the values obtained on polycrystalline samples. The prime line corresponds to the case when only the most densely populated plane occurs on the surface*

theoretical considerations,[4,7,80] because these values do not characterize unambiguously a particular metal.

When comparing the theoretical and experimental values of the work function, one has to bear in mind the exact physical meaning of the experimental value. In the introductory considerations the potential energy diagrams have been discussed for a semi-infinite metal with one plane interface (Figure 4). The problem arises if the metal is limited by several crystallographic planes, having different work functions. The Fermi level is unique for the whole crystal and its position with respect to V_∞ (Figure 4) (the electrochemical potential of the valence electrons) is given by the interplay of all the surface planes

$$-\mu_e = \phi_m = \sum_i \theta_i \phi_i \qquad (5)$$

where ϕ_m is the effective mean value of the work function and θ_i is the fraction of the surface having the work function ϕ_i. The potential of an electron in the surface region depends on its position with respect to individual crystallographic planes and their surface areas. The problem is no more a one-dimensional one. However, it is possible to construct several one-dimensional diagrams for different paths of the electron. The electron, passing through a given plane, 'feels' initially the field of that plane only. The influence of the other planes begins to play a role farther from the surface and finally the structure insensitive image potential region (common for any path of the electron) is reached (Figure 11). The experimental value ϕ_m, obtained by any non-emission technique, approaches the value ϕ_{hkl} for a particular plane, if this plane prevails in the surface [equation (5)]. The only technique

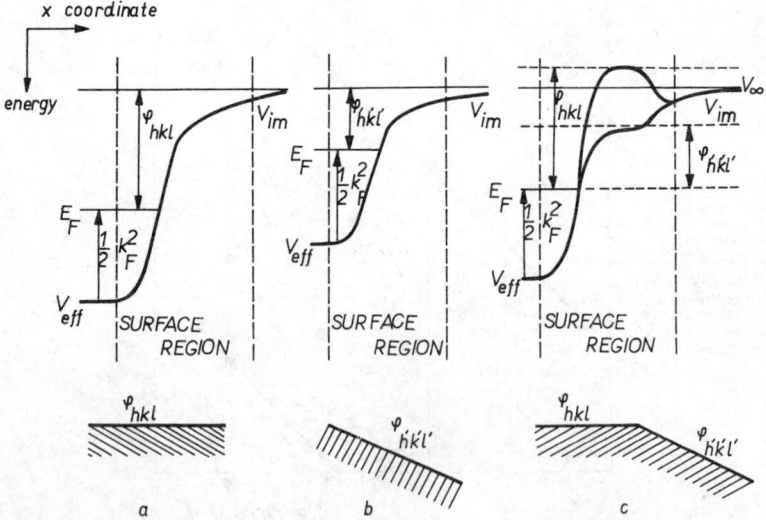

Figure 11 *Potential energy diagrams for the electron leaving the metal crystal, limited by:* (i) *one crystallographic plane, either* (hkl) *(having the work function ϕ_{hkl})* (a), *or* (h',k',l') *(having the work function $\phi_{h'k'l'}$)* (b); (ii) *by both of these planes simultaneously* (c)

[80] V. Heine and C. H. Hodges, *J. Phys.*, 1972, **C5**, 225.

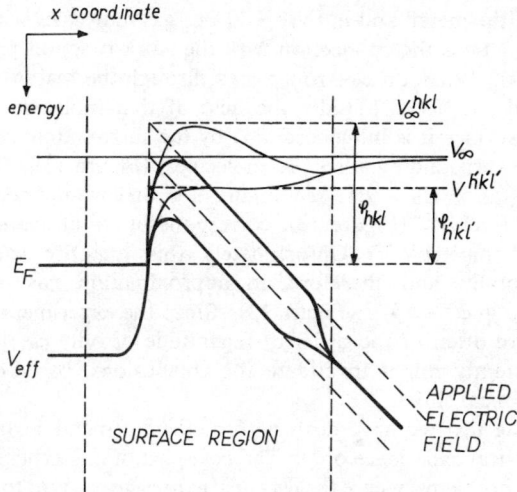

Figure 12 *Potential energy diagram for the electron leaving the metal in the FEM.* (V_{∞}^{hkl} and $V_{\infty}^{h'k'l'}$ are the effective levels of zero potential energy for the electron in the very neighbourhood of the plane (hkl) or (h'k'l'), respectively

providing us with some kind of information about the potential in the surface region, is the FEM (Figure 12). Any other technique, even when applied to a single crystal plane, supplies us with the work function value ϕ_m [equation (5)], which only approaches the theoretical value for a semi-infinite metal, for which the work function is a true physical constant, characteristic for one plane only.

The preceding considerations can be extended to the photoelectron spectroscopy of adsorbed species in connection with the challenging problem of the common reference level for both gas and adsorbed molecules.[81] The energy levels in the photoelectron spectra of the adsorbed species are measured with respect to the

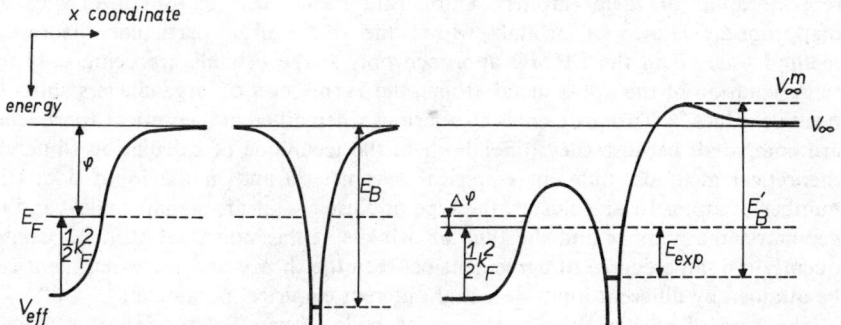

Figure 13 *Potential energy diagram for the electron leaving the surface through the region occupied by one adsorbed molecule* ($\Delta\phi$ = the WFC caused by the adsorbed molecule); E_{exp} = the experimental value of the binding energy of the electron in the adsorbed molecule; E_B and E'_B are the ionization potential of the gas and adsorbed molecules

[81] A. F. Carley, R. W. Joyner, and M. W. Roberts, *Chem. Phys. Letters*, 1974, **27**, 580.

Fermi level E_F of the metal, and in the spectra of gas molecules with respect to the vacuum level V_∞. Thus the connection with the work function problem becomes obvious (Figure 4). When an electron passes through the region occupied by the adsorbed molecule, it 'feels' initially the field of that molecule only and being farther from the surface it is influenced also by the surrounding region (either the bare metal or the surrounding adsorbed molecules) (Figure 13). For the chemical shift estimation ($\Delta E = E_B - E_B'$; see Figure 13) one would need to know the correct reference level V_∞^m (Figure 13), corresponding to effective 'vacuum level' of the adsorbed molecule.[82] Unfortunately this quantity cannot be determined experimentally and therefore an approximation has to be used, e.g. $\Delta E = E_B - (E_{exp} + \phi + \Delta\phi)$[82] (Figure 13). Since the experimental values of the chemical shifts are often of the order of magnitude of only ca. 1 eV, the above-mentioned uncertainty might invalidate the conclusions, based on the absolute values of the chemical shifts.

In the preceding text we have dealt with models of metal surfaces, completely based on the physical experience only. The next part of this article will be devoted to chemisorption problems, where the *chemical experience* proved to be fruitful. The surface chemical interactions, in complete analogy with the ordinary chemical reactions, usually concern the nearest neighbours only. Consequently one is tempted to construct the model of a metal crystal in the form of an individual metal atom or a small cluster of atoms. This seems to be an attractive approach, where quantum chemistry methods can be successfully applied. This type of model, which is probably appropriate for the chemical problems, need not, necessarily, be so for the physical problems of bulk crystals.[83] Nevertheless, these models have also been tested in the latter field and the results of Fermi level position have been compared with the bulk crystal values. One has to be careful in doing so and to bear in mind the preceding considerations, concerning the work function. The 'work function' values of clusters listed in Table 3 are mostly too high and they usually do not approach monotonically the bulk values even for very large clusters (Figure 14).

Another important problem is the number and the type of orbitals used for representation of metal atoms. Unrealistic results can be obtained with an inappropriate choice of orbitals within the frame of a particular theoretical method – e.g. if in the EHMO approach only *s*-type orbitals are being used for representation of the noble metal atoms, the Fermi level of large clusters shifts to positive values.[96] This problem is also serious when different theoretical approaches are compared, because they differ both in the technique of calculation (different theoretical methods, different empirical parameters) and in the input data (the number of atoms in one cluster, the type of cluster – linear, planar, bulk – and its geometry, the number and the type of orbitals at the individual atoms). Consequently the same degree of agreement between the theory and the experiment can be attained by different input data and different empirical parameters.

The type of cluster (linear, planar, or bulk clusters) is an important factor seriously influencing the theoretical results. The energy spectra of one-, two-, and three-dimensional clusters may considerably differ from each other – e.g. in the

[82] H. D. Hagstrum, *Surface Sci.*, 1976, **54**, 197.
[83] Y. W. Tsang and L. M. Falicov, *J. Phys.*, 1976, **C9**, 51.
[84] A. Madhukar and B. Bell, *Phys. Rev. Letters*, 1975, **34**, 1631.
[85] C. Kittel, 'Introduction to Solid State Physics', Wiley, New York, 1971.

Table 3 Theoretical Fermi level positions for the metal atom clusters with respect to the vacuum level

Metal	Ionization potential of the isolated atom/eV [85]	Number of atoms in cluster	Type of cluster	Type of orbitals	Fermi level/eV	Method	Ref.
Na	5.14	2	linear	$2p\ 3s$	−4.92	CNDO	86
W	7.98	9	3 dim.	$5p\ 5d\ 6s$	−8.47	EHMO	87
		12			−8.32		
		9	3 dim.	$5p\ 5d\ 6s$	−8.9	EHMO	88
Ni	7.63	8	3 dim.	$3d\ 4s$	−7.69	EHMO	89
		9			−7.66		
		10			−7.72		
		13			−7.64		
		54	3 dim.	$3d\ 4s\ 4p$	−7.13	EHMO	90
		9	3 dim.	$3p\ 3d\ 4s$	−9.1	EHMO	88
		8	linear	$3d\ 4s\ 4p$	−7.83	EHMO	92
		1		$3d\ 4s\ 4p$	−7.1	CNDO	91
		2	linear		−6.8		
		3			−6.53		
		4			−6.26		
		5			−6.26		
		6			−6.26		
		7			−5.99		
		3	planar		−7.10		
		5			−6.53		
		6			−7.89		
		7			−5.71		
		6	3 dim.		−7.62		
		9			−7.35		
		10			−6.72		
		13*			−6.26		
		13*			−7.35		
		8	3 dim.	$3d\ 4s\ 4p$	−4.62	SCF Xα SW	93
		13			−5.60		
		2			−4.86	SCF Xα SW	94
		4	3 dim.		−4.83		
		5			−4.83		
		6	3 dim.		−4.76	SCF Xα SW	95
Pd	8.33	2	linear	$4d\ 5s\ 5p$	−9.22	CNDO	86
		13	3 dim.		−5.98	SCF Xα SW	93
Pt	8.96	13	3 dim.		−6.66	SCF Xα SW	93

* Different geometrical structure.

[86] R. C. Baetzold, *J. Chem. Phys.*, 1971, **55**, 4355.
[87] L. W. Anders, R. S. Hansen, and L. S. Bartell, *J. Chem. Phys.*, 1973, **59**, 5277.
[88] A. B. Anderson and R. Hoffmann, *J. Chem. Phys.*, 1974, **61**, 4545.
[89] D. J. M. Fassaert, H. Verbeek, and A. van der Avoird, *Surface Sci.*, 1972, **29**, 501.
[90] D. J. M. Fassaert and A. van der Avoird, *Surface Sci.*, 1976 **55**, 291.
[91] G. Blyholder, *Surface Sci.*, 1974, **42**, 249.
[92] H. Itoh, *J. Phys.* (*F*), 1974, **4**, 1930.
[93] R. P. Messmer, S. K. Knudson, K. H. Johnson, J. B. Diamond, and C. Y. Yong, *Phys. Rev.*, 1976, **13**, 1396.
[94] I. P. Batra and O. Robaux, *J. Vacuum Sci. Technol.*, 1975, **12**, 242.
[95] N. Rösch and D. Menzel, *Chem. Phys.*, 1976, **13**, 243.

Table 3—*Contd.*

Metal	Ionization potential of the isolated atom/eV	Number of atoms in cluster	Type of cluster	Type of orbitals	Fermi level/eV	Method	Ref.
Cu	7.72	8	linear	3d 4s 4p	−8.61	EHMO	92
		13	3 dim.	4s 4p	−6.79	EHMO	96
		43			−5.12		
		79			−5.10		
		135			−4.70		
		2	linear	3d 4s 4p	−7.21	CNDO	86
		1		3d 4s	−4.76	SCF Xα SW	93
		2	linear		−4.90		
		8	3 dim.	3d 4s 4p	−5.59		
		13			−5.39		
		6	3 dim.	3d 4s 4p	−5.58	SCF Xα SW	95
Ag	7.57	13	3 dim.	5s 5p	−7.30	EHMO	96
		43			$\begin{cases} -6.47\dagger \\ -6.54\dagger \end{cases}$		
		43		4d 5s 5p	−6.35		
		79		5s 5p	$\begin{cases} -6.60\dagger \\ -6.55\dagger \end{cases}$		
		135			$\begin{cases} -6.27\dagger \\ -6.54\dagger \end{cases}$		
		2	linear	4d 5s 5p	−8.19	EHMO	97
		3			−6.81		
		4			−7.60		
		5			−6.63		
		6			−7.41		
		7			−5.81		
		8			−7.44		
		9			−6.91		
		10			−7.31		
		11			−6.91		
		12			−7.0		
		13			−6.91		
		14			−7.22		
		15			−6.91		
		16			−7.0		
		17			−6.91		
		18			−7.0		
		19			−6.81		
		20			−7.0		
		4	3 dim.	4d 5s 5p	−6.06	EHMO	97
		5			−6.81		
		6			−6.81		
		8			−7.66		
		9			−5.94		
		14			−6.50		
		18			−5.31		
		2	linear	4d 5s 5p	−7.23	CNDO	86

† Different orbital parameters.

[96] E. R. Davidson and S. C. Fain, *J. Vacuum Sci. Technol.*, 1976, **13**, 209.
[97] R. C. Baetzold, *J. Chem. Phys.*, 1971, **55**, 4363.

Table 3—Contd.

Metal	Ionization potential of the isolated atom/eV	Number of atoms in cluster	Type of cluster	Type of orbitals	Fermi level/eV	Method	Ref.
Ag		2	linear	4d 5s 5p	−7.23	CNDO	97
		3			−4.20		
		4			−6.85		
		5			−4.29		
		6			−6.94		
		7			−4.24		
		8			−6.04		
		6	3 dim.	4d 5s 5p	−5.03	SCF Xα SW	95
Au	9.22	13	3 dim.	6s 6p	−8.53	EHMO	96
		43			$\begin{cases} -7.31\dagger \\ -7.99\dagger \end{cases}$		
		43		5d 6s 6p	$\begin{cases} -6.86\dagger \\ -7.15\dagger \end{cases}$		
		79		6s 6p	$\begin{cases} -7.35\dagger \\ -8.00\dagger \end{cases}$		
		135			$\begin{cases} -6.89\dagger \\ -7.95\dagger \end{cases}$		
		2	linear	5d 6s 6p	−7.67	CNDO	86

type of the highest occupied orbitals.[91] This would, of course, change the 'chemical' properties of the 'surface' atoms of the cluster. Consequently three-dimensional models should be used, because only in this case it is possible to compare physically

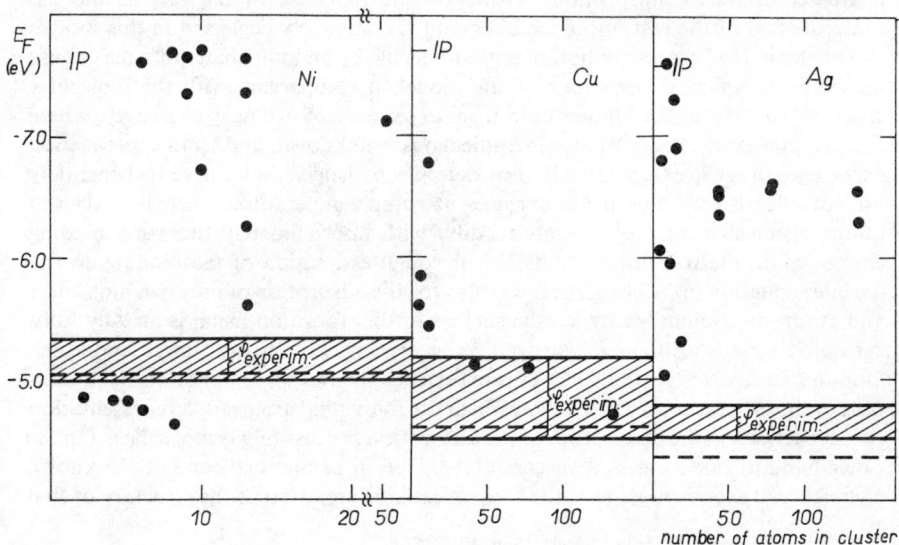

Figure 14 *The Fermi levels of the three-dimensional clusters (IP = ionization potential of the isolated atom; ϕ_{exp} denotes the range of the experimental work function values of different low index crystallographic planes; the dashed lines represent the experimental work function values of the polycrystalline samples*

analogous experimental and theoretical quantities. The inadequacy of one- and two-dimensional models for a description of the surface phenomena has already been shown in the case of the accommodation coefficient, where qualitatively different results from the ones observed on three-dimensional models have been obtained.[98] However, these oversimplified models have persisted in theoretical literature until recently, mainly because of their computational simplicity.

The statement, mentioned in one of the preceding paragraphs, concerning the inadequacy of the 'chemical' model (cluster model) for the description of the bulk metal properties probably also holds in the reciprocal way, *viz.* that the 'physical' model (the bulk crystal model, having ion lattice and delocalized valence electrons only) need not be the optimal one for solution of the chemical problems.[83, 84]

3 Chemisorption of Gas Molecules on Transition Metal Surfaces

Model Representation of the Gas–Metal System.—In this part the interaction of the gas molecule and the metal surface will be dealt with. Again, one has to decide the degree of simplification required in the theoretical representation of both of them.

Chemisorption of gas molecules results usually in the formation of surface compounds, resembling to some extent ordinary chemical compounds. Consequently the strongest interaction occurs predominantly among the nearest neighbours only. Obviously the most simplified model for this case would be the interaction of the two individual atoms: one representing the metal, the other one representing the gas molecule. This model would qualitatively show the localized character of the chemisorption. However, the influence of the rest of the gas molecule and of the rest of the metal crystal is completely neglected in this model.

The theoretical representation of a gas molecule by an individual atom may cause qualitatively different behaviour of the model in comparison with the molecular species. This statement follows both from experience of ordinary chemistry, where the exceptional reactivity of atoms (radicals) is well known, and from experimental experience in surface science itself. For example, hydrogen molecules experimentally do not adsorb on noble metal surfaces at room temperature, whereas hydrogen atoms chemisorb on these metals readily, with approximately the same binding energy as on the transition metals [99] [for rough estimation of the binding energy see later equation (6)]. The same holds also for the adsorption of nitrogen molecules and atoms on iridium.[100] When the surface of the transition metal is already fully saturated by adsorption of hydrogen molecules, it can still trap large additional amounts of hydrogen atoms.[101] This experience throws some doubt on the usefulness of those theoretical approaches, which use individual atoms for a representation of gas particles in the description of the adsorption process of gas molecules. On the other hand, if one's aim is a successful description of the final state of the known chemisorbed atomic layer (*e.g.* the theoretical interpretation of the structure of that

[98] G. Ehrlich, *Ann. Rev. Phys. Chem.*, 1966, **17**, 295.
[99] G. Ehrlich, in 'Proceedings of the Third International Congress on Catalysis', ed. W. M. H. Sachtler, G. C. A. Schuit, and P. Zwietering, North Holland, Amsterdam, 1965, Vol. 1, p. 113.
[100] Z. Knor and K. Kuchynka, unpublished results.
[101] V. Ponec, Z. Knor, and S. Černý, *J. Catalysis*, 1965, **4**, 485.

layer), when one ignores the problem of the initial interaction, then it is possible to study the system composed of atomic species only.

The necessity of working with molecules in theoretical considerations is a logical extension of the arguments presented in the preceding part, viz. that neither the metal nor the gas particle should be represented by an oversimplified model, which does not take account of (at least in a qualitative way) the basic properties of the real participant of the interaction. Basic properties are those properties which are substantial (characteristic) for the investigated effect (process), and need not be all the characteristic properties of the investigated system (e.g. high density of the nearly free electrons and the ion core lattice are the basic properties, sufficient for a discussion of the work function problem).

In the forthcoming text mainly transition metals are dealt with, because they exhibit the most pronounced chemisorption activity. Remembering the preceding considerations, one can ask what are the basic properties of the transition-metal surfaces which have to be included in the theoretical model for chemisorption of gas molecules.

It seems reasonable, at least in the present state of knowledge, to include in the model as many characteristic features of the real system (empirical input data) as possible, even allowing for the imprecise nature of the computational procedure. This unfortunately contradicts the common attitude of the theoretician, where he usually prefers to work with oversimplified models and to exploit sophisticated theoretical methods and higher order approximations. However, in these methods the empirical contribution is always included, but in a less obvious way, which is often less understandable in terms of the initial model; e.g. the empirical estimation of the diagonal and off-diagonal matrix elements in the EHMO and CNDO approaches,[86,88,89,102] the semi-empirical estimation of the exchange and correlation terms,[7,103] the empirical limits for the pseudopotential (r_c) in the density functional approach,[7,104,105] the semiempirical estimation of the averaged α values,[106–108] and the problem of non-overlapping or overlapping spheres in the cluster formation in the SCF Xα SW method,[95,109] etc.

When a gas molecule interacts chemically with a transition-metal surface, the perturbation of the metal is localized to the nearest neighbourhood of the adsorption complex. Therefore one is tempted, in analogy with ordinary chemistry, to use localized orbitals [110–112] to describe these processes. The question then arises, whether one can think of localized orbitals on transition-metal surfaces. Ample evidence for this kind of description of the surface interactions can be found in recent literature, e.g. the field ionization of inert gas atoms at metal surfaces was qualitatively described successfully by means of the overlap of the localized 'atomic-

---

[102] G. Blyholder, *J. Chem. Phys.*, 1975, **62**, 3193.
[103] D. W. Bullett, *J. Phys.*, 1975, **C8**, 2695.
[104] S. C. Ying, J. R. Smith, and W. Kohn, *Phys. Rev.*, 1975, **B11**, 1483.
[105] H. B. Huntington, L. A. Turk, and W. W. White, *Surface Sci.*, 1975, **48**, 187.
[106] I. R. Schrieffer, *J. Vacuum Sci. Technol.*, 1976, **13**, 335.
[107] K. H. Johnson and F. C. Smith, *Phys. Rev.*, 1972, **B5**, 831.
[108] J. C. Slater and K. H. Johnson, *Phys. Rev.*, 1972, **B5**, 844.
[109] K. H. Johnson and R. P. Messmer, *J. Vacuum Sci. Technol.*, 1974, **11**, 236.
[110] Z. Knor, *Adv. Catalysis*, 1972, **22**, 51.
[111] G. C. Bond, *Faraday Soc. Discuss.*, 1966, **41**, 200.
[112] D. A. Dowden, *J. Res. Inst. Catalysis, Hokkaido Univ.*, 1966, **4**, 1.

like' surface metal orbitals with the molecular (atomic) orbitals of the gas particles (the formation of the surface complex), either in 'one-step-ionization'[113] or in connection with field adsorbed particles.[114] The formation of a surface complex in the field ionization process has been verified by atom probe hole experiments.[115,116] This interpretation of the FIM images, which explained regional brightness of the fcc metals, alternating visibility of the rows of atoms in the fcc and hcp metal surfaces,[113] was based on the assumption that the surface metal orbitals at the very moment of the interaction with a gas particle are pointed towards the nearest missing neighbours.[113] This assumption has been later supported by recent work,[117,118] and also by a theoretical paper on the asphericity of the d-electron clouds in the transition metal surfaces[119] [the only exception was the hcp basal plane (0001), where surprisingly only two lobes were reported[119] instead of three lobes, corresponding to its trigonal symmetry]. Additional evidence, supporting the description of the surface interactions in terms of localized bond formation came from: ion neutralization spectroscopy (INS);[120] electron stimulated desorption (ESD)[121] (where the overlap of the gas particle orbitals with the spatially directed localized surface orbitals, as in FIM,[113] has been suggested[121] for the explanation of the spatial anisotropy of the ESD); angular anisotropy of the photoemission.[122–127] The localized orbitals (localized at the surface atoms) can be regarded as the 'chemical representation' of the surface states common in solid state physics.[128,129] These orbitals are probably hybridized on the bare metal surface; however, the interaction with a gas particle removes the original hybridization and the process can be described in the first approximation by means of 'atomic-like' orbitals.[110,130,131] This has been recognized recently in theoretical papers.[83,84,106]

CFSO–BEBO Approach.—Besides a number of theoretical papers, based in principle on the above model,[87–90,102,131] an interesting empirical procedure – CFSO–BEBO (Crystal Field Surface Orbitals Bond Energy Bond Order) method –

[113] Z. Knor and E. W. Müller, *Surface Sci.*, 1968, **10**, 21.
[114] E. W. Müller, 'Proc. 2nd Internat. Conf. on Solid Surfaces, 1974', *Japan J. Appl. Phys.*, 1974, Suppl. 2, part 2, p. 1.
[115] E. W. Müller, S. B. McLane, and J. A. Panitz, *Surface Sci.*, 1969, **17**, 430.
[116] E. W. Müller, *Quart. Rev.*, 1969, **23**, 133.
[117] F. Forstmann, W. Berndt, and P. Büttner, *Phys. Rev. Letters*, 1973, **30**, 17; J. E. Demuth, D. W. Jepsen, and P. M. Marcus, *Phys. Rev. Letters*, 1974, **32**, 1182.
[118] B. A. Hutchins, T. N. Rhodin, and J. E. Demuth, *Surface Sci.*, 1976, **54**, 419.
[119] M. C. Desjonquères and F. Cyrot-Lackmann, *J. Chem. Phys.*, 1976, **64**, 3707; *J. Phys. (F)*, 1975, **5**, 1368; *Surface Sci.*, 1975, **53**, 429.
[120] H. D. Hagstrum and G. E. Becker, *J. Chem. Phys.*, 1971, **54**, 1015.
[121] T. F. Madey, J. J. Czyzewski, and J. T. Yates, *Surface Sci.*, 1975, **49**, 465.
[122] J. W. Gadzuk, 'Proc. 2nd Internat. Conf. on Solid Surfaces 1974'; *Japan J. Appl. Phys.*, 1974, Suppl. 2, Pt. 2, p. 851.
[123] B. Feuerbacher, *Surface Sci.*, 1975, **47**, 115.
[124] J. Waclawski, T. V. Vorburger, and R. J. Stein, *J. Vacuum Sci. Technol.*, 1975, **12**, 301.
[125] T. B. Grimley and G. F. Bernasconi, *J. Phys. (C)*, 1975, **8**, 2423.
[126] A. Liebsch, *Phys. Rev.*, 1976, **B13**, 544.
[127] B. Feuerbacher and R. F. Willis, *Phys. Rev. Letters*, 1976, **36**, 1339.
[128] C. R. Brundle, *Surface Sci.*, 1975, **48**, 99; *J. Vacuum Sci. Technol.*, 1974, **11**, 212.
[129] M. Tomášek and P. Mikušík, *Phys. Rev.*, 1973, **B8**, 410.
[130] Z. Knor, *J. Vacuum Sci. Technol.*, 1971, **8**, 5.
[131] H. Kölbel and K. D. Tillmetz, *J. Catalysis*, 1974, **34**, 307.

has been suggested [132-137] for the description of the adsorption process in terms of the potential energy (E) versus reaction co-ordinate (or bond order – n) curves. The CFSO–BEBO method starts with the energy balance consideration: $E = \sum_i E_i - \sum_j E_j$, where E_i and E_j are the energies of the bonds which were split and newly formed in the investigated process. The dependence of the individual energy terms on distances between the interacting particles (reaction co-ordinate or bond order) follows from the simplified BEBO [138] correlations for the gas molecules: [132,134,136] $E_i = E_i^s n_i$, where E_i^s is the single-bond energy and n_i is the bond order related to the bond length R_i according to $R_i = R_i^s - 0.6 \log n_i$ (where R_i^s is the bond length of the single bond).[139] Since the BEBO correlations are available for a limited number of bonds only, the interrelation among various bond orders involved in the investigated process has to be found. The metal is in this method represented by 'diatomic molecules' with free valencies in the directions towards nearest missing neighbours, corresponding to a given crystallographic plane. The sequence of the bond splitting and new bond formation (in the gas and 'metal' molecule, in the surface complex) is speculatively postulated together with the geometry of the initial, transitional, and final state arrangements.[132-136] From these assumptions the desired interrelation of the bond orders follows so that the final equation of the potential energy for a given process usually rests on a single BEBO correlation only.[132,134,135] Moreover, this method depends on the reliability of the simplified version of Pauling's 'mixing rules' [139] (without taking into account the electronegativity corrections,[134,135,137] because the single-bond energies necessary for the BEBO correlations are seldom available in the literature. The complete potential energy curve is constructed with respect to the postulated reaction mechanism and the system configuration by discontinuous combination of intervals (postulated reaction steps) with monotonically increasing or decreasing functions (polynoms constructed from BEBO correlations), which intersect at the interval boundaries.[134,135,137] These intersections are considered in the CFSO–BEBO formalism to be maxima (related to the activation energies) and minima (related to the adsorption heats) and their co-ordinates directly follow from the postulated system configurations in the individual reaction steps. The absolute values of these extremes are uncertain [134,137] and thus the predictive power of this method is limited. Moreover, the only published potential energy versus reaction co-ordinate curve has an unrealistic shape (compare refs. 134 and 136 with ref. 140). All other results were published in the form of the potential energy versus bond order curves.

The Localized Free Electron Interplay Model.—It has been shown in the preceding considerations that the surface atoms preserve to some extent their individual character and that their chemical interaction with gas molecules concerns nearest

---

[132] W. H. Weinberg, *J. Catalysis*, 1973, **28**, 459.
[133] W. H. Weinberg, W. H. Lambert, R. M. Comrie, and J. W. Linnett, *Surface Sci.*, 1972, **30**, 299.
[134] W. H. Weinberg and R. P. Merrill, *Surface Sci.*, 1972, **33**, 493.
[135] W. H. Weinberg and R. P. Merrill, *Surface Sci.*, 1973, **39**, 206.
[136] W. H. Weinberg, *J. Vacuum Sci. Technol.*, 1973, **10**, 89.
[137] W. H. Weinberg, H. A. Deans, and R. P. Merrill, *Surface Sci.*, 1974, **11**, 312.
[138] H. S. Johnston, 'Gas Phase Reaction Rate Theory', Ronald Press, New York, 1966.
[139] L. Pauling, 'The Nature of the Chemical Bond', Cornell University Press, Ithaca, 1960.
[140] A. Clark, 'The Chemisorption Bond', Academic Press, New York, 1974.

neighbourhood only. This is also the consequence of the short screening length in metals – due to the large number of nearly free electrons. The surface chemical interaction thus implicitly contains both local (chemical) and delocalized (solid state) aspects.[141-143] Accordingly, it has proved useful [122] already in the previous theories to divide the valence electrons into two separate groups: localized d-type electrons and delocalized s- (sp-) type electrons. One can then consider the role of these two types of electron.

The interpretation of the surface chemical interactions in terms of localized electrons only neglects completely the high density of the nearly free electrons, which play a dominant role in the transport phenomena (electrical conductivity, Hall effect, etc.). This, however, contradicts the experimental evidence that chemisorption considerably influences the transport phenomena.[110] Additionally the selectivity of a given metal in heterogeneously catalysed reactions is usually lower than the selectivity of the reaction homogeneously catalysed by the complex ion of the same metal.[144] This effect may help us to understand the role of the two types of electrons in surface chemical reactions. Qualitatively the role of the nearly free s- (sp-) electrons can be understood in terms of Berlin's formulation of the Hellman–Feynman theorem.[145] The space around two atoms, bound together, can be divided into the binding region (between the two ion cores, where the valence electron density screens their mutual repulsion) and the antibinding region (outside the two ion cores). When the charge density in the antibinding region becomes non-zero, the bond between the two ion cores is weakened. This effect can qualitatively explain the 'dissolution' of those bonds in the adsorbed molecules, which are 'immersed' into the charge density of the nearly free electrons. Consequently the adsorbed molecule is perturbed not only by the localized bond(s) with the surface atom(s) (in the same way as it is perturbed by a homogeneous catalyst–single complex ion), but also by the screening effects of the high density of nearly free electrons (screening of the unfavourable distribution of the charges in the surface complex, polarization effects). This high density of nearly free electrons is available in the surface of the metal only. Thus the larger number of different products in heterogeneous reactions can be understood.

The role of the high density of nearly free electrons in the surface chemical interactions has been overemphasized by some authors,[104,146] who attempted to describe the chemisorption of gases on transition metals in terms of delocalized (nearly free) electrons only (using the jellium model of the metal), neglecting completely the empirically well known close connection between high chemisorption activity and the presence of partly occupied d-orbitals (exhibiting localized properties) in the transition metals.

Hence it is possible to construct the following model for the chemical interaction of a gas molecule with a transition metal surface. From this point of view the characteristic features of transition-metal surfaces, which have to be represented by the model are: (i) the presence of the partly occupied d-orbitals, localized at

[141] J. R. Schrieffer and P. Soven, *Physics Today*, 1975, **28**, 24.
[142] D. A. Shirley, *J. Vacuum Sci. Technol.*, 1975, **12**, 280.
[143] B. Bell and A. Madhukar, *J. Vacuum Sci. Technol.*, 1976, **13**, 345.
[144] R. Ugo, *Catalysis Rev. Sci. Eng.*, 1975, **11**, 225.
[145] B. M. Deb, *Rev. Mod. Phys.*, 1973, **45**, 22.
[146] N. D. Lang and A. R. Williams, *Phys. Rev. Letters*, 1975, **34**, 531.

the surface atoms, and (ii) the presence of high density of the nearly free s- (sp-) electrons. When a gas molecule approaches the transition-metal surface, the first electronic interaction probably involves the nearly free electrons and the outermost electrons of the gas molecule. The measure of the binding energy of the metal electrons is the work function (around 5 eV for transition metals, see Table 1) and for the outermost electrons of a gas molecule it is the ionization potential (10–15 eV for simple molecules – Table 4). Therefore one can expect that the

Table 4 *The ionization potentials of gas molecules and atoms* [147]

Gas particle	Ionization potential/eV
N_2	15.6
H_2	15.4
N	14.5
CO	14.0
CO_2	13.8
H	13.6
O	13.6
CH_4	13.0
O_2	12.1

nearly free metal electrons are initially 'blown' away due to the exchange and correlation forces, 'uncovering' thus the 'atomic-like' d-orbitals (in other words the surface hybridization is destroyed by the approaching molecule). The gas molecule itself is then perturbed and localized bonds are formed. As a consequence of the changes in the electron distribution in the original molecule, the nearly free electron density relaxes then to the new distribution (the trapped molecule being 'immersed' into the Fermi sea of then early free electrons) (Figure 15). The presence of partly occupied localized d-orbitals represents thus the necessary condition for trapping of the gas molecules. It is interesting to note here that from theoretical calculations [89, 148] emerges the minor role of the d-electrons in hydrogen atom adsorption, whereas it seems to be important for hydrogen molecule adsorption.[149]

From the point of view of heterogeneous catalysis the geometric configuration of the surface complex might be important for the additional activation (weakening) of the bonds in the original molecule [by 'immersing' the particular bond(s) into the Fermi sea of electrons]. The gas molecule has to stay on the surface long enough to enable: (i) the relaxation of the nearly free electron gas to the new density distribution; (ii) the interaction of the 'immersed' and thus weakened bond(s) of the reactant molecule with the neighbouring surface complexes or with another approaching molecule; (iii) the complete dissociation of the 'immersed' bond(s) so that the fragments could migrate apart, or react with other species, or recombine again, eventually in the isomeric form. The first of the above-listed processes is probably fast enough,[142] but any one of the others might be rate-determining.

The desired selectivity of a given catalyst can be attained if the particular bond(s) are: (i) perturbed by the formation of the localized surface bond(s) (in complete

---

[147] R. W. Kiser, 'Introduction to Mass Spectrometry and its Applications', Prentice Hall, Englewood Cliffs, New York, 1965.
[148] G. Blyholder, *J. Chem. Phys.*, 1975, **62**, 3193.
[149] R. C. Baetzold, *Surface Sci.*, 1975, **51**, 1.

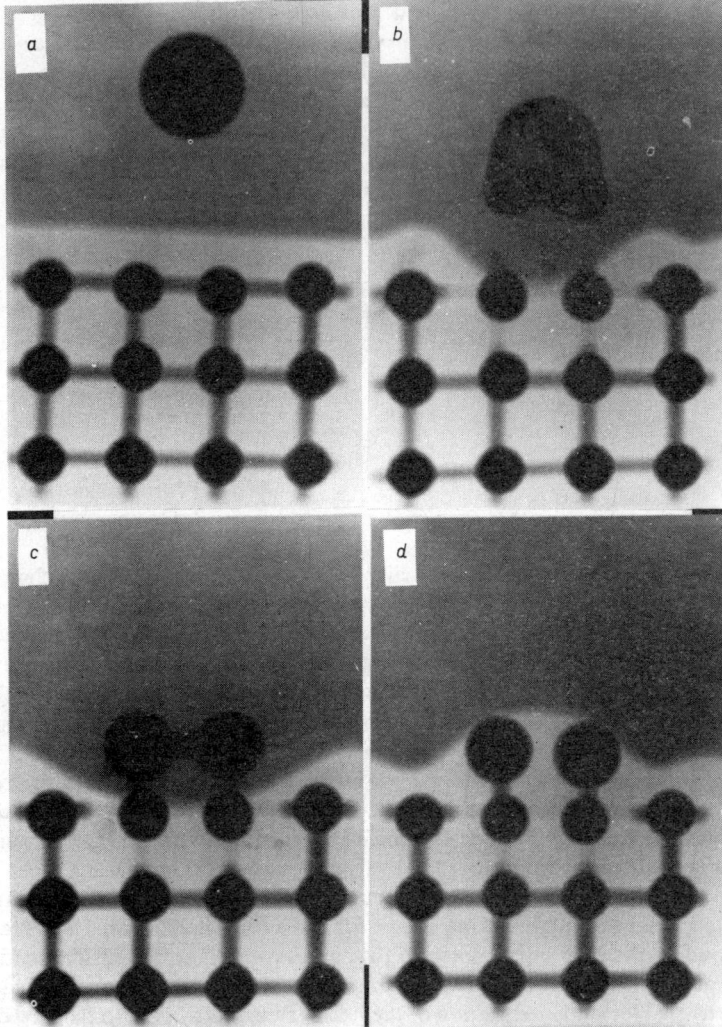

Figure 15 *A gas molecule, approaching the transition-metal surface* (a—d)

analogy to homogeneous catalysis); (ii) 'immersed' into the appropriate electron density region. The 'immersing' of the surface complex into the nearly free electron density (the embedding) can be theoretically simulated by applying to the surface complex an electric field, corresponding to a given charge density distribution in the metal surface.

Many catalytic effects can be understood in terms of this model: the crystallographic anisotropy of the catalytic properties (resulting from the different orientation and/or number of localized orbitals and from different nearly free electron density distributions on various crystallographic planes); the specific properties of 'stepped' surfaces [150,151] (where the smoothing effect might cause partial 'uncovering' of the localized orbitals together with increased nearly free electron density at the edges); the role of constituents in the bimetallic and alloy catalysts (as trapping centres and regulators of the nearly free electron density distribution); catalysis by EDA complexes [152] (cooperation of the 'free' electron reservoir with the proper d-electrons on the central ion); catalysis by graphite, loaded with alkali metals [152] (where the reactant molecules are held in the region of appropriate 'free' electron density by intermolecular forces between the graphite layers).

The transition metals are the only metals where the dual role of electrons (the formation of the localized bonds and the screening effects) can be realized by the two types of electron: partly occupied localized d-orbitals and nearly free s- (sp-) electrons. Thus their high activity in chemisorption and heterogeneous catalysis can be understood. This conclusion agrees well with the exceptional role of the frontier orbitals (highest occupied and lowest unoccupied orbitals), known from the field of ordinary chemical reactions.[153] Only the transition metals have the frontier orbitals of both d- and s- (sp-) type, because their Fermi level crosses the overlapping d- and s- (sp-) bands.

The Comparison of Theoretical and Experimental Results.—The comparison of theoretical and experimental results in the field of chemisorption normally concerns chemisorption heats, spectroscopic results (particularly u.v. photoelectron spectroscopy), and work function changes.

Selected experimental values of chemisorption heats on well-defined surfaces are presented in Table 5, from which the following conclusions can be drawn: (i) no calorimetric heats of adsorption on single-crystal planes are so far available; (ii) all the chemisorption heats on single-crystal planes have been obtained either as isosteric adsorption heats (by means of the Clausius–Clapeyron equation, based on the assumption of complete reversibility of the investigated process, which is, however, seldom fulfilled in chemisorption) or as activation energies of desorption (which need not be equal to the heat of adsorption); (iii) several adsorption states (or more exactly several desorption states) have often been postulated.[199,200] If one state prevails, having considerably higher activation energy of desorption than the other states, then it can be interpreted (in the case of biatomic molecules) as an atomic species and directly compared with theoretical results. When the population of different states is comparable, the interpretation is no more straightforward, particularly if the desorption heats differ considerably (see Table 5, systems H_2—W, CO—Mo, CO—W, CO—Pt). Several calorimetric results obtained on polycrystalline samples have been included in Table 5. They have in the context of this article illustrative value only, because the surface atomic structure of these samples is unknown.

[150] S. L. Bernasek, W. L. Sıckhaus, and G. A. Somorjai, *Phys. Rev. Letters*, 1973, **30**, 1202.
[151] G. A. Somorjai, *Surface Sci.*, 1973, **34**, 156.
[152] K. Tamaru, *Adv. Catalysis*, 1969, **20**, 327; *Amer. Scientist*, 1972, **60**, 474.
[153] R. G. Pearson, *Chem. Eng. News*, 1970, Sept., 66.
[199] M. Smutek, S. Černý, and F. Buzek, *Adv. Catalysis*, 1975, **24**, 343.
[200] D. A. King, *Surface Sci.*, 1975, **47**, 384.

Table 5 Experimental values of the adsorption heats[a]

Gas	Metal	Crystallographic plane	Adsorption heats/kJ mol⁻¹		Activation energy of desorption	Reference
			Calorimetric	Isosteric		
H_2	Nb	100		110.8		154
	Mo	110			2 adsorption states: 142, 117	155
		100			3 adsorption states: 113, 84, 67	155
		polycryst. film	125			156
	W	110			2 adsorption states: 134, 113	155
		100			2 adsorption states: 134, 109	155
					134, 105	157
				146		158
		211			2 adsorption states: 146, 67	159
				167		158
		111			4 adsorption states: 130, 105, 79, 50	160
					155, 125, 92, 58	161
				155		158
		polycryst. wire	263			162
	Ni	110		89.9		163
					123.3	164
		100		96.1		163
					96.1	165
		111		96.1		163
					94.9	166
		polycryst. film	66.9—71.1			167
		polycryst. wire	234			168
	Pd	110		102		169
		111		86.9		169

[a] The values of only one strongly bound state are listed, if the other states exhibit considerably lower adsorption heats; otherwise all the published states are included.

[154] D. J. Hagen and E. E. Donaldson, *Surface Sci.*, 1974, **45**, 61.
[155] M. Mahnig and L. D. Schmidt, *Z. phys. Chem.*, 1972, **80**, 71.
[156] S. Černý, *Surface Sci.*, 1975, **50**, 253.
[157] T. E. Madey and J. T. Yates, 'Structure et Propriètes des Surfaces des Solides', Editions du CNRS, Paris 1970, No. 187, p. 155.
[158] M. Domke, G. Jähnig, and M. Drechsler, *Surface Sci.*, 1974, **42**, 389.
[159] R. R. Rye, B. D. Barford, and P. G. Cartier, *J. Chem. Phys.*, 1973, **59**, 1693.
[160] T. E. Madey, *Surface Sci.*, 1972, **29**, 571.
[161] P. W. Tamm and L. D. Schmidt, *J. Chem. Phys.*, 1971, **54**, 4775.
[162] D. D. Eley and P. R. Norton, *Proc. Roy. Soc.*, 1970, **A314**, 301.
[163] K. Christmann, O. Schober, G. Ertl, and H. Neumann, *J. Chem. Phys.*, 1974, **60**, 4528.
[164] G. Ertl and D. Küppers, *Ber. Bunsengesellschaft phys. Chem.*, 1971, **75**, 1017.
[165] J Lapoujoulade and K. S. Neil, *Surface Sci.*, 1973, **35**, 288.
[166] J. Lapoujoulade and K. S. Neil, *J. Chem. Phys.*, 1972, **57**, 3535.
[167] G. Wedler and G. Fisch, *Ber. Bunsengesellschaft phys. Chem.*, 1972, **76**, 1160.
[168] D. D. Eley and R. P. Norton, *Proc. Roy. Soc.*, 1970, **A314**, 319.
[169] H. Conrad, G. Ertl, and F. E. Latta, *Surface Sci.*, 1974, **41**, 435.

Table 5—Contd.

Gas	Metal	Crystallographic plane	Adsorption heats/kJ mol^{-1}		Activation energy of desorption	Reference
			Calorimetric	Isosteric		
H_2	Pt	111			39.7	170
		polycryst. wire	103.3			171
		polycryst. film	87.8			172
O_2	Mo	100			489	173
		polycryst. sample	719			174
	W	100			568	173
		110			577	173
					1028	175
		polycryst. sample			585	176
		polycryst. film	811			174
	Pd	110		334		177
		100		251		177
		111		209—251		178
		polycryst. film	280			174
	Ag	100			167	179
N_2	Mo	110			339	155
		100			364	155
	W	110			330	155
					330	180
		100			326	155
					334	181
					307	182
		polycryst. wire	385			183
		polycryst. wire	433			184
	Fe	100			230	185
CO	Nb	110			289	186
	Mo	100			5 adsorption states: 337, 318 308, 288, 270	187
					4 adsorption states: 368, 318, 226, 71	188

[170] K. Christmann, G. Ertl, and T. Pignet, *Surface Sci.*, 1976, **54**, 365.
[171] R. P. Norton and P. J. Richards, *Surface Sci.*, 1974, **44**, 129.
[172] S. Černý, M. Smutek, and F. Buzek, *J. Catal.*, 1975, **38**, 245.
[173] N. P. Vasko, Yu. G. Ptushinskii, and B. A. Chuikov, *Surface Sci.*, 1969, **14**, 448.
[174] K. R. Lawless, *Reports Prog. Phys.*, 1964, **37**, 231.
[175] O. Engel, H. Niehus, and E. Baner, *Surface Sci.*, 1975, **52**, 237.
[176] A. E. Dabiri, V. S. Aramati, and R. E. Stickney, *Surface Sci.*, 1973, **40**, 205.
[177] G. Ertl and J. Koch, *Z. phys. Chem. (Frankfurt)* 1970, **69**, 323.
[178] G. Ertl and J. Koch, 'Adsorption and Desorption Phenomena', ed. F. Ricca, Academic Press, New York, 1972.
[179] G. Rovida, *J. Phys. Chem.*, 1976, **80**, 150.
[180] P. W. Tamm and L. D. Schmidt, *Surface Sci.*, 1971, **26**, 286.
[181] H. R. Han and L. D. Schmidt, *J. Phys. Chem.*, 1971, **75**, 227.
[182] L. R. Clavenna and L. D. Schmidt, *Surface Sci.*, 1970, **22**, 365.
[183] P. Kisliuk, *J. Chem. Phys.*, 1959, **31**, 1605.
[184] H. Yamazaki, T. Oguri, and I. Kanomata, *Jap. J. Appl. Phys.*, 1971, **10**, 1105.
[185] G. Ertl, M. Grunze, and M. Weis, *J. Vacuum Sci. Technol.*, 1976, **13**, 314.
[186] D. A. Degras, 'Battelle Colloquium Molecular Processes on Solid Surfaces', Kronberg, 1968.
[187] L. D. Mathews, *Surface Sci.*, 1971, **24**, 248.
[188] L. D. Schmidt, in 'Adsorption and Desorption Phenomena', ed. F. Ricca, Academic Press, London, 1972, p. 391.

Table 5—Contd.

Gas	Metal	Crystallographic plane	Adsorption heats/kJ mol^{-1}		Activation energy of desorption	Reference
			Calorimetric	Isosteric		
CO	W	110			4 adsorption states: 309, 259, 238, 92	188
					280	189
	Ni	111			117	190
					105	191
		100			125	190
					193	186
				125	106	192
					2 adsorption states: 116, 96	193
	Co	0001			193	186
	Pd	111		142		194
		100		159		177
				145		195
		110		231		186
				171		177
	Ir	110		155		196
	Pt	100			3 adsorption states: 288, 259, 184	197
		110			3 adsorption states: 130, 121, 105	198

The occurrence of several adsorption states is usually considered to be connected with different types of adsorbed species and/or adsorption sites on the metal surfaces. However, the identification of both the adsorbed species and their position on the surface is extremely difficult and has been to some extent achieved only in a limited number of simple adsorption systems.[117,118] Consequently it is difficult to decide which experimental value should be compared with a particular theoretical result. Moreover, even if it is possible to interpret the adsorption heat in terms of one adsorbed species, one has to bear in mind that chemisorption is a complex process, where, e.g., diatomic molecules are dissociated, the bonds in the metal surface have to be split (the surface hybridization being perturbed by the approaching molecule) and then the new bonds have to be formed. The energetic effects connected with the latter two processes are inaccessible to direct experimental estimation, and this again complicates the comparison of the experimental adsorption heats with the theoretical values. The binding energy E_{A-M} (the adsorption energy

[189] C. Kohrt and R. Gomer, *Surface Sci.*, 1971, **24**, 77.
[190] G. Doyen and G. Ertl, *Surface Sci.*, 1974, **43**, 197.
[191] J. Lapoujoulade, *J. Chem. Phys.*, 1971, **68**, 73.
[192] H. H. Madden, J. Küppers, and G. Ertl, *J. Chem. Phys.*, 1973, **58**, 3401.
[193] T. N. Taylor and P. J. Estrup, *J. Vacuum Sci. Technol.*, 1973, **10**, 26.
[194] G. Ertl and J. Koch, *Z. Naturforsch.*, 1970, **25a**, 1906.
[195] J. C. Tracy and P. W. Palmberg, *J. Chem. Phys.*, 1969, **51**, 4852.
[196] K. Christmann and G. Ertl, *Z. Naturforsch.*, 1973, **28a**, 1144.
[197] G. Kneringer and F. P. Netzer, *Surface Sci.*, 1975, **49**, 125.
[198] H. P. Bonzel and R. Ku, *J. Chem. Phys.*, 1973, **58**, 4617.

Table 6 Theoretical values of the adsorption energies for various atomically adsorbed species and molecularly adsorbed CO.

Gas	Metal	Representation of the metal	Adsorption energy of atomic species/kJ mol^{-1}	Adsorption heat/kJ mol^{-1}	Method	Reference
H_2	Ni	linear chain Ni_4		276	EHMO	201
		linear chain Ni_5		145	EHMO	201
H	Ni	3 dim. clusters				
		Ni_{10} [type (111)]	501^a	$568^{a,d}$	EHMO	89
			366^a	$298^{a,d}$		
		Ni_9 [type (100)]	520^a	$606^{a,d}$		
		Ni_8 [type (110)]	337^a	$240^{a,d}$		
			530	625^d		
H	Ni	up to Ni_{250},				
		5—6 layer crystal		-10.8—$66.9^{b,d}$	EHMO	90
		(100)	212—251^b	8.4—$104.5^{b,d}$		
		(110)	222—270^b	-30.1—$46.8^{b,d}$		
		(111)	202—241^b			
H_2	Ni	cluster Ni_{14} type (100)		116—135^c	EHMO	202
H_2	Ni	cluster Ni_5 type (100)		163	EHMO	203
		cluster Ni_4 type (110)		0.8		
		cluster Ni_3 type (111)		-37.2		
H	Fe	single atom	432	429^d	EHMO	204
	Co	single atom	407	379^d		
	Ni	single atom	299	163^d		

[201] H. Itoh, 'Proc. 2nd Internat. Conf. on Solid Surfaces 1974', *Jap. J. Appl. Phys. Suppl.*, 2, Pt 2, 1974, p. 497.
[202] D. J. M. Fassaert and A. van der Avoird, *Surface Sci.*, 1976, **55**, 313.
[203] D. Shopov, A. Andreev, and D. Petkov, *J. Catalysis*, 1969, **13**, 123.
[204] V. A. Zasukha and L. M. Roev, *Teor. i Eksp. Khim.*, 1971, **7**, 8.

Table 6—Contd.

Gas	Metal	Representation of the metal	Adsorption energy of atomic species/kJ mol^{-1}	Adsorption heat/kJ mol^{-1}	Method	Reference
H_2	Pd	cluster Pd_2	418	401^d	EHMO	149
		cluster Pd_9 linear	405	376^d		
		cluster Pd_9 planar fcc	433	431^d		
		cluster Pd_{13} 3-dim. fcc	385	378^d		
O	Fe	single atom	617	745^d	EHMO	205
	Co	single atom	280	70.2^d		
	Ni	single atom	116	−257^d		
H	W	single atom	434	−433^d	EHMO	87
		cluster W_9 type (100)	289^a	−144^{a,d}		
		cluster W_9 type (100)	166^a	−103^{a,d}		
		cluster W_{12} type (100)	264	669^a		
H	Ni	small influence of the cluster size (Ni_4—Ni_9)			CNDO	102
		type (111)		125		
		type (110)		75.2		
		type (100)		−37.6		
CO	Ni	bulk metal	adjusted to 125		Anderson formalism	190
		(100)	125			
		(110)	104			
		(111)				
	Pd	bulk metal	adjusted to 167			
		(110)	153			
		(100)	130			
		(111)				

H^+	Cu	bulk metal	(100) (110) (111)	adjusted to Ni values 76.9 81.9 69.0			
H^+	W	jellium		Ion desorption energy 867 corresponding adsorption energy 67.5	-297^d	Density functional theory	104
H	W	jellium		145	-107^d	Density functional theory	146
O	W	jellium		520	468^d		

[a] Different geometry of the adsorption complex; [b] Different number of metal layers; [c] Different adsorbate–adsorbent distance; [d] In these cases the adsorption heats have not been explicitly stated and therefore they were calculated by means of equation (6), using adsorption energy values (which approximate E_{A-M}) and the relevant dissociation energies $D_{H_2} = 434.7$ kJ mol^{-1} and $D_{O_2} = 489.1$ kJ mol^{-1}.

205 G. F. Kventsel and G. J. Golodets, *Zhur. fiz. Khim.*, 1970, **44**, 2142.

of the atomic species) for the diatomic molecules can be roughly approximated (under the assumption that there are free valences on the metal surfaces) by [99,157]

$$E_{A-M} = \tfrac{1}{2}(D_{A-A} + Q) \tag{6}$$

where D_{A-A} is the dissociation energy of the diatomic molecule and Q is the experimental adsorption heat. The theoretical value of the energy effect of the gas atom interaction with an individual surface atom (or with a small cluster) approximates the adsorption energy of atomic species, whereas the theoretical value of the energy effect of a gas molecule interaction with a large cluster (or with a bulk crystal) approximates, when appropriately treated, the adsorption heat.

Remembering all these restrictions, one has to consider that the direct comparison of theoretical values with experimental values of adsorption heats (particularly with those obtained on polycrystalline materials, or even on supported metal catalysts) is inappropriate. The difference between the adsorption heat and binding energy has sometimes been disregarded, and the theoretical results dealing with atomic adsorption have been directly compared with the activation energy of desorption of molecular gases.[146] The poor agreement between the theoretical and experimental values of adsorption heats (energies) (Tables 5 and 6) is not very encouraging. The theoretical values do not obey the empirically estimated correlation between the adsorption heats and the heats of formation of the highest oxide per atom.[206] In particular the negative values of the adsorption heats, resulting from some theoretical calculation, are physically meaningless (Table 6).

For a long time it was expected that the study of the work function changes, caused by adsorption, could directly elucidate the character of the chemisorption bond. The work function change (WFC) has been related to the dipole moment of the individual adsorption complexes by the Helmholtz equation [140] (based on the oversimplified model of the electrostatic double layer). WFC depends on the product of the distance and the electric charge. Its interpretation in terms of the charge distribution is mostly speculative because many questions, concerning the geometric configuration of the adsorption complex still remain open. For example, it is not known whether the dipole moment is concentrated in the adsorbed particle (this is probably the case in large undissociated molecules) or if the adsorbed particle forms one pole only. Depending on the position of the adsorbed particle with respect to the surface plane, one particle can simulate several types of particle, and several types of particle (even with reversed sign of the charge) can effectively behave as one type of particle (Figure 16). This again stresses how important it is to know the geometric configuration of the adsorption complex. Recently the closely related problem of the geometry of the adsorption site has been attacked theoretically by consideration of the donor and acceptor properties of different arrangements of surface atoms (the electronic affinity of the gas particle and the occupancy of the metal surface orbitals was taken into account).[119] A meaningful comparison of the theoretical values of the net charge on the adsorbed particle with the experimental WFC values would be possible, if the problem of the adsorption complex geometry and the screening problem were solved. Selected examples of the WFCs caused by the chemisorption of simple gases on individual crystallographic planes can be found in Table 7.

[206] K. Tanaka and K. Tamaru, *J. Catalysis*, 1963, **2**, 366.

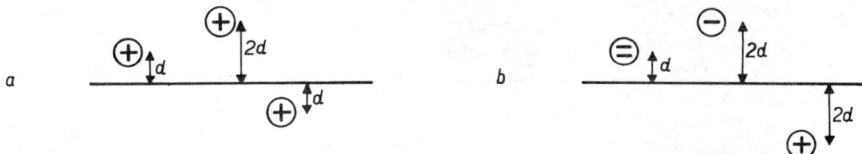

Figure 16 The charged particles at the metal surface (a) one particle, producing several dipole moment values. (b) several particles, producing one dipole moment only

Table 7 Work function changes of individual crystallographic planes caused by chemisorption of simple gases at 273–300 K ($\Delta\phi = \phi_{adsorb} - \phi_{clean}$)

Gas	Metal	Crystallographic plane	$\Delta\phi$/eV	Method[a]	Reference
H_2	Nb	100	+0.05	CPD	170
	Ta	110	+0.1	RP	34
	Mo	100	+0.68	FE	38
		110	+0.48	FE	38
		111	+0.42	FE	38
	W	100	+0.54	CPD	207
			+0.85	RP	208
			+0.84	RP	160
			+0.73	FE	38
		110	−0.14	CPD	207
			+0.14	FE	38
		111	+0.30	CPD	209
			+0.07	RP	160
		112	+0.65	FE	38
			+0.27	CPD	159
	Ni	111	+0.20	CPD, RP	163
		100	+0.08	CPD	210
			+0.17	CPD, RP	163
		110	+0.53	CPD, RP	163
			+0.5	RP	211
			+0.35	CPD	164
			+0.40	CPD	210
	Pd	110	+0.36	CPD	169, 212
		111	+0.18	CPD	169
	Pt	111	−0.23	CPD	170
O_2	Nb	111	+0.4	PE	32
	Ta	110	+0.8	RP	34
	Mo	100	+1.5	RP	213
		110	+1.35	RP	214
	W	100	+1.75	RP	215

[a] The same symbols have been used for the experimental methods of WFC estimation as in Table 1.

[207] B. J. Hopkins, S. Usami, and B. Williams, *Le Vide*, 1969, No. 139, 26.
[208] J. T. Yates and T. E. Madey, *J. Chem. Phys.*, 1971, **54**, 4969.
[209] B. T. Hopkins and R. R. Pender, *Surface Sci.*, 1966, **5**, 316.
[210] J. E. Demuth and T. N. Rhodin, *Surface Sci.*, 1974, **45**, 249.
[211] T. N. Taylor and P. J. Estrup, *J. Vacuum Sci. Technol.*, 1974, **45**, 61.
[212] H. Conrad, G. Ertl, and E. E. Latta, *J. Catalysis*, 1974, **35**, 363.
[213] R. Rivan, C. Guillot, and J. Paigne, *Surface Sci.*, 1975, **47**, 183.
[214] V. M. Zykov, D. S. Ikonnikov, and V. K. Tskhakaya, *Fiz. Tverd. Tela*, 1975, **17**, 274.
[215] H. Berndt and J. Völter, *Z. phys. Chem.*, 1973, **254**, 178.

Table 7—*Contd.*

Gas	Metal	Crystallographic plane	$\Delta\phi$/eV	Method[a]	Reference
			+1.6	CPD	216
			+1.4	RP	217
			+1.8	RP	218
			+1.5	RP	219
			+1.36	RP	220
			+1.5	CPD	221
		110	+1.2	RP	219
			+0.8	CPD	222
			+0.7	CPD, RP	
			+1.2	CPD	56
		111	+1.2	RP	160
			+1.5	FE	56
		112	+1.1	CPD	223
			+1.7	FE	56
	Ni	100	+0.36	CPD	210
		110	+0.42	CPD	210
		111	+0.7	CPD	210
	Ru	100	+0.8	CPD	224
	Pd	111	+0.61	CPD	178
		100	+0.18	CPD	178
	Cu	111	+0.13	CPD	61
		100	+0.39	CPD	61
		110	+0.68	CPD	61
	Ag	111	+0.4	PE	64
N_2	Ta	100	−0.36	FE	225
		111	+0.09	FE	225
	W	100	+0.06	CPD	207
			−0.57	RP	219
			−0.09	FE	49
			−0.6	CPD	226
			−0.06	CPD	41
			−0.31	FE	74
			−0.1—(−0.3)	CPD	227
		110	+0.25	CPD	207
			+0.03	RP	219
			+0.18	CPD	41
			0.0—(−0.3)	CPD	227
			0	FE	74
		111	+0.17	CPD	227

[216] B. J. Hopkins and R. R. Pender, *Surface Sci.*, 1966, **5**, 155.
[217] P. J. Estrup and J. Anderson, 'Proceedings of the 27th Annual Physics and Electronics Conference', MIT 1967, 47.
[218] Ya. P. Zingerman and V. A. Ischuk, *Fiz. Tverd. Tela*, 1966, **8**, 2999; 1967, **9**, 797.
[219] T. E. Madey and J. T. Yates, *Supplemento al Nuovo Cimento*, 1967, **5**, 483.
[220] T. E. Madey, *Surface Sci.*, 1972, **33**, 355.
[221] J. L. Desplat, 'Proc. 2nd Internat. Conf. on Solid Surfaces, 1974', *Jap. J. Appl. Phys.*, 1974, Suppl. 2, Pt. 2, p. 177.
[222] B. J. Hopkins, C. B. Williams, and P. C. Wilmer, *Surface Sci.*, 1971, **25**, 63.
[223] J. C. Tracy and J. M. Blakely, 'The Structure and Chemistry of Solid Surfaces', ed. G. A. Somorjai, Wiley, New York, 1969.
[224] T. E. Madey, H. A. Engelhardt, and D. Menzel, *Surface Sci.*, 1975, **48**, 304.
[225] T. Oguri, *Jap. J. Appl. Phys.*, 1970, **9**, 1461.
[226] D. L. Adams and L. H. Germer, *Surface Sci.*, 1971, **27**, 21.
[227] T. A. Delchar and G. Ehrlich, *J. Chem. Phys.*, 1965, **42**, 2686.

Table 7—Contd.

Gas	Metal	Crystallographic plane	$\Delta\phi$/eV	Method[a]	Reference
			+0.14	FE	74
			+0.38	CPD	41
			+0.3	FE	49
		112	+0.34	CPD	41
			+0.40	FE	74
CO	Mo	100	+0.5	RP	228
	W	110	+0.49	CPD	207
			+0.2	RP	219
			+0.45	CPD	40
		100	+0.41	CPD	207
			+0.48	RP	229
			+0.5	RP	219
			+0.4	CPD	230
			+0.49	PE	231
			+0.58	RP	208
		111	+0.80	CPD	40
		211	+0.68	RP	229
			+0.7	CPD	230
	Ni	100	+1.4	CPD	192
			+1.1	CPD	210
		110	+0.95	CPD	210
			+1.38	RP	193
		111	+0.7	CPD	210
	Ru	100	+0.6	CPD	232
	Pd	111	+1.0	CPD	178
			+1.0	CPD, RP	194
	Cu	111	+0.01	CPD	61
			−0.3	PE	233
		100	+0.11	CPD	61
		110	+0.02	CPD	61

Recently the photoelectron spectra of adsorption complexes became available, making possible the verification of the calculated energy levels of the electrons in the adsorbed particles. The simplest way of doing so is to compare directly the system of theoretical electron energy levels with the experimental photoelectron distribution curve.[94, 234–237] The fundamental problem of the common reference levels for the gas and the adsorbed molecules, together with the problem of the chemical shift estimation, is discussed in the section dealing with the work function. These problems are often formally overcome by the superposing the theoretical

[228] J. Lecante, R. Rivan, and C. Guillot, *Surface Sci.*, 1973, **35**, 271.
[229] R. A. Armstrong, *Canad. J. Phys.*, 1968, **46**, 949.
[230] B. J. Hopkins and S. Usami, *Supplemento al Nuovo Cimento* 1967, **5**, 535.
[231] T. V. Vorburger, D. R. Sandstrom, and B. J. Waclawski, *J. Vacuum Sci. Technol.*, 1976, **13**, 287.
[232] T. E. Madey and D. Menzel, 'Proc. 2nd Internat. Conf. on Solid Surfaces 1974', *Jap. J. Appl. Phys.*, 1974, Suppl. 2, Pt. 2, p. 229.
[233] H. Conrad, G. Ertl, J. Küppers, and E. E. Latta, *Solid State Comm.*, 1975, **17**, 613.
[234] R. P. Messmer, C. W. Tucker, and K. H. Johnson, *Surface Sci.*, 1974, **42**, 341.
[235] G. Blyholder, *J. Vacuum Sci. Technol.*, 1974, **11**, 865.
[236] K. H. Johnson and R. P. Messmer, *J. Vacuum Sci. Technol.*, 1974, **11**, 236.
[237] H. Conrad, G. Ertl, and J. Küppers, *Solid State Comm.*, 1975, **17**, 613.

and experimental spectra in such a way that most of the experimental and theoretical levels coincide, particularly the low-lying levels. However, it does not seem appropriate to fit the theoretical spectrum to the experimental one by tentative shifting of one level only.[88] Sometimes the theoretical energy levels are artificially broadened by using Gaussian functions (the broadening factor being usually 0.5 eV) and such a theoretical spectrum can be directly fitted to the experimental one.[93,95] An attempt to exploit quantitatively the experimental photoelectron spectra for the estimation of the adsorption heat from the experimental chemical shifts by using the Anderson formalism has appeared in the literature.[238] This procedure was criticized,[239] although the binding energy of electrons in the adsorbed molecules, obtained from photoelectron spectra, has been successfully correlated with the adsorption heats of molecular species.[240]

In the field of angular distribution of the photoemitted electrons the interplay of theory and experiment has proved to be a very fruitful one, contributing substantially to the solution of the above mentioned problem of the adsorption complex configuration.[122,126]

4 Conclusion

The problem of an appropriate choice of model for the theoretical description of metal surface phenomena has been discussed using two examples: work function and surface chemical interaction. It has been shown that in our present state of knowledge it appears useful to work preferentially with approximate theoretical methods, applied, however, to realistic models which might be different for various surface processes. This approach enables one to exploit fully experimental results. Moreover, it leaves less space for purely theoretical operations, which in fact sometimes only introduce empirically adjustable parameters and thus make it possible to achieve good agreement between the theoretical results and experimental data even if completely unrealistic models were employed (*e.g.* jellium model of the metal surface).

A meaningful comparison of theoretical and experimental results in the field of surface science is possible in only a limited number of cases so far. Some basic experimental data are still lacking (*e.g.* reliable chemisorption heats on single-crystal planes), and even the experimentally simple systems are often still too complicated from the theoretical point of view. Because of these difficulties theoretical and experimental quantities are sometimes compared, the physical meanings of which differ considerably from each other (*e.g.* activation energy of desorption being compared with the binding energy of the adsorbed particles, *etc.*). On the other hand, the rapidly growing number of both experimental and theoretical results will undoubtedly soon provoke some kind of synthesis of all the different achievements.

The author would like to thank Professor M. W. Roberts, Dr. S. Černý and Dr. J. Müller for their critical reading of the manuscript.

[238] J. E. Demuth and D. E. Eastman, *Phys. Rev. Letters*, 1974, **32**, 1123.
[239] K. Y. Yu, W. E. Spicer, I. Lindau, P. Pianetta, and S. F. Liu, *J. Vacuum Sci. Technol.*, 1976, **13**, 277.
[240] R. W. Joyner and M. W. Roberts, *Chem. Phys. Letters*, 1974, **29**, 447.

4
Angle-resolved Ultraviolet Photoelectron Spectroscopy of Clean Surfaces and Surfaces with Adsorbed Layers

BY D. R. LLOYD, C. M. QUINN, AND N. V. RICHARDSON

1 Introduction

The use of photoelectron spectroscopy as a monitor of the chemical and electronic nature of the adsorbed state and of the solid–vacuum interface has become widespread in recent years. To date, most of the attention in this field has been centred on the phenomenological aspects of the technique, and these have been discussed and summarized in reviews in this series [1] and elsewhere.[2-4] In the past few years it has been realized that a more sophisticated level of comparison with theory can be achieved if angular as well as energetic resolution is available, since the basic photoionization cross-section is necessarily calculated in angle-resolved form. Early angle-resolved measurements of photoelectric current were made by Gardner,[5] Hughes,[6] and Ives and his co-workers.[7] The technique was revived in 1964 by Gobeli, Allen, and Kane,[8] but unfortunately the vast majority of modern photoelectron spectrometers have been designed for maximum sensitivity and sample the photoemission over a large solid angle, generally with a wedge or hollow cone shape. Comparison with theory therefore requires integration over a very specific set of angles.[9]

The simple expedient of stopping down the aperture on such instruments to a narrow cone allows the study of angle-resolved photoelectron spectroscopy (ARPS), and such measurements have been carried out with considerable success.[10-13] However, the use of a fixed electron analyser means that the electron

---

[1] M. W. Roberts, in this series, 1972, Vol. 1, p. 144; C. R. Brundle, *ibid.*, p. 171; W. E. Spicer, K. Y. Lu, I. Lindau, P. Pianetta, and D. M. Collins, in this series, 1976, Vol. 5, p. 103.
[2] A. M. Bradshaw, L. Cederbaum, and W. Domcke, in 'Photoelectron Spectroscopy: Structure and Bonding', Springer-Verlag, Heidelberg, 1975.
[3] D. E. Eastman, in 'Vacuum Ultraviolet Radiation Physics', ed. E. E. Koch and R. Haensel, Pergamon, Vieweg, 1974.
[4] N. V. Smith, *CRC Crit. Rev. Solid State Sci.*, 1971, **2**, 45.
[5] W. Gardner, *Phys. Rev.*, 1916, **8**, 70.
[6] H. L. Hughes, *Phys. Rev.*, 1917, **10**, 5.
[7] H. E. Ives, A. R. Olpin, and A. L. Johnsrud, *Phys. Rev.*, 1928, **32**, 57.
[8] G. W. Gobeli, F. G. Allen, and E. O. Kane, *Phys. Rev. Letters*, 1964, **12**, 94.
[9] J. C. Fuggle, M. Steinkelberg, and D. Menzel, *Chem. Phys.*, 1975, **11**, 307.
[10] B. Feuerbacher and B. Fitton, *Phys. Rev. Letters*, 1972, **29**, 786.
[11] P. Heinmann, H. Neddemeyer, and H. F. Roloff, *Phys. Rev. Letters*, 1976, **37**, 775.
[12] U. Gerhardt and E. Dietz, *Phys. Rev. Letters*, 1971, **26**, 1477.
[13] J. Stohr, G. Apai, P. S. Wehner, F. R. McFeely, R. S. Williams, and D. A. Shirley, *Phys. Rev. (B)*, in press.

exit angle can only be changed by rotating the emitter,[14,15] with a corresponding change in photon incidence angle. The advent of angular dispersive electron spectroscopy (ADES) in which the electron analyser can be moved around the emitter has been an important advance. The first instruments of this kind used simple retarding grid analysers,[16,17] but more recently instruments with moving deflection analysers have become available.* The use of ARPS and ADES should lead to direct comparisons with theory.

This report attempts to summarize the relevant theories which have been developed for u.v. photoemission through clean surfaces or adsorbed layers and to indicate the degree of correspondence between this theory and angle-resolved u.v. photoelectron spectral data available to date. No attempt is made to cover angle-resolved spectroscopy in the X-ray region.[20,21]

2 Photoemission from Clean Surfaces

Theory.—In the external photoelectric effect the interaction between the photon field and the electron field gives rise to electron current in the region surrounding the emitter. The description of this general process at a quantum level capable of direct application is still a challenging problem, even though the basic photoemission experiment is one of the corner-stones of modern electron physics.

In general terms it is easy to describe. Photoemission corresponds to the development of a scattered state from an initial system of photons, electrons, phonons, *etc.* At the stage corresponding to the detection of electron current in the vacuum, photons have been annihilated and at least electrons and positive holes have been created above and below the Fermi level. Then, as most recently pointed out by Grimley and Bernasconi,[22,23] the transition rate from the initial state of the whole system to the state at which the current is measured follows directly from Lippmann's generalization [24,25] of a theorem by Ehrenfest.[26] The transition rate is given by equation (1) for a final state $|F\rangle$ developing from the

$$P_F \propto |\langle F|T|I\rangle|^2 \delta(E_f - \hbar\omega - E_i) \qquad (1)$$

initial state $|I\rangle$ conserving total energy as specified by the δ-function.

---

* The Birmingham group design is now available as Model ADES 400 spectrometer from V.G. Scientific Ltd.[18] A similar machine was described by Lindau and Hagström in 1971.[19] The acceptance cone of the ADES 400 spectrometer is 3°; on the prototype it is 1.5°.

[14] W. F. Egelhoff and D. L. Perry, *Phys. Rev. Letters*, 1975, **34**, 93; J. W. Linnett, D. L. Perry, and W. F. Egelhoff, jun., *Chem. Phys. Letters*, 1975, **36**, 331.
[15] R. H. Williams, J. M. Thomas, M. Barber, and N. Alford, *Chem. Phys. Letters*, 1972, **117**, 142.
[16] R. Y. Koyama and L. R. Hughey, *Phys. Rev. Letters*, 1972, **29**, 1518.
[17] N. V. Smith, M. M. Traum, and F. J. Di Salvo, *Phys. Rev. Letters*, 1974, **32**, 1241.
[18] D. R. Lloyd, C. M. Quinn, N. V. Richardson, P. M. Williams, K. Yates, and A. A. Jones, to be published.
[19] S. B. M. Hagström and J. Lindau, *J. Phys. (E)*, 1971, **4**, 936.
[20] C. S. Fadley, in 'Photoemission from Surfaces', ed. R. F. Willis, B. Feuerbacher, B. Fitton, and C. Backx, Wiley, New York, 1977.
[21] C. S. Fadley, *Progr. Solid State Chem.*, 1976, **11**, 265.
[22] T. B. Grimley, in 'Faraday Discussions of the Chemical Society', Cambridge, 1974.
[23] T. B. Grimley and G. F. Bernasconi, *J. Phys. (C)*, 1975, **8**, 2423.
[24] B. A. Lippmann, *Phys. Rev. Letters*, 1965, **15**, 11.
[25] B. A. Lippmann, *Phys. Rev. Letters*, 1966, **16**, 135.
[26] R. M. Eisberg, 'Fundamentals of Modern Physics', Wiley, New York, 1961.

In this entirely general form the T-matrix is that appropriate for the whole system allowing for any possible scattering events. It is simplified into more meaningful physical form by approximating for the nature of the component fields: as Ashcroft [27] has emphasized, considerable uncertainty surrounds the validity of this procedure.

With the exception of recent model calculations by Feibelman [28,29] and Kliewer [30,31] the photon field is treated classically at a macroscopic level. With one exception [31] only single photon dipole interactions have been considered for the range of photon intensities and the energies considered here. It is also desirable to keep separate the possible interactions between the Fermion fields of electrons and positive holes and also any interactions with the phonon field of the emitter lattice, because then the independent particle band theory of solids [32] can be carried over directly into the photoemission theory to provide the basis of a simple physical picture of the photoemission process.

Here again, however, is an area of uncertain validity. In band theory extended to include crystals with surfaces,[33-35] the surface geometry is normally assumed to be that corresponding to an idealized flat plane resulting from the truncation of an idealized perfect crystal lattice, midway between lattice sites. The attraction of the flat plane assumption rests with the simplifications introduced into the mathematics of the independent particle approximation for the electronic levels in solids with surfaces of this kind, and this can be carried over into the similarly based theories of photoemission. Surface roughness effects may however have profound significance in mechanisms for photoemission.[36-38]

It may be quite improper to attempt a realistic one-particle explanation of photoemission from solids. Electron–electron interactions and hole effects must be expected as a minimum. Electrons excited within a metal to levels only a few eV above the Fermi level exhibit lifetimes of only $\sim 10^{-16}$ s before deactivation by inelastic processes. In this time such electrons can travel at most about 10 Å. In contrast the radiation is absorbed over distances in the solid emitter up to three orders of magnitude greater than this typical electron escape depth.[27]

However, in spite of these strictures, it is within the basic framework of the independent particle approximation that most photoemission theory for solid emitters has been developed, and to a substantial degree many-body effects have

---

[27] N. W. Ashcroft, in 'Vacuum Ultraviolet Radiation Physics', Pergamon, Vieweg, 1974.
[28] P. W. Feibelman, *Phys. Rev. Letters*, 1975, **34**, 1092.
[29] P. W. Feibelman and D. E. Eastman, *Phys. Rev. Letters*, 1976, **36**, 234.
[30] K. L. Kliewer, *Phys. Rev. Letters*, 1974, **33**, 900; K. L. Kliewer, P. Rimbey, and R. Fuchs, in 'Photoemission from Surfaces', ed. R. F. Willis, B. Feuerbacher, B. Fitton, and C. Backx, Wiley, New York, 1977.
[31] I. Idawi, *Phys. Rev.*, 1964, **134**, A788.
[32] C. M. Quinn, 'An Introduction to the Quantum Chemistry of Solids', Clarendon Press, Oxford, 1972.
[33] R. O. Jones, in 'Surface Physics of Phosphors and Semiconductors', ed. C. A. Scott and C. E. Reed, Academic Press, New York, 1975.
[34] S. J. Davison and J. D. Levine, *Solid State Phys.*, 1970, **25**, 1.
[35] J. B. Pendry and S. J. Gurman, *Phys. Rev. Letters*, 1973, **31**, 637.
[36] N. W. Ashcroft and W. L. Schaich, 'Proceedings of the 3rd Materials Research Symposium on the Electronic Density of States', 1969, N.B.S. (U.S.) Special Publication No. 323.
[37] W. E. Spicer, *Phys. Rev.*, 1967, **154**, 395.
[38] J. Endriz, *Phys. Rev.*, 1973, **B7**, 346.

been allowed for empirically by the device of damping final state wavefunctions [39–42] or using a complex self-energy.[43,44] In these terms, equation (1) reduces to the Fermi Golden Rule form derived in several equivalent ways in the past 15 years.[31,39,40,42–48]

$$P_f = \frac{1}{\hbar} \sum_i |\langle f|V|i\rangle|^2 \delta(E_f - \hbar\omega - E_i) \qquad (2)$$

In equation (2) the $|i\rangle$ are one-electron states of the emitter below the Fermi level, the $|f\rangle$ are one-electron positive energy states of the emitter–vacuum system, perhaps damped into the emitter in attempts to take some account of many-body interactions, and chosen to exhibit photocurrent in the vacuum.[21] The introduction of a self-energy can confer some quasi-particle nature and account for electron–electron and hole–electron interactions and so deviations from Koopmans theorem.[49,50]

The V in equation (2) is the first term remainder of the T-matrix of equation (1), the non-relativistic approximation for the interaction between an electron of momentum P and a photon of vector potential A. With the gauge chosen to make the scalar potential zero,[51–54] we have equation (3). The roles and natures of the

$$V = A \cdot p + p \cdot A \qquad (3)$$

individual terms in this equation have been discussed explicitly.[28,39,52,55] For a simple plane wave approximation to the photon beam, the electric vector A is given by equation (4), and the wave propagates with frequency ω in the direction $\hat{q}$

$$A = A_0 \exp(-i\omega t + i\mathbf{q} \cdot \mathbf{r}) \qquad (4)$$

normal to the plane containing A ($A \cdot q = 0$). For u.v. photoemission the photon momentum is small and the spatial dependence of the plane wave approximation can be represented by the first term in its Taylor series expansion, yielding the dipole or acceleration forms for equation (3). The plane-wave approximation explicitly excludes the occurrence of a surface within this model for the system and

---

[39] P. W. Feibelman and D. E. Eastman, *Phys. Rev.*, 1974, **B10**, 4932.
[40] C. Caroli, D. Lederer-Rozenblatt, B. Roulet, and D. Saint James, *Phys. Rev.*, 1973, **B8**, 4552.
[41] J. B. Pendry, *J. Phys.* (C), 1973, **8**, 2413, 1975.
[42] J. B. Pendry, *J. Phys.* (C), 1975, **8**, 2413; *Surface Sci.*, 1976, **57**, 679; *Comm. Phys.*, 1977, in press.
[43] R. O. Jones and J. A. Strozier, *Phys. Rev. Letters*, 1969, **22**, 1186.
[44] J. F. Janak, A. R. Williams, and V. L. Moruzzi, *Phys. Rev.*, 1975, **B11**, 1522.
[45] N. W. Schaich and N. Ashcroft, *Solid State Comm.*, 1970, **8**, 1959.
[46] J. W. Gadzuk, *Phys. Rev.*, 1974, **B10**, 1011.
[47] G. D. Mahan, *Phys. Rev. Letters*, 1970, **24**, 1068.
[48] G. D. Mahan, *Phys. Rev.*, 1970, **B2**, 4334.
[49] L. Hedin and B. I. Lundquist, *J. Phys.* (C), 1971, **4**, 2064.
[50] W. Kohn and L. J. Sham, *Phys. Rev.*, 1966, **145**, 561.
[51] H. A. Bethe and E. E. Salpeter, *Handbuch der Phys.*, 1957, **35**, 88.
[52] B. Feuerbacher and R. F. Willis, *J. Phys.* (C), 1976, **9**, 169.
[53] A. Messiah, 'Quantum Mechanics', Vol. 2, North Holland, Amsterdam, 1969.
[54] J. Avery, 'The Quantum Theory of Atoms, Molecules and Photons', McGraw Hill, London, 1972.
[55] E. W. Plummer, in 'Topics in Applied Physics', ed. R. Gomer, Springer-Verlag, Berlin, 1974, Vol. 4, p. 143.

so reduces equation (2) to the form shown in equations (5), which are appropriate

$$P_f = \frac{1}{\hbar} \sum_i |\langle f|A_0 \cdot p|i\rangle|^2 \, \delta(E_f - \hbar\omega - E_i)$$

$$= \frac{1}{\hbar} \sum_i |\langle f|p|i\rangle \cdot A_0|^2 \, \delta(E_f - \hbar\omega - E_i)$$

$$= \frac{1}{\hbar\omega} \sum_i |\langle f|\nabla U|i\rangle \cdot A_0|^2 \, \delta(E_f - \hbar\omega - E_i) \qquad (5)$$

only in a potential field U of constant dielectric properties.

Discussion of the generation of photocurrent in these terms identifies the volume photoelectric effect, in which the primary photoexcitation in terms of equation (5) occurs within the volume of the emitter and increases with the bulk of the emitter illuminated by the radiation.

Volume photoemission was first discussed by Tamm and Schubin,[56] and then developed and amplified by various authors.[47,48,57–71] The essence of the mechanism is a three-step model of independent basic processes: electron excitation by photon absorption, transport of the excited electron to the surface of the emitter and then transmission into the vacuum. The first and last of these processes can be treated quantum-mechanically within one-electron theory in terms of the optical transitions of equation (5) and regional wavefunction matchings [33] or equivalently scattering across interfaces.[72] However, both the latter stage and the transport of the excited electron are generally accounted for in terms of semi-classical probability functions.[44,73–78] The photoexcitation step in the volume limit is thus built up to an expression for the total external photocurrent by expressions such as equation (6),[78] in which $T(E_f,K)$ is typically a combined overall escape factor

$$D(E,\hbar\omega) = \sum_{fi} \int_{B.Z.} d^3K\, P_f\, T(E_f,K)\, \delta(E - E_i) \qquad (6)$$

[56] I. Tamm and S. Schubin, *Z. Phys.*, 1931, **58**, 97.
[57] H. Y. Fan, *Phys. Rev.*, 1945, **68**, 43.
[58] H. Mayer and H. Thomas, *Z. Phys.*, 1957, **147**, 419.
[59] H. Puff, *Phys. Stat. Sol.*, 1961, **1**, 636.
[60] E. O. Kane, *Phys. Rev.*, 1962, **127**, 132.
[61] W. E. Spicer, *Phys. Rev.*, 1958, **112**, 114.
[62] W. E. Spicer and C. N. Berglund, *Phys. Rev.*, 1964, **136**, A1030.
[63] W. E. Spicer and C. N. Berglund, *Phys. Rev.*, 1964, **136**, A1044.
[64] W. E. Spicer, in 'Optical Properties and Electronic Structure of Metals and Alloys', ed. F. Abeles, North Holland, Amsterdam, 1966.
[65] W. E. Spicer, *Comments Solid State Phys.*, 1973, **5**, 105.
[66] W. F. Krolikowski and W. E. Spicer, *Phys. Rev.*, 1969, **185**, 882.
[67] R. N. Stuart, F. Wooten, and W. E. Spicer, *Phys. Rev. Letters*, 1964, **10**, 1.
[68] R. N. Stuart, F. Wooten, and W. E. Spicer, *Phys. Rev.*, 1964, **135**, A495.
[69] A. Meessen, *J. Phys. Radium*, 1961, **22**, 135.
[70] A. Meessen, *J. Phys. Radium*, 1961, **22**, 308.
[71] A. Meessen, *Phys. Stat. Sol.*, 1968, **26**, 125.
[72] J. B. Pendry and S. J. Gurman, *Surface Sci.*, 1975, **49**, 87.
[73] J. F. Janak, V. L. Moruzzi, and A. R. Williams, *Phys. Rev.*, 1973, **B8**, 2546.
[74] J. F. Janak, V. L. Moruzzi, and A. R. Williams, *Phys. Rev. Letters*, 1971, **28**, 671.
[75] J. F. Janak, D. E. Eastman, and A. R. Williams, *J. de Phys. Colloq.*, 1972, **3**, 131.
[76] R. Koyama and N. V. Smith, *Phys. Rev.*, 1970, **B2**, 3049.
[77] M. M. Traum and N. V. Smith, *Phys. Rev.*, 1974, **B9**, 1353.
[78] J. E. Rowe and N. V. Smith, *Phys. Rev.*, 1974, **B10**, 3207.

representing the probability that a photoexcited electron at energy E_f and with crystal momentum K can contribute to the external photocurrent.[78–88] Often this factor is simply an attenuation of final state structure by the group velocity of the excited state, for example,[37,44,80,84,86] but account can also be taken of the probability of many-electron scattering.[73,76] If $T(E_f,K)$ is assumed constant and the additional assumption is made that the moment P_f is also constant, then equation (6) is an expression for the energy distribution of the joint density of states (EDJDOS) at energy E.[76,86]

Equations like equation (6) are standard for the analysis of volume photoemission with total photocurrent monitoring* or in fixed angle instruments for randomly oriented polycrystalline emitters [78–85] when the photoexcitation process is direct, since all possible contributions to the photocurrent are accounted for by integration over the Brillouin zone of the emitter. In the direct volume limit photoexcitation occurs vertically between bands of the one-electron band structure of the emitter in the reduced zone scheme at given K, or in the extended zone scheme across the bands with K vectors related through a reciprocal lattice vector of the crystal structure.[47,48,52,58] This restriction provides within one-electron theory a mechanism for both energy and momentum conservation. Energy conservation follows from the δ-function, $\delta(E_f - \hbar\omega - E_i)$ in equation (4), while momentum conservation is achieved in the periodic field within the crystal volume during photoexcitation through scattering of an electron from a state of crystal momentum $\hbar K$ to a state $\hbar(K + G)$ with G a reciprocal lattice vector, the u.v. photon momentum being neglected.

Because this direct transition mechanism is a fairly restrictive condition, an indirect volume process has also been proposed in which conservation of momentum is achieved by exchange of momentum with the emitter lattice creating or annihilating phonons, thus facilitating electron excitation from any point on the energy-wave vector dispersion curve to any suitable excited level in the crystal conserving energy.[62,63] However, good agreement has been obtained in a fairly large number of cases between u.v. photoelectron spectra and a theoretical volume photoemission model based on the various approximations to equation (6) involving direct interband optical transitions,[44,83,86,89–95] and it does appear that the indirect model is less satisfactory unless unrealistic assumptions are made.[75]

---

* Christensen and Feuerbacher have pointed out that full angle measurement on a single crystal emitter is distinct from total photocurrent monitoring due to work-function and refraction effects.[86]

[79] N. E. Christensen, *Phys. Stat. Sol.*, 1972, **52**, 241.
[80] N. E. Christensen, *Phys. Stat. Sol.*, 1972, **54**, 551.
[81] N. E. Christensen and B. O. Seraphin, *Phys. Rev.*, 1971, **B4**, 3221.
[82] N. V. Smith and W. E. Spicer, *Opt. Comm.*, 1969, **1**, 157.
[83] N. V. Smith, *Phys. Rev.*, 1971, **B3**, 1862.
[84] J. F. Janak, D. E. Eastman, and A. R. Williams, *Solid State Comm.*, 1970, **8**, 271.
[85] N. V. Smith, *Phys. Rev.*, 1974, **B9**, 1365.
[86] N. E. Christensen and B. Feuerbacher, *Phys. Rev.*, 1974, **B10**, 2335.
[87] B. Feuerbacher and N. E. Christensen, *Phys. Rev.*, 1974, **B10**, 2349.
[88] G. F. Bassani, in 'Optical Properties of Solids', ed. J. Taue, Academic Press, New York, 1966.
[89] N. E. Christensen, *Phys. Letters*, 1971, **A35**, 206.
[90] P. O. Nilsson, C. Norris, and L. Wallden, *Solid State Comm.*, 1964, **7**, 1705.
[91] P. O. Nilsson, C. Norris, and L. Wallden, *Phys. Kondenseirten Mater.*, 1970, **11**, 220.
[92] D. E. Eastman and J. K. Cashion, *Phys. Rev. Letters*, 1970, **24**, 310.
[93] N. V. Smith, *Phys. Rev.*, 1972, **B4**, 1192.

The introduction of an interface-delimiting medium of different dielectric constant into the model for the photoemission experiment leads to the possibility of other contributions to the external photocurrent and identifies the other semi-classical limit in photoemission theory, surface photoemission.[31,38,96–102] In surface photoemission two terms dominate the transition moment in equation (2). Because the vector potential of the radiation is changed across the surface due to the change in dielectric permittivity at the surface, e.g. a step-like variation, the $p \cdot A$ term in equation (3) is not zero. Furthermore, the ∇U term in equation (5) contains a surface component corresponding to the change in potential in the surface region, so that the external photocurrent depends on the matrix element involving these extra contributions as well, viz.

$$|\langle f| A_0 \cdot \nabla U + \frac{i\delta A}{\delta Z} |i\rangle|^2 \qquad (7)$$

with $U = U_{\text{lattice}} + U_{\text{surface}}$.

However, no clear distinction can be made between surface and volume contributions in real systems, and the separation of photocurrent features into distinct surface and bulk contributions depends on the dominance of one mechanism in particular cases.[38,51,86,87] The escape depth in the u.v. region appears to be relatively small [103] and so emission originates in any case in the surface regions where all terms in (7) can be expected to contribute. Moreover, in some materials the difficulties of interpretation are increased because the density of electronic states in the surface region, the surface density of states, can be substantially different from that in the bulk because of the incompleteness of shells of neighbouring atoms in the surface layers of a crystal,[104–107] a quite separate feature to the occurrence of surface states.[33–35]

Finally, the direct transition conservation rule on the electron wave vector loses significance for the normal component of the wave vectors of initial and final states when the system is divided into a crystal and a vacuum region by the surface through which emission takes place. The one-particle wavefunctions of this total system for any energy E are the appropriately weighted superpositions of all regional wavefunctions in the crystal and the vacuum at this energy which are well behaved in slope and value across the interfacial plane. Viewed as a matching process [33,43] or as a multiple scattering process,[41] only the parallel components of the electron wave vector remain good quantum numbers, modulo a reciprocal lattice vector of the surface net for the electron states. Thus momentum conserva-

[94] N. V. Smith, *Phys. Rev. Letters*, 1969, **23**, 1452.
[95] N. V. Smith and M. M. Traum, *Phys. Rev. Letters*, 1970, **25**, 1077.
[96] K. Mitchell, *Proc. Roy. Soc.*, 1934, **146A**, 442.
[97] K. Mitchell, *Proc. Camb. Phil. Soc.*, 1935, **31**, 416.
[98] K. Mitchell, *Proc. Roy. Soc.*, 1935, **153A**, 513.
[99] R. E. B. Mackinson, *Phys. Rev.*, 1949, **75**, 1908.
[100] M. J. Buckingham, *Phys. Rev.*, 1950, **80**, 704.
[101] H. B. Huntington and L. Apker, *Phys. Rev.*, 1950, **80**, 704.
[102] H. B. Huntington, *Phys. Rev.*, 1953, **89**, 357.
[103] J. P. Goudonnet, G. Charbrier, and P. J. Vernier, in 'Vacuum Ultraviolet Radiation Physics', ed. E. E. Koch and R. Haensel, Pergamon, Vieweg, 1974.
[104] D. Kalkstein and P. Soven, *Surface Sci.*, 1971, **26**, 85.
[105] S. J. Gurman and M. J. Kelly, in this series, 1976, Vol. 5, p. 1.
[106] R. Haydock and M. J. Kelly, *Surface Sci.*, 1973, **38**, 139.
[107] F. Ducastelle and F. Cyrot-Lackmann, *J. Phys. Chem. Solids*, 1970, **31**, 1295.

tion in the optical excitation is evident only in these components of the electron wave vectors of the initial and final states.

This basic real situation is summarized in Figure 1. In Figure 1a an optical transition is envisaged to occur between initial and final states of the crystal with energies in the range of bands of Bloch functions of the idealized periodic model for the crystal. $K_\perp$ is not conserved since the wavefunction components normal to the surface are not Bloch-like and the photoemission intensity will include volume and surface terms in keeping with expression (7). The optical transition is direct only for $K_\parallel$, as in Figure 1b.

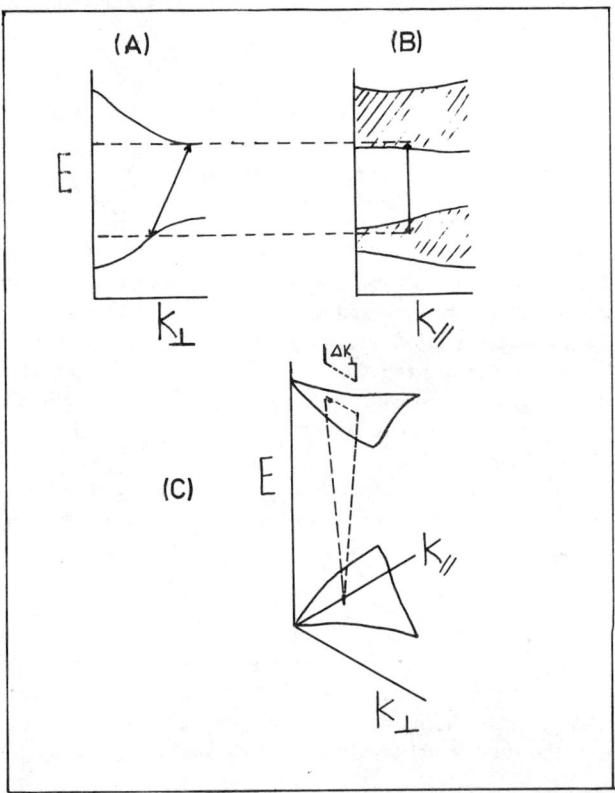

Figure 1 Because $K_\perp$ is not conserved in a semi-infinite crystal, the volume effect photo-excitation occurs as in (a) without $K_\perp$ conservation. $K_\parallel$ is a good quantum number and so the transition is direct as in (b) for this component of K. In (c) the degree to which total K conservation is maintained in a direct volume transition is represented as the broadening $\Delta K_\perp$

Although $K_\perp$ is not conserved in the general case, in practice there appears to be evidence that some degree of total K conservation persists in many cases. For transitions from states in the first Brillouin zone predominantly to final st⁻'es in some other Brillouin zone, the initial states exhibit crystal momentum K while the final states must exhibit momentum K' towards the surface and, after refraction in the surface, generate current in the detector. The conservation of $K_\perp$ requiring that these two momenta be related through a G of the emitter reciprocal lattice, as in the limiting three-step model, and the degree to which this is attained referenced to the band structure, can be represented as a momentum broadening in the final state as in Figure 1c.

In other than NFE metals, the contribution to photoemission intensity of the surface terms [38] in expression (7) is assumed often to be quite low. Even for the simplest metals, however, little real distinction can be made between bulk [76,108] and surface contributions [109–114] to the observed photocurrent, and it is only with the advent of angle-resolved spectroscopy for single-crystal emitters that the relative contributions of the terms in expression (7) can perhaps be unambiguously assessed. In tungsten, for example,[86] a band gap exists above the vacuum level, but photoemission corresponding to the final states in this band gap is apparent in normal emission spectra from the (110) face.[87] This photocurrent cannot originate in indirect volume transitions because of the absence of bulk final states. Nor can it originate in direct transitions in regions of the Brillouin zone other than the ΓN band structure and be observed along the (110) normal (ΓN) because on such low index planes [(100) and (110)] the required change in direction in the final state wave vector for this process cannot be accommodated by Umklapp theory,[47,48] owing to the absence of suitable reciprocal lattice vectors.[55,87]

Apart from these cases, the significance of the general inseparability of surface and volume contributions for angle-resolved measurements is underlined by the results (Figure 2) of a model calculation by Schaich and Ashcroft [45] for a Kronig–Penney lattice [32] with a surface. For an assumed escape depth of $L = 30$ Å, features well defined in energy and angle appear in the angle-resolved photocurrent in agreement with the direct transition model. As the escape depth is assumed to be less, the angular resolution is decreased until in the limit the distribution is closely similar to that expected for the pure surface photoelectric effect,[96–98] as the $K_\perp$ momentum uncertainty is broadened across the Brillouin zone and final state structure in the bands of the idealized periodic model becomes irrelevant.

For simple lattices Mahan [47,48] has shown that an escape depth of 5 Å corresponds to an uncertainty in K in the zone of about 20%, so that a considerable remnant of total K conservation remains, and it is more plausible that in general the absolute resolution of structure relates to initial state lifetimes.[114]

An alternative but entirely equivalent expression (8) for the external photocurrent follows from the rewriting of equation (2) in a form explicitly setting out the nature of the electronic structure in the surface regions [104–107] and the angular dependence of the measurement.[41,42,115]

[108] S. Methfessel, *Z. Phys.*, 1957, **147**, 442.
[109] S. A. Flodstrom and J. G. Endriz, *Phys. Rev. Letters*, 1973, **31**, 893.
[110] F. Wootten, T. Huen, and R. N. Stuart, in 'Optical Properties and Electronic Structure of Metals and Alloys', ed. F. Abeles, North-Holland, Amsterdam, 1966.
[111] J G. Endriz and W. E. Spicer, *Phys. Rev. Letters*, 1971, **27**, 570.
[112] N. V. Smith and W. E. Spicer, *Phys. Rev.*, 1969, **188**, 593; *Phys. Rev.*, 1971, **B3**, 3662.
[113] T. F. Gesell and E. T. Arakawa, *Phys. Rev.*, 1973, **B17**, 5141.
[114] J. B. Pendry, in 'Photoemission from Surfaces', ed. R. F. Willis, B. Feuerbacher, B. Fitton, and C. Backx, Wiley, New York, 1977.
[115] R. Haydock and R. K. McLean, unpublished results.

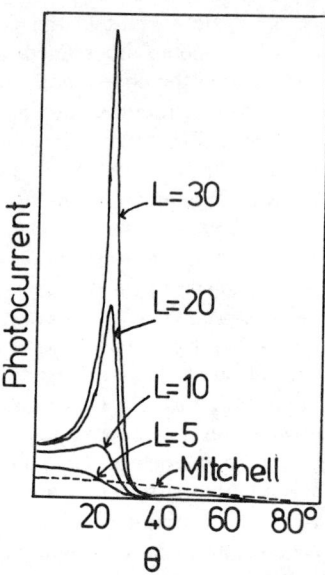

Figure 2 *The angular distribution of photocurrent from a modified Kronig–Penney lattice direct transition as a function of the escape depth L in Å. The dashed curve is the result for surface photoemission at the same energy* [96]
(Reproduced by permission from *Solid State Comm.*, 1970, **8**, 1959)

$$P_f = \frac{1}{\hbar} \sum_i |\langle f|V|i\rangle|^2 \, \delta(E_f - \hbar\omega - E_i) \qquad (2)$$

$$= \frac{1}{\hbar} \sum_i \langle f|V|i\rangle \, \delta(E_f - \hbar\omega - E_i) \langle i|V^\dagger|f\rangle$$

$$= \frac{-1}{\hbar\pi} \mathrm{Im} \sum_i \langle f|V|i\rangle \langle i|(E_f - \hbar\omega - E_i)^{-1}|i\rangle \langle i|V^\dagger|f\rangle$$

$$= \frac{-1}{\hbar\pi} \mathrm{Im} \, \langle f|V(E_f - \hbar\omega - E_i)^{-1}V^\dagger|f\rangle \qquad (8)$$

A Hamiltonian operator $(E - H)$ is diagonalized in the basis of its own eigenfunctions. Equally, the corresponding Greenian operator [32] $(E - H)^{-1}$ is diagonalized by the same eigenfunctions. Here, these correspond to the initial states $|i\rangle$ in the photoemission model of equations (2) and (8). For the diagonal elements $\langle i|(E - H)^{-1}|i\rangle$, H is equal to E_i, the eigenvalue for $|i\rangle$. These eigenvalues can be determined as the singularities of $\langle i|(E - E_i)^{-1}|i\rangle = G_{ii}(E_i)$ on the energy axis. The imaginary part of G_{ii}, $\mathrm{Im} G_{ii}$, is zero except at the singularity at $E = E_i$ where $\mathrm{Im} G_{ii} = -\infty$. From the contour integration result

$$\int_{-\infty}^{+\infty} dE \, \frac{1}{E - E_i} = \int_{-\infty}^{+\infty} dE \, \frac{1}{E - E_i} - i\pi$$

for $\int_{-\infty}^{+\infty}$ along the path ⌒·E_i⌒, it follows that this singularity in $I_m G_{ii}$ is a δ-function of strength $-\pi$. Thus

$$\sum_i \delta(E - E_i) = -\frac{1}{\pi} I_m \sum_i G_{ii}$$

and the transformation in equation (8) follows. For a particular final state $|f\rangle$, for example a plane wave, emitting into a particular direction away from the surface, equation (8) is the imaginary part of the diagonal element for the state $V|f\rangle$, calculable using recursion.[45] For photocurrent in $|f\rangle$ passing into the detector across a plane and determined by given parallel momentum at the emission energy, $|f\rangle$ can be constructed as a multiply-scattered time-reversed state of LEED theory.[41,42] This separation of the scattering events in the generation of initial and final states allows for the clear separation of lifetime effects in the analysis of angle-resolved photoemission. Moreover, the initial states can be represented now on a local orbital basis and so surface density of states effects are on an equal footing with bulk structure.

Comparisons with Angle-resolved Spectra.—The basic conservation of parallel momentum in electron emission across planar surface boundaries as proposed in early theory [60] was first demonstrated in silicon.[8] For the metal chalcogenide layer compounds the consequences of this conservation rule have been investigated in detail by Smith and Traum.[17,77,95,116–121] In these materials bonding is highly anisotropic, strong in the planes perpendicular to the crystal c-axis, but van der Waal's like along the c-direction. Thus, the band structure of these materials is predominantly two-dimensional and the energy-wave vector dispersion in the surface Brillouin zone of the cleavage planes is virtually the three-dimensional band structure. Photoemission data for such materials should relate more directly to this two-dimensional band structure when the emitter surface is a cleavage plane of the chalcogenide crystal, than in materials with more isotropic bonding. For any siting of the detector (Θ,ψ) with respect to the surface normal, the component $|K_\parallel|$ of wave vector of electrons detected with energy E is given by equation (9)

$$|K_\parallel| = (2mE/\hbar)^{\frac{1}{2}} \sin \Theta$$
$$= (E)^{\frac{1}{2}} \sin \Theta \qquad (9)$$

for energies in Rydbergs, and this permits the recovery of the two-dimensional band structure directly from the experiment. The assumption that the $K_\parallel$ of the final state in the optical transition is the same as that of the detected electrons requires that specular reflection occurs across the emitter surface. With the additional assumption that the optical transition is direct in $K_\parallel$, the connection with initial-state band structure is established. For external current at energy E in the

[116] N. V. Smith and M. M. Traum, *Phys. Rev. Letters*, 1975, **31**, 1247.
[117] N. V. Smith and M. M. Traum, *Phys. Rev.*, 1975, **B11**, 2087.
[118] M. M. Traum and N. V. Smith, *Phys. Letters*, 1975, **A54**, 439.
[119] N. V. Smith and M. M. Traum, *Surface Sci.*, 1974, **45**, 745.
[120] N. V. Smith and M. M. Traum, *Surface Sci.*, 1975, **53**, 121.
[121] N. V. Smith, M. M. Traum, and F. J. Di Salvo, *Solid State Comm.*, 1974, **15**, 211.

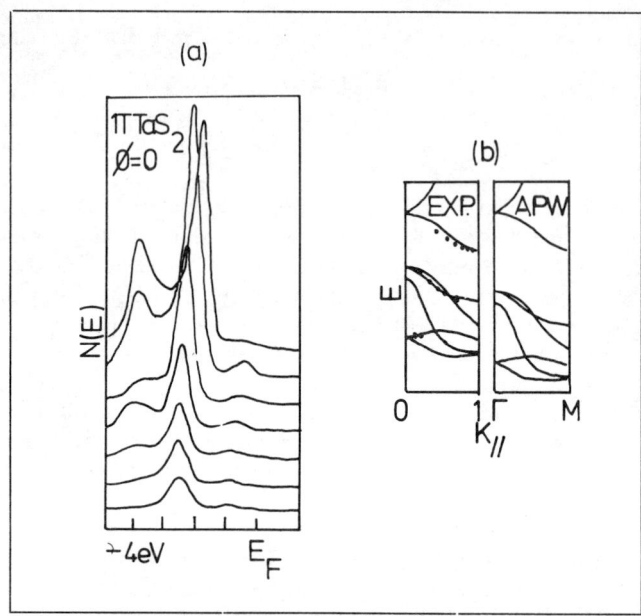

Figure 3 (a) *Angle-resolved spectra for* 1T–TaS$_2$ *at zero azimuthal angle setting of the detector for* 10.2 eV *photons. The polar angles are* 15, 25, 35, 45, 55, 65, *and* 75° *reading from the top of the figure.* (b) *Recovery of the two-dimensional band structure from these data as described in the text. The smooth curves drawn through the data are the APW calculated bands*
(Reproduced by permission from *Solid State Comm.*, 1974, **15**, 211)

electron detector with position defined by a $K_\parallel$, the photoelectrons originate in initial band states at $E - \hbar\omega - \phi$ with the same $K_\parallel$, where $\hbar\omega$ is the photon energy and ϕ is the emitter surface work function.

Figure 3a shows the angle-resolved photoemission from the cleavage face of a 1T–TaS$_2$ crystal for photons of energy 10.2 eV for analyser settings off the normal in the arc corresponding to ΓMLA in the three-dimensional Brillouin zone. For fixed θ in this diagram points taken in the same spectrum plotted as an $E/K_\parallel$ dispersion, as $(E - \hbar\omega - \phi)$ against $E^{\frac{1}{2}}$, lie on a parabola as in Figure 3b and map out the APW energy bands [122] quite well.

This particular example illustrates the deficiencies of partial angular resolution in this kind of study.[14,15,123] In the similar material MoS$_2$, partially resolved angular measurements [15] cannot be related directly to the band structure because the analyser azimuthal setting with respect to the emitter surface Brillouin zone is uncertain. In polycrystalline emitters the same direct relations are absent, even in fully resolved spectra,[16] because no surface Brillouin zone can be defined for the emitter surface and relations with the band structure can be made only by integration.[123]

The additional information in the ADES spectra of chalcogenide crystals which

[122] L. F. Mattheiss, *Phys. Rev.*, 1973, **B8**, 3719.
[123] F. Martino and K. F. Berggren, *Solid State Comm.*, 1976, **20**, 1057.

can be extracted from peak intensities and azimuthal angle and photon energy dependences also underlines basic theoretical features. From peak intensities and azimuthal angle dependences the basic surface symmetry can be established and in certain cases the electron escape depth estimated.[117] In 1T-TaSe$_2$ the Ta atoms are in trigonal antiprismatic co-ordination and sited in Se-Ta-Se sandwiches in the cleavage planes, exhibiting three-fold rotational symmetry about the normal to these planes. Peak intensities at fixed θ exhibit this basic symmetry when the dependence on azimuthal angle ψ is monitored.[119] 2H-TaSe$_2$ (D_{6h}^4) differs from the trigonal material (D_{3d}^3) in that the Ta atoms exhibit trigonal prismatic co-ordination. Thus while each Se-Ta-Se sandwich exhibits three-fold symmetry as before, six-fold symmetry can be attained with respect to the cleavage planes, if the rotation involves a non-symmorphic translation where necessary. However, because of the finite escape depth of the photoelectrons, a lowered emission intensity of the second-layer components in the azimuthal angle dependence of the spectra can be detected. For silicon emitters similar azimuthal measurements reveal the reconstruction of the (111) surface of the crystal.[124,125]

The fine detail in angle-resolved spectra for chalcogenide crystals at higher photon energies [120,126] reveal also the role of Umklapp processes in the photo-emission, since for higher photon energy the parallel components of the electron wave vector can lie outside the first Brillouin zone of the crystal structure. If an optical transition occurs at a point K, in the limit, in the three-dimensional reduced zone, or, at least, in the surface Brillouin zone, the final state, basically $K + G$ contains components related in wave vector to $K + G$ as $G_u = G + G_{hkl}$ where G_{hkl} is any reciprocal lattice vector of the structure or the surface.[47,48,57—60,120]

Since at positive energy each component of this final state inside the crystal can contribute to the external electron current, a given transition can generate current therefore in a number of specific directions. These current components at given energy, accounting for refraction in the surface, give rise to the primary and secondary cones detailed by Mahan.[47,48]

In Figure 4a angle-resolved spectra for 1T-TaS$_2$ and He(I) photons are given as a function of polar angle at constant azimuthal siting of the analyser.[120] For the Ta d-states contributing the spectral features -6 eV below the vacuum level, azimuthal-angle-dependent changes in the emission intensity due to these states are observed, as in Figure 4b at constant polar angle $\theta = 60°$. This corresponds to a sampling of the energy/wave-vector dispersion in the surface Brillouin zones on a circular path about the $\bar{\Gamma}$ point, as in Figure 4c. The occupied Ta d-states contributing to this emission intensity centre around the $\bar{M}$ points in the Brillouin zones. For small $|K_\parallel|$ only the d-state components within the first Brillouin zone contribute to the emission. As $|K_\parallel|$ is increased, and this depends on photon energy, contributions from the $\bar{M}$ points in other Brillouin zones can appear leading to the distortion of the simple azimuthal lobes apparent at lower photon energies.[117]

For clean, well-defined, single-crystal metal surfaces, the impetus in angle-

[124] M. M. Traum, J. E. Rowe, and N. V. Smith, *J. Vac. Sci. Technol.*, 1975, **12**, 298.
[125] J. E. Rowe, M. M. Traum, and N. V. Smith, *Phys. Rev. Letters*, 1974, **33**, 1333.
[126] D. R. Lloyd, C. M. Quinn, N. V. Richardson, and P. M. Williams, *Comm. on Phys.*, 1976, **1**, 11.

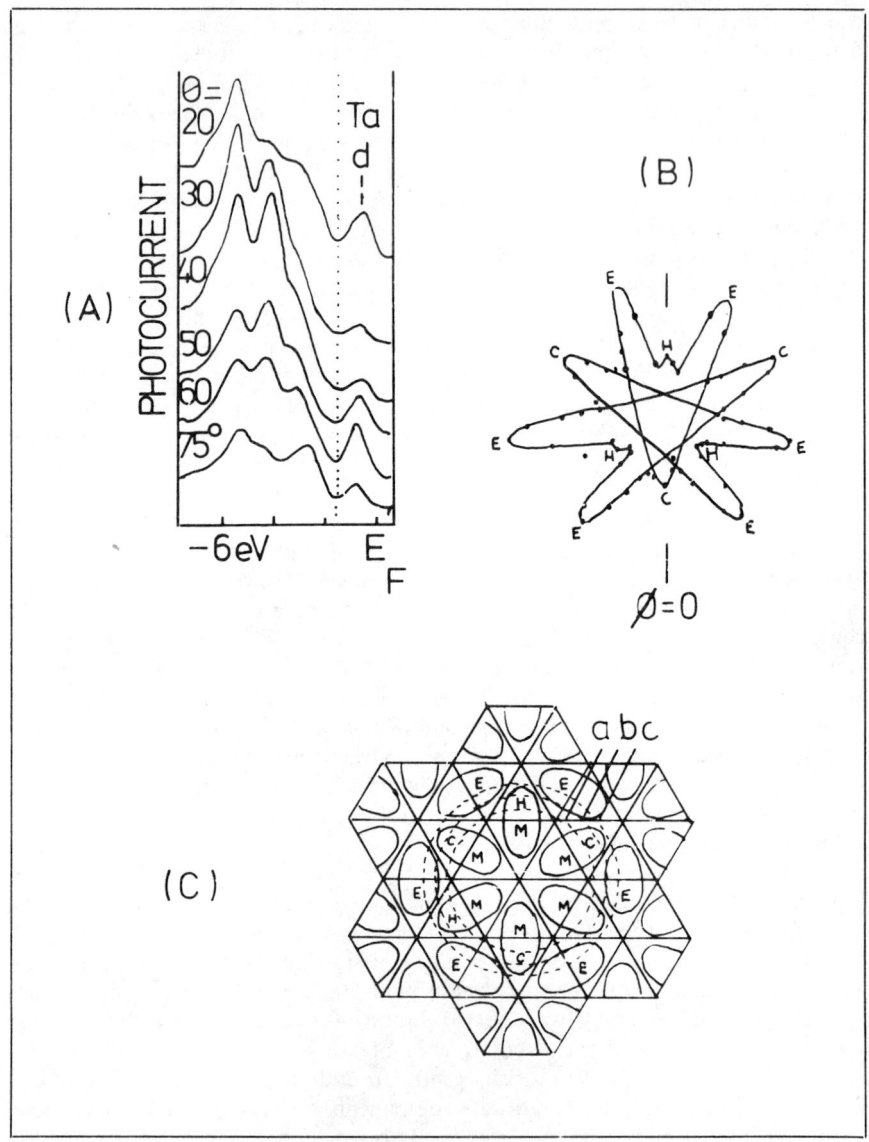

Figure 4 (a) *Angle-resolved spectra for* 1T–TaS$_2$ *and* He(I) *photons for different polar emission angles at constant azimuthal angle* $\phi = 275°$. (b) *Radial azimuthal dependence of the photocurrent signal from the Ta d-states in* 1T–TaS$_2$ *for* He(I) *photons and* $\theta = 60°$. (c) *Repeated zone scheme for the chalcogenide* 1T–TaS$_2$ *d-states in the elliptical regions around the* $\overline{M}$ *points. The broken circles are of radii corresponding to* $K_{\parallel}$ *in photocurrent signals at polar angles of* 40, 50, *and* 60° (a, b, *and* c, *respectively). The areas labelled* E, C, *and* H *are the regions of K-space corresponding to the azimuthal features so marked in* (b) *and lying outside the first Brillouin zone*
(Reproduced by permission from *Solid State Comm.*, 1974, **15**, 211)

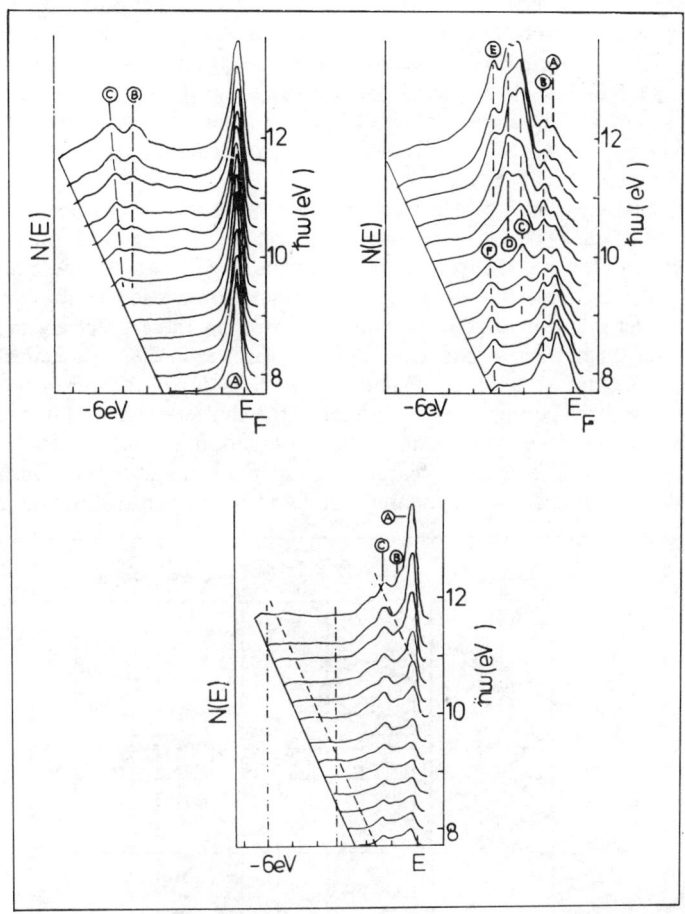

Figure 5 *Angle-resolved photoelectron spectra normal to* W(100), W(111), *and* W(110) *single-crystal surface for the photon energies marked and into a detector with acceptance cone of* 12°.[87] *In* W(100) *(upper left) the surface normal corresponds to the* ΓH *band structure in Figure 7. Feature* A *is a surface resonance. Features* B *and* C *are due to surface emission and a direct transition, respectively. In* W(110) *(upper right) the surface normal corresponds to the* ΓN *band structure in Figure 7. No final states are available between 6 and 11 eV above* E_F; *so in this range, designated by oblique dashed lines, surface emission only can occur. In agreement with this conclusion, all spectra in this range are identical, exhibiting no final-state modulation. The vertical dash–dot lines correspond to the* ΓN *band gap in the initial density of states. At the highest photon energies, the leading peaks may be due to direct transitions since the initial and final state energies lie within the crystal bands.* (Lower) W(111) *contributions to the spectra for states along* ΓP *symmetry line in Figure 7 and also possibly from* PH *due to Umklapp;* A + B *surface emission features;* E *and* F *possibly due to direct transitions; no obvious assignment for structures* C *and* D (Reproduced by permission from *Phys. Rev.*, 1974, **B10**, 2349)

resolved photoemission spectroscopy is also the testing of theory and, if possible, the recovery of the band structure from angle-resolved spectra. Indeed, the aim here is ultimately the checking and refining of these calculations. An immediate distinction with the chalcogenide materials can be made. The energy/vector dispersion for the normal component of wave vector is very different in metals and chalcogenides. So in projection on the surface Brillouin zone of an emitter, the band structure will appear generally as in Figure 1b, smeared out bands of allowed energies at each $K_{\parallel}$. Moreover, only for normal emission can refraction effects be ignored.[51,86,87]

Such normal emission spectra were the first reasonably well-defined measurements to be made on metal single-crystal surfaces. In Figure 5 photoelectron spectra [10,87,127] for an electron collection cone of 12° about the normals to single-crystal tungsten surfaces are given for the photon energies marked on the diagram. Interpretation of the individual features in these spectra can then be based on the assignment of particular spectral features in the hypothetical case in Figure 6. In Figure 6a a section of band structure is drawn for the Brillouin zone symmetry line corresponding to the normal of the emitter surface. For the photon energy $\hbar\omega$, and with the assumption of $K_{\perp}$ conservation, a direct optical transition can be assigned as the spectral feature shown in Figure 6b. The width of this feature relates to the lack of $K_{\perp}$ conservation and to the various lifetime effects in

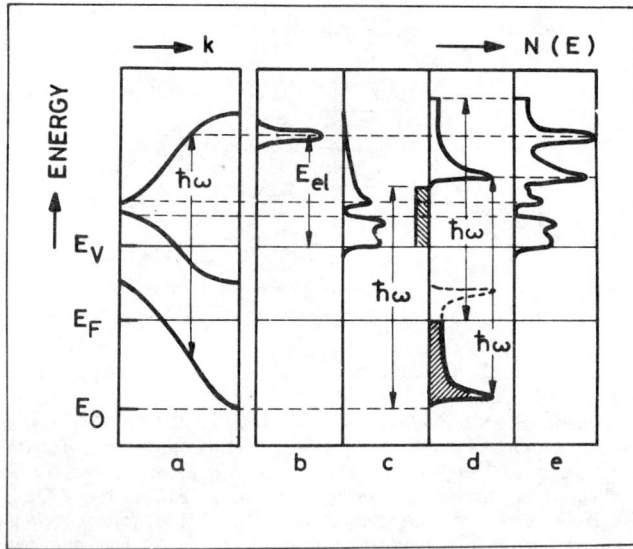

Figure 6 *A hypothetical band dispersion along the Brillouin zone symmetry line corresponding to the surface normal for the angle-resolved experiments in Figure 5. (a) Direct inter-band transition. (b) Direct transition related spectral feature. (c) Contribution to the observed spectrum from inelastically scattered photoelectrons reflecting in the spectra the final density of states along the normal direction in the B.Z. (d) Surface emission contribution reflecting the surface density of states. (e) Total observed angle-resolved spectrum* (Reproduced by permission from *Phys. Rev.*, 1974, **B10**, 2349)

[127] B. Feuerbacher and B. Fitton, *Phys. Rev. Letters*, 1973, **30**, 923.

the photoemission. The identification of such direct transition-related features can be made by observation of the movement of the initial state energy participating with varying photon energy and from good band structure calculations for the emitter. In this manner several of the principal features in the spectra in Figure 5 can be assigned to direct transitions along only the symmetry lines in the Brillouin zone corresponding to the normals to the surfaces, except in the case of the (111) surface. For this surface Umklapp processes in transitions in the HP band structure of the zone can cause emission contributions in the normal photocurrent from this surface, and thus provide a reason for the observation of photocurrent apparently from the gap region in the ΓP band structure for which no surface states have been observed.

This explanation is preferable for several reasons to one involving the indirect transition mechanism, proposed originally by Spicer.[62,63] For the (110) face, little photocurrent is observed from states between −2.5 and −5.0 eV. High densities of states in other areas of the Brillouin zone thus do not contribute in this part of this set of spectra, and the low photocurrent observed is in agreement with the band gap in the ΓN direction. Similarly in the (100) spectra there is no evidence of contributions from, for example, the high density of states around P in the band structure given in Figure 7.

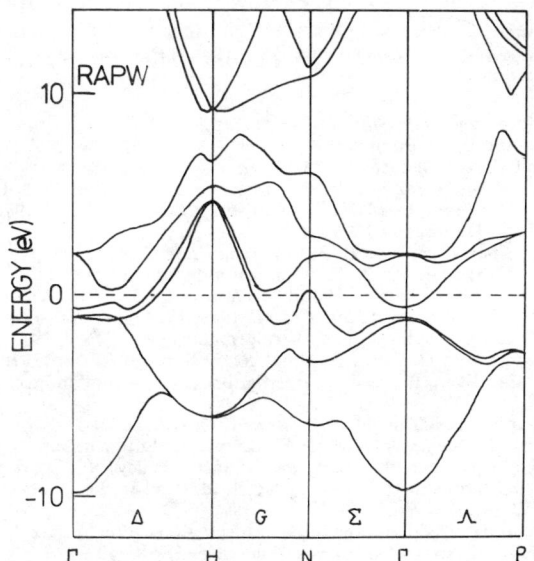

Figure 7 *Band structure of tungsten to positive energies calculated using the RAPW method* (Reproduced by permission from *Phys. Rev.*, 1974, **B10**, 2335)

Thus in the absence of indirect contributions to the photocurrent for the data in Figure 5, the remaining components in these spectra can be assigned to inelastically scattered photoelectron contributions reflecting, in the secondary tail at low kinetic energies, the final density of states as in Figure 6c, and to surface photoemission, Figure 6d, yielding information directly on the density of initial states in the surface used as a photoemitter.

The contribution of surface photoemission is particularly important in the spectra for the (100) surface. In this surface, theory [128] and other experiments [129] have led to the identification of a surface state resonance just at the initial-state energy of the large peak ~ 0.35 eV below E_f. In the photoemission measurements this feature in the spectra of the (100) surface is very sensitive to adsorbed gas on the surface but insensitive entirely to photon energy, observations entirely consistent with surface photoemission from a surface state.

The energy and angular distribution of photoelectrons emitted from the (110) and (111) surfaces of clean tungsten have also been measured by Turtle and Callcott [130] for photon energies between 7.7 and 10.7 eV. Then peaks in the emission from the two surfaces have been assigned to the same direct transition in K space on the basis of coincidence in energy and symmetry. For each such transition $K_\parallel$ in the particular surface can be obtained as described earlier. For the same transition observed in both surfaces at different angles to the normals the three-dimensional wave vector of these transitions can be obtained, and thus some of the band structure in the plane of the Brillouin zone defined by the (110) and (100) axes.

Similar heuristic analyses of angle-resolved photoelectron spectra have now been made for several clean single-crystal emitters [131-148] and in one case for a random substitutional alloy.[149] Correlations with direct transitions in the

[128] K. Sturm and R. Feder, *Solid State Comm.*, 1974, **14**, 1317.
[129] B. J. Waclawski and E. W. Plummer, *Phys. Rev. Letters*, 1972, **29**, 783.
[130] R. R. Turtle and T. A. Callcott, *Phys. Rev. Letters*, 1975, **34**, 86.
[131] D. R. Lloyd, C. M. Quinn, and N. V. Richardson, *J. Phys.* (C), 1975, **8**, L371.
[132] D. R. Lloyd, C. M. Quinn, and N. V. Richardson, *J.C.S. Faraday II*, 1976, **72**, 1036.
[133] D. R. Lloyd, C. M. Quinn, and N. V. Richardson, *Surface Sci.*, in press.
[134] R. F. Willis, B. Feuerbacher, and B. Fitton, *Solid State Comm.*, 1976, **18**, 1315.
[135] D. R. Lloyd, C. M. Quinn, and N. V. Richardson, to be published.
[136] P. M. Williams, D. Latham, and J. Wood, *J. Electron Spect.*, 1975, **7**, 281.
[137] P. O. Nilsson and L. Ilver, *Solid State Comm.*, 1976, **18**, 677.
[138] H. Becker, E. Dietz, U. Gerhardt, and H. Angenmuller, *Phys. Rev.*, 1975, **B12**, 2084.
[139] E. Dietz, H. Becker, and U. Gerhardt, *Phys. Rev. Letters*, 1976, **36**, 1397.
[140] P. O. Nilsson and L. Ilver, 'Proceedings International Symposium on Electron Spectroscopy', Kiev, 1975.
[141] P. O. Gartland and B. J. Slagsvold, *Phys. Rev.*, 1975, **B12**, 4047.
[142] P. Heinmann, H. Neddemeyer, and H. F. Roloff, in 'Photoemission from Surfaces', ed. R. F. Willis, B. Feuerbacher, B. Fitton, and C. Backx, Wiley, New York, 1977.
[143] P. O. Gartland, in 'Photoemission from Surfaces', ed. R. F. Willis, B. Feuerbacher, B. Fitton, and C. Backx, Wiley, New York, 1977.
[144] P. Heinmann and H. Neddemeyer, in 'Photoemission from Surfaces', ed. R. F. Willis, B. Feuerbacher, B. Fitton, and C. Backx, Wiley, New York, 1977.
[145] G. P. Williams and C. Norris, in 'Photoemission from Surfaces', ed. R. F. Willis, B. Feuerbacher, B. Fitton, and C. Backx, Wiley, New York, 1977.
[146] C. Guillot, Y. Ballu, G. Chauvin, J. Lecante, J. Paigne, R. Thiry, D. Dagneaux, Y. Petroff, R. Pinchaux, and R. Cinti, in 'Photoemission from Surfaces', ed. R. F. Willis, B. Feuerbacher, B. Fitton, and C. Backx, Wiley, New York, 1977.
[147] F. J. Himpsel and W. Steinmann, in 'Photoemission from Surfaces', ed. R. F. Willis, B. Feuerbacher, B. Fitton, and C. Backx, Wiley, New York, 1977.
[148] F. J. Himpsel and W. Steinmann, *Phys. Rev. Letters*, 1975, **35**, 1025.
[149] B. L. Gyorffy, G. M. Stocks, W. Temmerman, D. R. Lloyd, C. M. Quinn, and N. V. Richardson, to be published.

volume,[84,131,135,146] surface emission,[86,142] and surface state effects [52,132,134,141,142] have been made on the bases of available band structure, initial densities of states,[86] the sensitivity of spectra to adsorbed surface species,[86,132,150] and surface state calculations.[151-153]

For copper, clear evidence for surface state formation on the (111) surface in the band gap around the Fermi level has been obtained,[132,141,142] in agreement with calculations.[151-153] For the (100) surface the calculations suggest surface state formation in the s–d band gap,[35,151] but the experimental evidence is not so well defined for He(I) radiation.[131,137,140] At lower photon energies the coupling to the surface initial state on the (100) is greater, reflecting the general significance of the optical matrix element in the spectra.[154]

Strong dependences on photon angle and polarization within the matrix element can also be observed and this data can be used to evaluate particular mechanisms for the photoemission. Thus,[143] on a molybdenum crystal at low photon energies, it appears that surface emission dominates, in line with data on aluminium.[109,155] This conclusion is preferable to one involving the volume effect with increased intensity whenever emission occurs parallel to an important reciprocal lattice vector [48] because the spectral features are very sensitive to the presence of adsorbed species on the emitter surface.

For the copper (100) and (111) and palladium (100), (111), and (110) surfaces, many features in ADES at different photon energies correlate well with direct optical transitions in the volume.[131,135,140] For the Cu(100) surface the modulation of the optical matrix element by the photon incidence dependence is in agreement with the optical selection rules that can be set down using group theory.[135]

Other particular points of interest in the angle-resolved data, reported up to this time, are the observation of Mahan cone effects in angle-resolved spectra off normal for silver (110) surfaces;[156] the conclusion that in normal emission spectra from the same surface, only surface emission is important and that the spectra correlate very well with the initial one-dimensional density of states;[11,142] the use of the neck orbit in the Fermi surface around the normal to a Cu(111) surface; and the emission from the surface state band within that orbit as a function of photon energy to evaluate free-electron refraction theory.[132]

For the case of the random alloy crystal [149] the direct correspondence between the energy/wave vector dispersion along electron emission trajectories referred to the Brillouin zone and the angle-resolved spectra provides more convincing evidence for the validity of the coherent potential approximation [157,158] theory of random

---

[150] W. F. Egelhoff, jun., J. W. Linnett, and D. L. Perry, *Phys. Rev. Letters*, 1976, **36**, 98.
[151] S. J. Gurman, *Surface Sci.*, 1976, **55**, 93.
[152] S. P. Liebmann and C. M. Quinn, to be published.
[153] S. J. Gurman, in '13th Annual Solid State Physics Conference', Inst. of Physics, Manchester, 1976; *J. Phys. (C)*, 1976, **9**, L609.
[154] J. B. Pendry, in 'Photoelectric Emission', Proc. of Daresbury Study Weekend No. 8, 1976, p. 1.
[155] J. K. Grepstasl, P. O. Gartland, and B. J. Slagsvold, *Surface Sci.*, in press.
[156] G. V. Hansson, S. A. Flodstrom, and S. B. M. Hagström, in 'Photoemission from Surfaces', ed. R. F. Willis, B. Feuerbacher, B. Fitton, and C. Backx, Wiley, New York, 1977.
[157] R. J. Elliot, J. A. Krumhansl, and P. L. Leath, *Rev. Mod. Phys.*, 1974, **46**, 465.
[158] G. M. Stocks, E. S. Guiliano, B. L. Gyorffy, and S. Ruggeri, *Phys. Rev.*, in press.

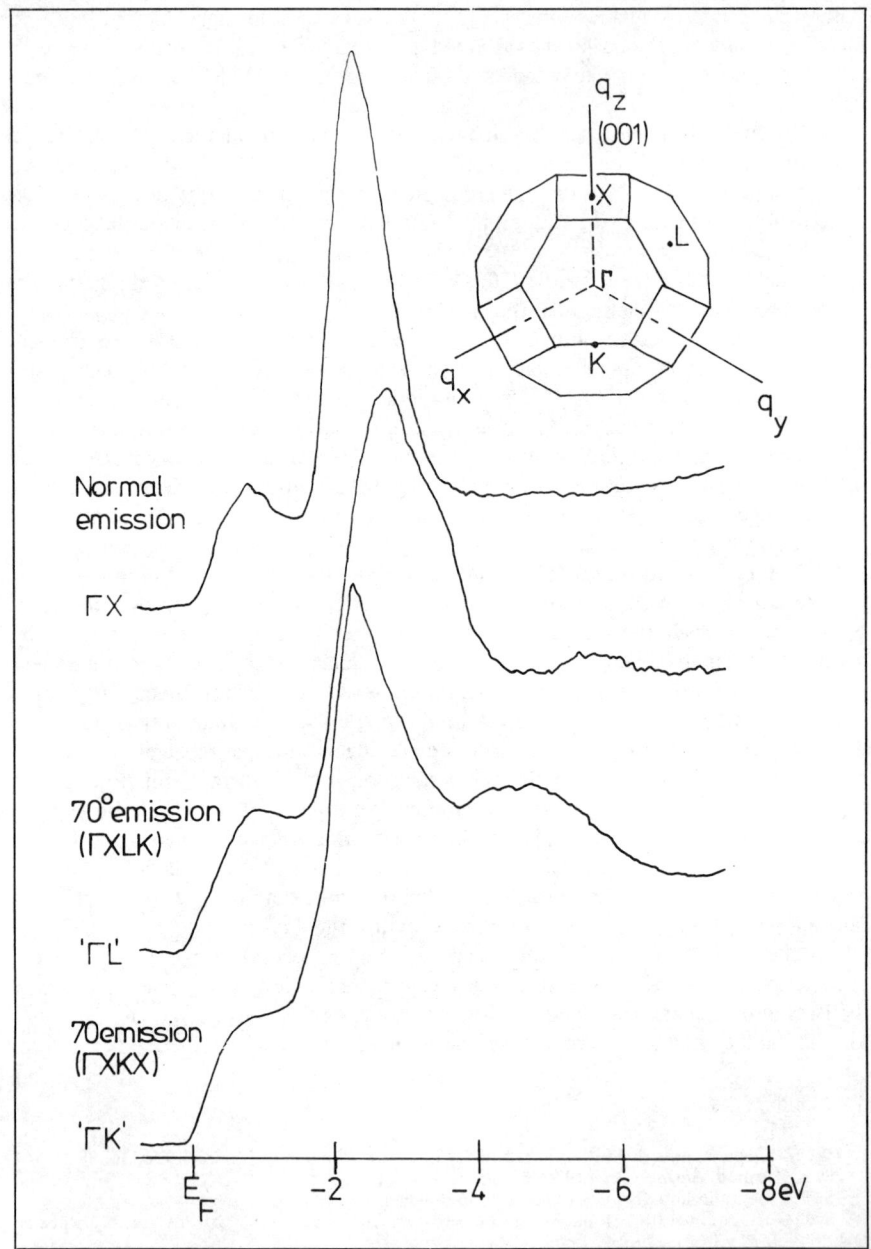

Figure 8 He(I) *angle-resolved photoemission spectra from the (100) surface of a 23% Ni/77% Cu random alloy. The incident photon angle was 20° off the surface normal in the plane it makes with the emission direction. The electron emission angle was also measured from the normal to the surface in the planes as indicated. Inset the Brillouin zone of the fcc crystal structure*

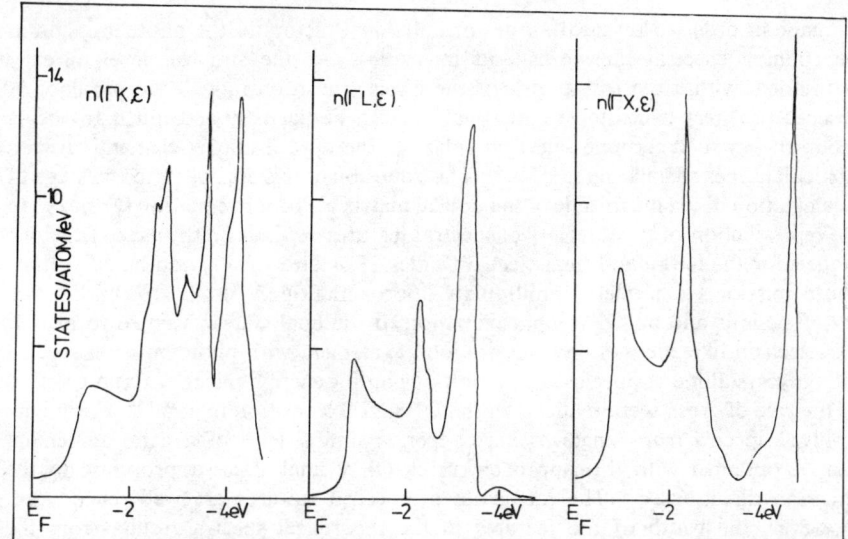

Figure 9 *The one-dimensional densities of states for* $Cu_{77}Ni_{23}$ *in the* ΓX, ΓL, *and* ΓK *directions as calculated from the KKR–CPA Bloch spectral functions* [149]

alloys than previous photoemission measurements.[159—165] Figure 8 shows He(I) angle-resolved spectra from the (100) surface of a 23% Ni/Cu alloy crystal. The one-dimensional densities for the alloy in the ΓX, ΓK, and ΓL directions as calculated from the KKR–CPA Bloch spectral functions are drawn in Figure 9. The nickel resonances below the Fermi level are found principally around the X-point of the Brillouin zone, and this is reflected in the decay of this feature for emission angles off the (100) surface normal towards ΓK and ΓL directions.

Away from the nickel resonances the remainder of the alloy spectra are closely similar to the equivalent spectra for a pure copper (100) surface. This finding agrees well with the predictions of the KKR–CPA [158] as reflected in the spectral functions.[149] The initial-state lifetimes are long except where the copper '*s*' band crosses the nickel impurity band when random hybridization gives rise to very short lifetimes. Thus, in the alloy, direct transitions appear to be meaningful for the main copper *d*-band complex in contrast to previous suggestions based on non-angle-resolved data.[159,160] For Ar(I) radiation a direct transition can be assigned from near the Fermi level in the pure copper angle-resolved spectrum taken normal to a (100) surface.[135] This feature is absent in the alloy spectrum for Ar(I) photons since the initial state in the alloy is of short lifetime.[149]

With the increasing availability of angle-resolved data for well-defined clean surfaces more detailed comparisons with the theory in equations (2) and (8) can

---

[159] D. H. Seib and W. E. Spicer, *Phys. Rev.*, 1970, **B2**, 937.
[160] D. H. Seib and W. E. Spicer, *Phys. Rev.*, 1970, **B2**, 1676.
[161] A. Kichon, *Phys. Rev.*, 1970, **B2**, 939.
[162] D. H. Seib and W. E. Spicer, *Phys. Rev. Letters*, 1969, **22**, 1711.
[163] D. H. Seib and W. E. Spicer, *Phys. Rev.*, 1970, **B2**, 1694.
[164] N. J. Shevchik and C. M. Penchina, *Phys. Stat. Sol.*, 1975, **70**, 619.
[165] G. P. Williams and C. Norris, *Comm. on Phys.*, 1976, **1**, 199.

be made in order to reveal the roles of individual factors in the photoemission in determining spectral linewidths and intensities. At the simplest level, already mentioned, within the framework of the direct transition model, the presence or absence of direct transition related features can be partially accounted for using group theory to determine selection rules for the optical matrix element between particular types of bulk state.[8,132,135] The extension of this approach to the stage of a calculation of the magnitude of the optical matrix element in equation (2) and so to the reproduction of the intensity of spectral features depends on the use of tractable models for the initial and final electron states. For direct transition based volume photoemission, a model Hamiltonian approximation[83,166—168] to initial-state wavefunctions and an OPW approximation to the final states assumed to exhibit free-electron like energies have led to good agreement with photoemission spectra for polycrystalline copper as a function of photon energy and to an appreciation of the role of cross-section effects in the UPS/XPS transition region.[169] For angle-resolved spectra from single-crystal copper, a similar level of general agreement can be obtained with this approach, using OPW final states appropriate to the emission directions.[135] The intensities of spectral features are well reproduced. However, the width of the features in the theoretical spectra results from the arbitrary broadening of the direct transitions to simulate escape depth and lifetime effects. Moreover, no account can be taken in such spectra of contributions from specifically surface effects, such as the surface local density of states and surface state bands.

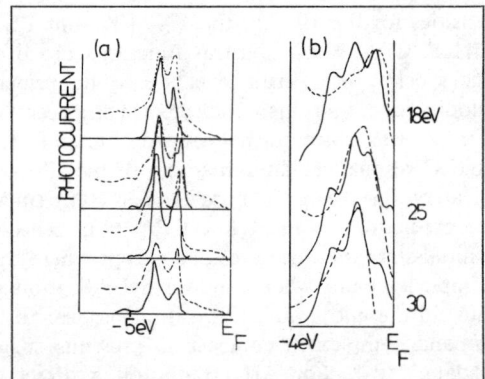

Figure 10 (a) (———) *Photocurrent from a copper* (100) *surface at various polar angles with azimuthal setting in the plane-containing* ΓKLX *for* He(I) *photons as calculated by Pendry.*[154] (---) *experimental data.*[131] *In the theoretical spectra inverse lifetimes of* 4 eV *for the final state electron and of* 0.05 eV *for the final state hole are used.* (b) (———) *Pendry's theoretical spectra*[154] *for* Ni(111) *at different photon energies for emission* 54° *off normal towards* ΓX; *inverse lifetimes of* 4 eV *and* 0.27 eV *are assumed as discussed in the text.* (---) *experimental data*[145]
(Reproduced by permission from ref. 154)

[166] L. Hodges, H. Ehrenreich, and N. D. Lang, *Phys. Rev.*, 1976, **152**, 505.
[167] F. M. Mueller, *Phys. Rev.*, 1967, **153**, 659.
[168] H. Ehrenreich and L. Hodges, *Methods Computational Phys.*, 1968, **8**, 108.
[169] J. Stohr, F. R. McFeely, G. Apai, P. S. Wehner, and D. A. Shirley, *Phys. Rev. (B)*, in press

Theoretical angle-resolved spectra in which all these possible factors are treated on an equal footing have been calculated for copper and nickel by Pendry,[154] based on equation (8). These spectra are compared with angle-resolved data for copper and nickel surfaces in Figure 10 and quite good agreement is apparent in many features of the spectra.

For the final state $|f\rangle$ in equation (8) Pendry has used time-reversed wavefunctions of LEED theory,[41,42,170,171] *viz.* equation (10), where $|\phi\rangle$ is a source of electron

$$\langle r|f\rangle = \langle r|G_2^-(E_i + \hbar\omega)|\phi\rangle \qquad (10)$$

current into the vacuum, defined by a certain z-value, of given parallel momentum. The attraction of this approach lies not only in the fact that all the computational procedures of LEED theory can be carried over directly into the photoemission theory, but also in that lifetime effects in initial and final states can be separately assigned as self-energies in the Green's functions on the basis of the width of spectral features. Moreover, any surface effects are accounted for in the local orbital calculation of the initial states.

The initial and final state inverse lifetimes (Σ_1, Σ_2) are related to the width of features (ΔE) in the spectra by equation (11).[172] For the copper and nickel data

$$\Delta E = \left[\sum_1^2 + \sum_2^1 \left(\frac{\partial K_2/\delta E}{\partial K_1/\delta E}\right)\right]^{\frac{1}{2}} \qquad (11)$$

presented it is apparent that considerable differences in lifetimes occur in the two materials. In copper, the density of states around E_F is low and thus electron holes exhibit relatively long lifetimes because they have a small recombination amplitude. Thus very sharp spectral features can be resolved; in our experiments FWHM values in the 100 mV range have been observed. In nickel, on the other hand, the Fermi level is within the flat *d*-bands, deactivation of electron holes is much more probable, and it appears that spectral features are much broader for this material.[145] The asymmetric broadening of the spectral features in nickel spectra can be attributed also to the rapid variation in the hole lifetimes across the nickel valence bands.[114]

In Figure 11 Pendry's theoretical Ne(I) spectrum [154] for an exit angle 45° off the normal of an (100) copper surface and 40° from the ΓX line is compared with the experimental spectrum measured by Ilver and Nilsson.[137] The surface state position on the theoretical spectrum and found in the measurement agree well with each other and with the independent wave-matching calculations of Gurman.[151]

Pendry's calculated spectra take account of refraction exactly as in LEED theory, but neglect the variation in photon field across the emitter surface [*i.e.* the term $\partial A_\perp/\partial Z$ in equation (7)]. Recent calculations by Haydock and McLean [115,173] take this variation into account for a simple step in dielectric permittivity at the

---

[170] J. B. Pendry, 'Low Energy Electron Diffraction', Academic Press, New York, 1974.
[171] C. B. Duke, *Adv. Chem. Phys.*, 1974, **27**, 1.
[172] J. B. Pendry, in 'Photoemission from Surfaces', ed. R. F. Willis, B. Feuerbacher, B. Fitton, and C. Backx, Wiley, New York, 1977.
[173] R. K. McLean, in 'Photoemission from Surfaces', ed. R. F. Willis, B. Feuerbacher, B. Fitton, and C. Backx, Wiley, New York, 1977.

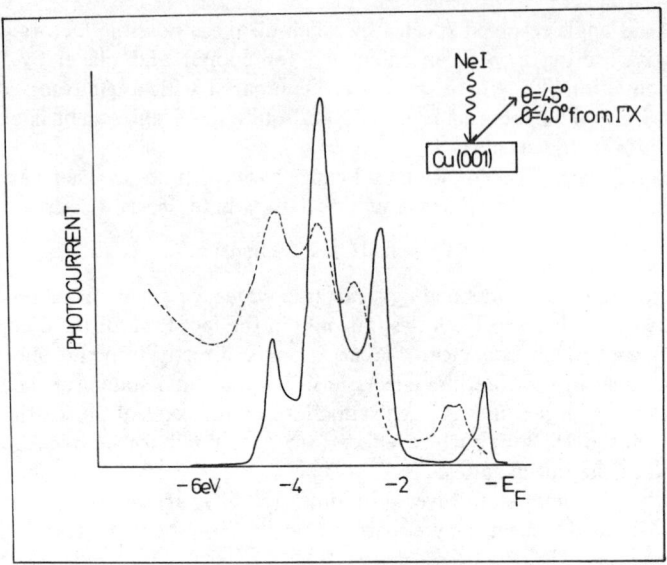

Figure 11 (———) *Theoretical ADES spectrum as calculated by Pendry* [154] *for* Ne(I) *photons generating photocurrent at 45° polar emission angle and azimuthal angle 40° from* ΓX *for a* Cu(100) *surface. Inverse lifetimes as in Figure 4(a).* (---) *experimental data* [137] (Reproduced by permission from ref. 154)

surface.[174] These calculations start from the same basis, equation (8), as the Pendry calculation, but generally involve only plane-wave final states, except when strong dependences on photon incidence angles have to be accounted for. Refraction at the surface is neglected while initial states are calculated only using the simplest tight binding approximation.[175]

The basic results of all these calculations are as follows. Photoemission from d-bands is largely controlled by matrix element effects. These vary substantially with photon energy, and the apparent invisibility of s-bands in He(I) angle-resolved spectra of copper and other transition metals can be traced to this effect. At least for palladium [133] there is evidence that s-band related features are present in the spectra. The differential sensitivity of surface state bands to emission can also change with photon energy. For example, Figure 12 shows a comparison of the contribution of the surface states on Cu(111) just below the Fermi level, to He(I), Ne(I),[132] and Ar(I) [135] normal emission angle-resolved spectra for this surface.

Complex relaxation processes [176] do not seem important in determining the angle-resolved spectra for copper, nickel, or palladium single-crystal emitters. In copper and nickel the more sophisticated calculations of angle-resolved spectra support the conclusion that the bulk local d-orbital model is substantially good up to the surface, which is surprising [177] in the light of accepted escape depths in the

[174] M. Born and E. Wolf, 'Principles of Optics', Pergamon Press, Oxford, 1970.
[175] J. C. Slater and G. F. Koster, *Phys. Rev.*, 1954, **94**, 1498.
[176] N. J. Shevchik and P. C. Kemeny, *Solid State Comm.*, 1975, **17**, 255.
[177] N. E. Christensen, in 'Photoemission from Surfaces', ed. R. F. Willis, B. Feuerbacher, B. Fitton, and C. Backx, Wiley, New York, 1977.

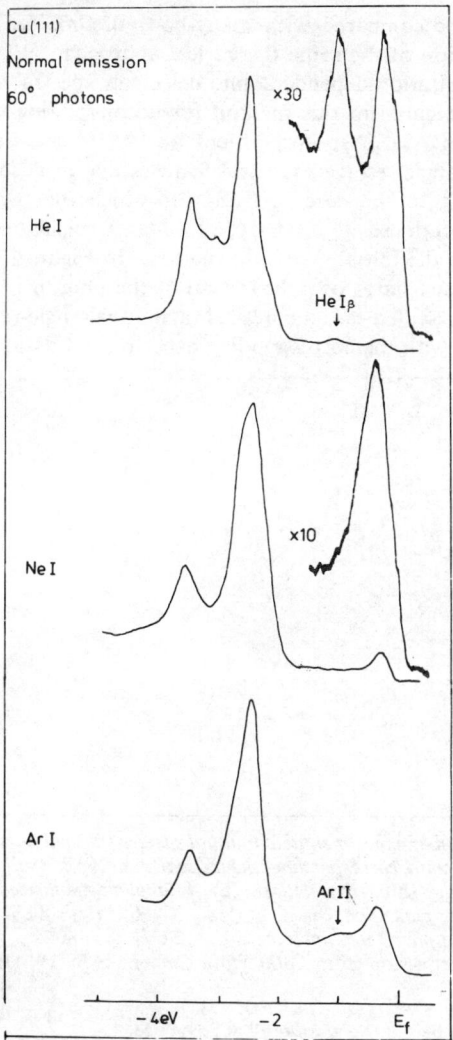

Figure 12 *Normal emission He(I), Ne(I),[132] and Ar(I) [135] ADES spectra for the region related to the copper d-band and the surface states in the gap near the Fermi level for a Cu(111) surface. The differential sensitivity of the surface state related feature to the photon energy is apparent*

u.v. region and the evidence for significant variations in surface electronic structure.[178—180] The use of optical potentials in the framework of equation (8) provides a sound theoretical foundation for the width of angle-resolved spectral

---

[178] P. Fulde, K. P. Bochnen, and H. Takayama, *Z. Phys.*, 1976, **B23**, 45.
[179] P. Fulde, A. Luther, and R. E. Watson, *Phys. Rev.*, 1973, **B8**, 440.
[180] M. C. Desjonqueres and F. Cyrot-Lackman, *J. Phys. (F)*, 1975, **5**, 1368.

features which can be compared with many-body lifetime formulations.[49,181—187]

A recent application of the same theoretical approach [188,189] to the analysis of angle-resolved polarization-dependent photoelectron spectra for TaS_2 [190] clearly demonstrates the potential in this method for deconvoluting the initial and final state information in ADES spectra. Liebsch's [191—193] one-electron formulation for the analysis of angle-resolved spectral features due to adsorbate levels can be carried over directly to the case of TaS_2 for which the two-dimensional-band structure is well reproduced by KKR-type thin-layer calculations, for example,[194] which thus provides the initial state information. In Figure 13 the results of this calculation for various states of polarization of the photon beam with respect to the surface are presented as theoretical azimuthal-angle-resolved spectra, in excellent agreement with the corresponding experimental data.

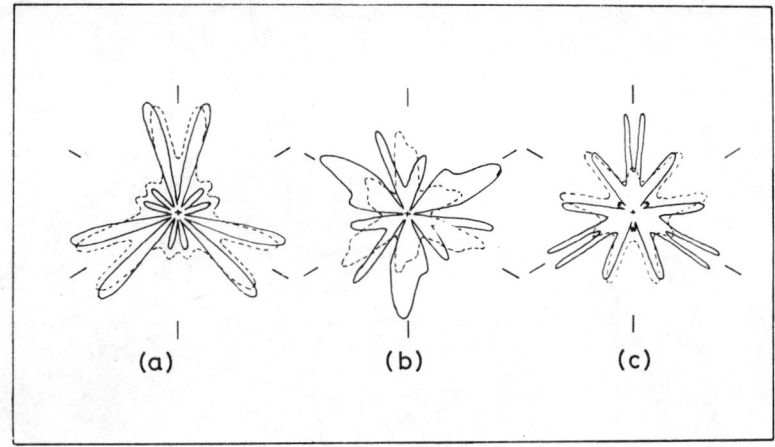

Figure 13 *Comparison of experimental azimuthal spectra (dashed lines) for TaS_2 synchrotron ADES spectra [190] with corresponding theoretical results [188] (solid curves) using actual initial and final state (arbitrary units).* (a) A in plane, polarized, $A_\perp k_{f_\parallel}$, $E_f = 10$ eV, $\theta_f = 42°$; (b) A in plane, polarized, $\angle(A,k_f) = 36°$, $E_f = 8$ eV, $\theta_f = 57°$ *(theoretical curve:* $48°$); (c) A in plane, unpolarized, $E_f = 15$ eV, $\theta_f = 60°$
(Reproduced by permission from *Solid State Comm.*, 1976, **19**, 1193)

[181] B. I. Lundquist, *Phys. Stat. Sol.*, 1969, **32**, 273.
[182] J. J. Quinn and R. Ferrell, *Solid State Phys.*, 1969, **23**, 1.
[183] K. Hedin, in 'Electrons in Crystalline Solids', IAEA, Vienna, 1973.
[184] L. Hedin, in 'X-ray Spectroscopy', ed. L. Azaroff, McGraw Hill, New York, 1974.
[185] S. Lundquist, in 'Elementary Excitations in Solids, Molecules and Atoms, Part A', NATO Adv. Study Inst. Series, Plenum Press, New York, 1974.
[186] S. Lundquist and G. Wedin, *J. Electron Spect.*, 1974, **5**, 513.
[187] J. W. Gadzuk in 'Electronic structure and Reactivity of Metal Surfaces', NATO Adv. Study Inst. Series, Plenum Press, New York, 1976.
[188] A. Liebsch, *Solid State Comm.*, 1976, **19**, 1193.
[189] N. V. Smith and A. Liebsch, to be published.
[190] N. V. Smith, M. M. Traum, J. A. Knapp, J. Anderson, and G. P. Lapeyre, *Phys. Rev.*, 1976, **B13**, 4462.
[191] A. Liebsch, *Phys. Rev. Letters*, 1974, **32**, 1202.
[192] A. Liebsch, *Phys. Rev.*, 1976, **B13**, 544.
[193] A. Liebsch, in 'Photoemission from Surfaces', ed. R. F. Willis, B. Feuerbacher, B. Fitton, and C. Backx, Wiley, New York, 1977.
[194] N. Kar and P. Soven, *Phys. Rev.*, 1975, **B11**, 3761.

3 Angle-resolved Spectra of Adsorbed Species

Theory.—The transition rate for the development of a final state from an initial state involving surface-adsorbed species illuminated by photons follows also from equation (1). Within the one-particle approximation again this equation can be simplified and expressed most compactly as in equation (8). But the Green's function matrix at the initial state energy is written now in the non-orthogonal mixed basis of the surface molecule model for the substrate–adsorbate binding,[195–216] just as in the simple molecular orbital description of the surface bond proposed over 40 years ago by Gurney.[217] It is an attractive starting point for a theory of photoemission from adsorbed species. The adsorbate-related features in spectra can be due to highly localized core levels on the adsorbate atoms only, the cluster-type levels of the surface molecules contained essentially within the adsorbate atoms and immediately surrounding neighbours, or the bands in the surface region formed by mixing with substrate valence bands. The mixed basis set of the surface molecule model, in principle, provides a complete description of all these states.

The final states within the one-particle formalism can be also so represented as positive energy states of the surface molecules in the presence of the semi-infinite solid. For isolated adsorbed species the limiting approximation to the final state is the plane wave,[22,23,46,218–220] or for the case of significant back-scattering from the substrate, a multiple scattered wavefunction.[188,190–193,221,222] For ordered layers of adsorbed species LEED overlayer theory is available.[173]

---

[195] D. M. Newns, *Phys. Rev.*, 1969, **178**, 1123.
[196] R. H. Paulson and J. R. Schrieffer, *Surface Sci.*, 1975, **48**, 329.
[197] T. B. Grimley, *J. Phys. (C)*, 1970, **3**, 1934.
[198] T. B. Grimley, in 'Dynamic Aspects of Surface Physics', ed. F. O. Goodman, Editrice Composition, Bologna, 1973.
[199] T. B. Grimley and C. Pisani, *J. Phys. (C)*, 1974, **7**, 2831.
[200] T. B. Grimley, in 'Batelle Colloquim; The Basis of Heterogeneous Catalysis', 1974.
[201] R. Gomer, *Crit. Rev. of Solid State Sci.*, 1974, **4**, 247.
[202] J. W. Gadzuk, in 'Surface Physics of Crystalline Materials', ed. J. M. Blakely, Academic Press, New York, 1975.
[203] J. W. Gadzuk, *Surface Sci.*, 1974, **43**, 44.
[204] J. W. Gadzuk, *Surface Sci.*, 1967, **6**, 133.
[205] J. R. Smith, in 'Topics in Applied Physics', ed. R. Gomer, Springer-Verlag, Berlin, 1975, Vol. 4.
[206] T. B. Grimley, *Adv. Surface Membrane Sci.*, 1975, **9**, 71.
[207] J. Appelbaum and D. R. Hamann, *Phys. Rev. Letters*, 1975, **34**, 806.
[208] N. D. Lang and A. R. Williams, *Phys. Rev. Letters*, 1975, **34**, 531.
[209] O. Gunnarson and H. Hjelinberg, *Physica Scripta*, 1975, **11**, 97.
[210] T. B. Grimley and M. Torrini, *J. Phys. (C)*, 1973, **6**, 886.
[211] J. R. Schrieffer, *J. Vac. Sci. Technol.*, 1972, **9**, 561.
[212] T. B. Grimley, *J. Vac. Sci. Technol.*, 1971, **8**, 31.
[213] R. Haydock and M. J. Kelly, *Surface Sci.*, 1973, **38**, 139.
[214] S. Lyo and R. Gomer, in 'Topics in Applied Physics', ed. R. Gomer, Springer-Verlag, Berlin, 1975, Vol. 4.
[215] B. J. Thorpe, *Surface Sci.*, 1972, **33**, 306.
[216] D. R. Penn, *Surface Sci.*, 1973, **39**, 333.
[217] R. W. Gurney, *Phys. Rev.*, 1935, **47**, 479.
[218] J. W. Gadzuk, *Surface Sci.*, 1975, **53**, 132.
[219] J. W. Gadzuk, *Jap. J. Appl. Phys.*, 1974, Suppl. 2, Pt. 2, p. 851.
[220] J. W. Gadzuk, *Solid State Comm.*, 1974, **15**, 1011; *Phys. Rev.*, 1974, **B10**, 5030.
[221] T. L. Einstein, *Surface Sci.*, 1974, **45**, 713.
[222] M. Scheffler, K. Kambe, and F. Forstmann, in 'Photoemission from Surfaces', ed. R. F. Willis, B. Feuerbacher, B. Fitton, and C. Backx, Wiley, New York, 1977.

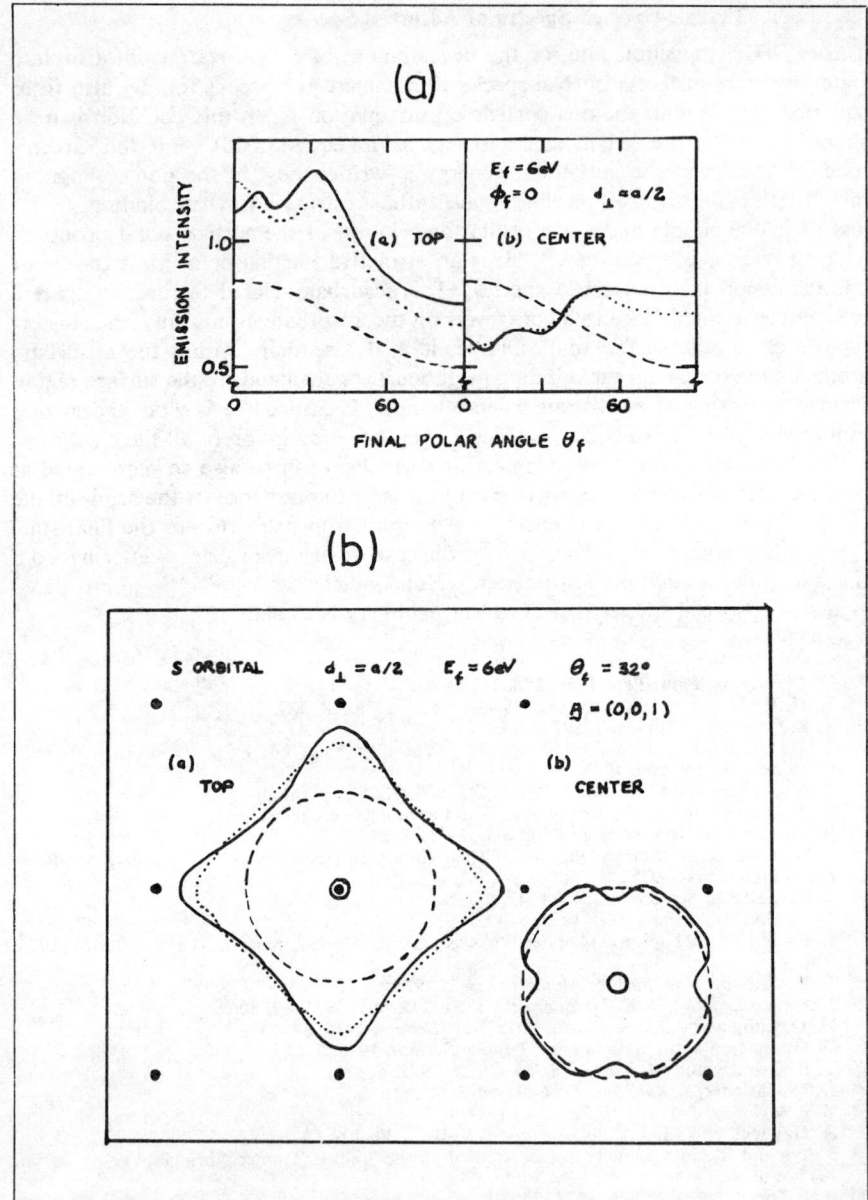

Figure 14 (a) *Predicted photocurrent intensity (arbitrary units) as a function of polar emission angle θ_f for photoemission from an s-orbital in an adsorbate atom sited* (a) *above a substrate atom of a square surface array, and* (b) *above the centre of the unit cell of this lattice.* (b) *Predicted azimuthal dependence of the photocurrent in these two cases for a polar emission angle of* $32°$
(Reproduced by permission from *Phys. Rev. Letters*, 1974, **32**, 1202)

Grimley and his co-workers [23,223] have emphasized the more crucial dependence likely in the photoemission signals from adsorbed species on many-body effects. For photoemission from a valence level in an isolated surface molecule the unrelaxed hole final state is shared mostly between the adsorbate and immediately surrounding surface atoms. The lifetime of the hole depends on many-body interactions, and a one-particle final state is not satisfactory. For photoemission from an adsorbate core level the final state electron–hole interaction will approximate closely to that in gas-phase photoemission.

Four different types of theoretical calculation have been made within the surface molecule framework. For localized adatom levels, final state angular anisotropies have been obtained for multiple scattered wavefunctions which reflect a strong dependence on the symmetry of the initial state orbital and the geometry of the adsorption site. In the case of an adsorbate s-orbital, for example, localized on two possible adsorption sites on a simple cubic lattice, the predicted polar and azimuthal angle dependences of the photoemission signals are as in Figures 14a and 14b.[191]* For cluster-type surface molecule levels photoemitting into plane-wave final states, strongly anisotropic effects also are to be expected, dependent on photon energy. For example, for an s-orbital in an adsorbate on the same cubic

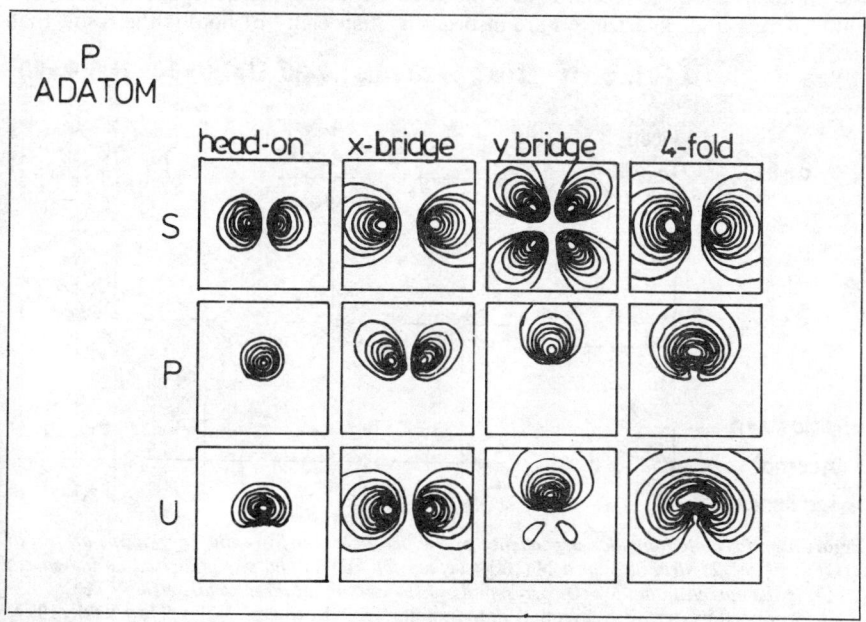

Figure 15 *Predicted azimuthal dependences for oriented p-orbitals with respect to the sitings identified in the inset for projection of the polar setting on a plane above the emitter surface.*[220] *The incidence angle of the photons is 45° and the ZY plane is the plane of incidence. Results are presented for s, p, and unpolarized radiation*
(Reproduced by permission from *Phys. Rev.*, 1974, **B10**, 1011)

[223] T. B. Grimley and G. M. Doyen, in Photoemission from Surfaces, ed. R. F. Willis, B. Feuerbacher, B. Fitton, and C. Backx, Wiley, New York, 1977.

lattice, but in this case, interacting with the in-plane d-orbitals of immediately surrounding neighbours, the azimuthal dependence in Figure 15 [220, 46] is to be expected.

In the third type of calculation, attempts [193, 222] have been made to treat the initial and final states similarly. For example, for the adsorption of oxygen on a (100) surface of nickel to give the $p(2 \times 2)$ structure, Scheffler, Kambe, and Forstmann [222] based their initial state wavefunctions on cluster calculations [224] but used time-reversed LEED states for the final states, allowing for refraction with free-electron theory. For the intensities of spectral features, results compatible with a group theoretical analysis [225] of equation (2) are obtained. Thus, from an initial s-state, grazing incidence light polarized perpendicular to the surface does not generate transitions except to p_z-type states, while a p_z-type initial state of the surface can couple only to s and d_{z^2}-type final states for the same radiation polarization.

The results of this more sophisticated calculation are summarized in Figures 16 and 17. In each Figure the first column of diagrams presents the polar angle dependence from the surface normal [$(\bar{1}00)$] with the analyser sitting azimuthally along the $(0\bar{1}0)$ direction. The remaining columns present the azimuthal dependence of the photoemission cross-section into the polar directions marked. The initial states, light energy, and polarization are also given. Especially of note is the result that

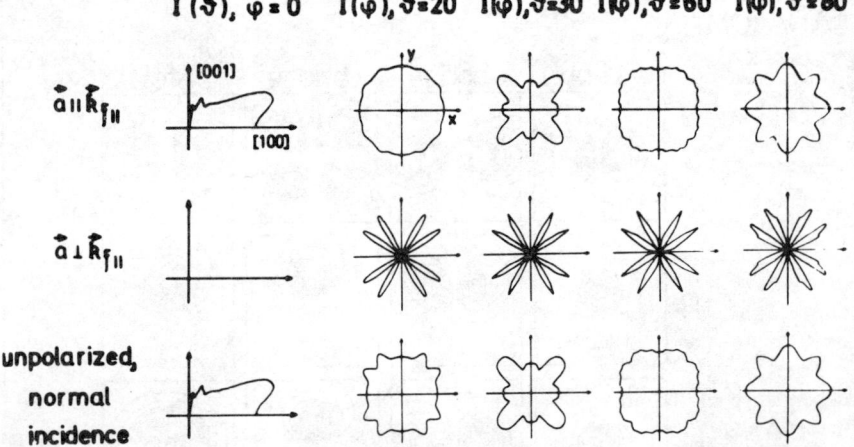

Figure 16 *Predicted angular dependence of the photocurrent from the oxygen $2p_z$ orbital of O/Ni $p(2 \times 2)$ structure on a Ni(100) surface for He(II) photons polarized as indicated. $I(\theta)$ is the current, and $\phi = 0$ corresponds to the azimuthal directional setting $\langle 110 \rangle$* (Reproduced by permission from 'Photoemission from Surfaces', Wiley, New York, 1977)

---

* The conclusion in this early calculation of Liebsch's that multiple scattering effects are not important in the photoemission final states is due to the simplicity of the model used (simple cubic crystal, s-wave only scatterers). In general this model is not adequate and Liebsch's later work shows that contributions due to multiple scattering can be about 40% of the signal (ref. 192 and A. A. Liebsch, *Phys. Rev. Letters*, 1977, **38**, 248).

[224] I. P. Batra and O. Robaux, *Surface Sci.*, 1975, **49**, 653.
[225] J. L. Birman, *Phys. Rev.*, 1966, **150**, 771.

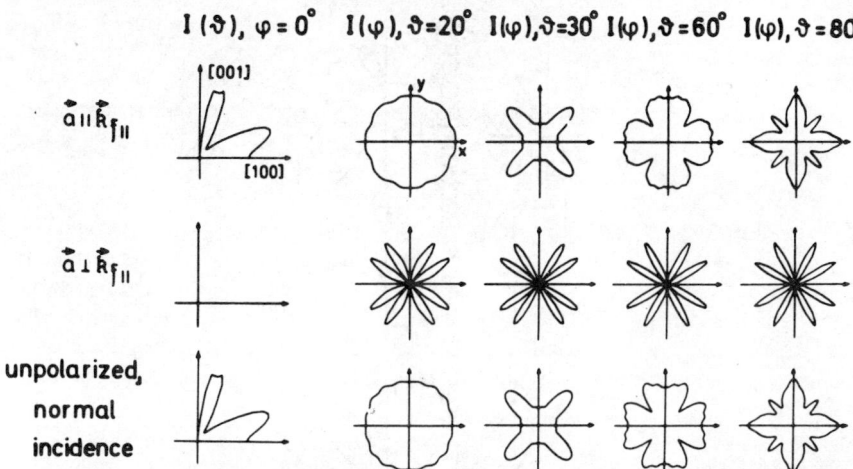

Figure 17 *Predicted angular dependence for the a_1 (mostly O 2s) orbital of the same surface molecule, as in Figure 16*
(Reproduced by permission from 'Photoemission from Surfaces', Wiley, New York, 1977)

the symmetry of the initial state alone is not necessarily of dominant importance in determining the angular distribution of the photoemission. For example, the s and p_z orbitals of an adsorbate atom on Ni($\bar{1}$00) exhibit the same azimuthal dependence around the surface normal, and yet the azimuthal angular dependence of the photoemission in these two cases are quite distinct.

In Figure 17, the predicted angular effects for other polarization states of the radiation again are in line with the group theory analysis. For emission into a direction lying in a mirror plane of the emitter the dipole matrix element in equation (2) must be invariant under reflection in the plane. For s-polarized radiation lying in such a plane, the initial and final states must exhibit the same parity.[9,222,226] Thus a distinction can be made between contributions to the photocurrent from the p_z orbital and the (p_x,p_y) degenerate pair in an adsorbate on such a surface.

For the $c(2 \times 2)$ structure on Ni(100) McLean[173] has calculated the angle dependence of the cross-section into time-reversed LEED states using overlayer theory and the recursion technique, *i.e.* with A lying in the surface plane. For p-polarization A lies in the plane containing the surface normal and the direction of polarization.

In the fourth type of calculation, the limiting case of the isolated surface molecule is invoked, and it is assumed that the photoemission can be discussed in terms only of individual adsorbate species with particular orientations. Two models have been proposed for species adsorbed molecularly *via* adsorbate p-orbital interaction. Grimley[22] has calculated the angular variations of photocurrent intensity due to the emission from oriented p-orbitals. In the model, it is assumed

---

[226] G. P. Lapeyre, J. Anderson, and R. J. Smith, in 'Photoemission from Surfaces', ed. R. F. Willis, B. Feuerbacher, B. Fitton, and C. Backx, Wiley, New York, 1977.

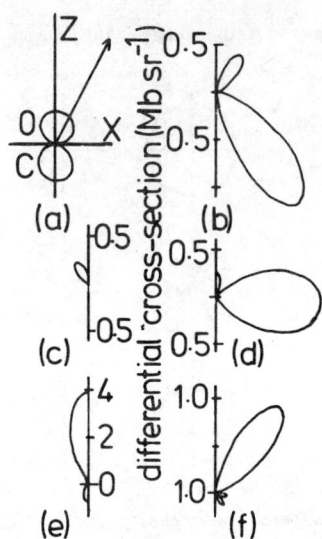

Figure 18 *Predicted angular distribution of photoemitted electrons from an oriented CO molecule at $\omega = 41$ eV. (a) Molecular orientation. (b)—(f) Polar plots of differential cross-section in Mb sr^{-1} for (b) the 5σ level with $\hat{A}$ along X axis, (c) the 1π level with $\hat{A}$ along Z, (d) the 1π level with $\hat{A}$ along X, (e) the 4σ level with $\hat{A}$ along Z, and (f) the 4σ level with $\hat{A}$ along X. For clarity, only half of each distribution is shown, but they are all symmetrical about the Z axis. Note the different scales*
(Reproduced by permission from *Phys. Rev. Letters*, 1975, **36**, 945)

that the final states are plane waves and that the radiation is polarized in the surface plane, *i.e.* incident along the normal. The principal result is that the relative emission intensity $p_z/(p_x,p_y)$ falls rapidly for photocurrent measured at greater angles to the surface normal.

For oriented N_2 and CO molecules specifically, Davenport [227] has calculated the angular and energy dependence of the photoemission cross-section using the scattered wave molecular photoionization method proposed by Dill and Dehmer.[228] In this technique the cluster final state wavefunctions are chosen to exhibit the asymptotic form of incident Coulomb waves plus incoming spherical waves, and so final state resonance effects, not obtainable with plane waves, can be determined in the energy dependence of the photoionization cross-section. The predicted angular dependence of the cross-section for oriented CO is given in Figure 18 for He(I) photons polarized with A in the surface plane or on the normal. The 5σ level is not represented in the diagram for A parallel to the normal ($\hat{z}$) because of the low cross-section values in this case. A distinction with the analysis for plane-wave final states is evident also in these angular distributions. For the electric vector in the surface plane the π level exhibits a large cross-section for emission along the normal, just less than that of the 4σ (p_z) orbital. For a plane-wave final state no photocurrent would be expected along the normal from this initial state [115]

[227] J. W. Davenport, *Phys. Rev. Letters*, 1975, **36**, 945.
[228] D. Dill and J. L. Dehmer, *J. Chem. Phys.*, 1974, **61**, 692.

since $A \cdot P$ is zero. This underlines clearly the need to include the effect of elastic scattering in the final state.

Comparisons with Experiment.—Direct comparison of much of the theory in the previous section with angle-resolved spectra for adsorbate-covered surfaces is limited at present because few angle-resolved experiments on adsorbate layers have been reported, and fewer still have used polarized radiation.

Carbon monoxide adsorption has been studied by several groups using angle-

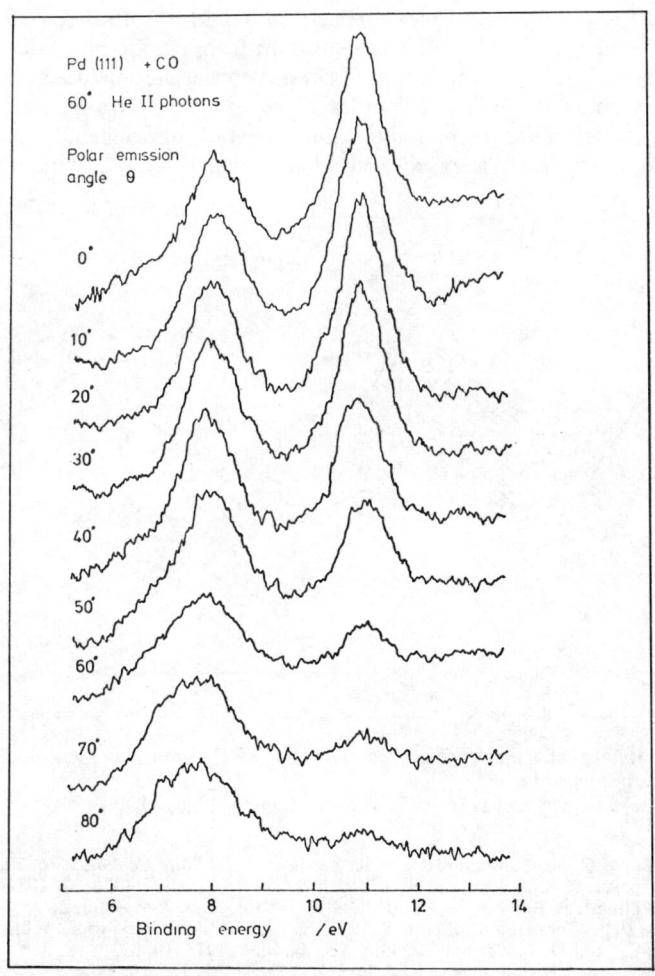

Figure 19 *The variation in photocurrent intensity of the CO related features in the ADES spectra at different polar emission angles for* He(II) *photons incident at* 60° *on a CO covered* Pd(111) *surface at* 25 °C
(Reproduced by permission from *Solid State Comm.*, 1976, **20**, 409)

resolving spectrometers.[226,229–231] Of these measurements only one series [230] has been carried out with polarized radiation. A non-angle-resolved measurement with polarized synchrotron radiation has also been carried out recently.[232]

Figure 19 shows the variation in intensity of the CO derived bands in He(II) spectra with polar emission angle for carbon monoxide adsorbed on a palladium (111) crystal surface.[229] In these measurements the radiation was incident at 60°. For He(I) and He(II) the less pronounced dependence on photon angle in the spectra is illustrated in Figures 20 and 21. Very similar spectra have been obtained for CO adsorbed on nickel (111) surfaces,[231] the difference in this system centring on the broadening of the CO related feature nearest the Fermi level.

The angle-resolved spectral measurements all conform with the general pattern observed in other studies of the photoemission from carbon monoxide adsorbed molecularly on transition metals.[233–239] Two chemisorption-induced bands, about 8 and 11 eV below the emitter Fermi level, are apparent. This contrasts with the gas-phase spectrum for carbon monoxide in which three bands are observed corresponding to clearly resolved 5σ (C lone pair), 1π (C—O bonding), and 4σ

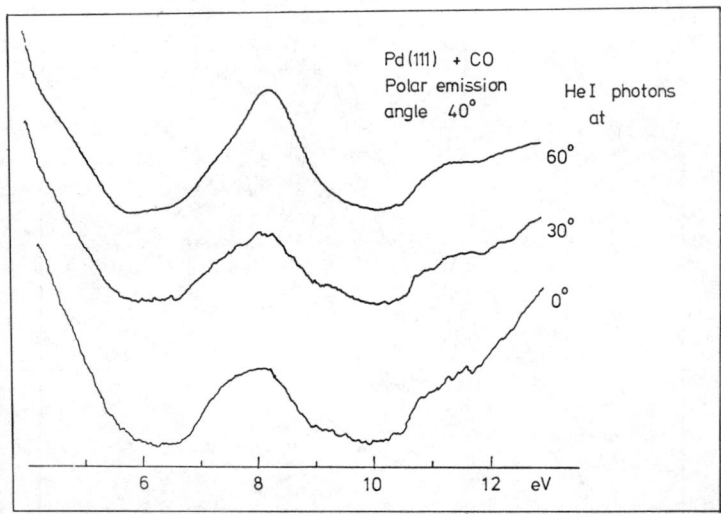

Figure 20 He(I) *photon incidence dependence in the ADES spectrum at* $\theta = 40$ *for the same surface as in Figure 19*
(Reproduced by permission from *Solid State Comm.*, 1976, **20**, 49)

---

[229] D. R. Lloyd, C. M. Quinn, and N. V. Richardson, *Solid State Comm.*, 1976, **20**, 409.
[230] G. P. Lapeyre, J. Anderson, and R. J. Smith, *Phys. Rev. Letters*, 1976, **37**, 1081.
[231] P. M. Williams, P. Butcher, S. Woods, and K. Jacobi, *Phys. Rev.*, in press.
[232] G. Apai, P. S. Wehner, J. Stohr, R. S. Williams, and D. A. Shirley, to be published.
[233] J. M. Baker and D. E. Eastman, *J. Vac. Sci. Technol.*, 1973, **10**, 233.
[234] G. E. Becker and H. D. Hagstrum, *J. Vac. Sci. Technol.*, 1973, **10**, 31.
[235] S. R. Atkinson, C. R. Brundle, and M. W. Roberts, *Chem. Phys. Letters*, 1974, **24**, 175.
[236] J. A. Clarke, I. D. Guy, B. Law, and R. Mason, *Chem. Phys. Letters*, 1975, **31**, 29.
[237] R. W. Joyner and M. W. Roberts, *J.C.S. Faraday I*, 1974, **70**, 1819.
[238] J. W. Linnett, D. L. Perry, and W. F. Egelhoff, jun., *Chem. Phys. Letters*, 1975, **36**, 331.
[239] P. G. Page, D. L. Trimm, and P. M. Williams, *J.C.S. Faraday I*, 1974, **70**, 1769.

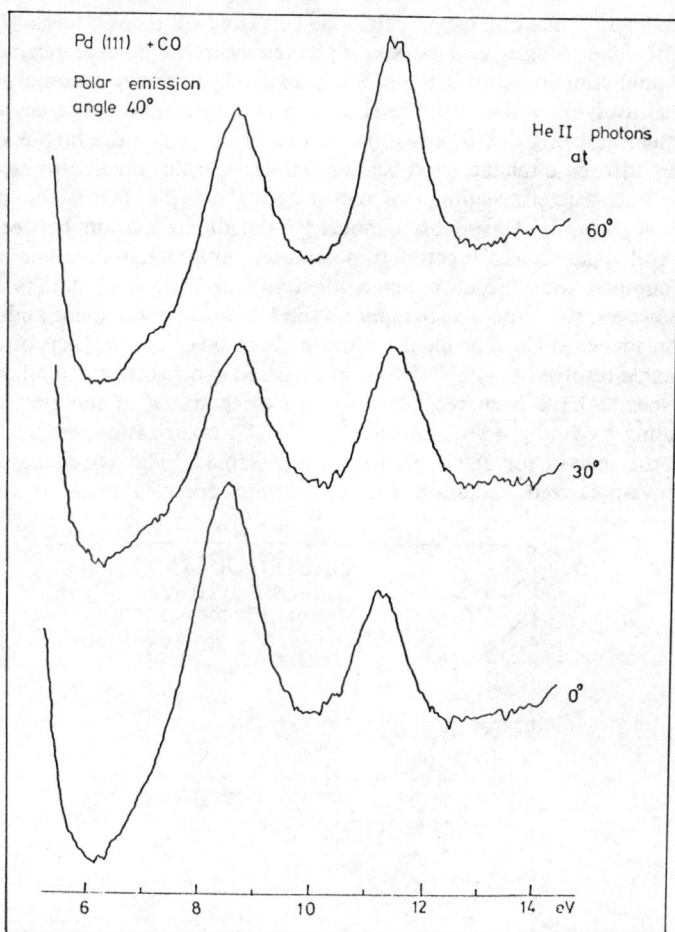

Figure 21 *Similar data for* He(II) *photons*
(Reproduced by permission from *Solid State Comm.*, 1976, **20**, 409)

(mainly O lone pair) contributions,[240] and there has been considerable controversy over the assignment of the adsorbate-induced features.[241] The angle-resolved spectra in conjunction with the theoretical angular dependence of the photoionization cross-sections for the individual orbital symmetry types aid the relating of the three gas-phase spectral features to the two bands observed for the adsorbed species.

In Grimley's model [22] the relative intensity of the $p_z/(p_x,p_y)$ type contributions to spectra should decrease rapidly with polar angle. The 5σ and 4σ orbitals of carbon monoxide exhibit substantial p_z character, and the 1π orbital is of pure

[240] A. D. Baker, D. Baker, C. R. Brundle, and D. W. Turner, 'Molecular Photoelectron Spectroscopy', Wiley, London, 1970, p. 49.
[241] D. R. Lloyd, C. R. Brundle, and E. W. Plummer, *Faraday Discuss. Chem. Soc.*, 1974, **58**, 136.

(p_x, p_y) character. Thus one expects that the contribution from σ levels should be most prominent in angle-resolved spectra taken near the surface normal, while π levels should contribute most to spectra taken well off the surface normal. The band further away from the Fermi level in both Ni and Pd spectra agrees well with its assignment as being due to emission on this basis. Since the high-energy side of the other adsorbate-related band behaves similarly to this it can also be assigned to be of σ character, leading to the ordering of the levels for adsorbed CO $1\pi > 5\sigma \gg 4\sigma$. In Davenport's model [227] the differentiation between σ and π levels is still apparent as a function of polar angle, although the angular variations are more complex than those predicted when a plane-wave final state is assumed. Overall, however, this model also supports the 1π, 5σ, 4σ assignment and provides a photon incidence angle dependence in the angle-resolved spectra as is observed.

In the angle-resolved study [226,230] with polarized synchrotron radiation normal emission spectra have been recorded for two orientations of the electric vector corresponding to s- ($\theta_A = 90°$) and p- ($\theta_A = 47.7°$) polarization, respectively. In Figure 22 the spectra for 28 eV photons are presented. The 4σ-related feature is absent for s-polarized radiation for all photon energies used in this work

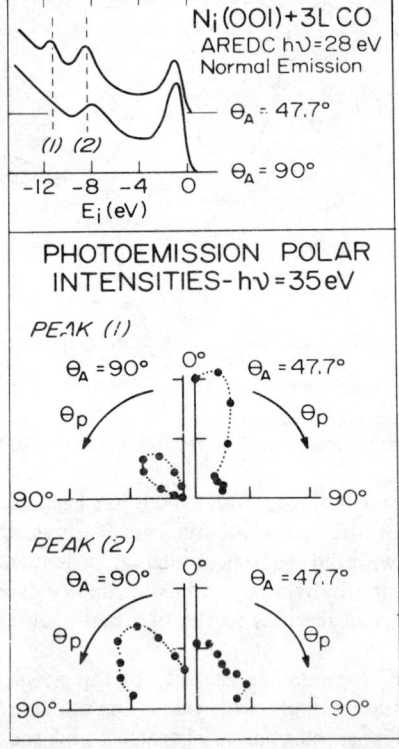

Figure 22 (a) *Angle-resolved EDC's (normal emission) for* Ni(001) + CO. (b), (c) *Polar dependent amplitudes of chemisorbed peaks* (1) *and* (2)
(Reproduced by permission from 'Photoemission from Surfaces', Wiley, New York, 1977)

($\hbar\omega \leqslant 40$ eV), while the combined peak is present in the spectra for both states of polarization of the photons. No azimuthal dependence was observed in the adsorbate-related spectral features. These results agree well with Davenport's calculation. For CO molecules bonded perpendicularly to the surface the absence of the 4σ peak is exactly as predicted for s-polarized radiation. A study of polar variations of intensity is also shown in Figure 22 (b) and (c). The data for $\theta_A = 90°$ should be directly comparable with Davenport's calculation (Figure 18), apart from a small difference in photon energy, and the agreement between the calculation for 4σ and the experimental data for peak (1) is striking. The data in the right hand side of Figure 22 are for an angle intermediate between the two sets of calculation, so comparison is less direct, but the agreement is probably still good. Since there is now substantial agreement from other sources that peak (2) does correspond to 4σ, these measurements also confirm the orientation of CO perpendicular to the metal surface, bonded through carbon. Further evidence is provided by the changes in intensity with photon energy variation which also show good agreement with the calculation.[242]

New features appear in similar angle-resolved synchrotron photoemission from a tungsten (100) surface with adsorbed hydrogen.[226,243] In Figure 23 the results

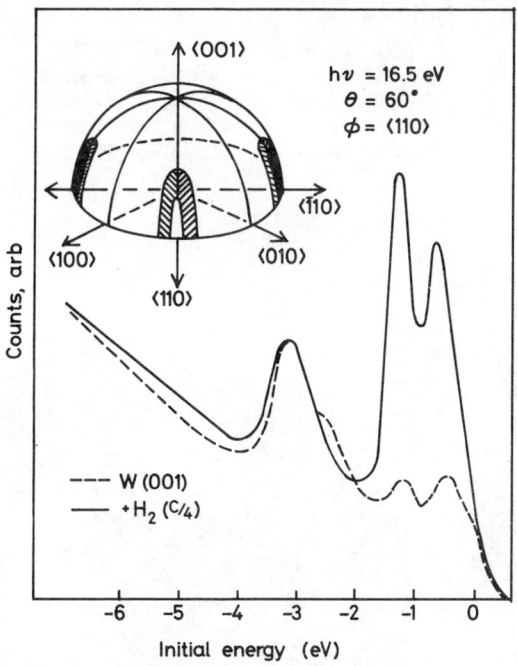

Figure 23 *AREDC's for clean* W(001) *(dashed line) and* W(001) + *quarter monolayer hydrogen (solid line). The inset is an emission hemisphere which shows, by shading, the angular properties of the hydrogen-induced structure*
(Reproduced by permission from *Phys. Rev. Letters*, 1976, **37**, 1081)

---

[242] T. Gustafsson, E. W. Plummer, W. Gudat, and D. E. Eastman, to be published.
[243] J. Anderson and G. J. Lapeyre, *Phys. Rev. Letters*, 1976, **36**, 376.

of angle-resolved photocurrent measurement for s-polarized radiation on clean W(100) and on W(100) plus one quarter monolayer of hydrogen are presented; the azimuthal angle of the analyser setting is given as the direction contained in that azimuth (*e.g.* $\langle 110 \rangle$). A strong four-fold symmetry of the major adsorbate-induced features with azimuthal angle is observed. It can also be demonstrated that these adsorbate-induced features obey selection rules for A perpendicular to or parallel to mirror planes of the substrate.[226]

From these observations it can be concluded that, in contrast to the situation with adsorbed molecular CO,[230] the energy bands formed by hydrogen chemisorption exhibit the mirror plane symmetry of the substrate. Also highly significant is the occurrence of two major spectral features for photon energies in the range $14 \leqslant \hbar\omega \leqslant 24$ eV, which are attributable to surface Umklapp involving the reciprocal lattice vector of length 1.4 Å^{-1} of the $c(2 \times 2)$ structure present at about one quarter monolayer hydrogen coverage on this surface plane.[244]

It is evident that at the time of writing, theory and experiment are progressing well in the study of angle-resolved photoemission from clean surfaces. Indeed, given the rough nature of real surfaces as opposed to the ideal surfaces of theory,

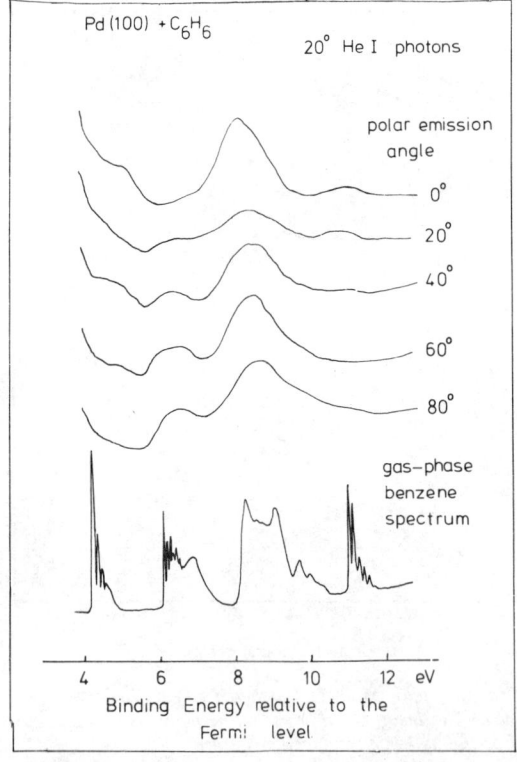

Figure 24 *ADES spectra for benzene layer on* Pd(111) *at* 25 °C

[244] D. C. Adams and L. H. Germer, *Surface Sci.*, 1970, **23**, 419.

it is very encouraging that such good agreement is being found. Experimental work on adsorbate-covered surfaces is being actively pursued by several groups, but very little of this has appeared so far in the literature, and in this sense theory at present is ahead of experiment. It may be appropriate here to comment that a theoretical ideal is often not easy to achieve in practice. Several theoretical studies have reported results for A perpendicular to a model surface. While this does indeed maintain a high symmetry, it is doubtful whether this can correspond to a physically realizable situation. Emitted signals fall very rapidly towards grazing photon incidence. More seriously, real surfaces are terraced frequently, so at high angles of incidence it is quite likely that there will be a majority of emission from the terraces rather than from the steps. Of course, limiting cases are of interest, but it is also highly desirable that calculations should also be carried out for intermediate situations which are reasonably easy to duplicate experimentally.

Finally, as an indication that experiments of the type discussed here are likely to be of significance in adsorption studies for more complex molecules, Figure 24 shows several ADES spectra for benzene adsorbed on a palladium (111) surface.[135]

5
Point Defects in Ionic Crystals

BY J. CORISH, P. W. M. JACOBS, AND S. RADHAKRISHNA

1 Introduction

This report deals with the nature, properties, and influence of point defects in ionic crystals, and the subject matter is as was defined in a previous report.[1] There are two main objectives. The first of these is to survey the literature for the years 1972—1975 inclusive (*i.e.* papers included in *Physics Abstracts* to July 1976) with coverage also of more recent literature which we know about and consider to be important. Thus the earlier sections, which deal with pure and doped ionic crystals, respectively, will essentially update the earlier report and we will not repeat the basic theoretical concepts and practical details. There have, we believe, also been exciting developments in the theoretical computation of defect parameters, both in the sophistication of the calculations themselves and in the model potentials used to simulate the crystals. Such results, which may now be of real assistance in the interpretation of experimental results, are reviewed in the third section.

The second objective is to extend the boundaries set by the previous article to encompass a number of significant topics which have emerged more forcibly recently. Since these have not been formally covered before they receive more thorough treatments which include basic concepts and details of experimental techniques where appropriate. In general, the defect types included arise because of a more adventurous choice of ions as dopants. Thus we discuss paraelectric centres, molecular impurities of various symmetries and complexity, and the optical properties of ions with the s^2 configuration. These fields were chosen for review because of the interest they are currently generating. We should have liked to enlarge this survey of point defects by the inclusion of rare-earth and transition-metal ion impurities but space, and time, did not permit an adequate treatment. The subject of fast ion conduction, which was previously touched upon only fleetingly, is now also given a more extended treatment with particular emphasis on those compounds exhibiting order–disorder transitions.

A number of books [2—5] have been published recently which contain important and useful sections on various aspects of point defects. Relevant new reviews include those by Steele and Dudley,[6] Hobbs,[7] who deals comprehensively with the theoretical and practical details of extended defects in the alkali halides, and

---

[1] J. Corish and P. W. M. Jacobs, in this series, 1973, Vol. 2, chap. 7.
[2] J. Hladik, 'Physics of Electrolytes', Vols. 1 and 2, Academic Press, New York, 1972.
[3] 'Crystals with the Fluorite Structure', ed. W. Hayes, Oxford, 1974.
[4] M. S. Seltzer and R. I. Jaffee, 'Defects and Transport in Oxides', Plenum Press, New York, 1974.
[5] J. H. Crawford, jun. and L. M. Slifkin, 'Defects in Solids', Plenum Press, New York, 1972.
[6] B. C. H. Steele and G. J. Dudley, in MTP International Review of Science, Inorganic Chemistry Series Two, Vol. 10, Butterworth, London, 1975.
[7] L. W. Hobbs, in this series, 1975, Vol. 4, chap. 6.

2 Intrinsic Defect Parameters

Self-diffusion.—For the ideal case of migration by a single defect species only the self-diffusion coefficient D and the ionic conductance σ are related by the modified Nernst–Einstein relationship (1), where q is the charge on the defect, N is the

$$D/\sigma = fkT/q^2N \tag{1}$$

number of ions per unit volume on the appropriate sub-lattice and f, the correlation factor, is typical of the migration mechanism.[1] In real crystals, however, the situation is complicated by the existence of contributions to ionic transport from more than one mechanism and by contribution to the diffusion (but not to the conductance) from neutral vacancy pairs.[1] The vacancy pair components of the self-diffusion may, in principle, be separated by measurements in crystals where bivalent doping on one sublattice has effectively reduced the concentration of single vacancies on the other. For the main part the low solubility of bivalent anions has confined such measurements to the anion vacancy pair component. At the lower temperatures anion diffusion may occur readily on grain boundaries or along dislocations and is then dominated by this 'extrinsic' component.[10] An exhaustive table of self-diffusion data prior to 1972 for the alkali halides is given by Fredericks.[8] Although there have been a reasonable number of measurements on a variety of crystals since then relatively few of these report separate data for vacancy pair and single vacancy components. The recent data are shown in Table 1 with the relevant temperature ranges and details of the contributing components where they have been specified.

In addition to their measurements on $PbBr_2$ (cf. Table 1), Rushbrook Williams and Barr[11] have used the sensitive isotope exchange technique to study anion self-diffusion in AgBr. They, as in the early work of Tannhauser,[12] found a continuously curved Arrhenius plot in the range $10^{-17} < D\, cm^2\, s^{-1} < 10^{-9}$ but by varying the bromine gas pressure concluded that a single anion vacancy mechanism was unlikely to operate. Anion self-diffusion in AgBr has also been studied recently by Batra and Slifkin,[13] who again found the curved Arrhenius plot. In spite of the reliability apparent in the primary data the effective energies and entropies appear to be very anomalously large, and Batra and Slifkin suggest that current theory cannot explain quantitatively the results at the very high defect concentration found in AgBr. The pressure dependence of ^{22}Na diffusion in NaCl has shown that the cation diffusion process is characterized by a pressure-

---

[8] W. J. Fredericks, in 'Diffusion in Solids: Recent Developments', ed. A. S. Nowick and J. J. Burton, Academic Press, New York, 1975, chap. 8.
[9] Proceedings of Conference 'Défauts de résau dans les cristaux ioniques', Marseille-Luminy 1973, Supplement C9 to *J. Phys. (Paris)*, 1973, **34**.
[10] K. Dawson and L. W. Barr, *Proc. Brit. Ceram. Soc.*, 1967, **9**, 171; R. G. Fuller, *Phys. Rev.*, 1966, **142**, 524; K. S. Sabharwal, J. Mimkes, and M. Wuttig, *J. Appl. Phys.*, 1975, **46**, 1839.
[11] S. Rushbrook Williams and L. W. Barr, *J. Phys. (Paris)*, 1973, **34**, C9-173.
[12] D. S. Tannhauser, *J. Phys. Chem. Solids*, 1958, **5**, 224.
[13] A. P. Batra and L. M. Slifkin, *J. Phys. (C)*, 1976, **9**, 947.

Table 1 *Recent self-diffusion data. D_0 and E_a are the pre-exponential and activation terms if the temperature-dependence of the self-diffusion coefficient D is expressed as $D = D_0 \exp(-E_a/kT)$. For contributing components v, i, and vp refer to vacancies, interstitials and vacancy pairs, respectively. A number of investigations which did not produce data suitable for inclusion in this table are discussed in the text*

Crystal	Diffusant	Temperature range/°C	D_0/cm² s⁻¹	E_a/eV	Contributing components (where specified)
NaCl[a]*	²²Na⁺	~550—760	8.44	1.96	vp
	³⁶Cl⁻	~500—750	1670	2.54	vp
NaBr[b]	Na⁺	569—760	0.06	1.29	(from isotope effect)
KF[c]	F⁻ (n.m.r.)	197—817		0.83	v(cation)
				1.35	v(anion)
KCl[a]	⁴³K⁺	616—747	5480	2.65	vp
	³⁶Cl⁻	543—747	133	2.39	vp
NH₄Cl[d]†	³⁶Cl⁻	127—171		0.6—0.8	
RbCl[e]	³⁶Cl⁻	576—700		1.88	v + vp
RbBr[f]	⁸⁶Rb⁺	525—675	55	2.01	v + vp
	⁸²Br⁻	600—675	32	1.99	v + vp
CsCl[g]	¹³⁷Cs⁺	360—460	1.10	1.35	v + vp
	³⁶Cl⁻	360—460	1.64	1.29	v + vp
				1.7 (estimated)	anion vp
AgCl[h]	³⁶Cl⁻	307—433	35	1.57	v
			133	1.66	vp
TlCl[i]	³⁵Cl⁻ (n.m.r.)	~40—111		0.73	
PbCl₂[j]	³⁶Cl⁻	97—352	1.03 × 10⁻³	0.79	v
PbBr₂[k]	⁸²Br⁻	50—350		0.31	v
TlBr[l]	²⁰⁴Tl⁺	300—420	1.316	1.19	
	⁸²Br⁻	250—420	1.227	0.77	
CaF₂[m]	⁴⁵Ca²⁺	987—1246	5.35 × 10³	4.15	
CdF₂[n]	¹⁸F⁻	270—640	1 × 10⁻⁶	0.39	v
BaF₂[o]	¹⁹F⁻ (n.m.r.)	RT—927		0.77	i
				0.61	v
TiO₂[p]‡	⁴⁴Ti	1197—1510	0.046	2.60	D_{zz}
			0.0024	2.11	D_{xx}

* These results refer to a reanalysis which includes ionic conductivity as well as cation and anion self-diffusion in pure and cation and anion bivalently doped crystals and which utilizes some data given in earlier publications by these authors. The free vacancy parameters are included in Table 2. † No definite assignment of activation enthalpy claimed. ‡ Diffusion in rutile is characterized by extreme anisotropy.

[a] M. Bénière, M. Chemla, and F. Bénière, *J. Phys. Chem. Solids*, 1976, **37**, 325;   [b] A. N. Murin, I. V. Murin, V. I. Portnyagin, and A. M. Andreev, *Soviet Phys. Solid State*, 1974, **15**, 1898;   [c] I. M. Hoodless, J. H. Strange, and L. E. Wylde, *J. Phys. (Paris)*, 1973, **34**, C9–21;   [d] J. D. Heneisen and A. L. Laskar, *Phys. Letters (A)*, 1974, **49**, 369;   [e] I. M. Hoodless and R. G. Turner, *Phys. Stat. Sol. (A)*, 1972, **11**, K55;   [f] H. L. Downing and R. J. Friauf, *Phys. Rev. (B)*, 1975, **12**, 5981;   [g] I. M. Hoodless and R. G. Turner, *J. Phys. Chem. Solids*, 1972, **33**, 1915;   [h] A. P. Batra and L. M. Slifkin, *J. Phys. (C)*, 1976, **9**, 947;   [i] G. L. Samuelson and D. C. Ailion, *Phys. Rev. (B)*, 1972, **5**, 2488;   [j] A. K. Pansare and A. V. Patankar, *Pramāna*, 1974, **2**, 282;   [k] S. Rushbrook Williams and L. W. Barr, *J. Phys. (Paris)*, 1973, **34**, C9–173;   [l] S. Kurihara, K. Fueki, and T. Mukaibo, *Chem. Letters*, 1972, 831;   [m] M. F. Berard, *J. Amer. Ceram. Soc.*, 1971, **54**, 144;   [n] P. Süptitz, E. Brink, and D. Becker, *Phys. Stat. Sol. (B)*, 1972, **54**, 713;   [o] D. R. Figueroa, A. V. Chadwick, and J. H. Strange, to be published;   [p] T. S. Lundy and W. A. Coghlan, *J. Phys. (Paris)*, 1973, **34**, C9–299.

dependent activation volume varying from $\sim 37\,\mathrm{cm}^3\,\mathrm{mol}^{-1}$ at low pressures to $\sim 25\,\mathrm{cm}^3\,\mathrm{mol}^{-1}$ at high pressures.[14]

The full potential for the study of diffusion processes by nuclear magnetic relaxation techniques has recently become apparent with the use of new experimental methods and advances in the theoretical treatment of the relaxation-time data which allow diffusion parameters to be evaluated. The early weak-collision theory [15] which treated the dipolar Hamiltonian as a perturbation on the Zeeman system was complemented by the strong-collision theory [16] appropriate to systems in which the external magnetic field was less than or comparable with the local dipolar field of the nuclei. Wolf and Jung [17] have developed a comprehensive perturbation formalism which applies to both the strong- and weak-collision cases. Wolf [18] in his 'encounter model' has also refined the random walk model to include the correlated motions of the nuclei which result from randomly migrating point defects. This new formalism has been applied to the cases of vacancy and interstitialcy mechanisms of diffusion on a fluorite lattice.[19] The results indicate an anisotropy in the relaxation times with respect to sample orientation in the field and this together with the shape of spin–lattice relaxation time minima are indicative of the diffusion mechanism. A comparison of ionic conductance and n.m.r. self-diffusion measurements interpreted in this way for BaF_2 show excellent agreement [20] (cf. Tables 1 and 2).

Data have been reported for cation self-diffusion in SrO and BaO,[21] in MgO [22] and in Li_2SO_4.[23] Matzke [24] has measured cation diffusion in the temperature range 1400—2200 °C in stoicheiometric and hyperstoicheiometric oxides MO_2 and MO_{2+x} (M = U or U + Pu), and the results indicate a vacancy mechanism. For MO_{2-x} an interstitialcy mechanism seemed most probable. Feit et al.[25] have discussed the problem of heterogeneous impurities present on the sample surface or in the tracer solution which may diffuse with the tracer in self-diffusion experiments on insulators which the tracer is applied as a thin source. They have shown that even small concentrations of impurity may cause marked deviations, and they present a formalism for a fairly precise determination of 'free' tracer diffusion.

Ionic Conductivity.—Extensive measurements of the variation in ionic conductivity with temperature over as wide a range as is feasible, and analyses of these data using computer techniques,[26] continue to provide most of the known intrinsic defect parameters. Results deduced in this way, however, are dependent on the

---

[14] G. Martin, D. Lazarus, and J. L. Mitchell, *Phys. Rev. (B)*, 1973, **8**, 1726.
[15] N. Bloembergen, E. M. Purcell, and R. V. Pound, *Phys. Rev.*, 1948, **73**, 679; H. C. Torrey, *Phys. Rev.*, 1953, **92**, 962; 1954, **96**, 690.
[16] C. P. Slichter and D. C. Ailion, *Phys. Rev.*, 1964, **135**, A1099.
[17] D. Wolf and P. Jung, *Phys. Rev. (B)*, 1975, **12**, 3596.
[18] D. Wolf, *Phys. Rev. (B)*, 1974, **10**, 2710, 2724.
[19] D. Wolf, D. R. Figueroa and J. H. Strange, *Phys. Rev. (B)*, 1977, **15**, 2545.
[20] D. R. Figueroa, A. V. Chadwick, and J. H. Strange, to be published.
[21] R. A. Swalin, *J. Phys. (Paris)*, 1973, **34**, C9-167.
[22] B. J. Wuensch, W. C. Steele, and T. Vasilos, *J. Chem. Phys.*, 1973, **58**, 5258; B. C. Harding and D. M. Price, *Phil. Mag.*, 1972, **26**, 253.
[23] A. Bengtzelius, A. Kvist, and A. Lundén, *J. Phys. (Paris)*, 1973, **34**, C9-199.
[24] Hj. Matzke, *J. Phys. (Paris)*, 1973, **34**, C9-317.
[25] M. D. Feit, J. L. Mitchell, and D. Lazarus, *Phys. Rev. (B)*, 1973, **8**, 1715.
[26] J. H. Beaumont and P. W. M. Jacobs, *J. Chem. Phys.*, 1966, **45**, 1496.

model assumed and even in the most frequently studied and apparently simple substances some anomalies remain (cf. detailed arguments of Fredericks [8]). Analyses should at least include data for both pure and doped crystals and, where possible, self-diffusion data also. Results for NaCl and KCl derived from such a variety of measurements, including bivalent anion doping, have recently been presented by Bénière et al.[27] In several careful studies of the alkali halides, dissatisfaction with analyses using the conventional model has been evident. This model, following the suggestion of Allnatt and Jacobs,[28] ascribes the curvature evident in the intrinsic Arrhenius plot to the onset of a contribution to the conductivity from an anion vacancy migration with a higher activation enthalpy. Various additional proposals, such as trivacancies [29] and the presence of cationic Frenkel defects,[30] have been made to account for discrepancies between theory and experiment. In KCl Jacobs and Pantelis [31] included models which allowed for the possibility of Frenkel defects on one or both sublattices and also an excess conductance model, but none was completely satisfactory. More recently data from experiments on pure and doped KBr, which included a study of the effects of uniaxial stress, have been analysed using a model which contained a dislocation contribution. Models which allowed for three different configurations of complexes and for the formation of clusters of impurity ions and cation vacancies were also tried, but again none of these was completely acceptable.[32] Downing and Friauf [33] have also reported an anomalous situation in RbBr. Anion and cation self-diffusion studies show almost equal values for the activation energies of their respective vacancy motions, but the Arrhenius conductivity plot for the same crystals is curved. The authors suggest that the most probable logical explanation is in the vacancy pair contributions to the diffusion, but recent theoretical calculations [34] may offer an alternative solution. These calculations report almost equal anion and cation vacancy migration enthalpies in each of the 16 alkali halides with the rock-salt structure, and indicate that the observed high-temperature curvature in conductivity plots is most likely due to transport of interstitials by the collinear interstitialcy mechanism. For the $NaCl:Sr^{2+}$ system Tallon [35] has suggested that impurity-vacancy dipole reorientation at internal and external surfaces may account for the small non-random discrepancies reported by Allnatt et al.[36]

Curvature in the high-temperature intrinsic Arrhenius conductivity plots of the silver halides has also been the subject of several studies. Corish and Jacobs [37] reinterpreted their earlier conductivity data for AgCl [38] using a number of new models which included a temperature-dependent enthalpy model. In this treatment the enthalpies of formation and of activation for the various mechanisms con-

[27] M. Bénière, M. Chemla, and F. Bénière, *J. Phys. Chem. Solids*, 1976, **37**, 525.
[28] A. R. Allnatt and P. W. M. Jacobs, *Trans. Faraday Soc.*, 1962, **58**, 1.
[29] R. G. Fuller and M. H. Reilly, *J. Phys. Chem. Solids*, 1969, **30**, 457.
[30] A. R Allnatt and P. Pantelis, *Solid State Comm.*, 1968, **6**, 309.
[31] P. W. M. Jacobs and P. Pantelis, *Phys. Rev.* (B), 1971, **4**, 3757.
[32] N. Brown and P. W. M. Jacobs, *J. Phys.* (Paris), 1973, **34**, C9–437.
[33] H. L. Downing and R. J. Friauf, *Phys. Rev.* (B), 1975, **12**, 5981.
[34] C. R. A. Catlow, J. Corish, K. M. Diller, P. W. M. Jacobs, and M. J. Norgett, *J. Phys.* (*Paris*), 1977, 1976, **37**, C7–253.
[35] J. L. Tallon, *J. Phys.* (C), 1975, **8**, 3801.
[36] A. R. Allnatt, P. Pantelis, and S. J. Sime, *J. Phys.* (C), 1971, **4**, 1778.
[37] J. Corish and P. W. M. Jacobs, *Phys. Stat. Sol.* (B), 1975, **67**, 263.
[38] J. Corish and P. W. M. Jacobs, *J. Phys. Chem. Solids*, 1972, **33**, 1799.

tributing to the conductivity were allowed to vary with temperature and this model was found to fit the data equally as well as the corresponding temperature-independent enthalpy model. Aboagye and Friauf [39] have analysed their data for both pure AgCl and AgBr using constraints on the fitting derived from earlier doped conductivity and tracer diffusion studies. The data were fitted and the parameters determined at intermediate temperatures using a Chi square test to define the range.[40] The differences between values extrapolated from these data and the experimental curve at higher temperatures are shown to exceed corrections provided by available defect interaction theories and are represented as a correction, $-\Delta g$, to the free energy of formation of Frenkel defects. The recent measurements of the diffusion of sodium in AgCl[41] show that the intrinsic cation vacancy concentration is given correctly by the excess free energy interpretation of Aboagye and Friauf. Re-examinations have been carried out of some early conductivity data for very heavily doped AgCl and AgBr using a model which includes a contribution to the conductivity from the formation and dissociation of vacancy–interstitial Frenkel pairs.[42] It was found to be unnecessary to assume the existence of a significant degree of vacancy–impurity association or that Coulombic forces seriously disturbed the randomness of defect distribution. The theory of spontaneous recombination of nearby Frenkel pairs has been developed in particular for metals[43] but has, more recently, been extended to some ionic crystals.[44] Spontaneous recombination radii of the order of 5 Å are evaluated.[45]

The formation and migration parameters for intrinsic defects are shown in Table 2. To present a reasonably comprehensive coverage within the acceptable size limits set for this report we have limited this Table to one entry per substance. For substances which have been studied extensively since our previous report this entry will be the latest well-analysed result. For substances for which no such recent work has been found we have given our biased view of the 'best' parameters available. In many cases, due to difficulties in preparation of well-characterized samples or in the interpretation of results, the parameters are of necessity less accurately known and we have made no attempt to signify the quality or reliability of the data in the Table. For further information the original papers should be consulted but we should emphasize that more extensive references for most of these substances and also references to earlier compilations of data will be found in Table 6 of reference 1. Since entropy values depend on the choice of vibrational frequency, original papers should be consulted before comparisons are made.

The ionic conductivity of single crystals of NH_4Cl and NH_4Br, and also of the complete range of their solid solutions from simple cubic through the structure transition to the fcc phase, has been measured.[46,47] The results in phase II for the

[39] J. K. Aboagye and R. J. Friauf, *Phys. Rev. (B)*, 1975, **11**, 1654.
[40] R. J. Friauf, *J. Phys. (Paris)*, 1973, **34**, C9–403.
[41] A. P. Batra and L. M. Slifkin, *Phys. Rev. (B)*, 1975, **12**, 3473.
[42] G. M. Fryer, *Phil. Mag.*, 1975, **32**, 173; 1976, **34**, 217.
[43] V. M. Koshkin, B. I. Minkov, L. P. Gal'chinetskii, and V. N. Kulik, *Soviet Phys. Solid State*, 1973, **15**, 87; K. Dettmann, G. Leibfried, and K. Schroeder, *Phys. Stat. Sol.*, 1967, **22**, 423, 433.
[44] V. M. Koshkin and V. M. Ekkerman, *Soviet Phys. Solid State*, 1975, **16**, 2426.
[45] K. Rössler, *J. Phys. (Paris)*, 1976, **37**, C.7–279.
[46] Y. V. G. S. Murti and P. S. Prasad, *Physica (B)*, 1975, **79**, 243.
[47] S. Radhakrishna and B. D. Sharma, *J. Appl. Phys.*, 1973, **44**, 3848.

Table 2* *Intrinsic defect formation and migration data. The symbols used to describe the disorder type and the nature of the mobile defects are as follows: a = anion; c = cation; e = excess conductivity; F = Frenkel; i = interstitial; ic = interstitialcy collinear; inc = interstitialcy non-collinear; v = vacancy. The data are mainly from conductance measurements except where designated by subscript as col = colour centre reactions, or n.m.r. = nuclear magnetic resonance techniques*

Crystal	Type of disorder	Enthalpy of formation H/eV	Entropy of formation S/k	Mobile defect	Migration enthalpy ΔH/eV	Migration entropy $\Delta S/k$
LiF	S	2.34^a	9.6^a	cv	0.65^b, 0.71$_{\text{n.m.r.}}^e$	
				av	0.67$_{\text{col}}^d$	
LiCl	S	2.12^b		cv	0.41^b	
LiBr	S	1.80^b		cv	0.39^b	
LiI	S	1.06^e	4.5^e	cv	0.43^e	
NaF	S	2.42^f	7.5^f	cv	0.95^f	0.9^f
				av	1.46^f	12.8^f
NaCl	S	2.44^g	9.8^g	cv	0.69^g	1.64^g
				av	0.77^g	1.38^g
NaClO$_3$	Fc	1.98^h	34.8^h	cv	0.46^h	
				ci	0.85^h	
NaBr	S	1.72^i	4.04^i	cv	0.80^i	3.58^i
				av	1.18^j	
NaI	S	1.84^k		cv	0.570^k	
				av	1.35^l	
KF	S	2.72^m		cv	0.84^m 0.83$_{\text{n.m.r.}}^m$	
				av	1.65^m 1.35$_{\text{n.m.r.}}^m$	
KCl	S	2.54^g	8.99^g	cv	0.73^g	2.40^g
				av	0.85^g	3.20^g
KBr	S	2.33^n	8.1^n	cv	0.66^n	0.86^n
				av	1.08^n	6.3^n
				e	1.35^n	‡
KI	S	2.21^o	8.88^o	cv	0.63^o	1.58^o
				av	1.29^o	9.34^o
KN$_3$	S	1.6^p		cv, $\parallel c$ axis	0.6^p	
				cv, $\perp c$ axis	0.7^p	
RbCl	S	2.12^g		cv	0.99^q	
RbI	S	2.1^r	5.8	cv	0.60^r	1.62
				av	1.6^r	1.51
RbN$_3$	S	1.7^p		cv, $\parallel c$ axis	0.8^p	
				cv, $\perp c$ axis	0.9^p	
CsCl	S	1.77^s	10.25^s	cv	0.63^s	19.0^s
				av	0.35^s	14.9^s
CsBr	S	1.80^t	5.6^t	cv	0.36^t	1.0^t
				av	0.51^t	4.3^t
CsI	S	1.828^u	5.3^u	cv	0.573^u	5.17^u
				av	0.287^u	0.154^u
AgCl	Fc	1.452^v	9.41^v	cv	0.275^v	−0.640^v
				cic	−0.014^v	−3.81^v
				cinc	0.104^v	−3.24^v
AgBr	Fc	1.134^v	6.55^v	cv	0.325^v	1.16^v
				cic	0.043^v	−3.18^v
				cinc	0.278^v	1.35^v
AgI (hexagonal)	Fc	0.60^w		cv, $\parallel c$ axis	0.50^w	
				$\perp c$ axis	0.39^w	
				ci, $\parallel c$ axis	0.61^w	
				$\perp c$ axis	0.29^w	

Table 2—cont.

Crystal	Type of disorder	Enthalpy of formation H/eV	Entropy of formation S/k	Mobile defect	Migration enthalpy ΔH/eV	Migration entropy ΔS/k
CaF$_2$	Fa	2.71[x]	5.53[x]	ai	0.79[x]	5.48[x]
				av	0.38—0.47[x]	1.1—2.0[x]
SrF$_2$	Fa	1.74[y]		av	0.94[y]	
				ai	0.94[y]	
CdF$_2$	Fa	4.5[z]		av	0.45[z]	
				ai	0.335[z]	
BaF$_2$	Fa	1.91,[aa]		free av	0.57[aa] 0.62$_{n.m.r.}$[aa]	
		1.80$_{n.m.r.}$[aa]		free ai	0.76[aa] 0.77$_{n.m.r.}$[aa]	
MgF$_2$	Fa	1.42[bb]		∥c axis	0.85[bb]	
MnF$_2$	Fa	1.44[bb]		av, ∥c axis	0.80[bb]	
				ai, ∥c axis	0.88[bb]	
TlCl	S	1.36[cc]	8[cc]	cv	0.4[cc]	
				av	0.09[cc]	
SrCl$_2$	Fa	1.60—1.82[dd]		av	0.34—0.46[dd]	
β-PbF$_2$	Fa			av	0.59—0.63$_{n.m.r.}$[ee]	
					0.68[ee]	
PbFCl	S			av Cl$^-$, ∥c axis	0.28[ff]	
				av F$^-$	0.53[ff]	
PbCl$_2$	S	1.55[gg]		av	0.38[gg]	
PbBr$_2$	S	1.80[hh]		av	0.23[hh]	
KSCN	S	<3.2[ii]§		cv	0.49[ii]§	
		1.8[ii]**		cv	0.74[ii]**	

[x] See text for source of references to earlier compilations.
‡ Excess conductivity contribution to σ of form $(B/T) \exp(-E_a/kT)$ with $10^{-6} B/\Omega^{-1} m^{-1} K = 3$.
§ Low-temperature phase.   ** High-temperature phase.

[a] T. G. Stoebe and P. L. Pratt, *Proc. Brit. Ceram. Soc.*, 1967, **9**, 181;   [b] Y. Haven, *Rec. Trav. chim.*, 1950, **69**, 1471, 1505;   [c] M. Eisenstadt, *Phys. Rev.*, 1963, **132**, 630;   [d] Y. Farge, M. Lambert, and R. Smoluchowski, *Solid State Comm.*, 1966, **4**, 333;   [e] B. J. H. Jackson and D. A. Young, *J. Phys. Chem. Solids*, 1969, **30**, 873;   [f] C. F. Bauer and D. H. Whitmore, *Phys. Stat. Sol.*, 1970, **37**, 585;   [g] M. Bénière, M. Chemla, and F. Bénière, *J. Phys. Chem. Solids*, 1976, **37**, 525;   [h] C. Ramasastry and K. Viswanatha Reddy, *Proc. Roy. Soc.*, 1973, **A335**, 1;   [i] H. Hoshino and M. Shimoji, *J. Phys. Chem. Solids*, 1967, **28**, 1169;   [j] H. W. Schamp and E. Katz, *Phys. Rev.*, 1954, **94**, 828;   [k] D. Kostopoulos, P. Varotsos, and S. Mourikis, *Canad. J. Phys.*, 1975, **53**, 1318;   [l] Ya. N. Pershits and V. L. Veisman, *Fiz. tverd. Tela*, 1970, **12**, 3175;   [m] I. M. Hoodless, J. H. Strange, and L. E. Wylde, *J. Phys. (Paris)*, 1973, **34**, C9-21;   [n] N. Brown and P. W. M. Jacobs, *J. Phys. (Paris)*, 1973, **34**, C9-437;   [o] S. Chandra and J. Rolfe, *Canad. J. Phys.*, 1970, **48**, 397; 412; *ibid.*, 1971, **49**, 2098;   [p] A. L. Laskar, in 'Transport Properties in Alkali Azides', Technical Report CUP101, U.S. Army Research Office, Durham, 1975; also *J. Phys. (Paris)*, 1976, **37**, C7-331;   [q] K. D. Misra and M. N. Sharma, *J. Phys. Soc. Japan*, 1974, **36**, 154;   [r] S. Chandra and J. Rolfe, *Canad. J. Phys.*, 1973, **51**, 236;   [s] I. M. Hoodless and R. G. Turner, *Phys. Stat. Sol. (A)*, 1972, **11**, 689;   [t] A. V. Chadwick, B. D. McNichol, and A. R. Allnatt, *Phys. Stat. Sol.*, 1969, **33**, 301;   [u] C. Nadler and J. Rossel, *Phys. Stat. Sol. (A)*, 1973, **18**, 711;   [v] J. K. Aboagye and R. J. Friauf, *Phys. Rev. (B)*, 1975, **11**, 1654;   [w] G. Cochrane and N. H. Fletcher, *J. Phys. Chem. Solids*, 1971, **32**, 2557;   [x] P. W. M. Jacobs and S. H. Ong, *J. Phys. (Paris)*, to be published;   [y] W. Bollmann, P. Gorlich, W. Hauk, and H. Mothes, *Phys. Stat. Sol. (A)*, 1970, **2**, 157;   [z] Y. T. Tan and D. Kramp, *J. Phys. Chem. Solids*, 1970, **53**, 3691; see also P. Müller, *ibid.*, 1971, **55**, 5144, and Y. Tan, *ibid.*, 1971, **55**, 5145 for discussion;   [aa] D. R. Figueroa, A. V. Chadwick, and J. H. Strange, to be published;   [bb] D. S. Park and A. S. Nowick, *J. Phys. Chem. Solids*, 1976, **37**, 607;   [cc] B. J. H. Jackson and D. A. Young, *Trans. Faraday Soc.*, 1967, **63**, 2246;   [dd] E. Barsis and A. Taylor, *J. Chem. Phys.*, 1966, **45**, 1154;   [ee] J. Schoonman, L. B. Ebert, C. H. Hsieh, and R. A. Huggins, *J. Appl. Phys.*, 1975, **46**, 2873;   [ff] A. F. Halff, J. Schoonman, and A. J. Eijkelenkamp, *J. Phys. (Paris)*, 1973, **34**, C9-471;   [gg] A. K. Pansare and A. V. Patankar, *Pramāna*, 1974, **2**, 282;   [hh] H. Hoshino, S. Yokose, and M. Shimoji, *J. Solid State Chem.*, 1973, **7**, 1;   [ii] A. V. Chadwick, A. J. Collins, and J. N. Sherwood, *J. Phys. (C)*, 1971, **4**, 584.

two salts are interpreted in terms of an anion vacancy mechanism in the pure crystals and extrinsic cation vacancy transport in the impure crystals. An excess conductivity observed at the transition is ascribed to proton release from freely rotating ammonium ions. Earlier Kessler and co-workers [48] had measured a.c. and d.c. conductivity and thermally stimulated depolarization on undoped, Ni^{2+}-doped and NH_3-saturated NH_4Cl crystals grown from solution. They noted that the a.c. and d.c. conductivity differed at low temperatures and were dependent on the amount of water adsorbed and attributed these complex effects to a charge transport mechanism that involved both vacancy and proton jumps. The differences between solution-grown and melt-grown crystals are due to the method of preparation and have been discussed in detail.[47]

In regard to technique Whitham and Calderwood [49] have compared a number of commonly used electrodes for d.c. conductivity measurements on NaCl. They suggest that superior performance below 500 °C is given by a silver-painted electrode and at higher temperatures by a composite silver paint over aquadag type. Pershits and Kalennikova [50] have demonstrated a spectrophotometric method, based on the study of the adsorption of aqueous halide solutions of heavy-metal salts, to determine the energy of formation of Schottky defects in the alkali halides. A recent reanalysis [51] of part of some earlier data for zone refined and 5N purity KCl [31] using two different non-linear minimization routines has emphasized the dangers, mentioned earlier, of basing such analyses on over-restricted data.

Table 3 *Recent data on the volume changes associated with the formation and migration of point defects in ionic crystals. The nomenclature for the type of disorder and the nature of the mobile defect is the same as that used in Table 2*

System	Type of disorder	Activation volume of formation ΔV_f/cm^3 mol^{-1}	Mobile defect	Activation volume for motion ΔV_m/cm^3 mol^{-1}
NaCl	S	55 ± 9^a	cv	7 ± 1^a
NaBr	S	44 ± 9^a	cv	8 ± 1^a
KCl	S	61 ± 9^a	cv	8 ± 1^a
KBr	S	54 ± 9	cv	11 ± 1^a
AgCl	Fc	22^b	cv	9.4^b
AgCl:Mn^{2+}	Fc		Mn	8.1^b
AgBr	Fc	14 ± 1.5^c	cv	5.5 ± 0.5^c
			ci	3.6 ± 0.4^c
NaCl*	S	28^d	cv	5^d

* Polycrystalline.

[a] D. N. Yoon and D. Lazarus, *Phys. Rev.* (*B*), 1972, **5**, 4935. [b] A. N. Murin, I. V. Murin, and V. P. Sivkov, *Soviet Phys. Solid State*, 1973, **15**, 98; [c] S. Lansiart and M. Beyeler, *J. Phys. Chem. Solids*, 1975, **36**, 703; [d] C. G. Homan, F. J. Rich, and J. Frankel, *Phys. Rev.* (*B*), 1976, **14**, 2672.

[48] A. Kessler, *Solid State Comm.*, 1973, **12**, 697; P. Berteit, A. Kessler, and T. List, *Z. Physik*, 1976, **B24**, 15; A. Kessler, *J. Electrochem. Soc., Solid State Sci. Tech.*, 1976, **123**, 1239.
[49] W. Whitam and J. H. Calderwood, *IEEE Trans. Elec. Insul.*, 1973, E1–8, 60.
[50] Ya. N. Pershits and T. A. Kalennikova, *Sov. Phys. Solid State*, 1975, **17**, 4.
[51] C. S. N. Murthy and P. L. Pratt, *J. Phys.* (*Paris*), 1976, **37**, C7–307.

Data obtained in recent investigations of the pressure dependence of ionic conductivity are shown in Table 3. Yoon and Lazarus have made a detailed analysis of their results which takes account of contributions from anions and extrinsic cation vacancies.[52]

3 Impurity Defect Parameters

Diffusion of Impurities.—A recent general treatment of impurity diffusion in the alkali halides,[8] which includes a detailed critical evaluation of experimental methods and techniques of data analysis, emphasizes the difficulties in making reliable measurements of this kind. The primary requirement is for a pure host crystal. The magnitude of the effect produced by an unsuspected impurity content depends on the difference in binding energy of the diffusant–impurity compound and the host–impurity compound. Allen and Fredericks[53] have discussed the $KCl:Hg^{2+}$ system containing OH^- which forms a complex set of impurity compounds. They deduced the diffusion coefficient of OH^- in KCl assuming that Cl_2 reacted with effusing OH^- at the surface, but Ikeda[54] has shown recently that reaction can occur deep in the crystal which suggests penetration by Cl_2. Anion vacancy trapping in KCl has been studied using the sensitivity of the internal vibration modes of molecular impurities to the presence of other defects in the nearest neighbour position.[55] Problems in impurity diffusion measurements also arise because of the chemical impurity of radioactive tracers and because tracers are often obtained in solution. For short-lived isotopes, decay to a different diffusant and subsequent simultaneous diffusion is an obvious source of error and radioactive tracers should be chemically purified before application.[56] Errors may also occur if the assumed boundary conditions are negated by effects such as reflection[57] or if the mathematical analysis employs an over-simplified model. Bénière et al.[58] have treated Fick's second law numerically for the initial conditions of a thin deposited layer and we have already mentioned the model developed by Feit et al.[25] for the simultaneous diffusion of a bivalent with a univalent diffusant. Krause and Fredericks[59] have interpreted the simultaneous diffusion of bivalent impurities.

Because of their importance in understanding the processes of substitutional and interstitial cation migration the diffusivities of Mn^{2+} and Cd^{2+} in AgCl have been carefully remeasured recently by Batra and Slifkin.[60] The segmented Arrhenius plots which had been reported earlier[61] and ascribed to the operation of an interstitial mechanism were not found, and the earlier result is now attributed to perturbation of the vacancy concentration by impurities in the tracer. The fact

---

[52] D. N. Yoon and D. Lazarus, *Phys. Rev.* (*B*), 1972, **5**, 4935.
[53] C. A. Allen and W. J. Fredericks, *Phys. Stat. Sol.* (*A*), 1970, **3**, 143; (*B*), 1973, **55**, 615.
[54] T. Ikeda, *Japan J. Appl. Phys.*, 1973, **12**, 1810.
[55] J. R. Zakis and V. P. Zeikats, *Phys. Stat. Sol.* (*B*), 1972, **51**, K63.
[56] H. Machida and W. J. Fredericks, *J. Phys.* (*Paris*), 1976, **37**, C7-385.
[57] C. A. Allen and W. J. Fredericks, *J. Solid State Chem.*, 1970, **1**, 205.
[58] F. Bénière, M. Bénière and M. Chemla, *J. Chem. Phys.*, 1972, **56**, 549.
[59] J. L. Krause and W. J. Fredericks, *J. Phys. Chem. Solids*, 1971, **32**, 2673; *J. Phys.* (*Paris*), 1973, **34**, C9-25.
[60] A. P. Batra and L. M. Slifkin, *J. Phys.* (*C*), 1975, **8**, 2911.
[61] A. L. Laskar and L. M. Slifkin, *J. Nonmetals*, 1972, **1**, 83; E. W. Sawyer and A. L. Laskar, *J. Phys. Chem. Solids*, 1972, **33**, 1149.

that Mn^{2+} and Cd^{2+} have therefore been shown to diffuse purely substitutionally means that of all the cations examined so far only the univalent noble metals now appear to diffuse interstitially in the silver halides. This suggests that transient covalent bonding between hybrid sp^3 orbitals of the solute and the four adjacent halide ions provides energetic stabilization. Chan and Van Sciver [62] have reported diffusion coefficients for copper in NaI which exceed the self-diffusion of Na^+ by factors of 10^3—10^4. At temperatures above 415 °C an interstitial mechanism is suggested with a value of 0.94 eV for the formation of a copper–vacancy Frenkel pair in NaI.

A number of isotope effect studies have been reported. Morimoto et al.[63] have studied cation electromigration in LiF, LiCl, LiBr, LiI, and KCl, and have elucidated a relationship between the mass effect and the activation energy for conductance. The diffusion of ^{22}Na and ^{24}Na in NaBr has been investigated by Murin et al.,[64] who obtained a value of $\Delta K = 0.84$ for the vacancy mechanism. The results of sodium isotope effect experiments in NaCl, KCl, and KBr [65] and ^{105}Ag and ^{111}Ag in AgBr and AgCl [66] have been summarized previously.[1] Chen and Peterson have reported cation isotope effect measurements in NiO, CoO, and $\langle FeO \rangle$.[67]

Impurity–Vacancy Complexes.—The kinetics of aggregation of bivalent impurity–vacancy complexes have been measured for some years now using a variety of techniques.[1] Cook and Dryden [68] interpreted their dielectric absorption data as due to an initial third-order process followed by equilibration between dipoles and trimers, which gave a plateau on the decay curve, and finally a second third-order process involving the formation of higher aggregates. Crawford [69] has suggested an alternative model involving a two-step process in which a dimer is formed initially and then a trimer by the addition of a third dipole. If the binding energy for the dimers is sufficiently low they will be in equilibrium with free dipoles and third-order kinetics will be observed. This theory suggests that dimer formation might be observed in crystals of low impurity content where the first step might be expected to be slower. More recently Unger and Perlman [70,71] have included dissociation as well as formation of dimers in a kinetic analysis and show that their own data for a number of systems as well as data from other workers, including those of Cook and Dryden, can be interpreted satisfactorily in terms of second-order kinetics. Dimer formation and dissociation occur until equilibrium is reached, and the formation of trimers and higher aggregates only becomes important beyond that plateau. Although Unger and Perlman do not find it necessary to consider the concept of barriers to the formation of stable dimers, as does Crawford, they regard their treatment as a simpler alternative model which does not negate the successive-reaction model. Detailed arguments have been presented recently in

[62] N. H. Chan and W. J. Van Sciver, *Phys. Rev. (B)*, 1975, **12**, 3438.
[63] T. Morimoto, I. Okada, and N. Saito, *Z. Naturforsch.*, 1971, **26A**, 300.
[64] A. N. Murin, I. V. Murin, I. Portnyagin, and A. M. Andreev, *Fiz. Tverd. Tela*, 1973, **15**, 2838.
[65] F. Nicolas, F. Bénière, and M. Chemla, *J. Phys. Chem. Solids*, 1974, **35**, 15.
[66] N. L. Peterson, L. W. Barr, and A. D. Le Claire, *J. Phys. (C)*, 1973, **6**, 2020.
[67] W. K. Chen and N. L. Peterson, *J. Phys. (Paris)*, 1973, **34**, C9–303.
[68] J. S. Cook and J. S. Dryden, *Austral. J. Phys.*, 1960, **13**, 260.
[69] J. H. Crawford, jun., *J. Phys. Chem. Solids*, 1970, **31**, 399.
[70] S. Unger and M. M. Perlman, *Phys. Rev. (B)*, 1974, **10**, 3692.
[71] M. M. Perlman and S. Unger, *J. Electrostatics*, 1975, **1**, 231.

support of third-order [72] and second-order [73] aggregation kinetics. Slifkin et al.[74] have studied the aggregation of Mn^{2+}–vacancy complexes in AgCl using e.p.r. at small concentrations and thermal depolarization techniques for more heavily doped samples. The results demonstrate dominant second-order kinetics under conditions of small initial super-saturation, but third-order kinetics under high initial super-saturation. Kessler [75] has re-examined the $NaCl:Ca^{2+}$ system and found a second-order aggregation process. Comparison with earlier data,[68] however, reveals that simultaneous 'precipitation' at dislocations most probably occurs.

The ultimate product of many of these aggregation processes is the Suzuki phase,[76] which is a precipitate of structure $6MX,M'X_2$ based on a superlattice of M'^{2+} impurity ions and cation vacancies in next nearest neighbour positions in a host crystal MX. Strutt and Lilley [77] have adopted a structural approach to the aggregation problem and have developed a detailed extension to the dimer–trimer model of Crawford.[69] The geometry of dimer and trimer configurations and of the Suzuki precipitate are considered in conjunction with the electrostatic constraints, and only one path leading to the Suzuki phase is apparent. The attendant kinetic equations for the initial decay of dipoles have the following three limiting cases: (i) simple third-order with equilibrium between dimers and dipoles; (ii) predominantly second-order at the start and eventually showing third-order; (iii) initial order less than two with no apparent barrier to reacting dipoles. The authors quote examples from the literature for each of these limiting cases.

Precipitation in the $NaCl:Cd^{2+}$ system with dopant level, c, in the range $5 \times 10^{-6} < c < 7 \times 10^{-3}$ has been studied using ionic conductance measurements.[78] The Suzuki phase is known to exist in this system, and the enthalpy and entropy for its dissolution in the temperature range 500—700 K are $\Delta H_d = 1.40$ eV and $\Delta S_d = 13.9$ k, respectively. Chapman and Lilley [79] have also made a solubility study of the $NaCl:Cd^{2+}$ as well as the $NaCl:Mn^{2+}$ systems using ionic conductance and dielectric loss measurements. The solubility was accounted for in terms of free bivalent cations and associated pairs and trimers, but the Cd^{2+} system showed a much greater tendency to form higher order clusters. Zierold et al.[80] have assumed the Suzuki form for the precipitate in heavily doped $AgCl:Cd^{2+}$ and presented a dipole association model to explain their observed conductivity data. They reported a solubility limit of 200 p.p.m. at room temperature. Simultaneous measurements of light scattering and ionic conductance in NaCl crystals doped with Pb^{2+}, Zn^{2+}, and Cd^{2+} show two light scattering peaks in the range 300—930 K.[81] The low-temperature maxima are attributed to the appearance of a new phase containing the bivalent impurity with release of cationic vacancies, while the high-temperature maxima are interpreted as decoration of charged

---

[72] J. S. Cook and J. S. Dryden, *Phys. Rev. (B)*, 1975, **12**, 5995.
[73] S. Unger and M. M. Perlman, *Phys. Rev. (B)*, 1975, **12**, 5997.
[74] L. Slifkin, J. Dutta, and D. Golopentia, to be published.
[75] A. Kessler, *J. Phys. (Paris)*, to be published.
[76] K. Suzuki, *J. Phys. Soc. Japan*, 1961, **16**, 67.
[77] J. E. Strutt and E. Lilley, *Phys. Stat. Sol. (A)*, 1976, **33**, 229.
[78] E. Laredo and D. Figueroa, *J. Phys. (Paris)*, 1973, **34**, C9–449.
[79] J. A. Chapman and E. Lilley, *J. Phys. (Paris)*, 1973, **34**, C9–455.
[80] K. Zierold, M. Wentz, and F. Granzer, *J. Phys. (Paris)*, 1973, **34**, C9–415.
[81] I. Baltog, C. Ghita, and M. Guirgea, *J. Phys. (C)*, 1974, **7**, 1892.

Table 4 Recent impurity defect data. The experimental methods used are given in parentheses following the references in the footnotes using the following schemes: c = ionic conductance; d = diffusion; d2 = simultaneous diffusion; e = e.s.r. measurements; f = internal friction measurements; n = n.m.r. measurements; o = optical techniques; r = dielectric relaxation; td = thermal depolarization. For an earlier compilation of these and other systems see Table 7 of reference 1. General references to compilations of diffusion data are given in the text

System	Association parameters		Activation parameters for orientation of complexes			Migration of impurities	
	$-H$/eV	$-S/k$	Type of jump (where specified)	$10^{-13}\tau_0^{-1}$/s^{-1}	ΔH/eV	D_0/cm^2 s^{-1}	ΔH/eV
NaCl:Mg^{2+}	*0.483^a			3.4	0.63^b		
					0.64^c		
NaCl:Ca^{2+}	0.752^d			12	0.695^b	1.14 × 10^{-3}	0.851^d
					0.70^c		
NaCl:V^{2+}	0.49^e	4.3	ω_1	30	0.7^e		
			ω_2	170	0.78^e		
NaCl:Cr^{2+}	0.66^f			1.97	0.62^f		
NaCl:Mn^{2+g}					0.65^c		
					0.67^b		
NaCl:Ni^{2+}	0.62^h	6.3		11	0.65^c		
NaCl:Zn^{2+}					0.65^c		
NaCl:Sr^{2+}	0.671^d				0.73^c	2.30 × 10^{-3}	0.925^d
	*0.515^a						
NaCl:Cd^{2+g}	0.46^i	0.48		1.5	0.65^b		
	†0.37^j				0.65^c		
NaCl:Ba^{2+}					0.77^c		
NaCl:Pb^{2+}			ω_2	0.16	0.81^k		
					0.65^c		
NaCl:Y^{3+}	1.17^l	9.5	ω_2	1.37	0.98^l		
NaCl:Pr^{3+}	0.76^m			0.13	0.68^m		
				449	0.88^m		

Point Defects in Ionic Crystals

NaCl:PO$_3^-$	0.44^n‡			
NaCl:Cd^{2+};				
CO$_3^{2-}$	0.25^o			
NaClO$_3$:Ca^{2+}	0.22^p			
NaI:Cu^{2+}			0.45^o	§0.545^q
				**0.957^q
				0.46^r
				1.60^s
KCl:He^r			0.49^t	3.9 × 10^{-5}
KCl:Li$^+$			0.64^c	46.5
KCl:Mg^{2+}		0.026	0.49^v	
KCl:Ca^{2+}	*0.42^u	0.016	0.46^t	
KCl:Mn^{2+}		0.0086	0.63^c	
KCl:Ni^{2+}	0.56^i	0.27	0.67^c	
KCl:Sr^{2+}	0.368^a			
KCl:Cd^{2+}	0.474^w	1.08	0.64^x	4.05 × 10^{-5}
KCl:Ba^{2+}	1.43		0.70^c	0.557^w

* Mean value. † Also see text. ‡ Report for (anion impurity)–(anion vacancy). § High-temperature interstitial mechanism. ** Low-temperature unknown mechanism. †† Interstitial impurity-vacancy complex. ‡‡ These data refer to rare-earth–interstitial complexes, and the details of the jump types will be found in the original references. For SrCl the ions studied were La, Ce, Nd, Gd, Tb, Dy, Ho, Tm, Yb, and Y and the results were within the ranges shown.

^{a} C. Nadler and J. Rossel, *Phys. Stat. Sol. (A)*, 1973, **18**, 711 (c); ^{b} P. Dansas, *J. Phys. Chem. Solids*, 1971, **32**, 2699 (td); ^{c} P. Varotsos and D. Miliotis' *J. Phys. Chem. Solids*, 1974, **35**, 927 (r); ^{d} H. Machida and W. J. Fredericks, *J. Phys. (Paris)*, 1976, **37**, C7–385. (d2); ^{e} V. Trnovcová, E. Mariani, and M. Lébl, *Phys. Stat. Sol. (A)*, 1975, **31**, K43 (c); ^{f} M. Hartmanová, M. Lébl, and E. Mariani, *Phys. Stat. Sol. (A)*, 1975, **31**, K85 (c); ^{g} J. A. Chapman and E. Lilley, *J. Phys. (Paris)*, 1973, **34**, C9–455 (c, r); ^{h} V. Trnovcová, E. Mariani, and K. Polák, *Czech. J. Phys.*, 1974, **24**, 765 (c); ^{i} M. Bénière, M. Chemla, and F. Bénière, *J. Phys. Chem. Solids*, 1976, **37**, 525 (c, d); ^{j} E. Laredo and D. Figueroa, *J. Phys. (Paris)*, 1973, **34**, C9–449 (c); ^{k} J. Grňo and V. Trnovcová, *Phys. Stat. Sol. (A)*, 1973, **18**, 303 (d); ^{l} F. Bénière and R. Rokban, *J. Phys. Chem. Solids*, 1975, **36**, 1151 (c, d); ^{m} S. Radhakrishna and B. D. Sharma, *Phys. Rev. (B)*, 1974, **9**, 2073 (c, d); ^{n} W. Whitam and J. H. Calderwood, *J. Phys. (D)*, 1975, **8**, 1133 (c); ^{o} V. Trnovcová and M. Pavlíková, *Crystal Lattice Defects*, 1974, **5**, 105 (c, d); ^{p} C. Ramasastry and K. Viswanatha Reddy, *Proc. Roy. Soc.*, 1973, **A335**, 1 (c); ^{q} N.-H. Chan and W. J. Van Sciver, *Phys. Rev. (B)*, 1975, **12**, 3438 (c, d, o); ^{r} A. Ya Kupryazhkin, P. V. Volobuev and P. E. Suetin, *Soviet Phys. Tech. Phys*, 1975, **19**, 1105 (d); ^{s} A. N. Murin, V. I. Portnyagin, and Yu. S. Shapkina, *Soviet Phys. Solid State*, 1973, **15**, 648 (d); ^{t} A. Brun and P. Dansas, *J. Phys. (C)*, 1974, **7**, 2593 (td); ^{u} A. Brun and P. Dansas, *J. Phys. (Paris)*, 1973, **34**, C9–61 (td); ^{v} A. Brun, P. Dansas, and F. Bénière, *J. Phys. Chem. Solids*, 1974, **35**, 249 (td, e); ^{w} J. L. Krause and W. J. Fredericks, *J. Phys. (Paris)*, 1973, **34**, C9–25 (d2); ^{x} R. E. Chaney and W. J. Fredericks, *J. Solid State Chem.*, 1973, **6**, 240 (td); ^{y} C. A. Allen and W. J. Fredericks, *Phys. Stat. Sol. (B)*, 1973, **55**, 615 (d); ^{z} N. Brown and P. W. M. Jacobs, *J. Phys. (Paris)*, 1973, **34**, C9–437 (c);

Table 4—cont.

System	Association parameters -H/eV	Association parameters -S/k	Type of jump (where specified)	$10^{-13}\tau_0^{-1}/s^{-1}$	ΔH/eV	D_0/cm^2 s^{-1}	ΔH/eV
KCl:Hg^{2+}	0.80—0.75^y					8.18×10^{-5}	0.57^y
KCl:Pb^{2+}	0.378^w	1.86				1.0×10^{-3}	0.908^w
KCl:Pr^{3+}	0.82^m			0.002	0.66^m		
				4.2	0.86^m		
KBr:Sr^{2+}	0.64^z	2.1			0.65^c		
KBr:Ba^{2+}					0.66^c		
KBr:Pr^{3+}	0.68^m			0.018	0.66^m		
				0.87	0.76^m		
KI:Ca^{2+}	0.778^a						1.05^{aa}
KI:Sr^{2+}	0.257^a						0.85^{bb}
KN$_3$:Na$^+$						7×10^{-3}	
RbCl:Sr^{2+}			ω_2	0.34	0.89^{cc}		
			ω_1	0.036	0.55^{cc}		
RbCl:Pb^{2+}				0.70	0.833^{dd}		
RbCl:O^{2-}					0.60^{ee}		
RbCl:Pr^{3+}	0.70^m			0.005	0.72^m		
				2.43	0.86^m		
RbBr:O^{2-}					0.56^{ee}		
RbI:Sr^{2+}	0.58^{ff}	2.55					
RbI:O^{2-}					0.5^{ee}		
CsCl:Pb^{2+}				0.005	0.64^{gg}		
CsBr:Ca^{2+}					~0.50^{gg}		
CsBr:Pb^{2+}	0.48^{gg}			0.009	0.65^{gg}		
CsI:Ca^{2+}	0.305^a						
CsI:Sr^{2+}	0.287^a						
CsI:Pb^{2+}	0.46^{gg}			0.002	0.62^{gg}		
AgCl:Na$^+$							0.49^{hh}

System					
AgCl:Mn²⁺†		ω_2	0.72^{ii}	3.5	1.19^{ii}
AgCl:Fe²⁺			0.36^{jj}	15.1	1.27^{kk}
AgCl:Zn²⁺	0.6			0.053	0.77^{ll}
AgCl:Cd²⁺†		ω_2	0.01	0.68^{ii}	4.1	1.17^{ll}
AgCl:Cs⁺				0.20	0.83^{mm}
NH₄Cl:Co²⁺	††0.55^{mm}				
CaF₂:Na⁺	0.71^{oo}		1.54	0.61^{nn}	
CaF₂:Ce³⁺‡‡		R_I	20	0.46^{pp}	
CaF₂:Gd³⁺‡‡		R_{II}	7.6 / 26.3	0.42^{qq} / 0.21^{qq}	
MnF₂:Y³⁺	0.56^{rr}				
MnF₂:Li⁺	0.73^{rr}				
SrCl₂:RE‡‡		I / II / III	3.3—0.1 / 3.3—0.01 / 3.3—0.0001	0.90—0.77^{ss} / 0.86—0.71^{ss} / 0.80—0.59^{ss}	
CdF₂:Eu³⁺‡‡			0.1	0.40^{tt}	
BaF₂:K⁺	0.74^{uu}				
BaF₂:La³⁺‡‡	0.21^{uu}				
BaF₂:O²⁻^{uu}					
PbBr₂:Na⁺	0.25^{vv}		16		

^{aa} A. L. Laskar, in 'Transport Properties in Alkali Azides', Technical Report CUP101, U.S. Army Research Office Durham, 1975 (d); ^{bb} G. P. Williams, jun. and J. W. Morton, *Phys. Stat. Sol. (A)*, 1973, **17**, 305 (r); ^{cc} G. P. Williams, jun. and D. Mullis, *Phys. Stat. Sol. (A)*, 1975, **28**, 539 (td, r); ^{dd} K. D. Misra and M. N. Sharma, *J. Phys. Soc. Japan*, 1974, **36**, 154 (t, c); ^{ee} Ch. Kokott and F. Fischer, *Z. Physik*, 1971, **249**, 31 (r). The activation energy for aggregation of complexes is given as 0.79 eV. ^{ff} S. Chandra and J. Rolfe, *Canad. J. Phys.*, 1973, **51**, 236 (c); ^{gg} S. Radhakrishna and K. P. Pande, *Phys. Rev. (B)*, 1973, **7**, 424 (c, d); ^{hh} S. Radhakrishna and S. Haridoss, *Solid State Comm.*, 1976, **18**, 1247 (td); ^{hh} A. P. Batra and L. M. Slifkin, *Phys. Rev. (B)*, 1975, **12**, 3473 (d); ^{ii} A. P. Batra and L. M. Slifkin, *J. Phys. (C)*, 1975, **8**, 2911 (d); ^{jj} P. Sladký, *Phys. Stat. Sol. (A)*, 1975, **27**, K81 (f); ^{kk} D. L. Foster and A. L. Laskar, *Phys. Stat. Sol. (A)*, 1975, **29**, 167 (D); ^{ll} A. P. Batra and L. M. Slifkin, *Phys. Stat. Sol. (A)*, 1973, **19**, 171 (d, below 350 °C); ^{mm} A. P. Batra and L. M. Slifkin, *J. Phys. Chem. Solids*, 1976, **37**, 967 (d); ^{nn} Y. V. G. S. Murti and C. S. N. Murthy, *J. Phys. (Paris)*, 1973, **34**, C9–337 (c, r); ^{oo} P. W. M. Jacobs and S. H. Ong, *J. Phys. (Paris)*, 1976, **37**, C7–331. (c); ^{pp} V. B. Campos and G. F. Leal Ferreira, *J. Phys. Chem. Solids*, 1974, **35**, 905 (r); ^{qq} A. D. Franklin, J. Crissman, and K. F. Young, *J. Phys. (Paris)*, 1973, **34**, C9–179 (e, r, f); ^{rr} D. S. Park and A. S. Nowick, *J. Phys. Chem. Solids*, 1976, **37**, 607 (c); ^{ss} E. J. Bijvank and H. W. Den Hartog, *J. Phys. (Paris)*, 1973, **34**, C9–75 (td); ^{tt} R. Capelletti, F. Fermi, and E. Okuno, *J. Phys. (Paris)*, 1973, **34**, C9–69 (td); ^{uu} D. R. Figueroa, A. V. Chadwick and J. H. Strange, to be published (c, n); ^{vv} H. Hoshino, S. Yokose, and M. Shimoji, *J. Solid State Chem.*, 1973, **7**, 1 (c).

dislocations. Harris [82] has pointed out that the low-temperature peaks may be ascribed to precipitate 'imprints' which are regions of local inhomogeneity left by dissolving precipitates.

Another aspect of impurity–vacancy complexes which has been studied extensively is their reorientation behaviour. A discussion of the experimental methods used and of the group-theoretical analysis of the relaxational modes expected for a complex which can adopt either the nearest neighbour (nn) or next-nearest neighbour (nnn) configurations has been given previously [1] and shows that the relaxation times are, in general, not simply related to a particular jump frequency. Brun and Dansas [83] have considered in detail the available data for KCl, where in certain cases evidence for two overlapping peaks has been observed in thermal depolarization studies. They concluded that in the majority of cases studied, the exceptions being Be^{2+}, Eu^{2+}, Sm^{2+}, and Yb^{2+}, the prominent peak observed is due to vacancy jumping from nnn to nn position. For $KCl:Mn^{2+}$ the second smaller peak has been assigned to aggregates.[84] A recent reinvestigation [85] of the $KCl:Sr^{2+}$ system, for which overlapping peaks had been reported earlier,[86] revealed that crystals annealed for 1 h at 623 K show one major relaxation peak with an activation energy of 0.74 eV. Extension of the annealing time to 15 h leads to the observation of two relaxation processes with activation energies of 0.68 and 0.74 eV. Further study of the dependence of the relaxation processes on annealing time has led to the assignment of the relaxation with $E = 0.68$ to M^{2+} cation vacancy complexes and that with $E = 0.74$ eV to dimers of these complexes. The thermal depolarization technique has been used extensively by Capelletti and co-workers to study the formation and solution of precipitates in doped alkali halide crystals.[87,88]

Recent defect data for the dissolution and diffusion of impurities in ionic crystals and for the behaviour of the resulting complexes are given in Table 4. Again no attempt has been made to signify the quality or reliability of the data. An earlier compilation for these and other systems will be found in Table 7 of reference 1. Extensive data for impurity diffusion in the alkali halides is given by Fredericks [8] and for ionic solids in general by Friauf.[89]

4 Theoretical Calculation of the Properties of Point Defects

Defect Energies: Methods.—The principal method employed in the calculation of defect energies has remained that based on the Mott–Littleton method.[90] This has been developed by the Harwell group into a sophisticated computer program HADES, standing for *H*arwell *A*utomatic *D*efect *E*valuation *S*ystem. The method of doing these calculations has been described in the literature [91–94] so that only

---

[82] L. B. Harris, *J. Phys. (C)*, 1975, **8**, L318; *J. Mater. Sci.*, 1976, **11**, 333.
[83] A. Brun and P. Dansas, *J. Phys. (C)*, 1974, **7**, 2593.
[84] A. Brun, P. Dansas, and F. Bénière, *J. Phys. Chem. Solids*, 1974, **35**, 249.
[85] A. M. Hor, P. W. M. Jacobs, and K. S. Moodie, *Phys. Stat. Sol. (A)*, 1976, **38**, 293.
[86] A. Brun, P. Dansas, and P. Sixou, *Solid State Comm.*, 1970, **8**, 613.
[87] R. Capelletti and E. Okuno, in 'Electrets Charge Storage and Transport in Dielectrics', ed. M. M. Perlman, The Electrochemical Society, Princeton, New Jersey, U.S.A. 08540, 1973, p. 13.
[88] R. Capelletti and A. Gainotti, *J. Phys. (Paris)*, 1976, **37**, C7–316.
[89] R. J. Friauf, in ref. 2, Vol. 2, p. 1103.
[90] N. F. Mott and M. J. Littleton, *Trans. Faraday Soc.*, 1938, **34**, 485.
[91] M. J. Norgett, Report AERE–7650, HM Stationery Office, London, 1974.

a brief summary of the principles of the calculation need be given here. The lattice is divided into two regions. In Region I, which contains the defect, the ions are described explicitly by specific interionic pair potentials and each ion is allowed to relax under the condition of zero force. Fast matrix minimization techniques [95,96] based on the Newton–Raphson procedure are used to obtain the minimum energy configuration. This inner region is embedded in an outer Region II, which is treated as a dielectric continuum. With proper matching at the boundary 100 ions in Region I are generally sufficient. The pair potentials must be chosen to satisfy not only the equilibrium and elastic properties of the perfect crystal but also the dielectric properties, so as to avoid any mismatch between the two regions. Most modern calculations employ the shell model to describe the dielectric displacements of the ions, in preference to the polarizable point ion (PPI) model.

Enthalpies of activation for the motion of interstitial ions in crystals with the rutile structure have been calculated [97] using the PPI model, and a method developed specifically for fast ion conductors by Flygare and Huggins [98] and applied by them to ionic motion in the α AgI structure. Essentially this method consists of calculating the energy of a small Region I (five unit cells in each direction were used in the TiO_2 and MgF_2 calculations) allowing for the polarizability of the ions but ignoring the ionic relaxations and dielectric continuity with the rest of the crystal which is not considered. The results of such calculations probably represent not much more than a useful qualitative guide but may serve to eliminate certain possible migration paths.

No molecular dynamics (computer simulation) studies of defect properties appear to have been made for ionic crystals although the migration of interstitials in a fcc structure, in which the pair interactions are described by a Lennard-Jones potential, has been studied.[99] A comparison of the dynamical [100] and equilibrium statistical mechanical theories [1,101] of the mobility of an idealized defect has been made by McCombie and Sachdev [102] using computer simulation techniques. The conclusion is that when the saddle point displays a large instability, the dynamical and statistical mechanical theories predict results that are nearly equal to one another and in very satisfactory agreement with the jump frequency determined from computer simulation.

Defect Energies: Results.—A summary of the results of defect energy calculations is given in Tables 5–14. The numerical values in the tables should be considered in the light of the following commentary which describes briefly the important

---

[92] A. B. Lidiard and M. J. Norgett, in 'Computational Solid State Physics', ed. F. Herman, N. W. Dalton, and T. R. Koehler, Plenum Press, New York, 1973, p. 385.
[93] I. M. Boswarva and A. B. Lidiard, *Phil. Mag.*, 1967, **16**, 805.
[94] C. R. A. Catlow and M. J. Norgett, *J. Phys. (C)*, 1973, **6**, 1325.
[95] R. Fletcher and M. J. D. Powell, *Computer J.*, 1963, **6**, 163.
[96] R. Fletcher, *Computer J.*, 1970, **13**, 317.
[97] O. B. Ajayi, L. E. Nagel, I. D. Raistrick, and R. A. Huggins, *J. Phys. Chem. Solids*, 1976, **37**, 167.
[98] W. H. Flygare and R. A. Huggins, *J. Phys. Chem. Solids*, 1973, **34**, 1199.
[99] L. B. Pederson, J. W. Martin, and R. M. J. Cotterill, *J. Phys. (C)*, 1972, **5**, 3296.
[100] C. P. Flynn, 'Point Defects and Diffusion', Clarendon Press, Oxford, 1972.
[101] G. H. Vineyard, *J. Phys. Chem. Solids*, 1957, **3**, 121.
[102] C. W. McCombie and M. Sachdev, *J. Phys. (C)*, 1975, **8**, L413.

Table 5 *Calculated values (in eV) of the formation energy of a Schottky defect and of cation and anion Frenkel defects in alkali halide crystals with the rock-salt structure*

	Schottky defect				Frenkel defects	
	BS	SR	CSR[a]	CSR[b]	SH[c]	SH[d]
LiF	2.72		2.28	2.17		
LiCl	1.72					
LiBr	1.53					
LiI	1.29					
NaF	3.24				3.53	3.40
NaCl	2.48		2.10	2.22	2.88	4.60
NaBr	2.23		1.87	1.92	2.59	4.84
NaI	1.87		1.62	1.63	2.01	5.15
KF	2.72				4.27	2.57
KCl	2.57				3.46	3.73
KBr	2.46				3.16	4.17
KI	2.28				2.73	4.30
RbF	2.34				4.70	2.35
RbCl	2.46	2.16			3.71	3.52
RbBr	2.41	1.97			3.35	3.68
RbI	2.29	1.86			2.70	3.89

BS = I. M. Boswarva and J. H. Simpson, *Canad. J. Phys.*, 1973, **51**, 1923.
SR = A. K. Shukla and C. N. R. Rao, *J.C.S. Faraday II*, 1974, **70**, 1628.
CSR = S. Chowdhury, S. K. Sen, and D. Roy, *Phys. Stat. Sol.* (*B*), 1973, **56**, 403. (*a*) without three-body effects; (*b*) including three-body contribution.
SH = P. D. Schulze and J. R. Hardy, *Phys. Rev.* (*B*), 1972, **6**, 1580. (*c*) cation sub-lattice; (*d*) anion sub-lattice.

features of the calculations. References to the original papers are given below the tables and are not, therefore, repeated in the text.

Boswarva and Simpson employed the Mott–Littleton (ML) method with 32 ions in Region I. They used the PPI model with the repulsive parameters in the short-range potential determined from the values of the relative permittivity ('dielectric constants') of the crystals. Three sets of van der Waals coefficients were used: from Mayer [103] and from Lynch,[104] with and without Lorentz local field corrections. Results are given in Table 5 for Schottky energies determined using Mayer van der Waals' coefficients and repulsive parameters that fit the permittivity, rather than the compressibility. The Lynch van der Waals' coefficients were used in conjunction with dielectric repulsion parameters in calculations with only six (nn) ions in region I and so the results will not be quoted here. Results are also presented in the original paper for Schottky energies calculated using repulsive parameters that fit the compressibility, but these values are not reliable for the reasons explained earlier.

Shukla and Rao have reported values for the Schottky formation energy in three rubidium halides. Again they use the ML method but Region I contained only the nn ions. It is not possible to evaluate the potentials used by Shukla and Rao since the source of their repulsive parameters is not explained. However, their

[103] J. E. Mayer, *J. Chem. Phys.*, 1933, **1**, 270.
[104] D. W. Lynch, *J. Phys. Chem. Solids*, 1967, **28**, 1941.

Schottky energy values are lower than those of Boswarva and Simpson given in the second column of Table 5, suggesting that the repulsive parameters may have been obtained from compressibilities.

Possible effects of three-body forces in the Schottky defect formation energy (E_S) have been sought by Chowdhury, Sen, and Roy. They used the ML method; in order to eliminate two-body effects in assessing their results we quote in columns 4 and 5 of Table 5 their values for E_S calculated without and with three-body contributions. While their values of E_S are probably low, because of the method of calculation employed and the potential used, the differences due to the inclusion of three-body forces are probably real and thus should be considered small but not negligible.

Formation energies of Frenkel defects on the cation and anion sub-lattices have been calculated by Schulze and Hardy using the method of lattice statics and the deformation dipole model, restricted to the cation–anion interaction. Short-range potential parameters were taken from Fumi and Tosi.[105] While the Fumi and Tosi potentials reproduce the cohesive energy of the alkali halides with good accuracy, they are known to give a less satisfactory representation of the elastic behaviour of these crystals.[106]

As we have emphasized there are two aspects to a defect calculation, the method of doing the calculation and the interionic potentials employed. Until recently, most emphasis has been on the former with comparatively little attention paid to the latter. An extremely thorough investigation of crystal potentials for the 16 alkali halides with the rock-salt structure has been performed recently by Catlow, Diller, and Norgett.[107] They derived two sets of potentials (designated I and II). In each set the second-neighbour repulsive parameters were calculated theoretically using the electron gas approximation,[108] and the nearest-neighbour interaction was assumed to be purely repulsive. In I the nn repulsive parameters A_{12}, ρ_{12}, and the van der Waals-like parameter C in the attractive $-C/r^6$ term of the short-range second-neighbour interactions, were determined by fitting bulk properties of the crystals. This attractive term was arbitrarily divided equally between the cation–cation (1–1) and anion–anion (2–2) interactions ($C_{11} = C_{22} = C$). In the fitting procedure the number of variables was reduced by writing $A_{12} = b \exp[(a_1 + a_2)/\rho_{12}]$ and constraining $a_1 + a_2 = r_0$, the nn separation. These potential parameters were adjusted simultaneously so as to reproduce the stability condition, the cohesive energy, and the elastic constants for the bulk crystals. In the second stage of the fitting procedure the shell-model parameters (shell charges and shell spring constants) were determined by fitting the high- and low-frequency permittivities and the frequency of the transverse optic mode. In potential II individual second neighbour attractive parameters (C_{11}, C_{22}) were chosen to fit the elastic constants. The short-range and long-range behaviour of the second-neighbour interaction being now defined, they were joined to the (adjustable) minimum by two spline functions. Thus in potential I, the second-neighbour interactions are described by an equation of simple form, but with

---

[105] F. G. Fumi and M. P. Tosi, *J. Phys. Chem. Solids*, 1964, **25**, 31; *cf.* also M. P. Tosi, in 'Solid State Physics', ed. F. Seitz and D. Turnbull, Vol. 16, Academic Press, New York, 1964, p. 1.
[106] F. H. Ree and A. C. Holt, *Phys. Rev. (B)*, 1973, **8**, 826.
[107] C. R. A. Catlow, K. M. Diller, and M. J. Norgett, *J. Phys. (C)*, 1977, in the press.
[108] P. T. Wedepohl, *Proc. Phys. Soc.*, 1967, **92**, 79.

different parameters for each substance. In potential II there is a *unique*, but complex, potential for each second-neighbour interaction. Both potentials gave excellent representations of the equilibrium, elastic and dielectric properties of the crystals.

Table 6 *Defect energies (in eV) for* NaCl, KCl, *and* KBr *calculated from potentials* CDNI *and* CDNII.[a] *The origin and form of these potentials are described in the text.* E_S *is the Schottky defect formation energy,* ΔE_{cv} *and* ΔE_{av} *are the activation energies for the migration of cations and anions, respectively.* E_{Fc} *and* E_{Fa} *are the Frenkel defect formation energies on the cation and anion sub-lattices, respectively*

	E_S	ΔE_{cv}	ΔE_{av}	E_{Fc}	E_{Fa}
CDNI					
NaCl	2.32	0.68	0.72	3.21	3.85
KCl	2.50	0.69	0.69	3.24	3.41
KBr	2.28	0.62	0.63	2.75	3.11
CDNII					
NaCl	2.54	0.66	0.71	3.50	4.33
KCl	2.56	0.71	0.69	3.61	3.71
KBr	2.45	0.67	0.66	3.40	3.58

[a] C. R. A. Catlow, J. Corish, K. M. Diller, P. W. M. Jacobs, and M. J. Norgett, *J. Phys. (Paris)*, 1976, **37**, C7-253.

Defect energies calculated from the Catlow, Diller, and Norgett potentials I and II (CDNI, CDNII) have so far been published only for NaCl, KCl, and KBr and these results are reproduced in Tables 6—8. Several important conclusions can be drawn. First the calculated values for the Schottky defect formation energy, the migration energy of cation vacancies, and the Arrhenius energy for migration of vacancy pairs are in rather good agreement with experimental data (Tables 1 and 2). Secondly, the activation energies for migration of anion vacancies

Table 7 *Defect energies (in eV) for* NaCl, KCl, *and* KBr *calculated from potential* CDNI *(see text).*[a] ΔE_{cic} *and* ΔE_{aic} *are the activation energies for the migration of interstitial cations and anions, respectively, by the collinear interstitialcy mechanism.* E_{mcv} *and* E_{mav} *are the theoretical Arrhenius energies for the migration of cation and anion vacancies, respectively.* E_{mci} *and* E_{mai} *are the Arrhenius energies for the migration of cation and anion interstitials, respectively, by the collinear interstitialcy mechanism, in crystals containing predominantly Schottky defects*

	ΔE_{cic}	ΔE_{aic}	E_{mcv}	E_{mav}	E_{mci}	E_{mai}
NaCl	0.29	0.16	1.84	1.88	2.34	2.85
KCl	0.38	0.28	1.94	1.94	2.37	2.44
KBr	0.37	0.24	1.76	1.77	1.98	2.21

[a] C. R. A. Catlow, J. Corish, K. M. Diller, P. W. M. Jacobs, and M. J. Norgett, *J. Phys. (Paris)*, 1976, **37**, C7-253.

Table 8 *Defect energies (in eV) for vacancy pairs in NaCl, KCl, and KBr, calculated from potential CDNI (see text).[a]* E_p *is the formation energy of the vacancy pair (a cation vacancy and an anion vacancy on adjacent lattice sites). E_{bp} is the binding energy of the pair* $= -(E_p - E_S)$. ΔE_{cp} *and* ΔE_{ap} *are the activation energies for the jump of a cation or anion, respectively, into the corresponding vacancy in a bound pair.* E_{mp} *is the Arrhenius energy for the migration of vacancy pairs (i.e. the lower of the two values for cation or anion jumps)*

	E_p	E_{bp}	ΔE_{cp}	ΔE_{ap}	E_{mp}
NaCl	1.42	0.90	0.90	0.89	2.32
KCl	1.53	0.97	0.79	0.80	2.33
KBr	1.42	0.86	0.71	0.71	2.13

[a] C. R. A. Catlow, J. Corish, K. M. Diller, P. W. M. Jacobs, and M. J. Norgett, *J. Phys. (Paris)*, 1976, **37**, C7–253.

(ΔE_{av}) are very close to those for cation vacancies and thus significantly lower than values derived from ionic conductivity measurements on pure crystals and crystals doped with bivalent cations. Values of ΔE_{av} derived from anion diffusion experiments or from anion-doped crystals are closer to, but still somewhat larger than, the theoretical values. This similarity in the calculated values of the anion and cation vacancy migration energies, and the close agreement of the cation values with experiment, make it unlikely that the observed curvature in conductivity plots is due to the sum of two exponential terms, one an Arrhenius energy for cation vacancies (E_{mcv}) and the other the corresponding quantity for anion vacancies (E_{mav}) (see Table 7). Catlow *et al.* suggested that this curvature is due instead to a contribution to the total ionic transport from the collinear interstitialcy mechanism.

The calculations described are quasiharmonic ones and so do not include the effects of thermal expansion. It was pointed out a long time ago by Jost [109] that the defect energies could depend on temperature because of the temperature-dependence of the lattice constant. Although this idea has been mentioned in the literature from time to time,[110—115] we know of only one attempt [37] to apply it quantitatively in the interpretation of conductivity data, although a similar treatment has been applied to diffusion data.[116] There are thus a number of possible contributing factors to the conductivity curve. However, the calculations described allow one to eliminate contributions from trivacancies in NaCl, KCl, and KBr and to confirm the role of vacancy pairs in anion and cation diffusion.

Similar conclusions have been reached for KCl by Ramdas, Shukla, and Rao, although their calculations must be considered less accurate than those of Catlow *et al.* because of restriction of Region I to the defect and its nearest neighbours

[109] W. Jost, *Z. phys. Chem. (A)*, 1934, **169**, 129.
[110] P. W. M. Jacobs and F. C. Tompkins, *Quart. Rev.*, 1952, **6**, 238.
[111] N. F. Mott and R. W. Gurney, 'Electronic Processes in Ionic Crystals', 2nd edn., Chapter 2, Clarendon Press, Oxford, 1948.
[112] P. Müller, *Phys. Stat. Sol.*, 1967, **21**, 693.
[113] G. J. Dienes, *Acta Metall.*, 1965, **13**, 433, 1215.
[114] G. B. Gibbs, *Acta Metall.*, 1964, **12**, 1302; 1965, **13**, 926.
[115] L. M. Levinson and F. R. N. Nabarro, *Acta Metall.*, 1967, **15**, 785.
[116] A. Seeger, *J. Phys. (F)*, 1973, **3**, 248.

Table 9 Defect energies (in eV) for KCl.[a] All the symbols are defined in the captions in Tables 6—8 except ΔE_{ci} and ΔE_{ai}, which stand for the activation energy for the motion of interstitial cations and anions, respectively, by the direct interstitial mechanism

	Formation energies			
	E_S	E_{Fc}	E_{Fa}	E_p
KCl	2.00	3.02	3.13	1.24

	Migration energies							
	ΔE_{cv}	ΔE_{av}	ΔE_{ci}	ΔE_{ai}	ΔE_{cic}	ΔE_{aic}	ΔE_{cp}	ΔE_{ap}
KCl	0.79	0.87	1.02	1.09	0.18	0.34	1.07	1.14

[a] S. Ramdas, A. K. Shukla, and C. N. R. Rao, *Phys. Rev.* (*B*), 1973, **8**, 2975.

and the use of what are essentially Tosi [105] potential parameters fitted to the cohesive energy and compressibility. Their results are summarized in Table 9.

There have been three sets of calculations on defects in Cs halides in the period under review.[117—119] We do not tabulate the results, however, those being only of qualitative interest because of the potentials used, which were of the traditional Born–Mayer form with the two repulsive parameters adjusted to give the correct lattice spacing and isothermal compressibility using an appropriate equation of state. The shortcomings of this procedure have already been stressed when describing calculations on the alkali halides with the NaCl structure. A better approach has been outlined by Müller and Norgett,[120,121] who have used their potentials mainly in calculations on rare gas diffusion in Cs halides. In the course of these calculations they found that $E_S = 1.80$ eV and $E_p = 1.40$ eV for CsCl.

Table 10 Defect energies (in eV) for AgCl and AgBr.[a] Symbols are defined in the captions to Tables 6—9, except ΔE_{cinc} which denotes the activation energy for the motion of cation interstitials by the non-collinear interstitialcy mechanism

	E_S	E_{Fc}	ΔE_{cv}	ΔE_{av}	ΔE_{ci}	ΔE_{cinc}
AgCl	1.65	1.48	1.04	1.09	1.17	2.20
AgBr	1.50	1.18	0.99	1.13	1.12	1.75

[a] A. K. Shukla, S. Ramdas and C. N. R. Rao, *J. Solid State Chem.*, 1973, **8**, 120.

The transport properties of the silver halides continue to be an intriguing problem. Defect properties of AgCl and AgBr have been calculated by Shukla, Ramdas, and Rao and their results are reproduced in Table 10. They used the ML method with Region I confined to the defect and its nearest neighbours. The usual Born–Mayer short-range potential was employed with an Ag$^+$ ion radius of

[117] Y. V. G. S. Murti and C. S. N. Murthy, *J. Phys.* (*C*), 1972, **5**, 401.
[118] A. K. Shukla, S. Ramdas, and C. N. R. Rao, *J.C.S. Faraday II*, 1973, **69**, 207.
[119] S. Chaudhuri, D. Roy, and A. K. Ghosh, *Phys. Stat. Sol.* (*B*), 1975, **70**, K33.
[120] M. Müller and M. J. Norgett, *J. Phys.* (*C*), 1972, **5**, L256.
[121] M. Müller and M. J. Norgett, *J. Phys.* (*Paris*), 1973, **34**, C9-159.

1.2 Å and the constants b, ρ adjusted to satisfy the equation of state and its volume derivative. Mayer [103] van der Waals coefficients were used, the use of the Lynch [104] values giving large positive deviations in computed lattice energies and 'unreliable' defect formation energies. It is difficult to avoid the conclusion that the not unreasonable value obtained for the Frenkel defect formation energy is somewhat fortuitous. Both the method of calculation and the potentials used may be criticized on the grounds of the small Region I and the disregard of a satisfactory description of the dielectric properties. The calculated migration energies for vacancies and for interstitials by the non-collinear interstitialcy mechanism are much larger than experimental values;[37,39] nevertheless such calculations represent an encouraging start at a difficult problem.

Table 11 *Defect energies (in eV) for the alkaline earth fluorides calculated by Catlow and Norgett,[a] Keeton and Wilson,[b] and Catlow, Norgett, and Ross.[c] For symbols see captions to Tables 6—9. ΔE_d is the activation energy for the direct exchange of two anions, without the assistance of defects*

	Catlow and Norgett[a]				Keeton and Wilson[b]			
	E_S	E_{Fa}	ΔE_{av}	ΔE_{ainc}	E_{Fa}	ΔE_{av}	ΔE_{ai}	ΔE_{ainc}
CaF_2	7.00	2.63	0.20	0.69	1.74	0.3	1.3	0.6
SrF_2	6.90	2.39	0.80	0.69				
BaF_2	6.32	1.92	0.37	0.69	1.95	0.4	2.3	0.8

	Catlow, Norgett, and Ross[c]					
	E_S	E_{Fc}	E_{Fa}	ΔE_{av}	ΔE_{ainc}	ΔE_d
CaF_2	5.75	8.00	2.75	0.35	0.91	2.29
SrF_2	5.92	7.57	2.38	0.43	0.80	1.93
BaF_2	5.64	7.22	1.98	0.46	0.72	1.80

[a] C. R. A. Catlow and M. J. Norgett, *J. Phys.* (*C*), 1973, **6**, 1325;   [b] S. C. Keeton and W. D. Wilson, *Phys. Rev.* (*B*), 1973, **7**, 834;   [c] C. R. A. Catlow, M. J. Norgett, and T. A. Ross, AERE, Harwell Report TP673, 1976.

Considerable attention has been paid recently to crystals with the fluorite structure both because of their intrinsic interest and as models for UO_2. Catlow and Norgett employed the shell model in their calculations on Ca, Sr, and Ba fluorides. For the anion–anion second-neighbour interaction they used the results of a Hartree–Fock calculation by Catlow and Hayns [122] the remaining potential parameters being fitted to bulk properties. Their results appear in Table 11. Also given in Table 11 are results of calculations by Keeton and Wilson, who used a very large Region I (up to 592 ions) and obtained the total defect energy either from an extrapolation procedure or by adding the contribution from Jost's simple continuum approximation (2),[123] where q is the virtual charge of the defect r_I, the

$$E_J = \frac{q^2}{(4\pi\epsilon_0)2r_I}\left(1 - \frac{1}{\epsilon_r}\right) \quad (2)$$

[122] C. R. A. Catlow and M. R. Hayns, *J. Phys.* (*C*), 1972, **5**, L237.
[123] W. Jost, *J. Chem. Phys.*, 1933, **1**, 466.

radius of Region I, and ϵ_r the static relative permittivity. Keeton and Wilson used the electron gas approximation [108] to calculate the repulsive potentials and a PPI model in the defect calculations. In more recent calculations Catlow, Norgett, and Ross again used the shell model but with a splined F^--F^- potential of similar form to that developed for the alkali halides.[107] Cation–cation interactions were neglected and the nearest neighbour interactions were described by an exponential repulsive term with A and ρ fitted to elastic constants. Their results show greatly improved agreement with experiment for the anion vacancy activation energies (which were too low in the earlier calculations) and once again emphasize the importance of utilizing the best possible second-neighbour interaction in defect calculations.

Table 12 *Defect energies (in eV) for some oxides calculated by Catlow, Faux, and Norgett (MgO),[a] Dienes, Welch, Fischer, Hatcher, Lazareth, and Samberg (Al_2O_3),[b] and Catlow and Lidiard (UO_2).[c] Symbols are defined in the captions to Tables 6, 8, and 11. Parentheses indicate preliminary results*

	E_S	ΔE_{cv}	ΔE_{av}	E_p	E_{Fc}	E_{Fa}	ΔE_{ainc}
MgO	7.5	1.9	2.0	5.2			
Al_2O_3	5.7	(3.8)	(2.9^d)		10.0	7.0	
UO_2	10.3	5.6	0.25		18.5	5.0	0.6

[a] C. R. A. Catlow, I. D. Faux, and M. J. Norgett, *J. Phys. (C)*, 1976, **9**, 419; [b] G. J. Dienes, D. O. Welch, C. R. Fischer, R. D. Hatcher, O. Lazareth, and M. Samberg, *Phys. Rev. (B)*, 1975, **11**, 3060; [c] C. R. A. Catlow and A. B. Lidiard, 'Thermodynamics of Nuclear Materials', International Atomic Energy Agency, Vienna, 1975, Vol. II, p. 27; [d] In the basal plane.

There have been three sets of calculations on oxides (Table 12): on MgO, Al_2O_3, and UO_2. For MgO, Catlow, Faux, and Norgett used the shell model in its normal form and in the 'breathing' modification. The calculated Schottky energy is so large that no intrinsic diffusion would be observable in MgO. The activation energies for vacancy migration are in harmony with an extrinsic vacancy mechanism for ionic transport in this material. The vacancy formation volumes were also calculated and found to be negative, in agreement with experimental results [124] for MgO. However, negative formation volumes appear to be a general feature of the ionic model, which is not in agreement with experimental data on the alkali halides, for instance. Catlow and Lidiard also used the shell model for UO_2. Their calculations prove that the intrinsic defects in UO_2 are Frenkel defects on the anion sublattice; ionic transport occurs by the migration of anion vacancies in stoicheiometric UO_2 and by oxygen interstitials (non-collinear interstitialcy mechanism) in oxidized UO_2. Catlow and Lidiard also investigated the stability of hole–interstitial clusters of the type originally proposed for fluorite structures on neutron diffraction evidence.[125] Numerical values for the binding energies are given in Catlow and Lidiard's paper. There are other reported calculations on UO_2[126]

[124] B. Henderson, D. H. Bowen, A. Briggs, and R. D. King, *J. Phys. (C)*, 1971, **4**, 1496.
[125] B. T. M. Willis, *Proc. Brit. Ceram. Soc.*, 1964, **1**, 9.
[126] E. Van der Voort, *J. Phys. (C)*, 1974, **7**, L395.

but the lattice model used has been criticized.[127] The calculations on Al_2O_3 also used the shell model but refer to a crystallite containing 1855 ions. For the formation of an Al^{3+} interstitial, for example, this leads to an error in the coulombic energy of 0.3%. The Schottky quintet appears to be the favoured intrinsic defect. Preliminary results for the migration of anion vacancies suggest that this may be the mechanism operating in anion diffusion [128] and in the annealing of radiation damage.[129]

In the period under review calculations of defect energies have also been made for γ-CuCl,[130] $NaNO_3$,[131,132] KN_3,[133] and NH_4ClO_4.[134] Because of the specialized rather than general interest of these materials, results are not quoted here and the original papers should be consulted.

Comparatively little attention has been paid to calculations on the defect properties of ionic crystals containing impurity ions. Some univalent impurities have been investigated by Catlow et al.[34] with a view to establishing which impurity ions occupy an off-centre configuration. It was concluded that the shell-model potentials currently employed, while sufficiently accurate to locate off-centre energy minima, may not be able to distinguish between different configurations of very similar energies. A more optimistic viewpoint is expressed by Pandey and Shukla,[135] who used the PPI model. Coupling between the motion of the impurity and the lattice phonons has been investigated by Shore and Sander.[136]

The energy of solution of the bivalent impurities Ca^{3+}, Sr^{2+}, Ba^{2+}, Cd^{2+}, and Pb^{2+} in alkali halide crystals has been calculated by Bowman [137] using the ML method. Dienes et al.[138] have investigated a number of neutral manganese centres in NaCl with a view to testing the proposals of Ikeya and Itoh [139,140] based on e.p.r. studies.

Defect Interactions.—The principal theoretical development in the treatment of defect interactions in the period since our last report [1] has been the physical cluster theory of Allnatt and co-workers.[141—145] The Allnatt and Cohen theory,[146,147] which is based on Mayer ionic-solution theory, is not very well adapted for the

[127] C. R. A. Catlow, A. B. Lidiard, and M. J. Norgett, *J. Phys.* (C), 1975, **8**, L435.
[128] Y. Oishi and W. D. Kingery, *J. Chem. Phys.*, 1960, **33**, 480.
[129] R. S. Wilks, *J. Nuclear Mater.*, 1968, **26**, 137.
[130] J. Laine, *J. Phys.* (C), 1973, **6**, 637.
[131] P. Cerisier and P. Gaune, *J. Solid State Chem.*, 1971, **3**, 473.
[132] P. Cerisier and P. Gaune, *J. Solid State Chem.*, 1973, **6**, 74.
[133] A. Danemar, D. O. Welch, and B. S. H. Royce, *Phys. Stat. Sol.* (B), 1973, **55**, 201.
[134] M. Goldstein and A. G. Keenan, *J. Solid State Chem.*, 1973, **7**, 286.
[135] G. K. Pandey and D. K. Shukla, *Phys. Rev.* (B), 1972, **5**, 3362.
[136] H. B. Shore and L. M. Sander, *Phys. Rev.* (B), 1975, **12**, 1546.
[137] R. C. Bowman, *J. Chem. Phys.*, 1973, **59**, 2215.
[138] G. J. Dienes, R. D. Hatcher, O. W. Lazareth, B. S. H. Royce, and R. Smoluchowski, *Phys. Rev.* (B), 1973, **7**, 5332.
[139] M. Ikeya and N. Itoh, *J. Phys. Soc. Japan*, 1970, **29**, 1295.
[140] M. Ikeya, *Phys. Stat. Sol.* (B), 1972, **51**, 407.
[141] A. R. Allnatt and E. Loftus, *J. Chem. Phys.*, 1973, **59**, 2541.
[142] A. R. Allnatt and E. Loftus, *J. Chem. Phys.*, 1973, **59**, 2550.
[143] A. R. Allnatt and P. S. Yuen, *J. Phys.* (C), 1975, **8**, 2213.
[144] A. R. Allnatt and P. S. Yuen, *J. Phys.* (C), 1975, **8**, 2199.
[145] A. R. Allnatt and P. S. Yuen, *J. Phys.* (C), 1976, **9**, 431.
[146] A. R. Allnatt and M. H. Cohen, *J. Chem. Phys.*, 1964, **40**, 1860.
[147] A. R. Allnatt and M. H. Cohen, *J. Chem. Phys.*, 1964, **40**, 1871.

discussion of the association of point defects. One needs not only the pair correlation function but higher order correlation functions as well, and when the correlation functions are expressed as cluster expansions [146,147] the convergence of these expansions is unacceptably slow over much of the interesting concentration range. The advantage of the Mayer cluster expansions is that the discreteness of lattice sites can be allowed for. This advantage is retained in the physical cluster theory. The concentrations and activity coefficients of defect aggregates (the 'physical clusters') are given by general mass action expressions, and the activity coefficients are then obtained from exact formal cluster expansions in terms of the concentrations of the physical clusters. The method has been applied to AgCl doped with $CdCl_2$,[142] to pure NaCl,[143] and to $Y^{3+}F_i^-$ complexes and to 2:2:2 clusters [3,148] in CaF_2 doped with YF_3.[144,145] The general conclusion is that deviations from the Lidiard–Teltow law of mass action [149,150] because of interactions between the physical cluster are not very pronounced, even at surprisingly high concentrations; consequently the generally used mass action expressions with Debye–Hückel activity coefficients are likely to be sufficiently accurate for use in most analyses of conductivity [37,151] or diffusion data.

Allnatt and Cohen [146,147] also gave a method for calculating the concentration of defect pairs from the defect radial distribution functions which can be calculated from Mayer cluster theory. In this approach certain higher order distribution functions are approximated by functions of the pair correlation functions.[152] The validity of this procedure, which is also inherent in the theory of Ford and Fong [153,154] would be difficult to check. The treatment of Ramdas and Rao [155] is more simplistic being based only on the pair correlation function.

The binding energy for various pairs of defects have been calculated by the ML method and results are given in Table 13. For M^{2+} cation vacancy complexes, or A^{2-} anion vacancy complexes, in the NaCl structure, the nn, nnn, and third nearest neighbour ($j = 1, 2, 3$) configurations (on the same sublattice) are of

Table 13 *Binding energies (in eV) for defect complexes. $j = 1, 2, 3$, denotes nearest neighbour, next nearest neighbour and third nearest neighbour configurations*

System	$j = 1$	$j = 2$	$j = 3$
vacancy pairs in KCl[a]	0.695	0.349	0.362
Sr^{2+}-cation vacancy complexes in KCl[a]	0.332	0.512	0.210
SO_4^{2-}-anion vacancy complexes in KCl[b]	0.18	0.61	0.13
K^{2+}-cation vacancy complexes in KCl[c]	0.61	0.58	0.37

[a] S. Ramdas, A. K. Shukla, and C. N. R. Rao, *Chem. Phys. Letters*, 1972, **16**, 14;   [b] P. K. Swaminiathan, A. K. Shukla, and C. N. R. Rao, *ibid.*, 1973, **23**, 318;   [c] That is an electrostatic perturbation of the KCl lattice: C. R. A. Catlow, *Chem. Phys. Letters*, 1976, **39**, 497.

[148] C. R. A. Catlow, *J. Phys. (C)*, 1973, **6**, L64.
[149] A. B. Lidiard, *Phys. Rev.*, 1954, **94**, 29.
[150] J. Teltow, *Ann. Phys. (Leipzig)*, 1949, **5**, 71.
[151] A. R. Allnatt, J. Corish, and P. W. M. Jacobs, *J. Phys. Chem. Solids*, 1975, **36**, 1233.
[152] A. R. Allnatt, *Adv. Chem. Phys.*, 1967, **11**, 1.
[153] R. L. Ford and F. K. Fong, *J. Chem. Phys.*, 1971, **55**, 2532.
[154] F. K. Fong, R. L. Ford, and R. H. Heist, *Phys. Rev. (B)*, 1970, **2**, 4202.
[155] S. Ramdas and C. N. R. Rao, *Chem. Phys. Letters*, 1975, **31**, 37.

symmetry C_{2v}, C_{4v}, and C_s, respectively, whereas for defects on different sublattices they are of C_{4v}, C_{3v}, and C_s symmetry respectively.

Finally, an interesting series of papers on the formation of the Suzuki phase 6MCl,M'Cl$_2$ has been published by Lilley and co-workers.[77,156,157] The lattice energy and anion displacements have been calculated for 6NaCl,M'Cl$_2$. Energetic considerations show, in agreement with experimental observations, that the Suzuki phase tends to form if $r_{M'^{2+}}/r_{Na^+} < 1.2$ but not for larger values. The calculated anion displacements for M' = Cd or Mn are in fair agreement with experimental values which are not known very accurately at present.[156] The second paper [157] deals with the morphology of the precipitating phase, and the third one [77] with the mechanism by which dimer aggregation leads to the nucleation of precipitates of the Suzuki phase.

Rare Gas Diffusion in Ionic Crystals.—The motion of rare gas atoms and their interaction with defects in CaF$_2$, in CsCl, and in K and Rb halides, have been investigated by Keeton and Wilson,[158] by Müller and Norgett,[120,121] and by Heinisch and Jaswal,[159] respectively. Keeton and Wilson find that He, Ne, and Ar can move interstitially in CaF$_2$ with activation energies ca. 1.4 eV and that anion vacancies can trap these atoms with binding energies of 1.7, 2.1, and 2.5 eV, respectively. For Xe in CsI, Müller and Norgett [120] show that the diffusion mechanism involves the motion of Xe atoms bound to vacancy pairs. In α-CsCl, Xe also diffuses by the vacancy pair mechanism but in β-CsCl, which has the NaCl structure, an interstitial mechanism is preferred for the diffusion of Ar, Kr, and Xe.[121] The lattice statics method has been used by Heinisch and Jaswal [159] to calculate the migration energies of Ar in K halides and of Kr in Rb halides. If anharmonic corrections are applied the results are similar to earlier ML calculations by Norgett and Lidiard [160] (except for the fluorides, where negative migration energies were obtained), but without these corrections the migration energies are much too large.

Defect Entropies.—Agrawal and Garg have applied Theimer's method [161] to the calculation of the entropy of Schottky defect formation in KCl, KBr, and KI. The elastic displacements were calculated by methods proposed by Brauer [162] and by Boswarva and Lidiard [163] respectively, and allowed for anharmonic effects in the calculation of the displacements. Their results are given in Table 14. Also given in this Table are results of a calculation by Bénière, who used an Einstein approximation and assumed that the vibrational frequency of the ions adjacent to the vacancy could be equated to that for a liquid state vibrational mode.

The entropy of binding between vacancies and solute atoms in linear and square (harmonic) lattice have been calculated from frequency spectra obtained from

[156] A. I. Sors and E. Lilley, *Phys. Stat. Sol.* (*A*), 1975, **27**, 469.
[157] A. I. Sors and E. Lilley, *Phys. Stat. Sol.* (*A*), 1975, **32**, 533.
[158] S. C. Keeton and W. D. Wilson, *Phys. Rev.* (*B*), 1973, **7**, 834.
[159] H. L. Heinisch and S. S. Jaswal, *Phys. Rev.* (*B*), 1974, **9**, 2754.
[160] M. J. Norgett and A. B. Lidiard, *Phil. Mag.*, 1968, **18**, 1193.
[161] O. Theimer, *Phys. Rev.*, 1958, **112**, 1857.
[162] P. Brauer, *Z. Naturforsch.*, 1952, **7a**, 372.
[163] I. M. Boswarva and A. B. Lidiard, *Phil. Mag.*, 1967, **16**, 805.

Table 14 *Entropy of formation of a Schottky defect pair (S_S) in alkali halides. S_S is given in units of k, B means the elastic displacements were calculated by Brauer's method and BL that they were calculated by the method of Boswarva and Lidiard*

	Agrawal and Garg[a,b]		Bénière[c]
	B	BL	
NaCl			12.6
KCl	6.6	6.8	12.3
KBr	6.7		10—12
KI	8.2	2.0	11.6
CsCl	8.0		
CsBr	8.3		
CsI	8.2		

[a] V. K. Agrawal and H. C. Garg, *Phys. Rev.* (*B*), 1973, **8**, 843; [b] V. K. Agrawal and H. C. Garg, *Indian J. Pure Appl. Phys.*, 1974, **12**, 168; [c] F. Bénière, *J. Phys.* (*Paris*) *Letters*, 1975, **36**, L9.

computer calculations.[164] The entropy change associated with the introduction of a Na^+, Rb^+, or Ag^+ ion into KCl has been calculated by Govindarajan, Jacobs, and Nerenberg [165] by a Green's function method.

5 Fast Ion Conduction

Extensive interest has been manifested over the past few years in a class of materials which have come to be known as fast ion conducting solids (FIC). This is because of their many laboratory and commercial functions in thermodynamic, kinetic, and monitoring systems and more especially because of their potential uses in alternative energy technology. A measure of the activity can be assessed from the number of major publications which have appeared since our earlier report [1] when we mentioned these substances only briefly. The proceedings of conferences at Belgirate [166] and Eindhoven [167] have been published; Kummer [168] has reviewed earlier work on β-alumina electrolytes, McGeehin and Hooper [169] have made a comprehensive materials survey, and Whittingham [170] has reviewed transport mechanisms in these compounds. Several chapters in the two volume work edited by Hladik [2] serve as a general introduction, while a recent review by van Gool [171] is extremely informative on both early and recent literature pertaining to all aspects of the subject. Some of the opportunities for practical applications of FIC have been described by Bénière.[172]

After early overemphasis on the searching out of new FIC, presumably brought

---

[164] F. Nakamura, J. Takamura, and M. Chikasaki, *J. Mater. Sci.*, 1973, **8**, 385.
[165] J. Govindarajan, P. W. M. Jacobs, and M. A. Nerenberg, *J. Phys.* (*C*), 1976, **9**, 3911.
[166] 'Fast Ion Transport in Solids', ed. W. van Gool, Proceedings NATO Summer School, Belgirate, Sept. 1972, North Holland, Amsterdam, 1973.
[167] International Society of Electrochemistry, Extended abstracts, 24th meeting, Eindhoven, Sept. 1973.
[168] J. T. Kummer, *Prog. Solid State Chem.*, 1972, **7**, 141.
[169] P. McGeehin and A. Hooper, Report R8070, AERE Harwell, 1975.
[170] M. S. Whittingham, *Electrochim. Acta*, 1975, **20**, 575.
[171] W. van Gool, *Ann. Rev. Mater. Sci.*, 1974, **4**, 311.
[172] F. Bénière, *La Recherche*, 1975, **6**, 36.

The Ultrastructure of Minerals

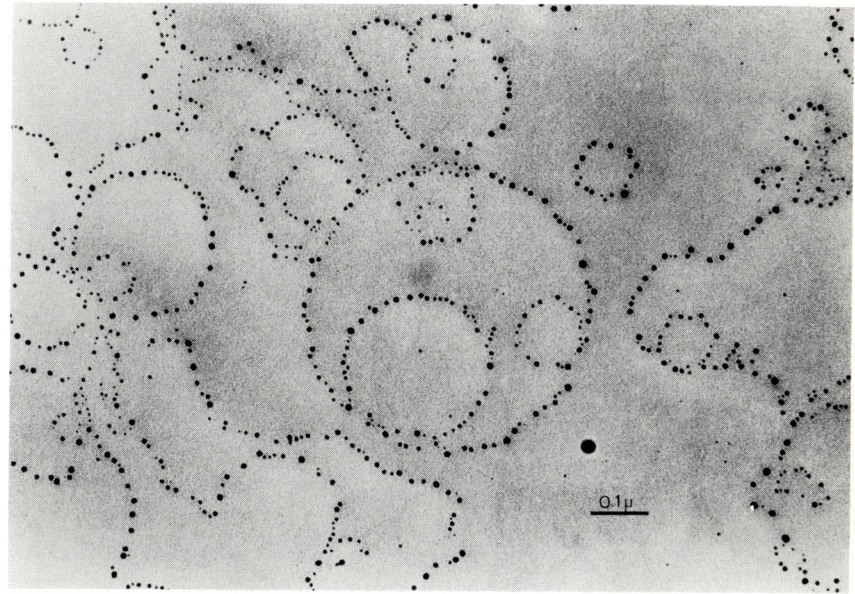

Plate 1 *Typical transmission micrograph showing tangential depressions of monatomic height rendered visible by the electron-opaque gold crystallites*

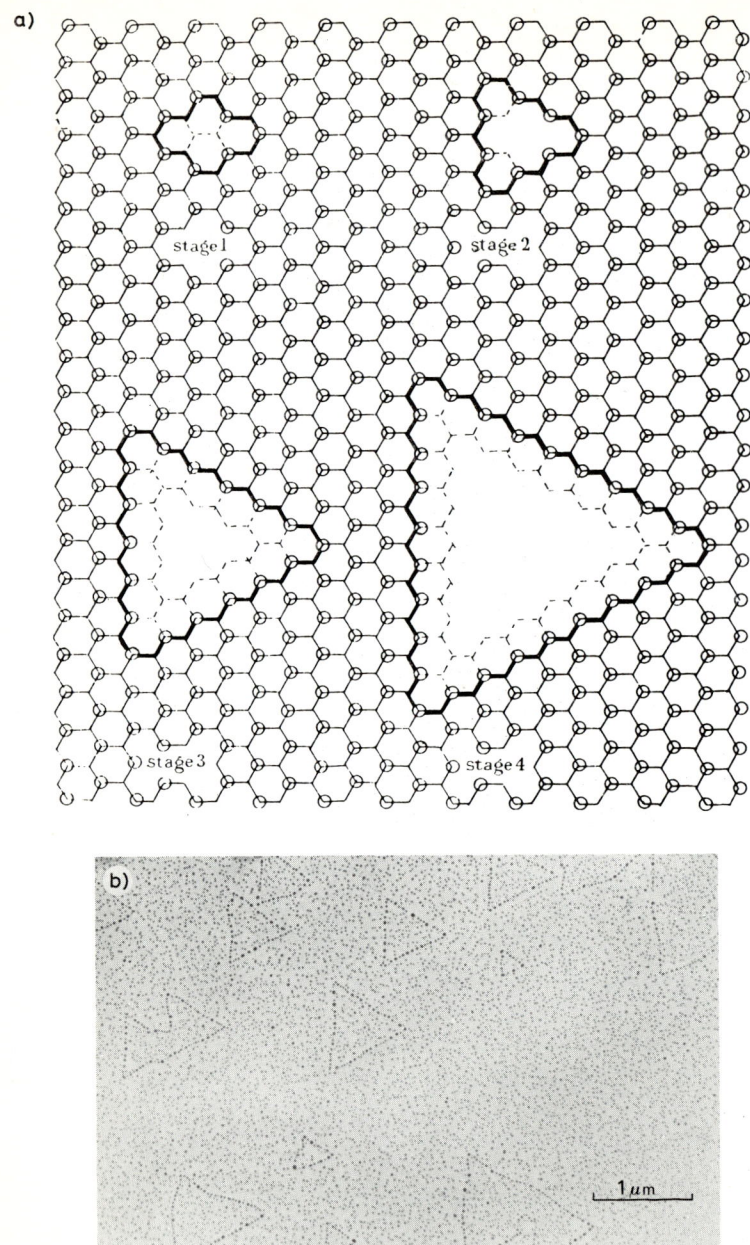

Plate 2 (a) Illustration of the production of a triangular depression possessing the observed orientation [see below]. Circles denote positions of sulphur atoms in one layer: the molybdenum atoms have, for clarity, been omitted. (b) Triangular depressions, decorated by gold particles on a surface which had been previously oxidized

The Ultrastructure of Minerals

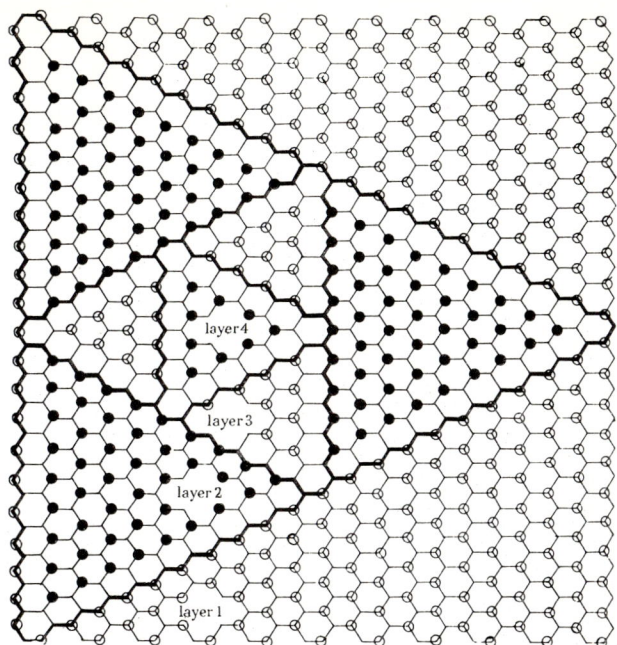

Plate 3 *Illustration of reorientation of the depressions as each successive layer plane is exposed. Empty circles denote positions of sulphur atoms in first and third layer, dark circles the positions of the sulphur atoms in the second and fourth layers*

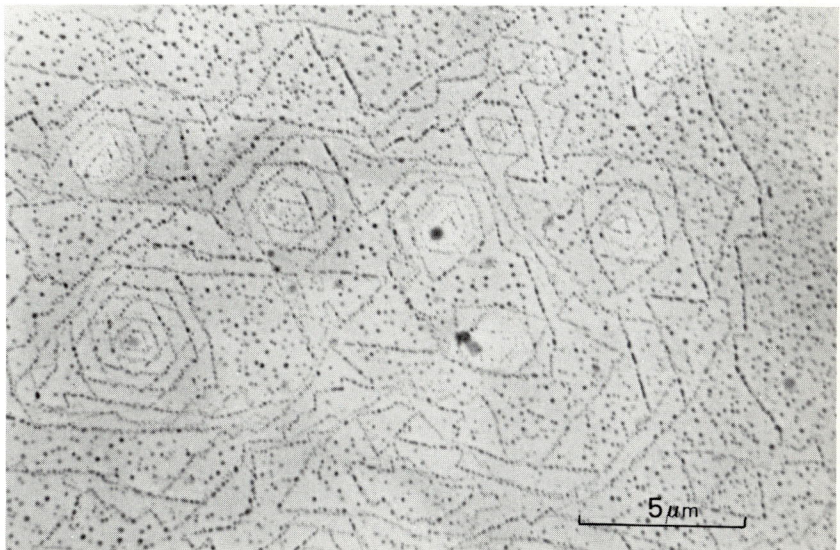

Plate 4 *Triangular depressions (of monatomic height), decorated by gold, on ten consecutive layers. Since the 60° reorientation is, as predicted in Plate 3, for the top 60—65 Å the hexagonal sequence is regular and there are no stacking faults*
(Reproduced by permission from *Proc. Roy. Soc.*, 1968, **A306**, 53)

Surface and Defect Properties of Solids

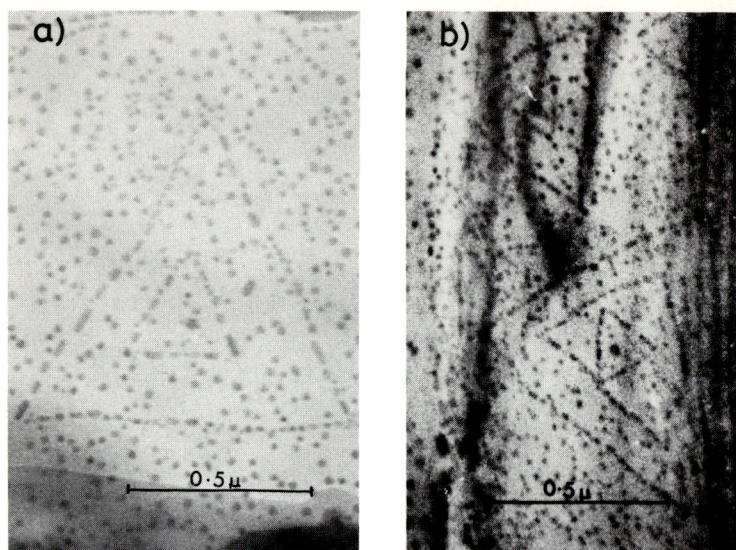

Plate 5 *Gold-decorated triangular depressions of the same orientation on successive sheets of molybdenite. This means that for both* (a) *and* (b), *the hexagonal stacking sequence is no longer maintained, and stacking faults of type* I *or* II *(see Table* 1*) are present. Note that, in* (b), *it may be seen that the second and third sheets from the outermost surface revert to the normal hexagonal sequence*
(Reproduced by permission from *Proc. Roy. Soc.*, 1968, **A306,** 53)

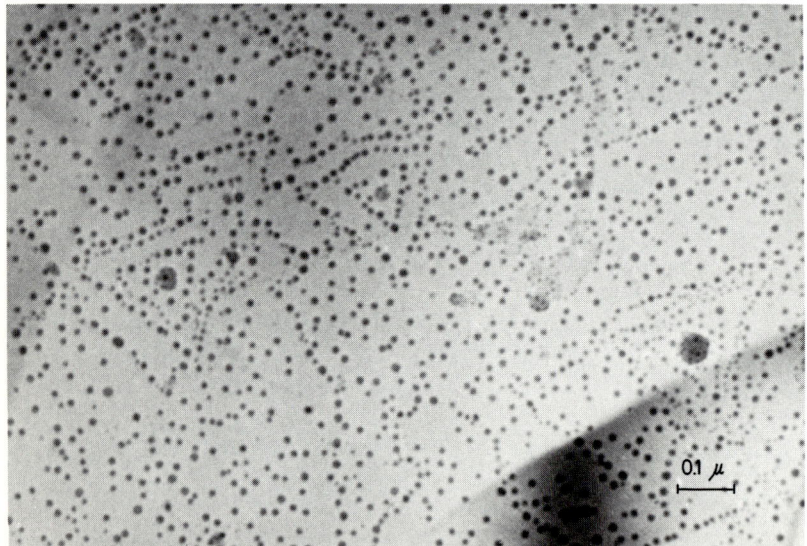

Plate 6 *Triangular depressions (decorated by gold) of the same orientation as in Figure* 3 *on successive layers of oxidized rhombohedral* MoS_2
(Reproduced from *Trans. Faraday Soc.*, 1968, **64,** 3354)

The Ultrastructure of Minerals

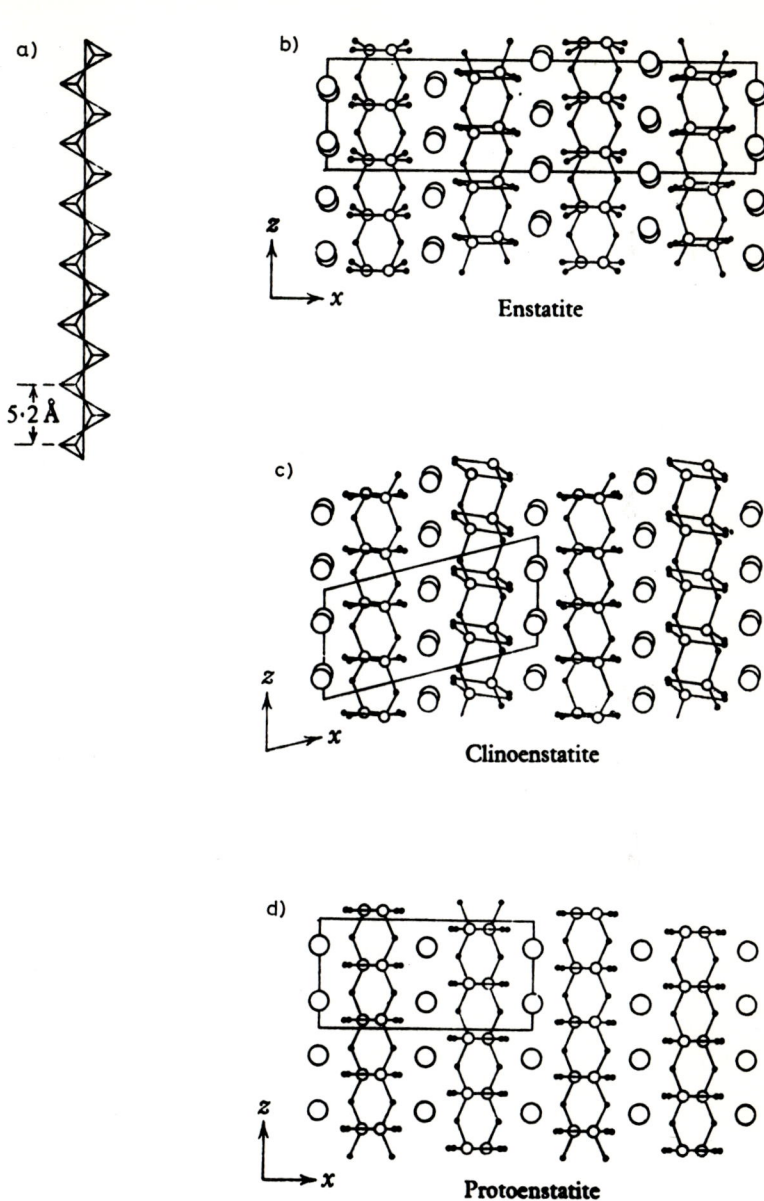

Plate 7 *The structures of the pyroxene polymorphs.* (a) *The basic pyroxene chain.* (b) *Orthoenstatite.* (c) *Clinoenstatite.* (d) *Protenstatite*

Surface and Defect Properties of Solids

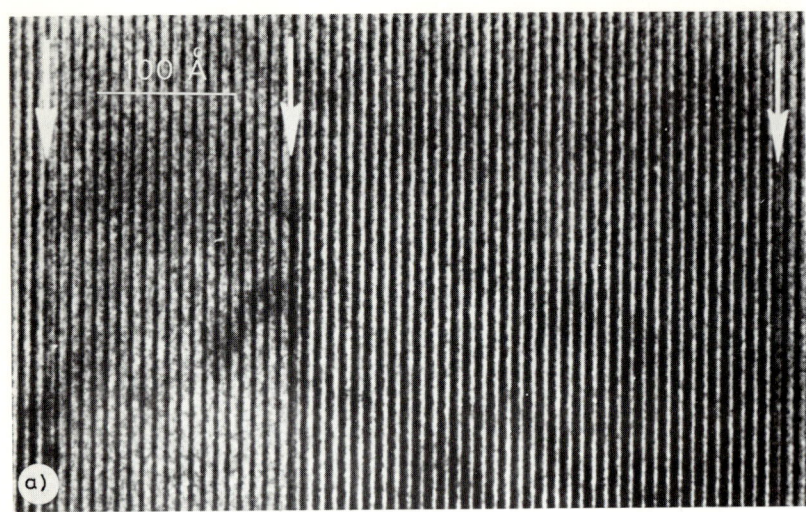

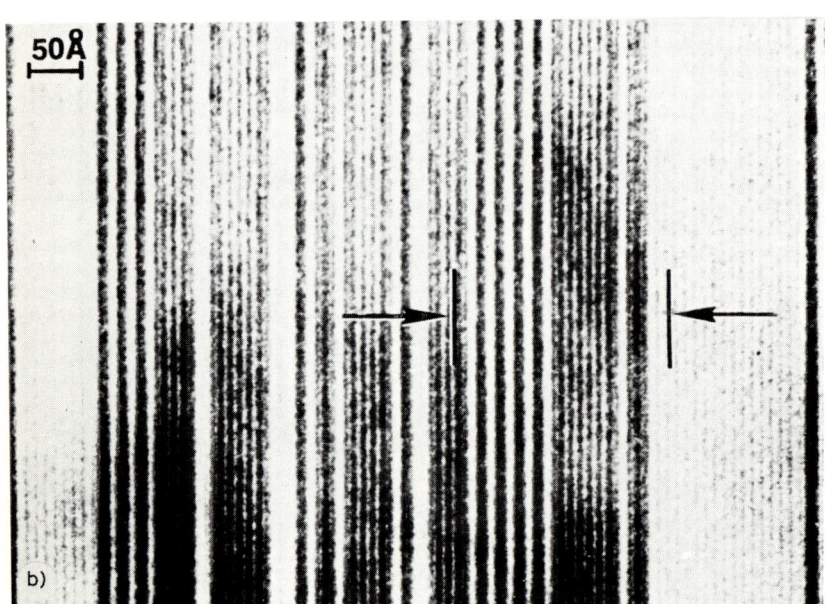

Plate 8 (a) *Relatively large twin lamellae in an enstatite which has been heated and quenched. The electron beam in this case is parallel to [010]*. (b) *The same specimen and similar heat-treatment, but in this case the lamellae are considerably finer*
(Reproduced by permission from 'Electron Microscopy in Mineralogy', Springer-Verlag, Berlin, 1976, p. 320, and *Amer. Mineral*, 1975, **60,** 762)

The Ultrastructure of Minerals

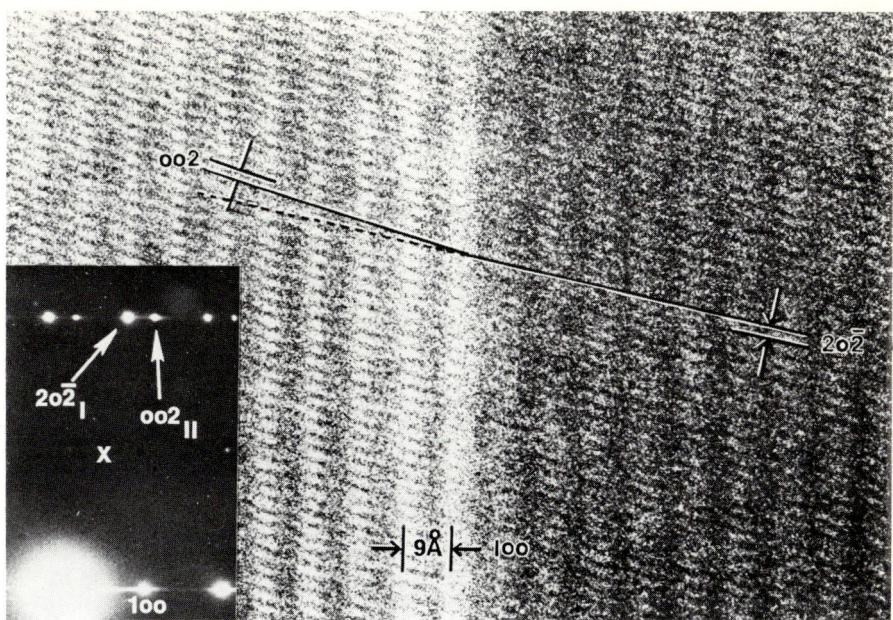

Plate 9 *High magnification image obtained with tilted illumination, showing* (002) *and* (20$\bar{2}$) *fringes in two of the twin individuals*
(Reproduced by permission, from 'Electron Microscopy in Mineralogy', Springer-Verlag, Berlin, 1976, p. 321)

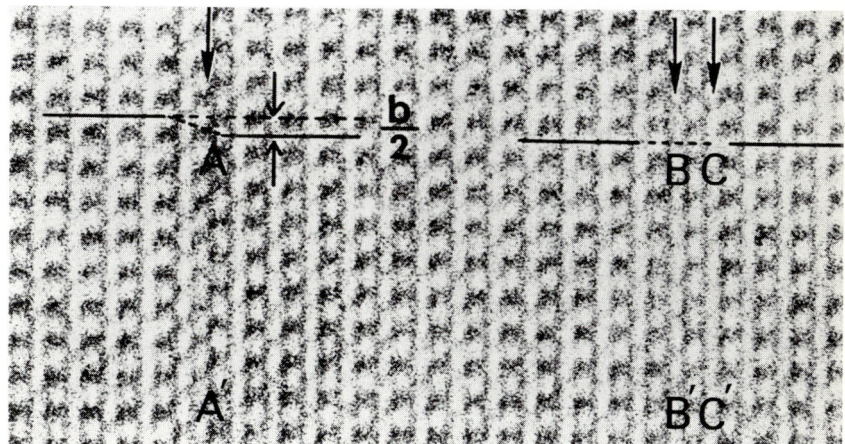

Plate 10 *Lattice image of the specimen of Plate 9, taken with beam parallel to* [001]. *The effect resulting from glide twinning is indicated* (AA')
(Reproduced by permission from *Amer. Mineral*, 1975, **60**, 763)

Surface and Defect Properties of Solids

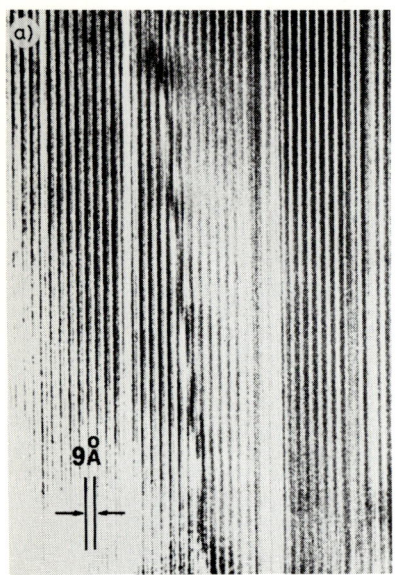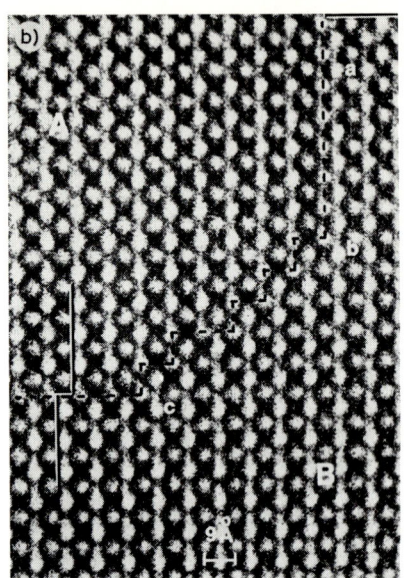

Plate 11 (a) *A 'stepped' twin composition plane, exhibited by a heated and quenched enstatite, which also shows other, normal twin planes. Electron beam parallel to* [010]. (b) *An alternative form of non-planar twin interface, viewed this time in projection down* [001] (Reproduced by permission from *Amer. Mineral*, 1975, **60**, 774, 776)

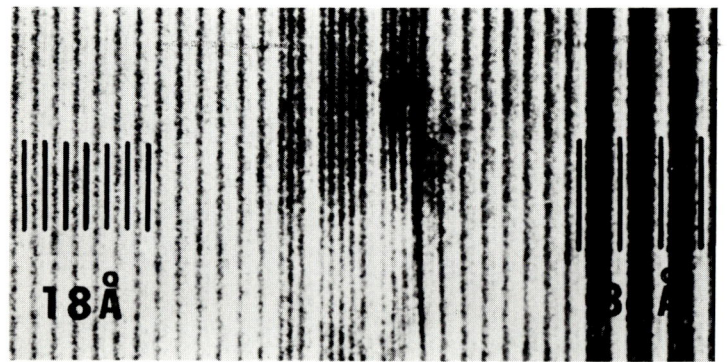

Plate 12 *More complex polytypes of the enstatite structure. Regions of 9, 18, and 36 Å periodicities can be observed. Electron beam parallel to* [010] (Reproduced by permission from *Amer. Mineral*, 1975, **60**, 773)

The Ultrastructure of Minerals

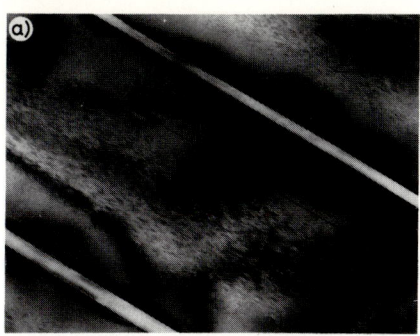

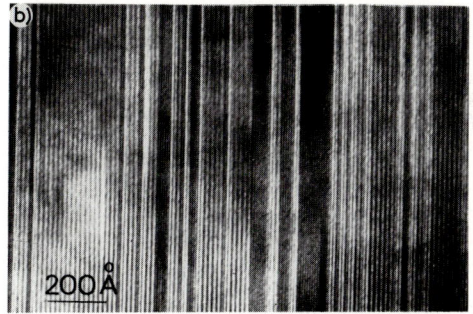

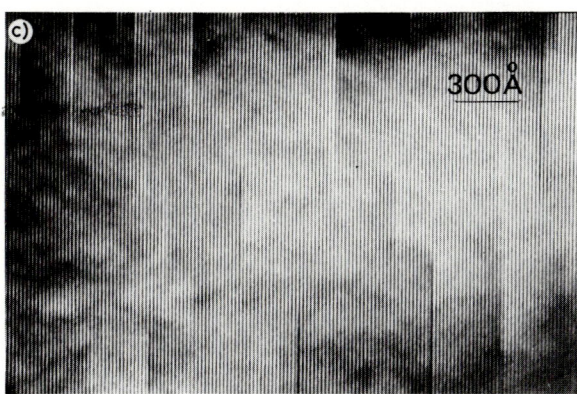

Plate 13 (a) *Relatively low magnification image of augite exsolution in an orthopyroxene, with the electron beam parallel to* [010]. (b) *Higher magnification image of exsolution in a different specimen, showing primitive lamellae (characterized by 9 Å spacing) in an 18 Å matrix. Electron beam parallel to* [010]. (c) *High-resolution micrograph of the GP zones*
(Reproduced by permission from *J. Mater. Sci.*, 1973, **8**, 469, and *Phil. Mag.*, 1974, **30**, 360, 363)

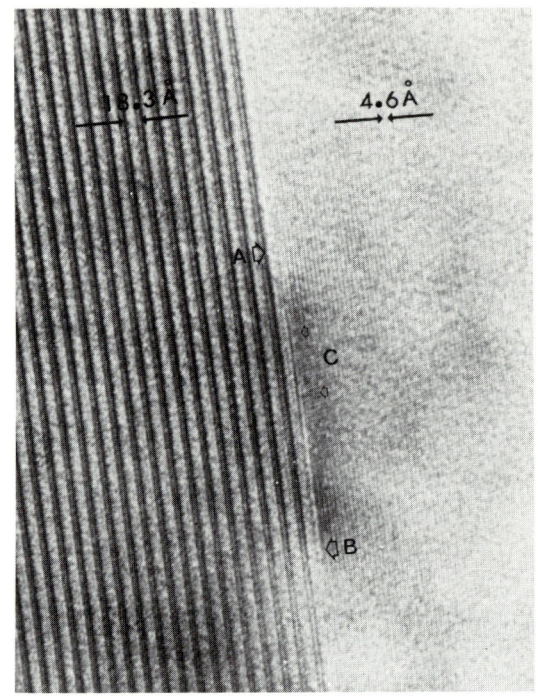

Plate 14 *Ledges at an augite-orthopyroxene interface during exsolution. Electron beam parallel to* [010]
(Reproduced by permission from *Phil. Mag.*, 1974, **29**, 1047)

The Ultrastructure of Minerals

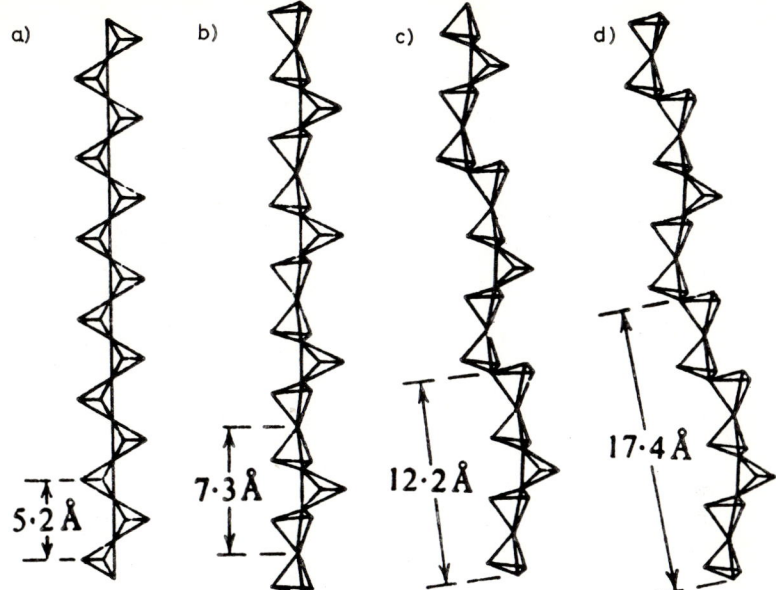

Plate 15 *Schematic illustration of the various types of pyroxenoid chain, with that of the pyroxenes for comparison. (a) Pyroxene type (two tetrahedra repeat). (b) Wollastonite type (three tetrahedra). (c) Rhodonite type (five tetrahedra). (d) Pyroxmangite type (seven tetrahedra)*

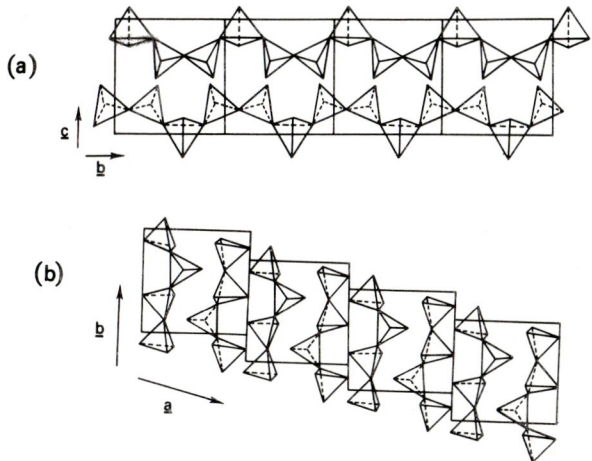

Plate 16 *The structure of the triclinic form of wollastonite, showing the arrangement of tetrahedra. (a) Projection on (100). (b) Projection along [001]*

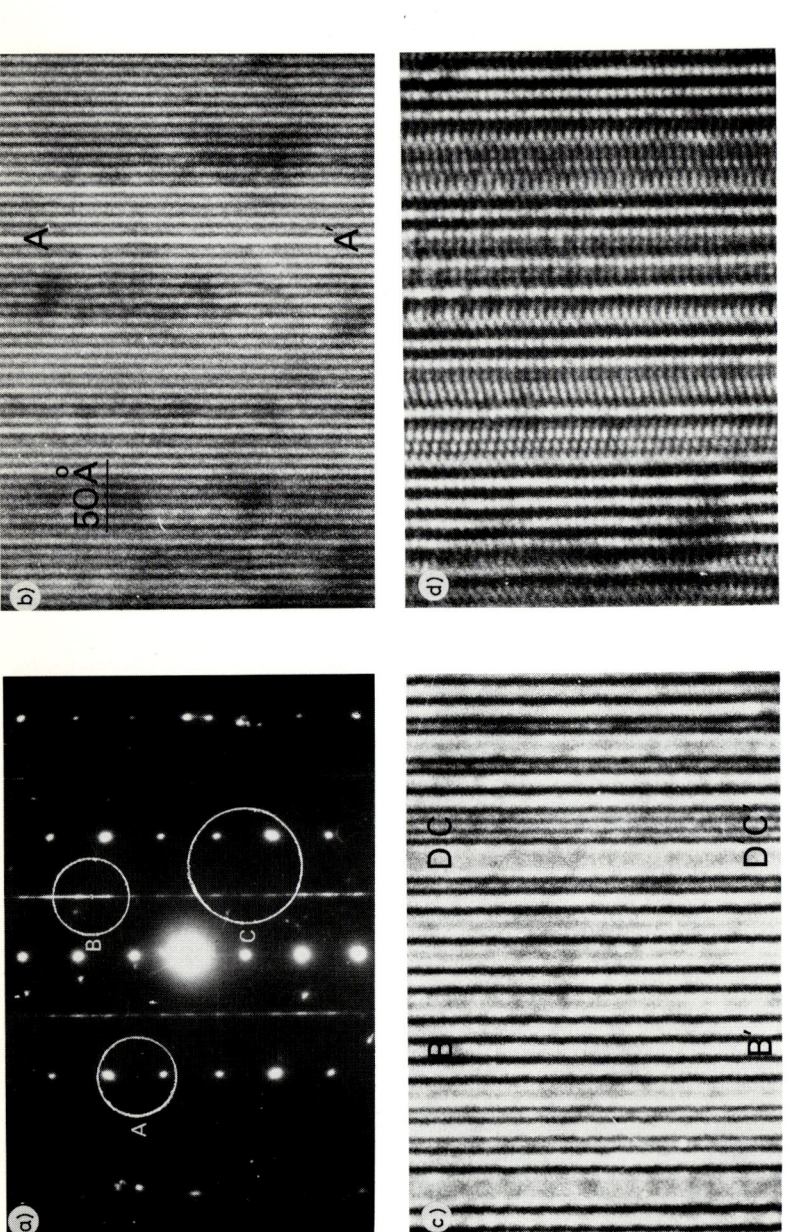

Plate 17 (a) The (hk0) electron diffraction pattern of a disordered Californian wollastonite, showing severe streaking of rows with $k = 2n + 1$. (b) Lattice image formed with an aperture placed around two of the spots on a $k = 2n$ row (position A). (c) Lattice image formed with the same aperture placed around a portion of the streak on a $k = 2n + 1$ row (position B). (d) Lattice image formed with a slightly larger aperture placed around a portion of both $k = 2n$ and $k = 2n + 1$ rows (position C) (Reproduced by permission from *Mater. Res. Bull.*, 1975, **10**, 764)

The Ultrastructure of Minerals

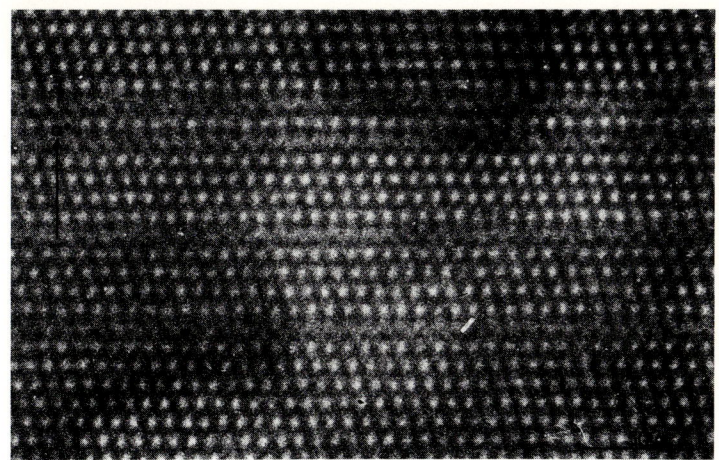

Plate 18 *Bright-field lattice image of a disordered wollastonite from New York State*
(Reproduced by permission from *Cont. Min. Pet.*, 1976, **55**, 306)

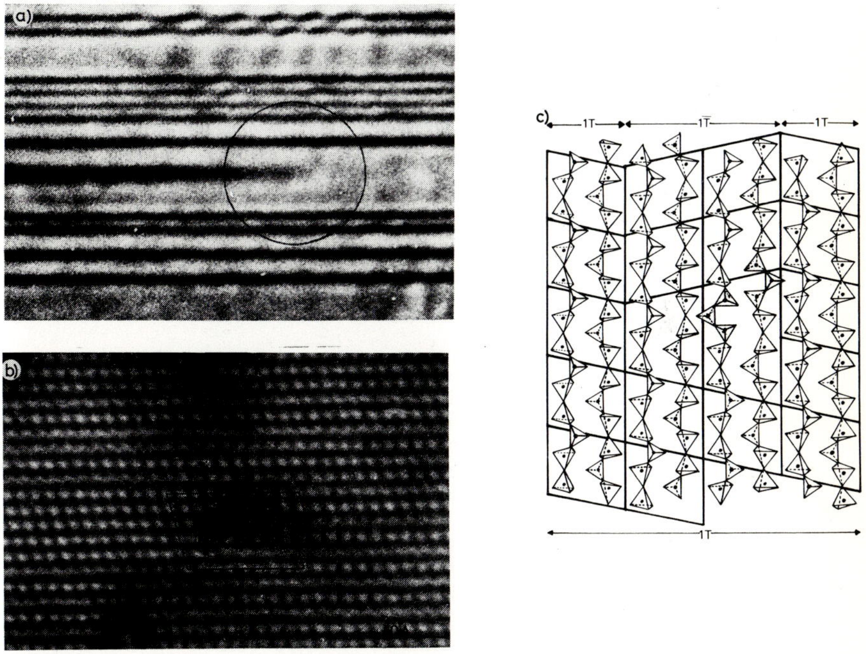

Plate 19 (a) *Dark-field image of a disordered Californian wollastonite, with a region of one polytype terminating within another (circled).* (b) *A similar type of defect, viewed in bright field in a wollastonite from New York State.* (c) *Schematic illustration of the manner in which a short length of pyroxene chain can be used to preserve a coherent interface at such a defect*
(Reproduced by permission from 'Developments in Electron Microscopy and Analysis', Academic Press, 1975, p. 277, and *Cont. Min. Pet.*, 1976, **55**, 308)

Surface and Defect Properties of Solids

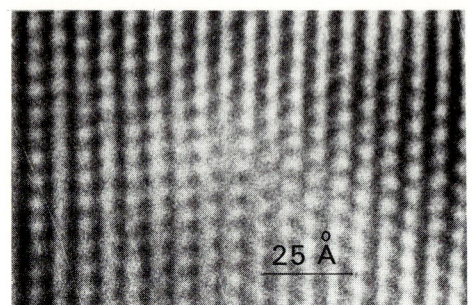

Plate 20 *An alternative type of defect, where an extra 'slab' of structure has apparently been inserted into the crystal. This specimen is a monoclinic wollastonite from Tanzania*

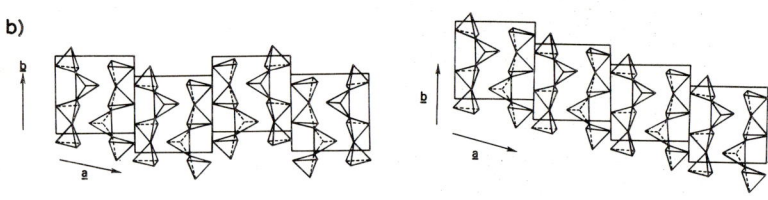

reflection twin   unfaulted structure

Plate 21 (a) *An unusual and abrupt contrast change across a crystal of wollastonite from Finland, accompanied by no apparent stacking fault.* (b) *Proposed model for the type of fault in (a), consisting of a (100) reflection twin with* **b**/4 *shift, and then continued stacking with* −**b**/4 *shifts at each layer*

The Ultrastructure of Minerals

Plate 22 *A (h0l) lattice image of nephrite (actinolite) from New Zealand, showing twinning on* (100)

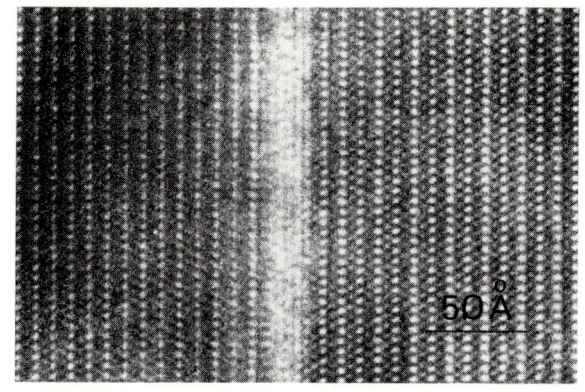

Plate 23 *A (010) Wadsley defect in New Zealand nephrite, identified as a six-fold chain strip in a normal (double chain) matrix. The electron beam is parallel to* **a**

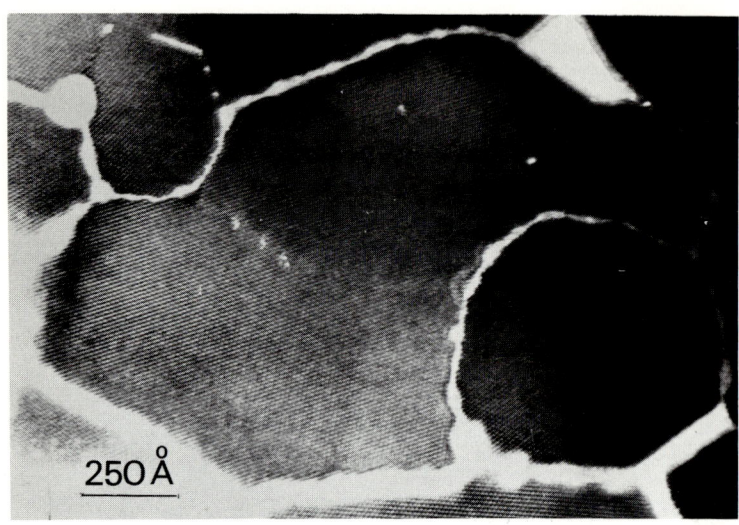

Plate 24 *A (hk0) lattice image of crocidolite fibrils, thinned as cross-sections normal to the fibre axis, showing highly irregular fibril boundaries*
(Reproduced by permission from *Nature*, 1977, **226**, 520)

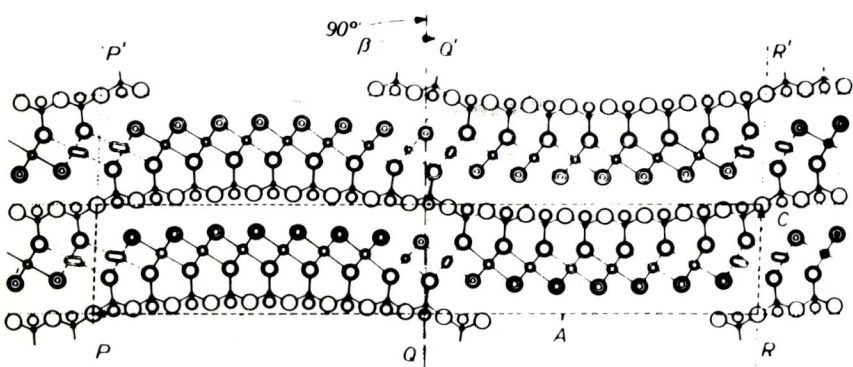

Plate 25 *The structure of the antigorite form of serpentine, showing the curvature of the layers and the resulting 'wave-like' structure*
(Reproduced by permission from *Z. Krist.*, 1956, **108**, 82)

The Ultrastructure of Minerals

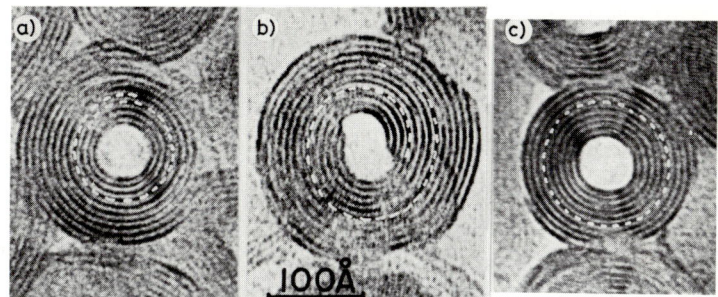

Plate 26 *Lattice images of cross-sections of chrysotile (Transvaal), showing* (a) *single-layer spiral growth;* (b) *multiple-layer growth and* (c) *concentric layer growth*
(Reproduced by permission from *Acta Cryst*, 1971, **A27**, 659)

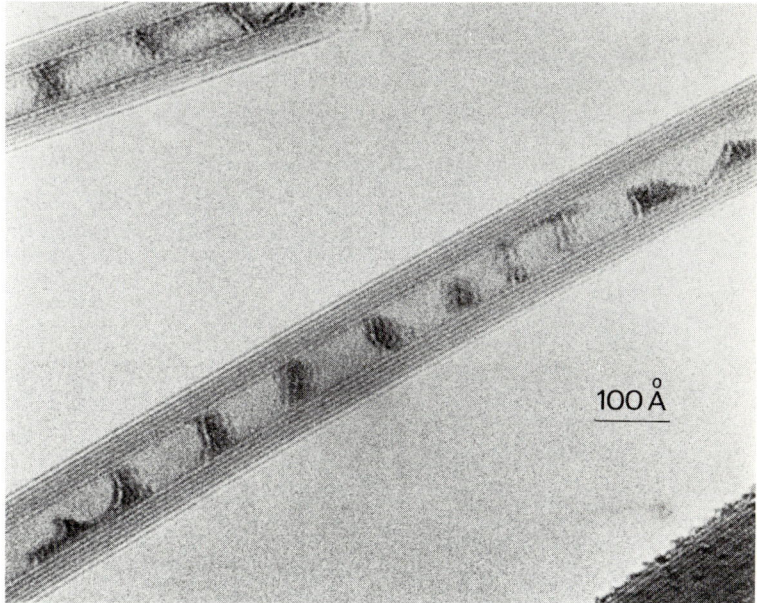

Plate 27 *Thin fibril of chrysotile (Clinton) containing blobs of amorphous silica*

Surface and Defect Properties of Solids

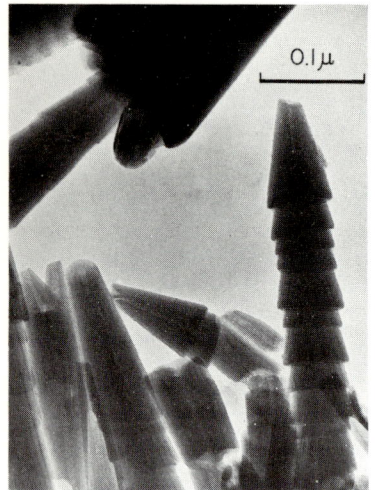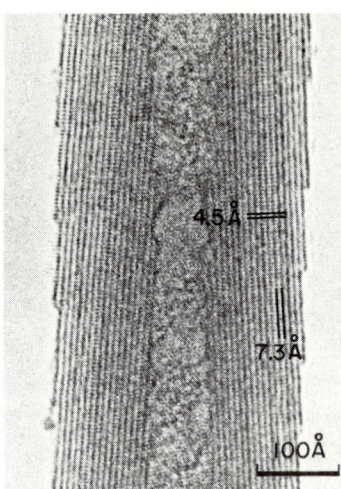

Plate 28 *'Cone-in-cone' formation of synthetic chrysotile*
(Reproduced by permission from 'Eighth International Congress on Electron Microscopy (Canberra)' 1974, Vol. 1, p. 494)

Plate 29 *A (hk0) lattice image of antigorite, displaying a* 46 Å *repeat along a, in addition to the 'serpentine' substructure*

The Ultrastructure of Minerals

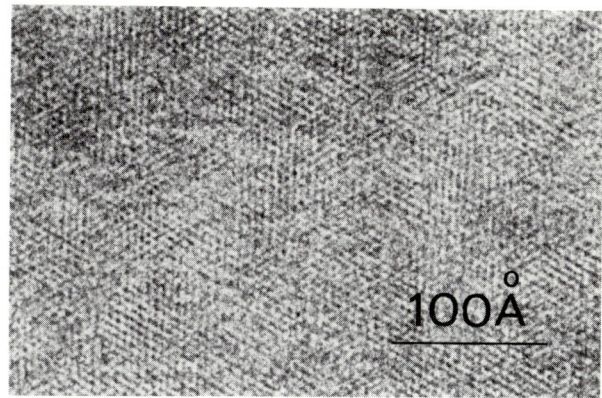

Plate 30 *A (hk0) lattice image of a synthetic lizardite, showing considerable disorder in the form of dislocations and domain structures*
[Reproduced by permission from 'Eighth International Congress on Electron Microscopy (Canberra)' 1974, Vol. 1, p. 494]

Plate 31 *Ion-thinned section of chrysotile, showing an incomplete fibre of 'polygonal chrysotile'. in addition to several smaller fibrils*
(Reproduced by permission from *Canad. Mineral*, 1976, **14**, 307)

Surface and Defect Properties of Solids

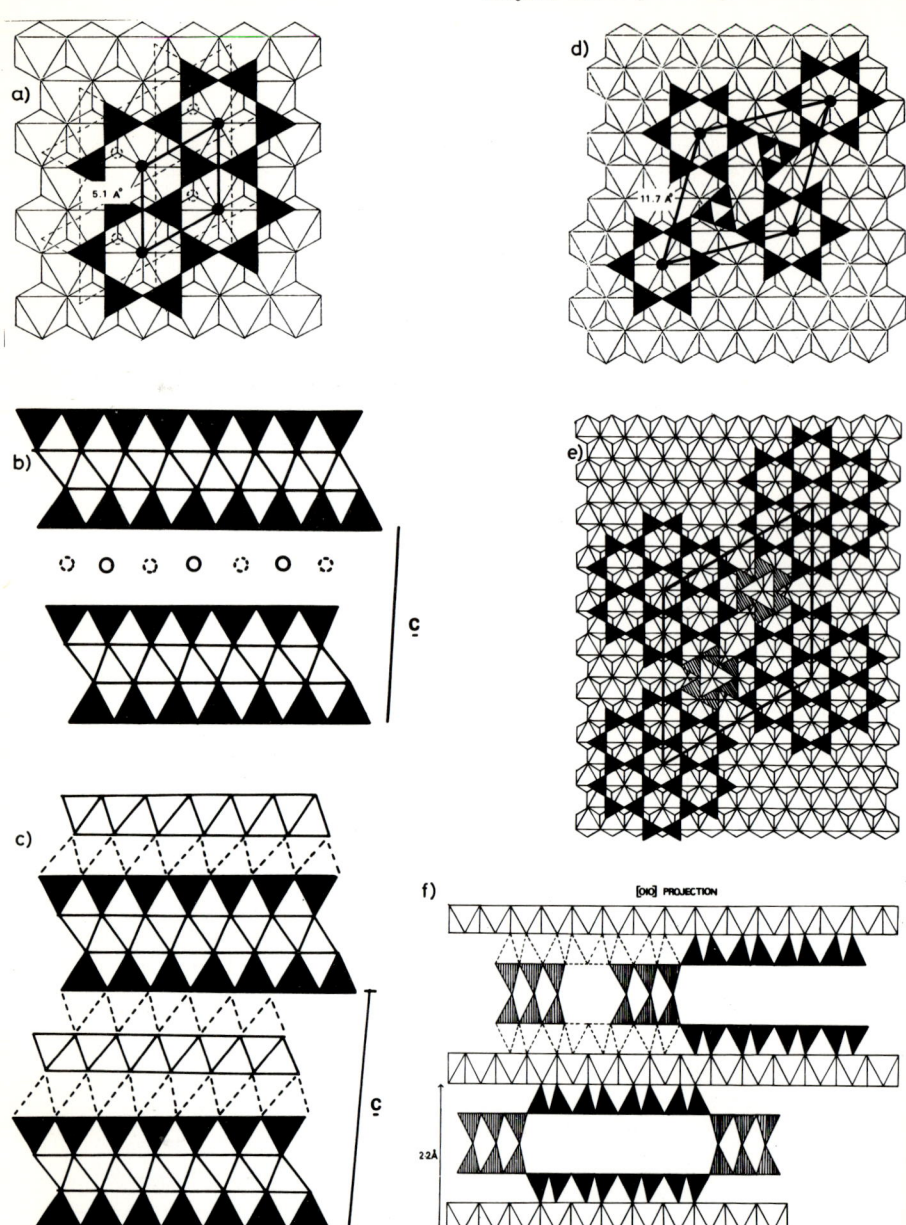

Plate 32 *Idealized structures of the layered silicates considered in the text.* (a) *Mica structure projected onto* (001). (b) *Mica structure viewed along* [010]. (c) *Chlorite structure viewed along* [010]. (d) *Zussmanite structure projected onto* (001). (e) *Stilpnomelane structure projected onto* (001). (f) *Stilpnomelane structure viewed along* [010]. (*Projections of the zussmanite structure along* [010] *or* [100] *do not illustrate the structural arrangement clearly and are therefore omitted*)

The Ultrastructure of Minerals

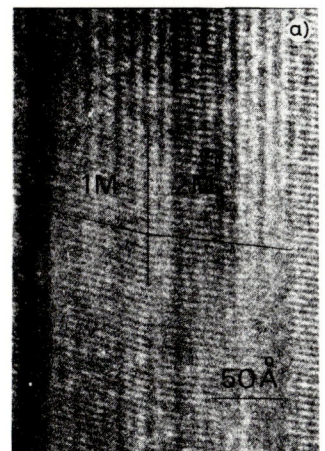
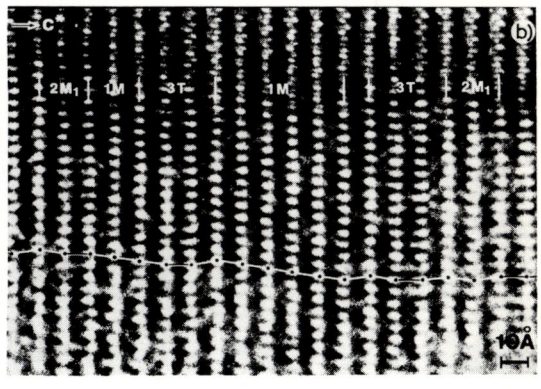
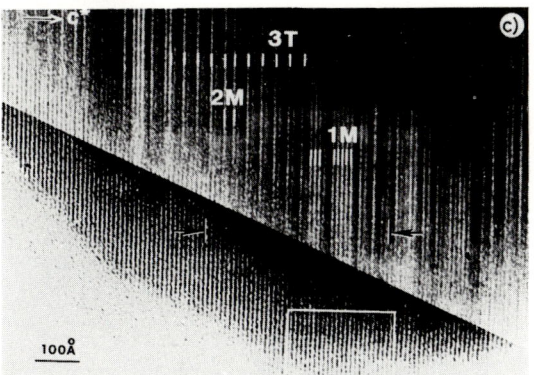

Plate 33 (a) *High-resolution lattice image of a muscovite flake after ion-thinning, with the electron beam parallel to [100]. Regions of the different polytypes are indicated.* (b) *Improved image of a microtome-cut section of a different sample, but in the same orientation, again showing clearly the different polytypes.* (c) *High-resolution image of the same sample as in (b) with the electron beam parallel to [010]. At the crystal edge (indicated) the various polytypes are not discernible, their existence becoming apparent at greater crystal thicknesses as a result of multiple scattering*
(Parts (b) and (c) reproduced by permission from *Acta Cryst.*, 1977, **A33**, (in press))

Surface and Defect Properties of Solids

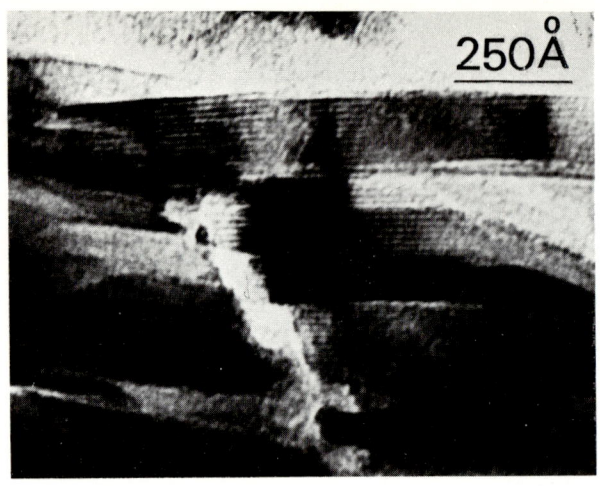

Plate 34 *High-resolution lattice image of a microtome-cut section of chlorite* (Reproduced by permission from *Clays and Clay Minerals*, 1973, **21**, 3)

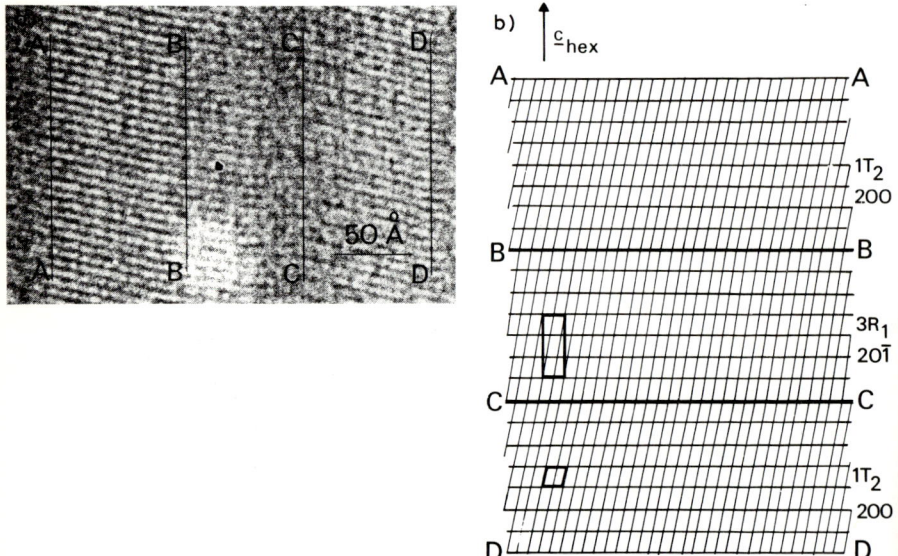

Plate 35 (a) *High-resolution lattice image of (h0l) fringes in zussmanite, with different polytypes shown by the alteration in fringe direction,* (b) *Schematic illustration of part of* (a), *with an intergrowth of rhombohedral* (3R) *and two-layer triclinic* (2Tc) *polytypes* (Reproduced from *J.C.S. Faraday II*, 1974, **70**, 1692)

The Ultrastructure of Minerals

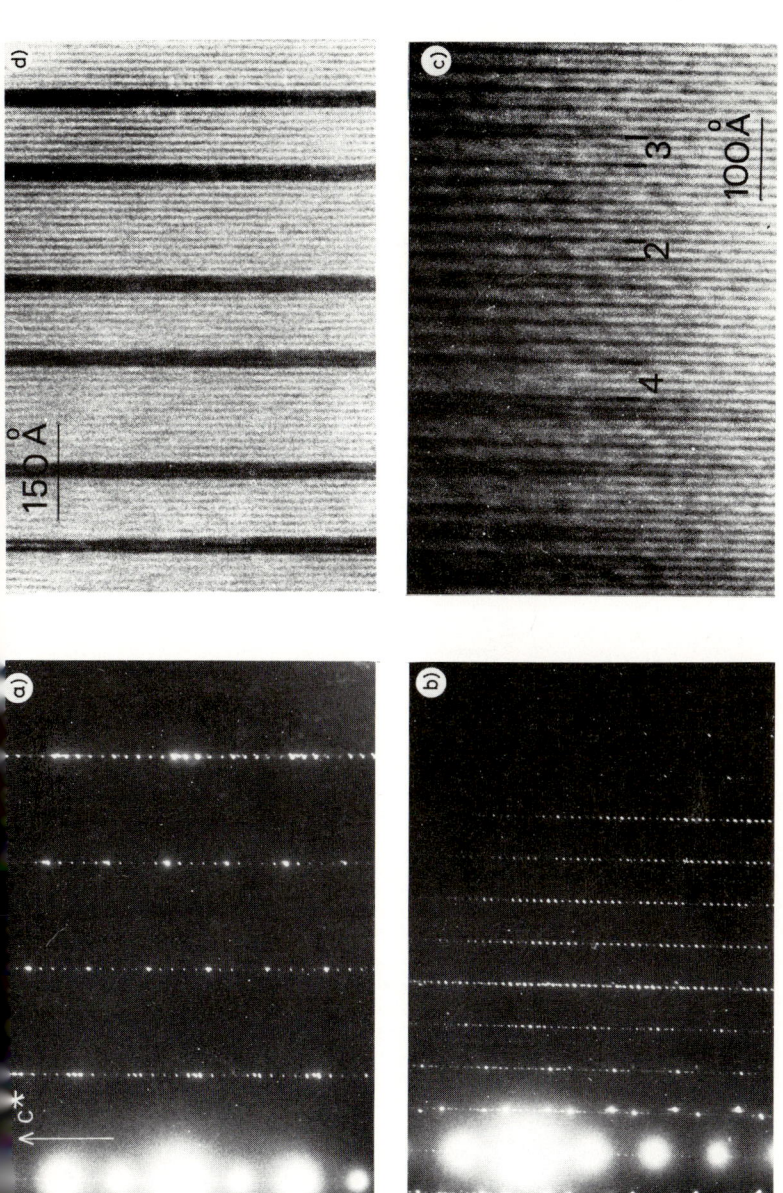

Plate 36 (a) (3h,h,l) electron diffraction pattern for a five-layer polytype of stilpnomelane. (b) (h0l) electron diffraction pattern from a nine-layer variant of the same mineral. (c) Three- and four-layer polytypes of stilpnomelane interspersed in a crystal where the stacking sequence is predominantly of the two-layer trigonal (2T) type. (d) Regular repetition of nine- and fourteen-layer stilpnomelane polytypes. The images at the crystal edges indicate only one-layer repeats, in the same manner as Plate 33(a) (Reproduced by permission from *Acta Cryst.*, 1977, **A33** (in press))

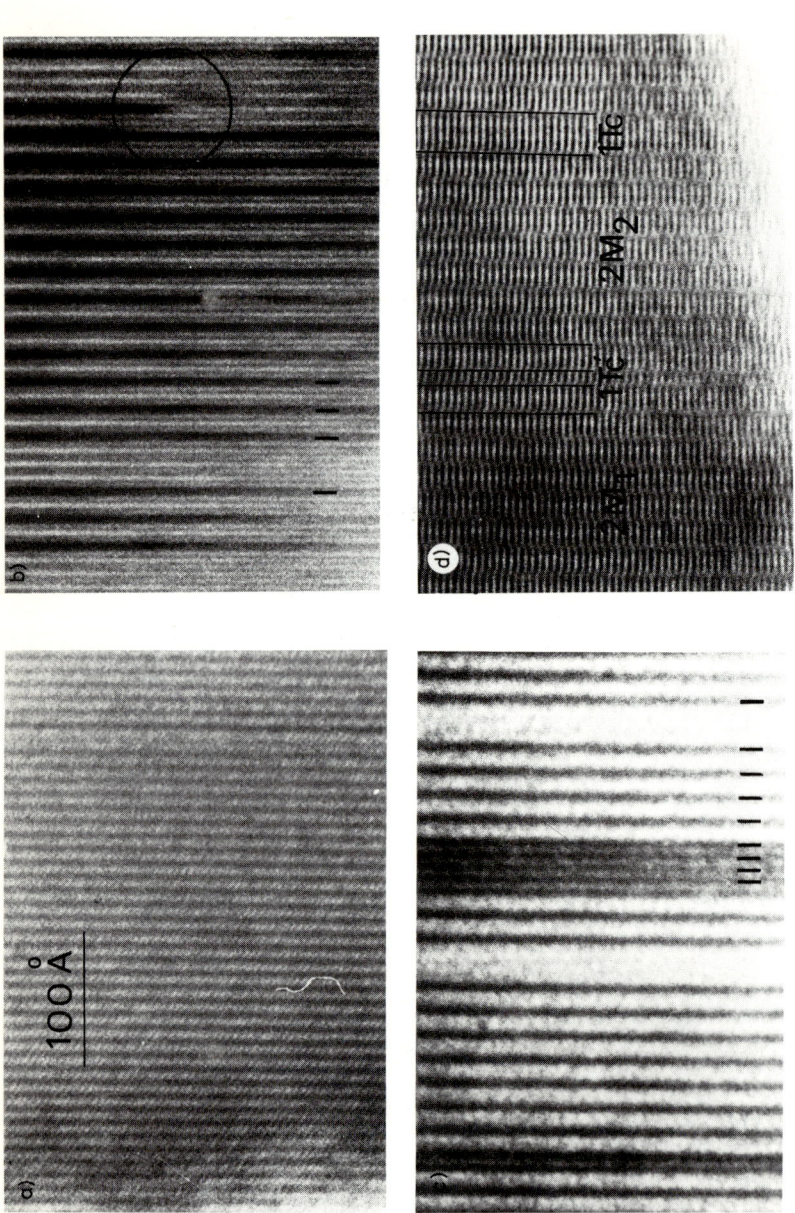

Plate 37 *High-resolution lattice images of chloritoid.* (a) Triclinic structure, showing an isolated monoclinic fault. Electron beam parallel to [110]. (b) Monoclinic 2M₂ structure, with considerable faulting. Electron beam parallel to [100]. (c) Alternative 2M monoclinic form, showing two types of fault. Electron beam parallel to [100]. (d) Two-dimensional lattice image of 2M₂ form, showing an intergrowth with 2M, and

The Ultrastructure of Minerals

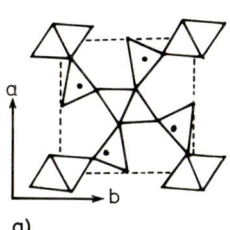 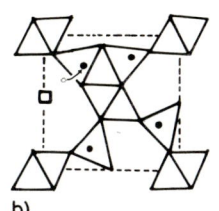 

a) b) c)

Plate 38 (a) [110] *projection of sillimanite, shown as a network of octahedra and tetrahedra. Tetrahedrally co-ordinated silicon or aluminium atom sites are shown by black dots.* (b) [110] *projection of mullite, showing a possible 'O_3' vacancy (□) and the shift of an associated tetrahedral cation (see text).* (c) [010] *projection of mullite, idealized as linked tetrahedra. Octahedral aluminium atoms are shown as black dots, 'O_3' vacancies as □. The lower half of the figure shows silliminate type structure with no vacancies*

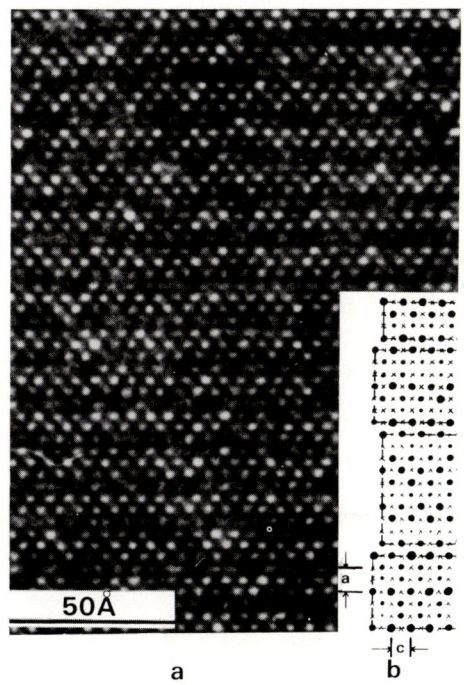

Plate 39 (h0l) *lattice image of mullite in which the array of white dots is related to rows containing 'O_3' vacancies (□ in Plate 38c)*
(Reproduced by permission from *Proc. Japan. Acad.*, 1975, **51**, 173)

Surface and Defect Properties of Solids

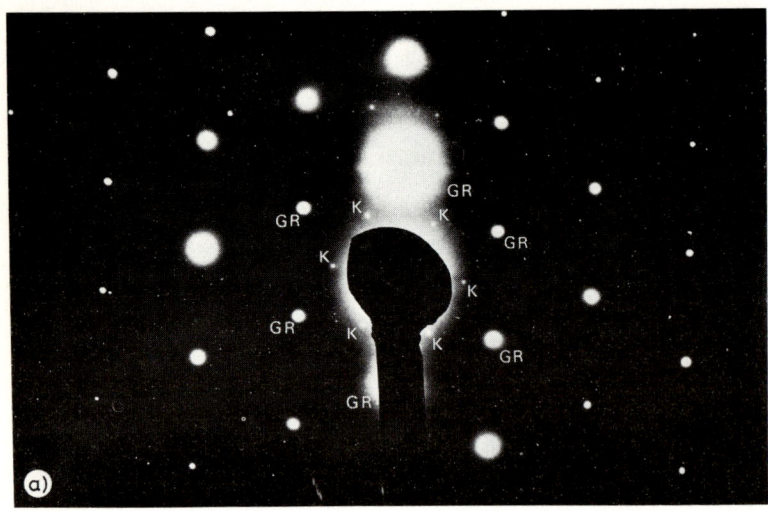

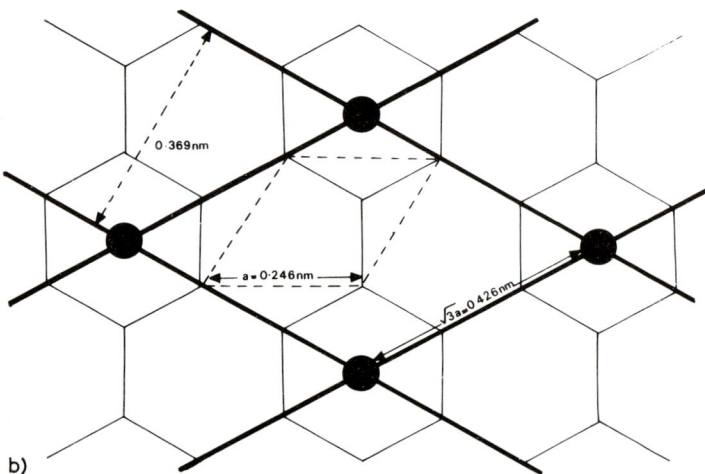

Plate 40 (a) *Selected area diffraction pattern of graphite–potassium intercalate showing that the potassium is ordered with a superlattice cell equal to 3a of graphite.* (b) *Schematic representation of possible arrangement of potassium on the graphite lattice that would produce the observed superlattice seen in* (a).
(Reproduced by permission from *J. Solid State Chem.*, 1975, **4**, 99)

The Ultrastructure of Minerals

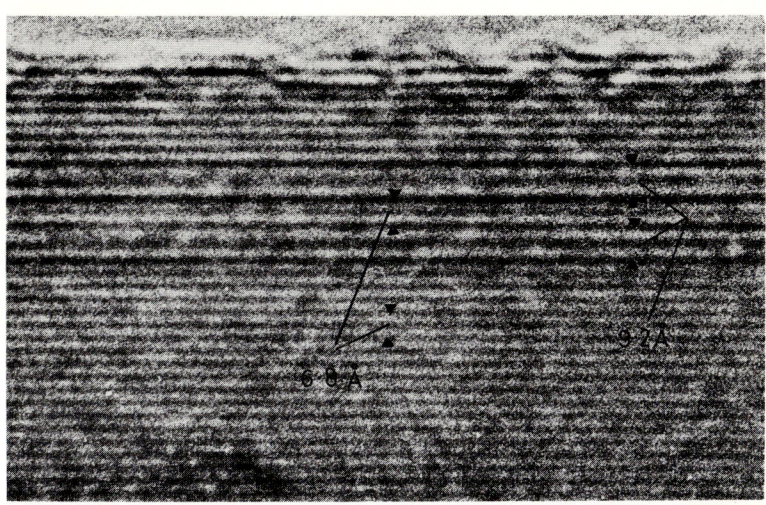

Plate 41 *Bright-field lattice image indicating the presence of two strips of first-stage graphite–ferric chloride intercalate (spacing 9.2 Å) in coherent contact with and separated by graphite (spacing 6.8 Å)*
(Reproduced from *J.C.S. Dalton*, 1976, 2443)

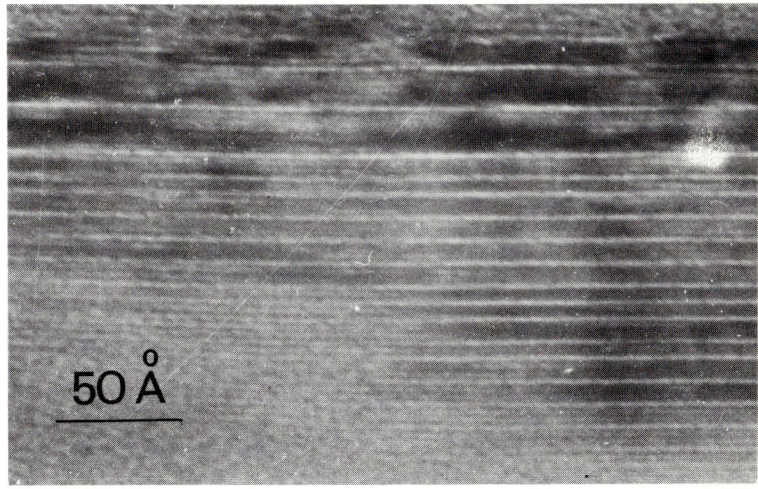

Plate 42 *Bright-field lattice image of a 210 Å thick slab of graphite–$FeCl_3$ intercalate with average composition $C_{12}FeCl_3$, corresponding to a second-stage intercalate. The light striations on the right of the micrograph delineate the sheets of $FeCl_3$ octahedra, and from their separation and the simple formula given in the text the particular state (i.e. whether sixth or first) of the various intercalates which are coherently intergrown, may be readily identified*
(Reproduced from *J.C.S. Dalton*, 1976, 2443)

Surface and Defect Properties of Solids

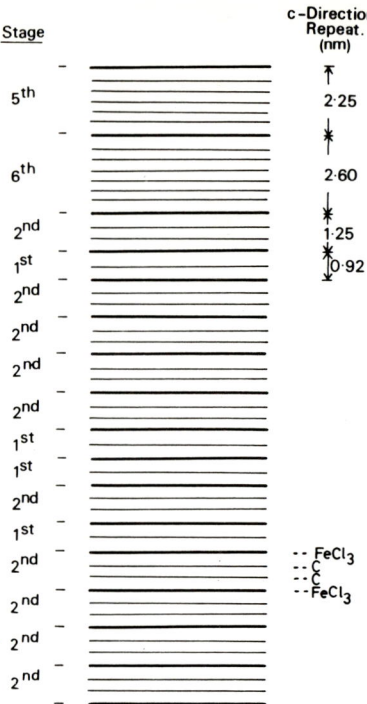

Plate 43 *Schematic diagram illustrating the interpretation of Plate 41 in terms of various stages of coherently grown graphite–$FeCl_3$ intercalates*

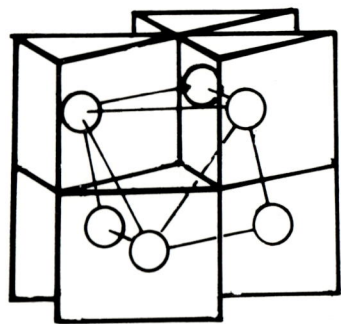

Plate 44 *Idealization of the NiAs structure type, showing arsenic atoms (circles) in trigonal prismatic co-ordination; nickel atoms are at prism vertices*

The Ultrastructure of Minerals

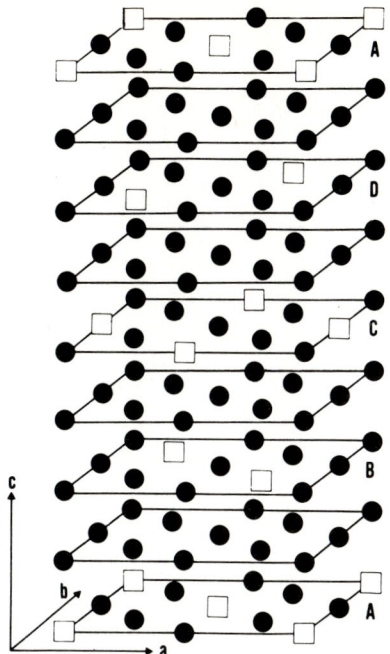

Plate 45 *Idealized structure of 4C pyrrhotite, showing only the iron layers. Occupied iron atom sites are shown by solid circles, vacant sites as squares*

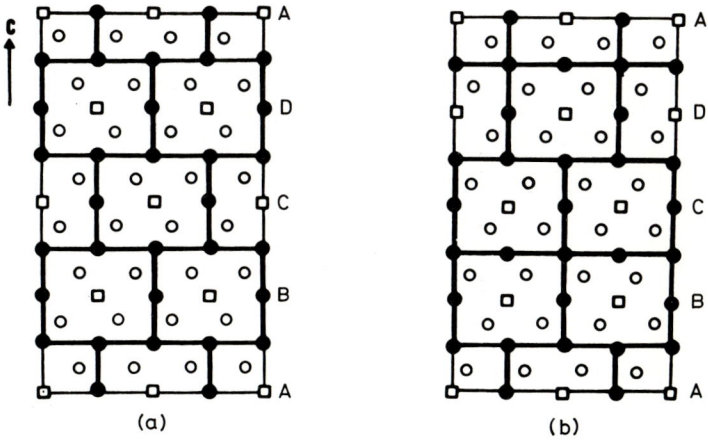

Plate 46 *Projections of 4C pyrrhotite, (a) along [010] and (b) along [110]. Open circles represent sulphur atoms*

Surface and Defect Properties of Solids

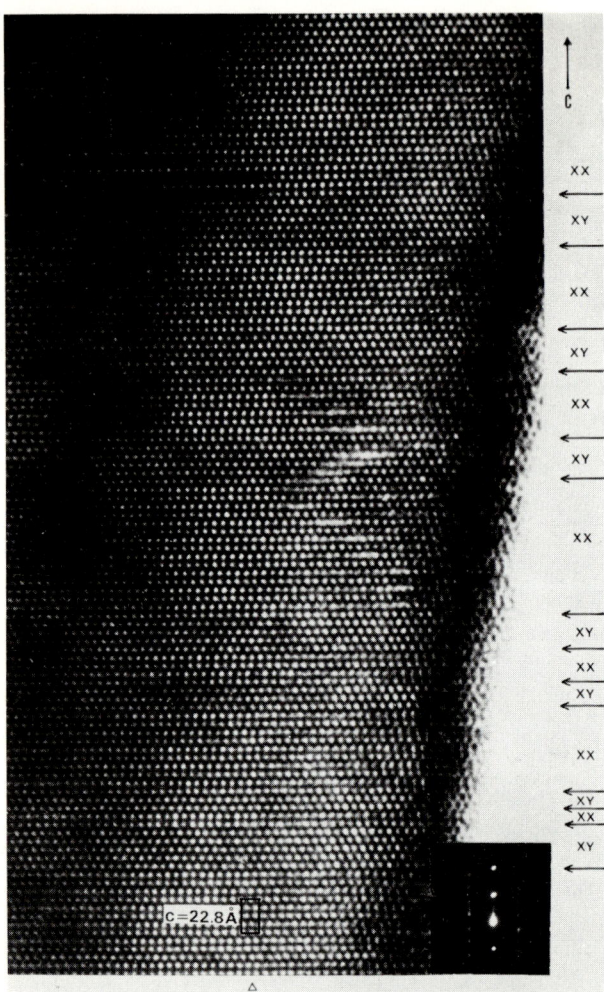

Plate 47 (*hk*0) *lattice image of* 4*C pyrrhotite, showing rotation twinning on* (001). *The electron beam is parallel to* [010] *in regions* 'XY'; *parallel to* [110] *in regions* 'XX'
(Reproduced by permission from *Amer. Mineral.*, 1975, **60**, 359)

The Ultrastructure of Minerals

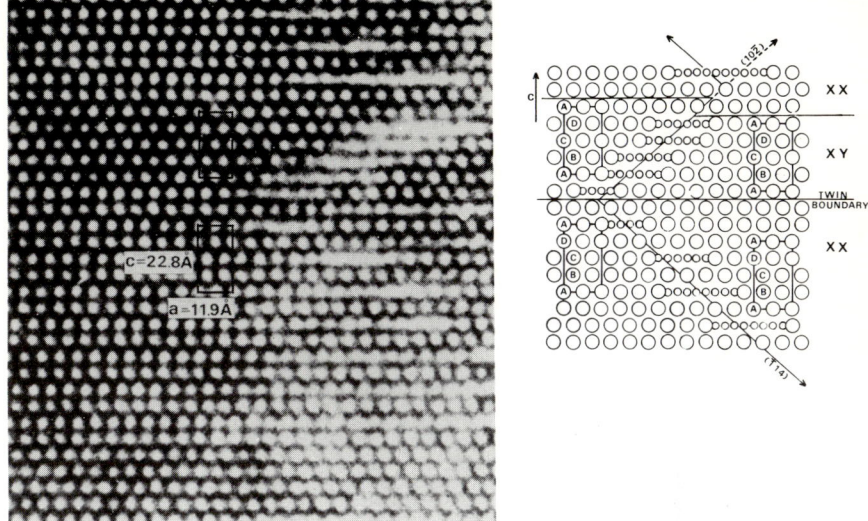

Plate 48 *Enlarged image of 4C pyrrhotite, showing details of 'out of step' boundaries (see text) with white streaks corresponding to overlap of twin domains*
(Reproduced by permission from *Amer. Mineral.*, 1975, **60,** 359)

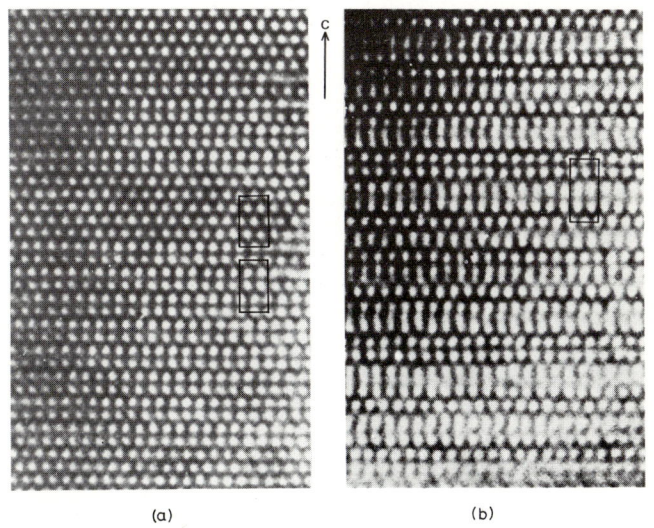

Plate 49 (a) *(hk0) lattice image of (twinned) 4C pyrrhotite, at same scale as* (b). (b) *(hk0) lattice image of nC pyrrhotite (n = 5) showing disorder between layers*
(Reproduced by permission from *Proc. Japan. Acad.*, 1974, **50,** 756)

Surface and Defect Properties of Solids

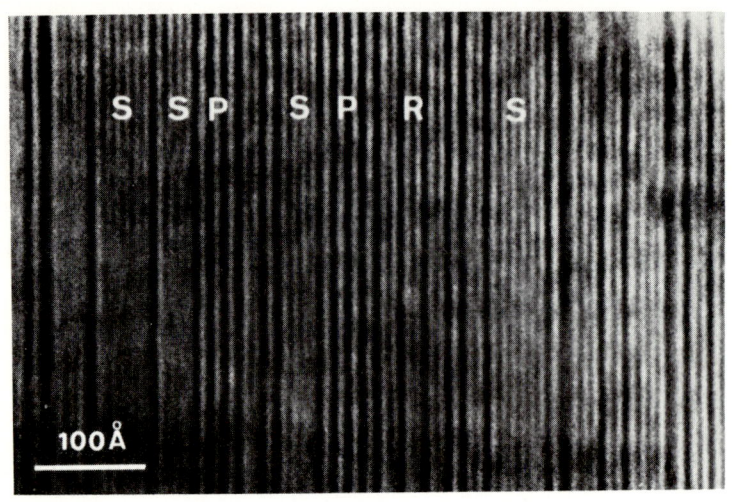

Plate 50 *'Systematics' lattice fringe image showing domains of parisite* **(P)**, *roentgenite* **(R)**, *and synchistite* **(S)** *randomly, but coherently, intergrown*
(Reproduced by permission from *Amer. Mineral*, 1975, **60**, 351)

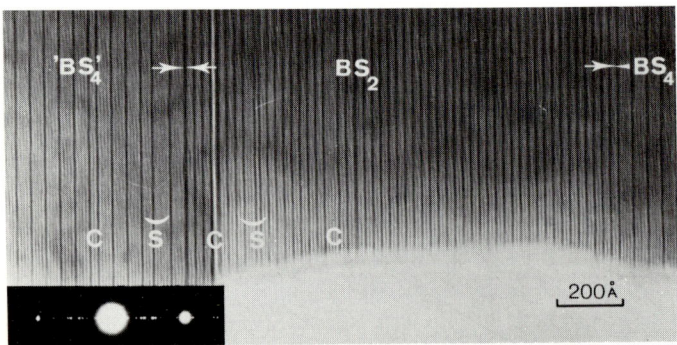

Plate 51 *Lattice fringe image showing intergrowth between roentgenite* **(BS$_2$)** *with* '**BS$_4$**'. '*C*' *and* '*S*' *indicate compositional and sequential faults, respectively*
(Reproduced with permission from *Amer. Mineral.*, 1975, **60**, 351)

Point Defects in Ionic Crystals

about by commercial pressure, there are now corresponding theoretical developments which allow an understanding of the atomistic nature of the operative transport mechanisms.[170,171] In what follows we shall survey briefly the nature of fast ion conduction and also current knowledge concerning the transport mechanism in the various classes of compounds which exhibit this property. In particular we shall endeavour to see the limits of common ground between the interpretations of fast ion and normal ionic conduction and also where it has been necessary to invoke new concepts.

The contribution of a single species to the specific conductance of an ionic conductor may be written [1] as equation (3), where q and u are the charge and

$$\sigma = |q|nu \qquad (3)$$

mobility of the species, respectively, and n is the number per unit volume. Since σ depends on both number and mobility it is essential to remember that an anomalously high conductivity might result either from a very large number of conducting species, or from extremely mobile species (due perhaps to the onset of a new transport mechanism with change in temperature) or from a suitable combination of both of these factors. Theoretical estimates [173—175] have been made of maximum values for σ and for the diffusion coefficient, D, and lead to suggested criteria for fast ion conductors. Both van Gool [171] and McGeehin and Hooper [169] point out that the activation energy for ion motion, E_a, should be in the range 0.1—0.5 eV, *i.e.* substantially less than the energy required to form an isolated point defect. If E_a is at the lower end of this range, then typical ambient temperature values of D and σ are $\sim 10^{-6}$ cm^2 s^{-1} and 10^{-2} ohm^{-1} cm^{-1}, with somewhat lower values also acceptable.[171] Some materials may have higher E_a values and thus show relatively high conductivities at higher temperatures. Values of this sort for the transport parameters ensure that many of the ions are active in the process and with mobilities close to the maximum. Although some 300 compounds are listed in the 1972 proceedings,[166] it has been noted [169] that FIC has been firmly established in relatively few of these. In this respect the difficulties sometimes present in the preparation of standard specimens and in their adequate testing have been pointed out, and both fast scanning techniques and accurate confirmation testing methods have been summarized.[176,177]

Ag$^+$ and Cu$^+$ as Fast Ion Conductors.—AgI undergoes a first-order structural phase transition, with an enthalpy change comparable to normal melting, from a lower temperature crystalline phase in which the ions are ordered to a high-temperature α-phase with a disordered cation state, in which the Ag$^+$ ions are arranged in a bcc configuration and show, using the Rickert 'region' concept,[178] only a one-third occupancy. α-AgI is the prototype for a large number of Ag$^+$ and Cu$^+$ fast ion conductors, and attempts to stabilize the high-temperature form at lower tempera-

---

[173] M. O'Keeffe, in ref. 166, p. 233.
[174] H. Rickert, in ref. 166, p. 3.
[175] Y. Haven, in ref. 166, p. 35.
[176] M. S. Whittingham, in ref. 166, p. 429.
[177] J. B. Wagner, jun., in ref. 166, p. 489.
[178] H. Rickert, *Z. phys. Chem.*, 1960, **E24**, 418.

tures have been successful when by substitution of either anions or cations the effective structure of the lower temperature phase has been altered. Perhaps the best-known examples are in the MAg_4I_5 class of compounds,[179] but Owens has reviewed the factors influencing stability[180] and a comprehensive list of these compounds will also be found in reference 169. Not unnaturally, attempts to explain the anomalously high conductivity of these compounds have been linked to the special features of their structures. These features are briefly as follows: (i) the cations are distributed over an excessive number of symmetrically equivalent available sites; (ii) there is a small activation energy for the motion of these ions resulting in a high mobility and the description of their behaviour as 'liquid-like'; and (iii) the compounds often exhibit continuous face-sharing of octahedra or tetrahedra through which ions may diffuse easily.

Whittingham[170] lists pre-exponential and activation parameters for conductivity and diffusion in a number of these compounds and also values for the Haven ratio,[1,181] which is useful in the determination of the transport mechanism. In an analysis based on concepts such as those used in treating high-temperature conductivity in $AgCl$[37] and $AgBr$[39] he shows these data to be consistent with an interstitialcy mechanism where there is no differentiation between regular and interstitial lattice sites but in which there are a proportion of co-operative jumps. The jump frequencies are typical of a regular lattice vibration, and there has been evidence in the literature for some time of a broad low-frequency absorption which has been associated with large amplitude vibrations of the cations in small potential wells.[182—185] A phenomenological theory has been developed in which it is suggested that the interaction of interstitial cation defects in the low-temperature β-phase with the strain field they induce plays a crucial role in the phase transition. It is concluded that anomalously large free volumes of cation interstitial formation below the transition temperature and large polarizabilities of both conducting and static ions are also likely to be important factors.[186] Recent inelastic neutron scattering experiments in β-AgI at room temperature show a low-lying optical mode at ~ 2 meV with a flat dispersion and an unusually high amplitude of vibration, which favours the movement of a cation to an interstitial position and is considered essential for the β- to α-phase transition.[187]

van Gool[171] stresses that with large fractions of mobile ions, spatially cooperative phenomena may be expected to occur, and he points to the lack of definite information in the liquid-like model regarding the relative times spent by ions on or between sites. He, like Roth,[188] uses a structural approach, and with Bottelberghs[189] has formulated a domain model in which the movement of domain walls gives the diffusion process. For α-AgI he suggests the presence of tetragonal

---

[179] J. N. Bradley and P. D. Greene, *Trans. Faraday Soc.*, 1967, **63**, 2516.
[180] B. B. Owens, *Adv. Electrochem. Electrochem. Eng.*, 1971, **8**, 1; also in ref. 166, p. 593.
[181] K. Compaan and Y. Haven, *Trans. Faraday Soc.*, 1956, **52**, 786; 1958, **54**, 1498.
[182] W. Jost, K. Funke, and A. Jost, *Z. Naturforsch.*, 1970, **25a**, 983.
[183] K. Funke and A. Jost, *Ber. Bunsengesellschaft phys. Chem.*, 1971, **75**, 436.
[184] K. Funke and R. Hackenberg, *Ber. Bunsengesellschaft phys. Chem.*, 1972, **76**, 883.
[185] G. Eckold and K. Funke, *Z. Naturforsch.*, 1973, **28a**, 1042.
[186] M. J. Rice, S. Staessler, and G. A. Toombs, *Phys. Rev. Letters*, 1974, **32**, 596.
[187] W. Bührer and P. Brüesch, *Solid State Comm.*, 1975, **16**, 155.
[188] W. L. Roth, *J. Solid State Chem.*, 1972, **4**, 60.
[189] W. van Gool and P. H. Bottelberghs, *J. Solid State Chem.*, 1973, **7**, 59.

domain structures [190] which is supported by X-ray evidence.[191,192] A structural approach to anomalous conductivity by Ag^+ ions has also been taken by Wiedersich and Geller [193] with particular emphasis on the networks and passages resulting from face-sharing by the various anion polyhedra.[194] Armstrong et al.,[195] also using structural principles, conclude that large concentrations of mobile cations are likely to exist at moderate temperatures only where those cations are univalent and stable in four-, three-, and even two-co-ordinated arrangements, preferably with anions which have some $d\pi$-bonding ability. These criteria are admirably fulfilled by Ag^+ and Cu^+ ions, and the partial covalent character of the bonds which they form may be most important in reducing the energy during the transition states. The silver ion, because of its polarizability, can apparently occupy sites ranging from 0.7 to 1.3 Å in radius,[170] this being a unique and valuable property where size-compatibility is essential. Even in the normal ionic conductor AgCl the vacancy migration enthalpy for Ag^+ (0.275 eV) is much less than that for the smaller univalent sodium ion (0.49 eV).[41]

Fast Conduction by Alkali Ions.—Because of their high electronegativities and since they are relatively cheap, there are sound scientific and commercial reasons which favour alkali ion fast conductors. The β-alumina class of compounds, which are used in sodium–sulphur cells, have been most widely studied,[171] and McGeehin and Hooper [169] include a separate recent bibliography. β-Alumina compounds are non-stoicheiometric, may be written as $M_{1+x}Al_{11}O_{17}$, and have a layered structure. Densely packed spinel-type layers, which are four oxygen positions thick, are separated from each other by the conduction planes so that ionic conduction is two-dimensional. The conduction planes which are perpendicular to the c-axis contain Al–O–Al groups and also abundant sites for the univalent conducting ions. Beevers and Ross [196] established that in the $P6_3/mmc$ space group there were two positions (now called the 'Beevers Ross', B-R, and 'anti-Beevers Ross', aB-R, sites) which the univalent ions might occupy. Recent X-ray work has shown excess Na^+ ions [197] to be distributed on B-R sites and on positions midway between B-R and aB-R sites, while in silver β''-alumina [188] the Ag^+ ions were found to occupy B-R, aB-R, and the mid point. In β-alumina, which is usually stabilized by MgO and is an even better conductor, there is a slight modification of the structure which makes the conducting layer no longer a mirror plane and allows all the cations to reside on identical sites. For β-alumina it has been shown [198] that the conductivity is proportional to the excess of cations over the stoicheiometric composition and pressure studies have confirmed the importance of ion size.[199]

[190] W. van Gool, in ref. 166, p. 201.
[191] L. W. Strock, *Z. phys. Chem.* (*Leipzig*), 1934, **B25**, 441; 1935, **B31**, 132.
[192] G. Burley, *Acta Cryst.*, 1967, **23**, 1.
[193] H. Wiedersich and S. Geller, in 'The Chemistry of Extended Defects in Non-metallic Solids', ed. L. Eyring and M. O'Keefe, North Holland, Amsterdam, 1970, p. 629.
[194] S. Geller, in ref. 166, p. 607.
[195] R. D. Armstrong, R. S. Bulmer, and T. Dickinson, in ref. 166, p. 269.
[196] C. A. Beevers and M. A. Ross, *Z. Krist.*, 1937, **97**, 59; C. A. Beevers and S. Brohult, *Z. Krist.*, 1932, **95**, 472.
[197] C. Peters, M. Bettman, J. Moore, and M. Glick, *Acta Cryst.*, 1971, **B27**, 1826.
[198] A. F. Sammells and J. H. Kennedy, *J. Electrochem. Soc.*, 1972, **119**, 1609.
[199] R. H. Radzilowski and J. T. Kummer, *J. Electrochem. Soc.*, 1971, **118**, 714.

Following measurements of transport properties Whittingham proposed that an interstitialcy mechanism, based on point defect concepts, operates in β-alumina, and he has shown the Haven ratio calculated for this mechanism to be in excellent agreement with the experimental value.[170,200] Recent calculations of potential energy curves for various carrier ions support this view. They show that although there is a substantial (~ 2 eV) difference between the B-R and aB-R sites in a stoicheiometric crystal, in the real (*i.e.* non-stoicheiometric) case carrier ions, when considered as interstitialcy pairs, yield activation energies compatible with the experimental values.[201] The predicted occupancy of sites is in good agreement with experiment for Na^+ and the poorer agreement for Ag^+ is attributed to the covalent bonding between silver and oxygen at the aB-R site. It should be noted that these calculations are rather less sophisticated than those now available for normal ionic conductors [92] in that they assume all the ions except the carriers to be fixed at equilibrium sites. The domain model has also been applied to β-alumina [188] with areas of stoicheiometric material joined to areas containing twice the normal number of cations. Such alkali-rich regions have also been suggested in the interpretation of X-ray data.[202] Electrostatic calculations on the application of the domain model to the stoicheiometric material have shown that the average additional energy necessary for the promotion of a sodium ion to an aB-R site decreases to reasonable values as the sizes of the domains increase.[189] Whittingham has pointed out that the ensuing diffusion mechanism is probably similar to that for his point model,[170] although now confined to the areas of the domain walls. The predicted Haven ratio and the dependence of σ on the excess of alkali ions would also be expected to be consistent. Two other stuctural approaches which have been used are the path probability method of Kikuchi,[203] which is based on a vacancy mechanism, and the free ion model of Rice and Roth.[204] The latter, being a macroscopic theory, is less likely to be useful and has also been criticized [205] on the basis that it is not clear how it differs from a simple hopping model.

Fast alkali ion conduction has also been observed in a number of compounds which have incompletely filled channels or tunnels in their structure. Notable among these are the tungsten and vanadium bronzes.[206] The general structural requirement for this type of conductor is the joining of simple octahedra by corner-, edge-, or face-sharing with the particular structure depending on the relative sizes of the participating ions. The criteria for FIC are met by the resulting channels in which the ions may move and the non-stoicheiometric ranges which lead to excess of available sites for the conducting ions. Whittingham and Huggins have reviewed a number of these systems and found diffusion for cations including ammonium and potassium in the hexagonal structure.[207] Singer *et al.* have investigated the

---

[200] M. S. Whittingham and R. A. Huggins, *J. Electrochem. Soc.*, 1971, **118**, 1.
[201] J. C. Wang, M. Gaffari, and S. Choi, *J. Chem. Phys.*, 1975, **63**, 772.
[202] Y. Le Cars, R. Comes, L. Deschamps, and J. Thery, *Acta Cryst.*, 1974, **A30**, 305.
[203] H. Sato and R. Kikuchi, *J. Chem. Phys.*, 1971, **55**, 677; R. Kikuchi and H. Sato, *ibid.*, p. 702.
[204] M. J. Rice and W. L. Roth, *J. Solid State Chem.*, 1972, **4**, 294.
[205] W. C. Haas, *J. Solid State Chem.*, 1973, **7**, 155.
[206] P. G. Dickens and M. S. Whittingham, *Quart. Rev.*, 1968, **22**, 30; P. Hagenmuller, *Progr. Solid State Chem.*, 1971, **5**, 71.
[207] M. S. Whittingham and R. A. Huggins, in ref. 166, p. 645.

hollandite and tetratungstate structures.[208] Roth and Muller,[209] in an extensive study, have concluded that conduction given by one-dimensional tunnel structures as described above may be expected to be improved in compounds which have two- or three-dimensional interconnected tunnels. McGeehin and Hopper[169] list experimental results for two systems of this type [$A_2BFe(CN)_6$ and $KSbO_3$] as well as for some 45 other compounds or classes of compounds which are alkali ion conductors.

Anionic Fast Conduction.—Doping of the Group IVB oxides (ZnO_2, HfO_2, CeO_2, and ThO_2) with rare earth or alkaline earth oxides can produce a stable fluorite phase with charge-compensating oxygen vacancies in the anion sublattice. At high temperatures (in the region of 1000 K) these systems are fast oxygen ion conductors with activation enthalpies typically in the range 0.6—1.1 eV. There have been a number of reviews of these systems,[210-212] with that of Etsell and Flengas[211] being especially comprehensive. These systems have found extensive use in oxygen monitors and for thermodynamic and kinetic measurements. Possible conduction mechanisms have been discussed by several authors, and Whittingham[170] considers the jump of a single ion along a string of vacancies as most likely to satisfy the available experimental data.

More recently there has been considerable interest in fluoride ions as possible fast conductors. The thermodynamics and kinetics of point defects in crystals with the fluorite structure have been dealt with extensively by Lidiard.[213] The special features of these crystals include a broad thermal anomaly which has been interpreted as an order–disorder transition in which the anions become distributed randomly between their normal positions and interstitial octahedrally co-ordinated sites.[214] This transition has been studied by neutron diffraction[215] and light-scattering techniques.[216] The crystals also exhibit a remarkable capacity for dissolving tervalent metal fluorides with the foreign cations entering the lattice substitutionally and the extra anions taking up interstitial positions. Accurate ionic conductivity measurements on fluorides at high temperatures are extremely difficult to perform because of reactivity of the crystals, their low resistance, and their volatility.[20] Rather rough measurements have been made over long temperature ranges including the melting points of the conductance in a series of alkaline earth halides in order to classify their behaviour in terms of possible order–disorder transitions.[217] $SrCl_2$ appears to pass through the temperature of its specific heat maximum without any anomalous change in σ, but for PbF_2 the ionic conductivity,[218] while normal at low temperatures, reaches values typical of a molten salt

---

[208] J. Singer, H. E. Kautz, W. L. Fielder, and J. S. Fordyce, in ref. 166, p. 653.
[209] W. L. Roth and O. Muller, NASA Final Report, N74 26498, 1974.
[210] B. C. H. Steele and G. J. Dudley, in MTP International Review of Science, Inorganic Chemistry Series Two, Vol. 10, Butterworth, London, 1975.
[211] T. H. Etsell and S. N. Flengas, *Chem. Rev.*, 1970, **70**, 339.
[212] C. Deportes, G. Robert, and M. Forestier, *Electrochim. Acta*, 1971, **16**, 1003.
[213] A. B. Lidiard, in ref. 3, p. 101.
[214] A. S. Dworkin and M. A. Bredig, *J. Phys. Chem.*, 1968, **72**, 1277.
[215] M. W. Thomas, *Chem. Phys. Letters*, 1976, **40**, 111.
[216] R. T. Harley, W. Hayes, A. J. Rushworth, and J. F. Ryan, *J. Phys. (C)*, 1975, **8**, L530.
[217] C. E. Derrington, A. Linder, and M. O'Keeffe, *J. Solid State Chem.*, 1975, **15**, 171.
[218] C. E. Derrington and M. O'Keeffe, *Nature (Phys. Sci.)*, 1973, **246**, 44.

at ~750 K and then increases only slightly up to its melting temperature (1102 K). Recently, high-temperature heat content measurements on PbF_2 have shown that the region of rapid increase in σ corresponds exactly to the heat capacity anomaly and is thus associated with fluoride ion disorder.[219] Recent σ measurements on BaF_2[220] in its transition temperature range, although hampered by volatility and other problems, indicate similar results but it is not clear that this is not due to the onset of a new conduction mechanism. Nagel and O'Keeffe[221] have reported a number of highly conducting fluorides including LaF_3, YF_3, and some MF_2–$M'F_3$ solid solutions. Other compounds in which fast ion conduction has been indicated by n.m.r. measurements are $CsPbF_3$[222] and compounds with the scheelite structure (*e.g.* $NaYF_4$ and $NaLaF_4$) which have a transition to the fluorite structure at high temperatures.[209]

6 Paraelectric Impurities

Since their discovery little more than a decade ago [223,224] there have been both a tremendous interest and a corresponding rapid growth in the study and understanding of point defects which produce polarizable permanent electric dipoles in ionic crystals. These systems which have become known as paraelectric centres (PEC) are of two main types: substitutional molecular defects which have intrinsic dipole moments (*e.g.* CN^- and OH^-), and substitutional small ions such as Li^+ and Cu^+ which may occupy off-centre positions in a host lattice. In a report of this size we cannot attempt comprehensive coverage: the essential elements of the theories used to understand the behaviour of PEC are included in an early review by Narayanamurti and Pohl,[225] while Lüty[226] presents a most useful survey. For a recent complete and critical review of the theoretical and experimental aspects see the article by Bridges;[227] Grimes[228] has considered some more general aspects of crystal structures with asymmetric and off-centre atoms.

Static and Dynamic Behaviour of PEC.—The static properties of these dipolar defects are characterized by an electric dipole vector and by an elastic dipole tensor, and may be studied by a number of techniques under applied static fields or stress. The orientation of the dipole depends on the nature of the multi-well potential which results from its electric and elastic interaction with the lattice. Dynamic relaxation processes have been observed to occur by both classical and tunnelling mechanisms,[226] and are also suitable for investigation by a variety of methods. Two principal theoretical models, namely the Devonshire model and the multi-well potential tunnelling model, provide a reasonable understanding of observed PEC

[219] C. E. Derrington, A. Navrotsky, and M. O'Keeffe, *Solid State Comm.*, 1976, **18**, 47.
[220] V. M. Carr, A. V. Chadwick, and D. R. Figueroa, *J. Phys. (Paris)*, 1976, **37**, C7–337.
[221] L. E. Nagel and M. O'Keeffe, in ref. 166, p. 165.
[222] V. M. Bouznik, Yu. N. Moskvich, and V. N. Voronov, *Chem. Phys. Letters*, 1976, **37**, 464.
[223] U. Kuhn and F. Lüty, *Solid State Comm.*, 1964, **2**, 281.
[224] G. Lombardo and R. O. Pohl, *Phys. Rev. Letters*, 1965, **15**, 291.
[225] V. Narayanamurti and R. O. Pohl, *Rev. Mod. Phys.*, 1970, **42**, 201.
[226] F. Lüty, *J. Phys. (Paris)*, 1973, **34**, C9–49.
[227] F. Bridges, *CRC Crit. Rev. Solid State Sci.*, 1975, **5**, 1.
[228] N. W. Grimes, *Contemp. Phys.*, 1976, **17**, 71.

behaviour and we shall describe these qualitatively and very briefly and also outline some of their recent modifications.

Devonshire Model.—Based on the original hindered rotor model [229] a static cubic crystal potential is superimposed on a free rotor. Initially the expansion of the potential in spherical harmonics was extended only to the second term resulting in a V_4 potential which had minima along the $\langle 100 \rangle$ and $\langle 111 \rangle$ directions. Inclusion of the next higher order harmonic V_6 can provide $\langle 110 \rangle$ minima also, and Beyeler [230] has given a detailed treatment of the potential shape. Multiple expansions of the potential have also been applied to the Devonshire model [231] and have been extended to allow an off-centre position for the impurity ion.[135] With these refinements the model attains considerable success but is still unable to explain several observed features.[227]

Multi-well Potential Tunnelling Model.—Although some PEC, *e.g.* off-centre Cu^+,[232] show classical rate behaviour during relaxation between two possible orientations many others, such as substitutional OH^-, reorientate by quantum mechanical tunnelling.[233] The well-known tunnelling picture for a two-well potential has been extended to cover the expected six $\langle 100 \rangle$, eight $\langle 111 \rangle$, and twelve $\langle 110 \rangle$ direction minima in cubic systems.[234, 235] Many of the off-diagonal matrix elements of the crystal field Hamiltonian are equal by symmetry and a small number of parameters suffice to describe these systems. The energy level splitting in the presence of an applied electric field is treated by including the dipole Hamiltonian, and the energy level diagrams have been worked out.[234, 235] The splitting of the energy levels by application of stress has been considered by several authors [227] and the problem formulated in terms of principal values of the stress tensor.[236, 237] More recently, Shore and Sander [136] have given an equivalent reformulation of the elastic interaction in terms of the stress-splitting coefficients, a_i, which are the quantities directly deducible from paraelastic measurements.[226, 238] Transitions between the localized states, which arise when field or stress interactions remove the equality of the multi-well potential, occur by phonon-assisted tunnelling processes.[239] Their rates at very low temperatures where single phonon processes dominate are predicted [240, 241] to show a linear dependence on temperature. Such behaviour has been observed [233] and the phonon-induced nature of the tunnelling confirmed by measuring the variation of the relaxation rate with the field. The T^4 dependence found at higher temperatures is interpreted in terms of multi-phonon relaxation processes.[240, 241] To complete our understanding of this type of tunnelling

[229] A. F. Devonshire, *Proc. Roy. Soc.*, 1936, **A153**, 601.
[230] H. U. Beyeler, *Phys. Stat. Sol. (B)*, 1972, **52**, 419.
[231] S. Chandra, G. K. Pandey, and V. K. Agrawal, *J. Phys. Chem. Solids*, 1969, **30**, 1644.
[232] M. S. Li, M. de Souza, and F. Lüty, *Phys. Rev. (B)*, 1973, **7**, 4677.
[233] S. Kapphan, *J. Phys. Chem. Solids*, 1974, **35**, 621.
[234] H. B. Shore, *Phys. Rev. Letters*, 1966, **17**, 1142.
[235] M. Gomez, S. P. Bowen, and J. A. Krumhansl, *Phys. Rev.*, 1967, **153**, 1009.
[236] A. S. Nowick and W. R. Heller, *Adv. Phys.*, 1963, **12**, 251.
[237] H. Härtel and F. Lüty, *Phys. Stat. Sol.*, 1965, **12**, 347.
[238] R. V. Jimenez and F. Lüty, *Phys. Rev. (B)*, 1975, **12**, 1531.
[239] J. A. Sussmann, *Phys. Kondens. Materie*, 1964, **2**, 146.
[240] B. G. Dick and D. Strauch, *Phys. Rev. (B)*, 1970, **2**, 2200.
[241] L. M. Sander and H. B. Shore, *Phys. Rev. (B)*, 1971, **3**, 1472; 1972, **6**, 1551.

it is also necessary to consider the distorting influence of the dipole on the surrounding lattice and, as this distortion must also rotate, the dipole is considered to be 'dressed' and its reorientation effectively hindered.[234,242] This polaron or elastic-dipole dressed tunnelling model has been tested in the off-centre RbCl:Ag$^+$ and RbBr:Ag$^+$ systems where the large preference for reorientation by tunnelling between next-nearest neighbours (90°) instead of nearest neighbours (60°) dipole states can be explained by the very different way in which the T_{2g} and E_g elastic-dipole distortions hinder these two motions.[238] This type of elastic dressing can also account for the major part of the observed trend in the variation of the OH$^-$ relaxation time with the size of the host ions.[233]

Experimental Techniques and Data.—The experimental techniques used to study the dielectric and elastic relaxation of classical aliovalent impurity–vacancy complexes have been described previously in these reports.[1] Provided that these techniques can be adapted for use at low and ultra-low temperatures they may also provide useful information on the analogous paraelectric centres. The reorientation of the off-centre Cu$^+$ defect in potassium halides follows classical rate theory and has been studied by ionic thermocurrents (ITC).[232] Dielectric loss methods [243,244] and elastic loss measurements [245] have also been used. We shall describe here only those new techniques which are particularly applicable to paraelectric centres.

Electro-optical Technique. Both static and dynamical behaviour [223,246] of all dipoles with measurable optical or u.v. absorptions may be studied by this technique. Since such absorptions are expected to be different perpendicular ($\perp$) and parallel ($\parallel$) to the dipole axis, the application of a field or stress which causes dipole alignment will give rise to characteristic absorption changes. The extent of these changes will vary for light polarized $\perp$ or $\parallel$ to the applied perturbation and hence the orientation of the dipoles may be established. Measurement of the time dependence of the dichroism after rapid changes in the applied field or stress allow the dipole relaxation to be studied.[233,246]

Electrocaloric Method. This method is based on the rate of exchange of energy between the lattice and a system of dipoles during electric polarization or depolarization. For thermal equilibrium to exist between these systems at every moment the rise and decay time of the field, τ_E, must be long compared to the dipole–lattice relaxation time τ. If the opposite is true and $\tau_E < \tau$ then, under adiabatic field increase or decrease, a heating or cooling of the lattice is observed.[233] The electrocaloric effect is closely related to dielectric loss, and the net energy dissipation (heating) for an a.c. cycle as a function of the frequency, ω, follows directly the Debye curve for $\varepsilon''(\omega)$. It has been shown [247] that below approximately 6 K, where

---

[242] P. Gosar and R. Pirc, Proceedings, Fourteenth Colloque Ampère, Ljublÿana 1966, ed. R. Blink, North-Holland, Amsterdam, 1968, p. 636; also R. Pirc and P. Gosar, *Phys. Kondens. Materie*, 1969, **9**, 377.
[243] W. Känzig, H. R. Hart, and S. Roberts, *Phys. Rev. Letters*, 1964, **13**, 543.
[244] W. Knop, G. Pfister, and W. Känzig, *Phys. Kondens. Materie*, 1968, **7**, 107.
[245] N. E. Byer and H. S. Sack, *Phys. Stat. Sol.*, 1968, **30**, 569; *J. Phys. Chem. Solids*, 1968, **29**, 677.
[246] S. Kapphan and F. Lüty, *J. Phys. Chem. Solids*, 1973, **34**, 969.
[247] G. A. Reymann and F. Lüty, *Phys. Stat. Sol. (A)*, 1973, **16**, 561.

the heat capacity is very small, the caloric technique is superior to the standard loss angle bridge method.

Specific Heat Measurements. A technique involving short heat pulses is used [248] and the time dependence of the crystal temperature is monitored. After application of a pulse the initial rise in temperature is found to decay exponentially because of a flow of heat to the weakly coupled dipole system. This decay directly measures the coupling time constant between the lattice and dipole baths and represents the dipole–lattice relaxation time. As the heat capacity of the auxiliary dipole bath is extremely small, very low temperatures (<0.1 K) are necessary to allow useful sensitive measurements.

Paraelectric Resonance (PER). When the orientational dipole levels are split by a suitable electric field, microwave-induced transitions may occur [249,250] giving rise to PER, which is the electric analogue of paramagnetic resonance. In suitable circumstances PER can be a most versatile technique, and may be used to measure all the parameters of the dipole system. The variation of the resonance frequency with applied electric or stress fields gives the electric dipole moment and stress tensor parameters. The dipole orientation may be probed by observing the changes in the positions of the resonance lines as the orientation of the crystal is changed relative to the electric field. The tunnelling parameters are established in the $E \to 0$ limit, and the relaxation rate is found from the saturation of the inhomogeneously broadened low-temperature spectra or by their life-time broadening at higher temperatures.[251]

The Kerr Effect Method. Electrobirefringence is caused by an electric-field-induced anisotropy in the refractive index.[252] In systems where only inaccessible optical absorptions exist (*cf.* electro-optical technique) these will introduce Kramers–Kronig related refractive index changes which extend into the transparent crystal range. This anisotropic refractive index will change linearly polarized light into elliptically polarized light, and measurements are made using sensitive techniques and a suitable Kerr-effect geometry. For OH^- dipoles the Kramers–Kronig relationship of the Kerr effect to the electrodichroism of the OH^- u.v. band has been confirmed,[252] and most recently successful Kerr-effect studies have been made of a number of off-centre F^- systems in the alkali halides.[253]

In addition to these specific techniques, use has also been made of various spectroscopic methods. I.r. and far-i.r. absorption [254,255] allow the study of transitions from ground to excited states and, in some cases, measurement of zero-field splittings of the ground-state multiplet. Magnetic resonance techniques [256]

---

[248] W. D. Seward, V. Reddy, and J. W. Shaner, *Solid State Comm.*, 1972, **11**, 1569.
[249] G. Feher, I. W. Shepherd, and H. B. Shore, *Phys. Rev. Letters*, 1966, **16**, 500.
[250] R. A. Herendeen and R. H. Silsbee, *Phys. Rev.*, 1969, **188**, 645.
[251] R. Osswald and H. C. Wolf, *Phys. Stat. Sol. (B)*, 1972, **50**, K93.
[252] G. Zibold and F. Lüty, *J. Nonmetals*, 1972, **1**, 1.
[253] A. Góngora and F. Lüty, *Bull. Amer. Phys. Soc.*, 1975, **20**, 469.
[254] R. D. Kirby, A. E. Hughes, and A. J. Sievers, *Phys. Rev. (B)*, 1970, **2**, 481; *Phys. Letters*, 1968, **A28**, 170.
[255] F. Lüty, *J. Phys. (Paris)*, 1967, **28**, C4–120.
[256] D. L. Hagrman and W. D. Ohlsen, *Phys. Rev. (B)*, 1971, **3**, 1918.

also yield information on dipole orientation and electric- and stress-field coupling constants.

Relaxation studies can be used to estimate off-centre positions of suitable nuclei. Raman spectroscopy,[257,258] although somewhat insensitive, has been used to determine the energy levels of PEC defect systems and, in conjunction with polarized sources and detectors, to reveal dipole orientations. A list of paraelectric systems with their orientations and uncorrected dipole moments is given in Table 15. Also given are the vibrational frequencies of the molecular PEC, CN^- and OH^-.

Table 15* *Experimental electric dipole parameters for paraelectric systems*

System	Orientation	Vibrational frequency/ cm^{-1} [ee]	Uncorrected dipole moment (eÅ)
$NaCl:^7Li^+$	$\langle 111 \rangle^{a\dagger}$		
$NaCl:OH^-$	$\langle 100 \rangle^{b,c}$	3654.5	$0.9^{c,d}$
$NaCl:CN^-$	$\langle 100 \rangle^e$	2106.8	
$NaBr:F^-$	$\langle 110 \rangle^{f,g}$		$0.7—1.0^f$
$NaBr:OH^-$	$\langle 110 \rangle^b$	3626	
$NaBr:CN^-$		2087	
$NaI:CN^-$		2074	
$KCl:^7Li^+$	$\langle 110 \rangle^{h-m}$		$1.1,^j\ 1.14,^{n,o,d}\ 1.13^p$ $1.15,^q\ 1.16,^h\ 1.25^i$
$KCl:^6Li^+$	$\langle 111 \rangle^{h,q}$		$1.16,^h\ 1.14^d$
$KCl:Cu^+$	$\langle 111 \rangle^r$		$1.49,^r\ 2.63^s$
$KCl:OH^-$	$\langle 100 \rangle^c$	3641	$0.82,^c\ 0.92,^t\ 0.79^u$ $0.69,^v\ 0.94^w$
$KCl:CN^-$	$\langle 100 \rangle^x$ $\langle 111 \rangle^e$	2087	0.104^u
$KCl:H_2O^-$	$\langle 111 \rangle^y$		0.31^y
$KCl:NO^-$	$\langle 110 \rangle^z$		0.044^{aa}
$KBr:Cu^+$			1.92^r
$KBr:OH^-$	$\langle 100 \rangle^{c,d}$	3618	$0.9—1.0^c$
$KBr:CN^-$	$\langle 100 \rangle^x$ $\langle 111 \rangle^e$	2078	0.063^{aa}
$KBr:SH^-$	$\langle 100 \rangle^{bb}$		1^{bb}
$KI:F^-$	$\langle 110 \rangle^g$		
$KI:Cu^+$			2.6^r
$KI:OH^-$	$\langle 110 \rangle^{c,cc,dd}$	3603	$1.35, 1.40,^c\ 1.33^{ee}$
$KI:CN^-$	$\langle 100 \rangle^x$ $\langle 111 \rangle^e$	2067	0.063^{aa}
$KI:NO_2^-$	$\langle 110 \rangle^z$		0.2^{aa}
$RbCl:Cu^+$	off-centre[r]		
$RbCl:Ag^+$	$\langle 110 \rangle^{d,ff}$		$0.78,^{gg}\ 0.78,^{ff}\ 0.1^d$
$RbCl:OH^-$	$\langle 100 \rangle^c$	3632	$1.4,^u\ 0.87, 1.0^c$
$RbCl:CN^-$	$\langle 100 \rangle^{hh}$	2081	0.102^{hh}
$RbBr:Cu^+$	off-centre[r]		

[257] J. G. Peascoe and M. V. Klein, *J. Chem. Phys.*, 1973, **59**, 5.
[258] J. G. Peascoe, W. R. Fenner, and M. V. Klein, *J. Chem. Phys.*, 1974, **60**, 4208.

Lüty [226] has discussed the factors which influence the trends evident in the classical and/or tunnelling behaviour of dipole systems. Recently, in a series of elegant experiments, Holland and Lüty [259] have reported hydrostatic pressure tuning from classical to tunnelling motion and from off- to on-centre behaviour for Ag^+, Cu^+, and Li^+ defects in various hosts. Recently also [253] a comprehensive survey has been concluded of substitutional defects in alkali bromides and iodides using Kerr-effect measurements. This confirms the $\langle 110 \rangle$ off-centre $NaBr:F^-$ system [260] and reports two new systems $KI:F^-$ and $RbI:F^-$ which are similarly

Table 15—cont.

System	Orientation	Vibrational frequency/ cm^{-1} [ee]	Uncorrected dipole moment (eÅ)
$RbBr:Ag^+$	$\langle 110 \rangle$ [ff]		0.95 [ff]
$RbBr:OH^-$	$\langle 100 \rangle$ [c]	3610	1.0 [c]
$RbBr:CN^-$		2070	
$RbI:F^-$	$\langle 110 \rangle$ [g]		
$RbI:OH^-$	$\langle 100 \rangle$ [c,d]	3601	1.0 [c]
$RbI:CN^-$		2063	
$CsCl:CN^-$		2078	
$CsBr:OH^-$	$\langle 111 \rangle$ [c]	3628	1.25 [c]
$CsBr:CN^-$		2066	
$CsI:CN^-$		2053	

* For more detailed information including orientational parameters on these systems and a list of systems which have failed to show paraelectric behaviour see ref. 227. † This result is in disagreement with a number of other investigations which have failed to find paraelectric behaviour.[227]

[a] T. L. Estle and Y. G. Garnes, Bull. Amer. Phys. Soc., 1969, 14, 346; [b] W. Heinicke and F. Lüty, ibid., 1972, 17, 143; [c] S. Kapphan and F. Lüty, J. Phys. Chem. Solids, 1973, 34, 969; [d] R. D. Kirby, A. E. Hughes and A. J. Sievers, Phys. Rev. (B), 1970, 2, 481; [e] A. Diáz-Góngora and F. Lüty, Solid State Comm., 1974, 14, 923; Bull. Amer. Phys. Soc., 1974, 19, 650; [f] R. J. Rollefson, Phys. Rev. (B), 1972, 5, 3235; [g] A. Diáz-Góngora and F. Lüty, Bull. Amer. Phys. Soc., 1975, 20, 469; [h] R. A. Herendeen and R. H. Silsbee, Phys. Rev., 1969, 188, 645; [i] D. Blumenstock, R. Osswald, and H. C. Wolf, Phys. Stat. Sol. (B), 1971, 46, 217; [j] A. V. Frantsesson, O. F. Dudnik, and V. B. Kravchenko, Soviet Phys. Solid State, 1970, 12, 126; [k] D. W. Alderman and R. M. Cotts, Phys. Rev. (B), 1970, 1, 2870; [l] N. E. Byer and H. S. Sack, J. Phys. Chem. Solids, 1968, 29, 677; [m] T. L. Estle, Phys. Rev., 1968, 176, 1056; [n] G. Lombardo and R. O. Pohl, Phys. Rev. Letters, 1965, 15, 291; [o] R. D. Kirby and A. J. Sievers, Bull. Amer. Phys. Soc., 1969, 14, 301; [p] S. Kapphan and F. Lüty, Solid State Comm., 1968, 6, 907; [q] J. P. Harrison, P. P. Peressini, and R. O. Pohl, Phys. Rev., 1968, 171, 1037; [r] M. S. Li, M. de Souza, and F. Lüty, Phys. Rev. (B), 1973, 7, 4677; [s] R. Sittig, Phys. Stat. Sol., 1969, 34, K189; [t] I. W. Shepherd, J. Phys. Chem. Solids, 1967, 28, 2027; [u] A. T. Fiory, Phys. Rev. (B), 1971, 4, 614; [v] W. E. Bron and R. W. Dreyfus, Phys. Rev., 1967, 163, 304; [w] R. Sittig, Phys. Stat. Sol., 1967, 21, K175; [x] W. D. Seward and V. Narayanamurti, Phys. Rev., 1966, 148, 463; [y] W. Rusch and H. Seidal, Phys. Stat. Sol. (B), 1974, 63, 183; [z] V. Narayanamurti, W. D. Seward, and R. O. Pohl, Phys. Rev., 1966, 148, 481; [aa] H. S. Sack and M. C. Moriarty, Solid State Comm., 1965, 3, 93; [bb] W. E. Bron and R. W. Dreyfus, Phys. Rev. Letters, 1966, 16, 165; [cc] F. Bridges, Solid State Comm., 1973, 13, 1877; [dd] K. Knop, G. Pfister, and W. Kanzig, Phys. Kondens. Materie, 1968, 7, 107; [ee] F. Bridges, CRC Crit. Rev. Solid State, 1975, 5, 1; [ff] S. Kapphan and F. Lüty, Phys. Rev. (B), 1972, 6, 1537; [gg] F. Bridges, ibid., 1972, 5, 3321; [hh] R. W. Dreyfus, J. Phys. Chem. Solids, 1968, 29, 1941; 1969, 30, 1903.

[259] U. Holland and F. Lüty, J. Phys. (Paris), 1976, 37, C7–272.
[260] J. Wahl and F. Lüty, Bull. Amer. Phys. Soc., 1975, 20, 469.

oriented. Dielectric and stress studies [259] are in agreement with these findings and the orientation behaviour is interpreted in terms of elastic dressing effects.

7 Molecular Impurities

Considerable attention has been devoted to the study of molecular impurities in ionic crystals. Several powerful experimental techniques can be utilized to obtain information about the host crystals by using molecular impurities as probes, and they provide a fund of information which is not easy to obtain from other methods. In a similar way the influence of the host crystal on the molecular impurity is also of interest. In studying these impurities embedded in ionic crystals the information that one seeks is (i) the orientation of the molecule, (ii) changes in vibrational frequencies from those characterizing a free ion, (iii) activation of forbidden modes in Raman or i.r. measurements, and (iv) splitting of degenerate modes due to lowering of symmetry. For simple diatomic molecules (iii) and (iv) are not relevant as there are no degenerate or forbidden modes. These features, however, become observable with more complex molecules. The molecular impurities that have been studied can be classified in the following categories.

Diatomic Molecular Ions.—Homonuclear diatomic molecules, *e.g.* S_2^-, Se_2^-, O_2^-, and N_2^-, may be incorporated very conveniently into host crystals like the alkali halides. Single crystals doped with S_2^- or Se_2^- can be prepared by heating the crystal in the presence of the corresponding vapour. These impurities have only one vibrational frequency which is Raman active and i.r. inactive. Such impurities are paramagnetic and they also have characteristic emission and absorption bands. Early e.p.r. experiments [261,262] showed that these molecular ions are oriented with the molecular axis directed along $\langle 110 \rangle$ directions. Subsequently [263,264] S_2^- was introduced into a large number of crystals, and from e.p.r. studies it was shown that the centre is characterized by a very high *l*-anisotropy and that the orbital contribution to the magnetic moment is very large. Since ^{33}S has a spin $I = 3/2$, $^{33}S^{32}S^-$ gives a four-line contribution to the spectrum while $^{33}S_2^-$ gives rise to a manifold of seven lines with an intensity ratio of $1:2:3:4:3:2:1$. In all crystals studied the minimum field for resonance was obtained with H_0 parallel to $\langle 110 \rangle$, demonstrating that the S—S bonds were parallel to one of the six $\langle 110 \rangle$ directions in the crystal. Se_2^- and SSe_2^- in alkali halides have also been investigated [265] and in these cases also the bond direction was found to be parallel to one of the $\langle 110 \rangle$ directions. However, in NaI, KBr, and KI the S—S bond is oriented along $\langle 100 \rangle$. This situation may be compared with the case of O_2^- where the *p* orbitals of O_2^- are parallel to $\langle 100 \rangle$ in sodium salts but are parallel to $\langle 110 \rangle$ in rubidium and potassium salts.[261]

In the process of doping a crystal with S_2^- ions, one cannot avoid the presence of S_3^- ions also. S_3^- is a bent triatomic molecule and has three vibrational frequencies.

---

[261] W. Kanzig and M. H. Cohen, *Phys. Rev. Letters*, 1959, **3**, 509.
[262] J. R. Morton, *J. Chem. Phys.*, 1965, **43**, 3418.
[263] L. E. Vannotti and J. R. Morton, *Phys. Rev.*, 1967, **161**, 282.
[264] J. R. Morton, *J. Phys. Chem.*, 1967, **71**, 89.
[265] L. E. Vannotti and J. R. Morton, *J. Chem. Phys.*, 1967, **47**, 4210.

Table 16 *Observed Raman frequencies^{a,b} in cm^{-1} of diatomic molecular ion impurities in alkali halide crystals*

	O_2^-	N_2^-	S_2^-	Se_2^-	SSe^-
NaCl	1144				
NaBr	1131		610 ± 2		
NaI			592	333^c	462
KCl	1145	1836 ± 3			
KBr	1135	1821 ± 3	612 ± 2		
KI	1123	1870 ± 2	594	325	464
RbCl	1141				
RbBr	1132		611		
RbI			598		
CsBr			602	328	

^{a} All frequencies are accurate to ±1 cm^{-1} unless otherwise indicated; ^{b} Data are from W. Holzer, W. F. Murphy, and H. J. Bernstein, *J. Mol. Spectroscopy*, 1969, **32**, 13, except those for O_2^-, which are from W. Känzig and M. H. Cohen, *Phys. Rev. Letters*, 1959, **3**, 509, and for CsBr, which are unpublished work; ^{c} Se^{77} isotope.

Table 17 *Spectroscopic constants of diatomic molecular ions in KI crystals, obtained from fluorescence measurements at 4.2 K, together with calculated values of dissociation energies D_e, internuclear distances r_e, and force constants k*

	v_{00}/ cm^{-1}	v_{00}/ eV	ω_e/ cm^{-1}	$\omega_e x_e$/ cm^{-1}	D_e/ eV	C/ cm^2	r_e/ Å	k/ mdyn Å^{-1}
O_2^-	26 520^a	3.29	1134	8.7	4.58^b	9 619	1.33^c	6.06
S_2^-	20 035	2.48	600.8	2.53	4.42	16 940	1.98	3.40
SeS$^-$	17 300	2.14	465.6	1.63	4.12	15 278^d	2.09	2.89
Se_2^-	16 011	1.98	329.3	0.75	4.48	13 617	2.25	2.55

^{a} J. Rolfe, W. Holzer, W. F. Murphy, and H. J. Bernstein, *J. Chem. Phys.*, 1968, **49**, 963; ^{b} Dissociation energies calculated from the formula $D_e = \omega_e^2/4\omega_e x_e$; ^{c} F. Halverson, *J. Phys. Chem. Solids*, 1962, **23**, 207, gives an experimental value $r_e = 1.33$ Å; ^{d} Mean of S_2 and Se_2 values.
All other data are from J. Rolfe, *J. Chem. Phys.*, 1968, **49**, 4193.

The S_2^- and S_3^- ions have characteristic optical absorption bands. The band ascribed to S_2^- is centred at 4000 Å while that associated with S_3^- is located at about 6100 Å. The presence of these optical absorption bands can be put to good use in the study of resonance Raman scattering by using a laser beam of wavelength close to the absorption band maximum.[266] An important feature of this work is that a large number of overtones, associated with the fundamental Raman line, have been observed. This is primarily because of the high intensity that resonance Raman scattering provides. The Raman frequencies observed for diatomic molecules are given in Table 16.

A considerable amount of work has also been done on emission spectra from diatomic molecules like O_2^-, S_2^-, Se_2^- and SeS$^-$ embedded in ionic crystals. The

[266] W. Holzer, W. F. Murphy, and H. J. Bernstein, *J. Mol. Spectroscopy*, 1969, **32**, 13.

early work on luminescence of O_2^-,[267] S_2^-,[268] and Se_2^- [269] was done using conventional methods. Recently, improved experimental techniques have been used to obtain more detailed results.[270,271] Figure 1 shows a typical emission spectrum obtained from S_2^- ions in KI crystals. Emission from Se_2^- in KI crystals is excited most efficiently by light of wavelength 5120 Å and lies mainly in the i.r. The spectrum appears complicated because of the large number of isotopes of nearly equal abundance. From the emission data several useful spectroscopic constants can be calculated. The data obtained for bivalent impurities in a typical host crystal are given in Table 17.

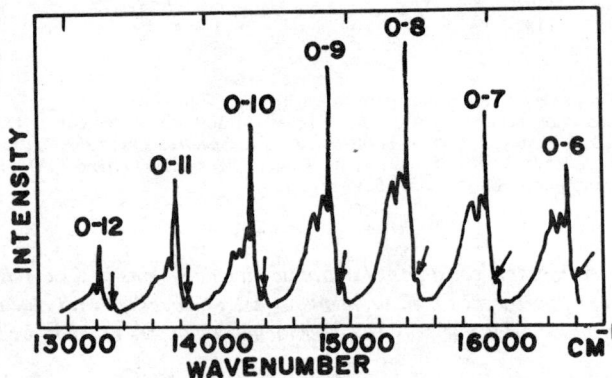

Figure 1 *The most intense part of the emission spectrum of S_2^- dissolved in a KI crystal, measured at 4.2 K. Arrows mark ($^{34}S^{32}S^-$) emission peaks*
(Reproduced by permission from *J. Chem. Phys.*, 1968, **49**, 4193)

Recently [270] several zero-phonon lines have been investigated in alkali halides doped with O_2^-, S_2^-, Se_2^-, or SeS^-. Several relevant spectroscopic constants were calculated from the experimental data, including $\Delta G''_{\frac{1}{2}} = \omega''_0 - \omega''_0 x''_0$ (where ω''_0, $\omega''_0 x''_0$ are the vibrational constants for the ground state). $\Delta G''_{\frac{1}{2}}$ corresponds to the fundamental Raman frequency so that the computed values of $\Delta G''_{\frac{1}{2}}$ could be compared with directly observed Raman frequencies. By a study of the effect of uniaxial stress on the zero-phonon lines it has proved possible to determine the orientation of the diatomic molecular ion centre. Owing to a lifting of the electronic or orientational degeneracy by the applied stress the zero-phonon lines split into one or more components. In this way two categories of local symmetries in alkali halides doped with diatomic molecular impurities were identified.[270] In the first category, where the molecular axis is oriented along the ⟨110⟩ direction, the symmetry is orthorhombic. In the second category the molecular axis is oriented along ⟨111⟩ and the symmetry is trigonal. The species O_2^- in NaCl, KBr, and KI was found to assume a ⟨111⟩ orientation and in all other crystals it was found to have a ⟨110⟩ orientation. The zero-phonon lines, which are due to vibronic

[267] J. Rolfe, F. R. Lipsett, and W. J. King, *Phys. Rev.*, 1961, **123**, 447.
[268] J. H. Schulman and R. D. Kirk, *Solid State Comm.*, 1964, **2**, 105.
[269] J. Rolfe, *J. Chem. Phys.*, 1968, **49**, 4193.
[270] M. Ikezawa and J. Rolfe, *J. Chem. Phys.*, 1973, **58**, 2024.
[271] G. J. Vella and J. Rolfe, *J. Chem. Phys.*, 1974, **61**, 41.

transitions within the molecular ion, are accompanied by phonon sidebands produced by the interaction of even-parity lattice modes with the allowed $\Pi_g \leftarrow \Pi_u$ transition which is responsible for the observed emission. For O_2^- dissolved in NaCl, KCl, KBr, and KI most of the sideband structure can be explained [272] by coupling with A_{1g} lattice modes perturbed by relatively small changes in nn force constants. In two cases, KCl and RbCl, better agreement of calculated sidebands with experiment is obtained either by including T_{2g} modes or by allowing for changes in other force constants.[272] Te_2^{3-} and Se_2^{3-} diatomic centres have been identified in AgBr by studying and analysing the hyperfine interaction observed in the e.p.r. spectrum.[273] These experiments have also shown that the direction of the molecular ion is $\langle 110 \rangle$. The centre consists of two impurity ions at lattice sites with a common trapped hole.

Molecular impurities like OH^- and CN^- are of special interest in view of the fact that they have an intrinsic dipole moment. These centres are dealt with in Section 6.

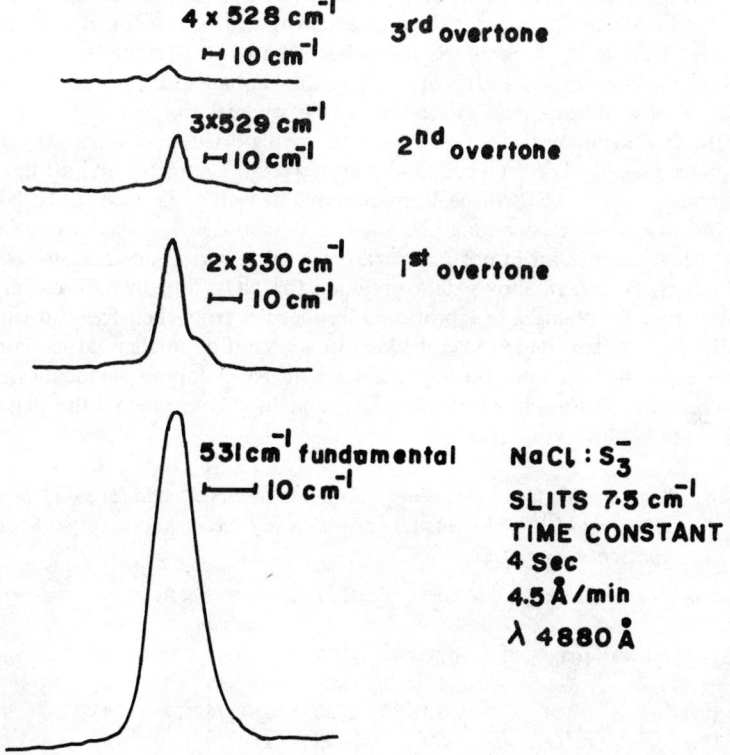

Figure 2 *Resonance Raman effect of S_3^- in* NaCl
(Reproduced by permission from *J. Mol. Spectroscopy*, 1969, **32**, 13)

[272] J. Rolfe, M. Ikezawa, and T. Timusk, *Phys. Rev.* (*B*), 1973, **7**, 3913.
[273] M. Hohne and M. Stasiw, *Phys. Stat. Sol.*, 1967, **20**, 657, 667.

Triatomic Molecular Ions.—The S_3^- ion is a typical example of a triatomic molecular ion. This impurity has a characteristic optical absorption band and most information has been obtained [266] by the use of resonance Raman scattering. Figure 2 shows the resonance Raman spectrum observed for the S_3^- ion in NaCl. The fundamental frequency of vibration, which is Raman active, occurs at 531 cm^{-1} and the three overtones observed are also shown. From these data several spectroscopic constants have been calculated. Other triatomic impurities which have been studied are NO_2^-, NH_2^-, ND_2^-, and H_2O.[274,275] These molecules have three modes of vibration, all of which are Raman as well as i.r. active, and the primary interest with these molecules has been to observe the change in frequency from that characteristic of the ion in its free state.

Planar Molecular Ions.—With a NO_3^- ion impurity no charge-compensating vacancy is required. The symmetry of the free NO_3^- ion is D_{3h} and it has characteristic vibrational modes belonging to the representations A_1', A_2'', and $2E'$. The A_1' and $2E'$ modes are Raman active while A_2'' and $2E'$ modes are i.r. active. However, the A_1' mode, which is i.r. inactive, appears in the i.r. spectrum of KBr crystals doped with NO_3^- ions.[276] Similarly it has been shown [277] that the A_2'' mode which is Raman inactive in the free ion appears in the Raman spectrum of NO_3^- doped crystals. A possible explanation for this is that whereas the free NO_3^- is a planar ion with D_{3h} symmetry, in the crystal it is distorted in such a way that it is no longer planar. The molecule will then have C_{3v} symmetry and all the vibrational modes ($2A_1 + 2E$) will be Raman active as well as i.r. active. In addition to the forbidden mode becoming activated, a shift in the frequencies is observed; this frequency shift is an important parameter for it gives information about the surroundings. Table 18 shows data obtained for NO_3^- ions in different crystals. It appears that the changes in vibrational frequencies from their free-ion value are generally larger when the ion is embedded in a crystal of smaller lattice constant. As there is no charge-compensating vacancy with NO_3^- doping the local symmetry is C_{3v} (because of non-planar distortions), and in this symmetry the degenerate modes are not split.

Table 18 Raman shifts (in cm^{-1}) observed for the free NO_3^- and for NO_3^- in various alkali halides. The modes are labelled according to the irreducible representations of D_{3h}[a]

Mode	Free NO_3^-	NaCl	KCl	KBr Raman	KBr i.r.	KI
A_1' (ν_1)	1048	1075	1070	1065	1054.8	1060
A_2'' (ν_2)	834	850	845	840	841.3	838
E' (ν_3)	1380	1380	1380	1375	1383.2	1375
E' (ν_4)	714	720	720	720	720	725

[a] I.r. data for KBr are from K. Kato and J. Rolfe, *J. Chem. Phys.*, 1967, **47**, 1901; other data from S. Radhakrishna and A. M. Karguppikar, *J. Nonmetals*, 1973, **2**, 75.

[274] A. R. Evans and D. B. Fitchen, *Phys. Rev. (B)*, 1970, **2**, 1074.
[275] C. Ruhenbeck, *Z. Phys.*, 1967, **207**, 446.
[276] K. Kato and J. Rolfe, *J. Chem. Phys.*, 1967, **47**, 1901.
[277] S. Radhakrishna and A. M. Karguppikar, *J. Nonmetals*, 1973, **2**, 75.

Other planar molecules which have been studied are CO_3^{2-} and BO_3^{3-}.[278] With these ions the presence of charge-compensating vacancies lowers the local symmetry and the degenerate modes are split.

Molecular Ions of Tetrahedral Symmetry.—Impurities like SO_4^{2-} and PO_4^{3-} have tetrahedral symmetry in their free state and therefore vibrational modes of A_1, E, and T_2 symmetry. A_1, E, and $2T_2$ are Raman active, and the two triply degenerate modes are i.r. active. When incorporated in a crystal one or more charge-compensating vacancies are required. These impurities are usually incorporated substitutionally at anion sites and the presence of the charge-compensating vacancy reduces the local symmetry. Much of the early work on SO_4^{2-} doped crystals was confined to i.r. studies,[279-281] but more recently Raman scattering measurements have been performed.[282] These experiments show (i) that the forbidden modes are activated, (ii) that the degeneracy of the degenerate modes is lifted, and (iii) that there is a shift in the position of the i.r. or Raman lines. This is because of the change in frequency of vibration brought about by changes in the force constants when the impurity is incorporated in the crystal lattice.

Impurities like CrO_4^{2-} and MnO_4^- have yet another advantage in that they introduce strong optical bands in the visible region. Early molecular orbital calculations[283] for CrO_4^{2-} showed that the ground state is $^1A_1(t_1^6)$ and that transitions to the immediately overlying $^1T_2(t_1^5e^1)$ and $^1T_2(t_1^5t_2^1)$ states would be allowed. Two strong bands are observed for CrO_4^{2-} in the spectral regions 360—380 nm and 260—280 nm are ascribed to these transitions.[284,285] The first band shows a characteristic vibronic structure representing transitions from the ground electronic state to the vibrational overtones (for total symmetric oscillations with frequency 800 cm^{-1}) of the nearest electronic states $^1T_2(t_1^5e^1)$.[284,285] CrO_4^{2-} and $Cr_2O_7^{2-}$ have similar molecular orbital diagrams and therefore essentially the same kind of optical spectrum.[286] Figure 3 shows typical spectra observed for $Cr_2O_7^{2-}$ and CrO_4^{2-} ions in KBr crystals. In KBr:MnO_4^- there are two strong bands at 520 and 336 nm which are ascribed to the allowed transitions $^1A_1 \rightarrow {}^1T_2(t_1^5e^1)$ and $^1A_1 \rightarrow {}^1T_2(t_2^5e^1)$.[287] Weaker bands at 600 and 390 nm, respectively, are assigned to transitions from the ground state to $^1T_1(t_1^5e^1)$ and $^1T_1(t_2^5e^1)$, which are forbidden in T_d symmetry. The rich vibronic structure of the 520 nm band is attributed to coupling of the $^1T_2(t_1^5e^1)$ state with both the symmetric stretching mode of MnO_4^- and also with lattice modes at 65 and 130 cm^{-1}, where there are peaks in the phonon density of states for KBr.

CrO_4^{2-} ions like SO_4^{2-} have T_d symmetry in their free state but because of charge-compensating vacancies the symmetry inside the crystal is lowered to C_s. As a

---

[278] S. C. Jain, A. V. R. Warrier, and H. K. Sehgal, *J. Phys. (C)*, 1973, **6**, 189.
[279] J. C. Wilson, E. H. Coker, and G. C. Brenna, *Spectrochim. Acta*, 1963, **19**, 1281.
[280] T. L. Maksimova, *Phys. Stat. Sol.*, 1969, **33**, 547.
[281] D. N. Mirlin and I. I. Reshina, *Soviet Phys. Solid State*, 1968, **10**, 895.
[282] S. Radhakrishna, *Phys. Rev. (B)*, 1971, **4**, 1382.
[283] C. J. Ballhausen and A. D. Liehr, *J. Mol. Spectroscopy*, 1958, **2**, 342.
[284] S. C. Jain, A. V. R. Warrier, and S. K. Agarwal, *Chem. Phys. Letters*, 1972, **14**, 211; *J. Phys. Chem. Solids*, 1973, **34**, 209.
[285] S. Radhakrishna and K. P. Pande, *Chem. Phys. Letters*, 1972, **13**, 62.
[286] S. Radhakrishna and B. D. Sharma, *J. Chem. Phys.*, 1974, **61**, 3925.
[287] S. C. Jain, D. Pooley, and R. Singh, *J. Phys. (C)*, 1972, **5**, L307.

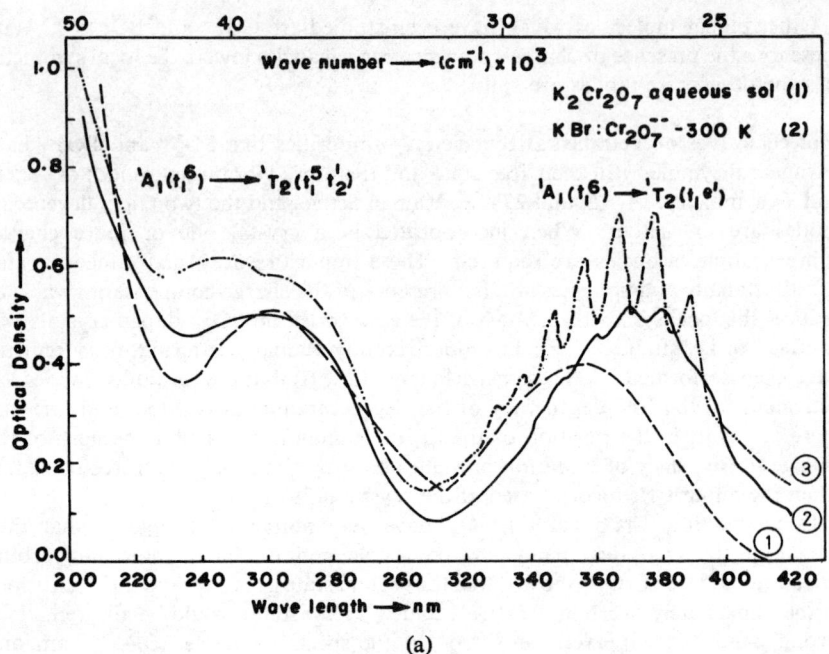

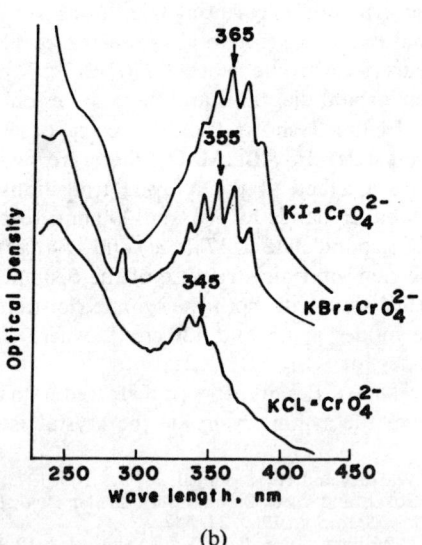

Figure 3 (a) *Optical absorption spectrum of* $K_2Cr_2O_7$ *in aqueous solution at* 300 K (curve 1) *and of* $KBr:Cr_2O_7$ *(curves 2 and 3 at 300 and 77 K, respectively).* (b) *Optical absorption spectra of* CrO_4^{2-} *ions in KCl, KBr and KI measured at 77 K* (Reproduced by permission from *J. Chem. Phys.*, 1974, **61**, 3925 and *J. Phys. Chem. Solids*, 1973, **34**, 209)

Table 19 *Vibrational frequencies of CrO_4^{2-} ions in KCl and KBr, labelled according to the irreducible representation of T_d. $\square^-$ denotes an anion vacancy. R = Raman spectrum,[a] i.r. = infrared spectrum[b]*

System	Local symmetry	$\nu_1 (A_1)$	$\nu_2 (E)$	$\nu_3 (T_2)$	$\nu_4 (T_2)$	Temperature, activity
Free ion	T_d	847	348	884	368	
KCl:CrO_4^{2-}; $\square^-$	C_s	860		939		
				896		300 K, i.r.
				887		
KCl:CrO_4^{2-}; Ca^{2+}	C_{2v}	860		944		
				929		300 K, i.r.
				881		
KCl:CrO_4^{2-}	T_d			913		300 K, i.r.
KBr:CrO_4^{2-}; $\square^-$	C_s	856		927		
				894		300 K, i.r.
				888		
KBr:CrO_4^{2-}; Ca^{2+}	C_{2v}	855		936	433	
				924	416	77 K, i.r.
				880	399	
KBr:CrO_4^{2-}; Ba^{2+}	C_{2v}	849		940	427	
				921	414	77 K, i.r.
				872	401	
KBr:CrO_4^{2-}	T_d			908		300 K, i.r.
KBr:CrO_4^{2-}; $\square^-$		851	341	932	380	R
				890	375	77 K
				883	365	

[a] W. B. Grant and S. Radhakrishna, *Solid State Comm.*, 1973, **13**, 109; [b] S. C. Jain, A. V. R. Warrier, and S. K. Agarwal, *J. Phys. Chem. Solids*, 1973, **34**, 209; V. P. Dem'yanenko, Yu. P. Tsyashchenko, and E. M. Verlan, *Soviet Phys. Solid State*, 1970, **12**, 417; P. J. Miller, G. L. Cessac, and R. K. Khanna, *Spectrochim. Acta*, 1971, **27A**, 2019.

consequence the degenerate modes are split and the split components are observed. As for SO_4^{2-}, all the four modes A_1, E, and $2T_2$ are Raman active while only the two triply degenerate modes are i.r. active. Table 19 shows the results obtained for CrO_4^{2-} in KCl and KBr. If, however, charge compensation is achieved by a bivalent cation in one of the six nearest neighbour cation sites the local symmetry is reduced from T_d to C_{2v}. Even in this case the degeneracy is completely lifted although now the splittings are slightly different. The extent of the splitting is governed by the strength of the local field inside the crystal while the nature of the splitting is governed by the local symmetry. By careful work it has been possible to show that in a typical crystal the local symmetry is lowered to both C_s and C_{2v} resulting in a complicated i.r. or Raman spectrum. By controlled experiments it is, however, possible to separate the spectra characteristic of these different symmetries.[284]

The experimental data for several tetrahedral ions in ionic crystals have been summarized.[288] Some of the other ions which have been studied are MnO_4^-,[284,289]

[288] S. C. Jain, A. V. R. Warrier, and S. K. Agarwal, NBS data book, NSRDS–NBS Publication No. 52, July, 1974.
[289] P. Manzelli and G. Todder, *J. Chem. Phys.*, 1969, **51**, 1484.

NH_4^+,[290] BH_4^-,[291] BF_4^-,[292] SeO_4^-,[293] and BeF_4^{2-};[294] NH_4^+, ND_4^+, and BH_4^- incorporated in bcc structures like CsCl or CsBr have also been studied.[290,291] In such lattices the symmetry is reduced to D_{2d} so that the degeneracy of the degenerate modes is only partially lifted.

Complex Molecular Ions.—Two other impurities which have been studied in some detail are the ferricyanide ion, $Fe(CN)_6^{3-}$, and the cobalticyanide ion, $Co(CN)_6^{3-}$.[295,296] The metal ion occupies a cation site and the six (CN^-) ions go into the six neighbouring anion sites. Three types of spectrum may be studied. The first category corresponds to charge-transfer transitions from the metal to the ligand. The second kind corresponds to crystal field transitions of the d^5 (Fe) configuration. In addition one can observe C—N stretch frequencies in the i.r. region. The first two kinds of transition give rise to bands in the region 200—500 nm, and a set of 10 unresolved bands have been reported and explained satisfactorily. From a careful study of the i.r. spectra two sets of lines in the frequency region 2131—2110 and 2073—2047 cm^{-1} have been reported. The former is interpreted as the stretch frequency of C—N in the $Fe(CN)_6^{3-}$ complex while the latter is ascribed to C—N stretch in the $Fe(CN)_6^{4-}$ molecule. Spectra characteristic of local D_{2h}, C_s, and C_{2v} symmetry have been reported.[295] In the i.r. spectra of $Co(CN)_6^{3-}$ several components arising from at least two local site symmetries, D_{2h} and C_s, have been reported.[296]

Molecular centres like MoO^{3+} and VO^{2+} have been studied by both e.p.r. and optical techniques.[297-299] To prepare crystals doped with the molybdenyl ion the

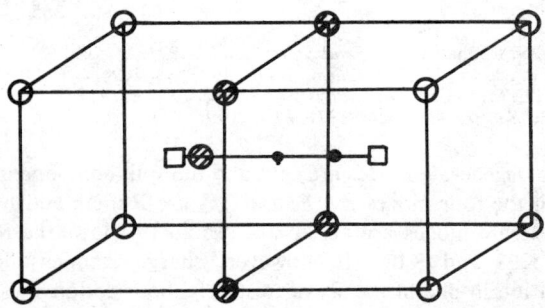

Figure 4 *Proposed model for the orientation of the molecule in the NH_4Cl lattice.*
● Mo^{5+}; ● O^{2-}; ○ and ⊘ Cl^-; □ NH_4^+ vacancy
(Reproduced by permission from *Chem. Phys. Letters*, 1975, **30**, 231)

---

[290] W. C. Price, W. F. Sherman, and G. R. Wilkinson, *Proc. Roy. Soc.*, 1960, **A255**, 5.
[291] E. H. Coker and D. E. Hofer, *J. Chem. Phys.*, 1968, **48**, 2713.
[292] H. Bonadeo and E. Silberman, *J. Mol. Spectroscopy*, 1969, **32**, 214.
[293] V. P. Dem'yanenko, Yu. P. Tsyashchenko, and E. M. Verlan, *Phys. Stat. Sol.* (*B*), 1971, **48**, 737.
[294] Results for $v_3(T_2)$ quoted in reference 288.
[295] S. C. Jain, A. V. R. Warrier, and H. K. Sehgal, *J. Phys.* (*C*), 1973, **6**, 193.
[296] S. C. Jain, A. V. R. Warrier, and H. K. Sehgal, *J. Phys.* (*C*), 1972, **5**, 1511
[297] S. Radhakrishna, B. V. R. Chowdari, and A. K. Viswanath, *Chem. Phys. Letters*, 1975, **30**, 231; *J. Chem. Phys.*, 1976, in the press.
[298] M. D. Sastry and P. Venkateswarlu, *Mol. Phys.*, 1967, **13**, 161.
[299] S. Radhakrishna, B. V. R. Chowdari, and A. K. Viswanath, *J. Mag. Resonance*, 1974, **16**, 199.

starting material used is $(NH_4)_2MoOCl_5$. In the process of growth the ion which enters the crystal is $MoOCl_5^{2-}$. This however depends on the lattice. In NH_4Cl or NH_4Br where a square-planar configuration of ligand ions is available, Mo enters the face centre with a $\langle 100 \rangle$ orientation for the O—Mo—Cl group (Figure 4). The four remaining Cl^- form a square-planar configuration.[297] The Mo^{5+} ion has an outer electron configuration of $3d^1$ and the unpaired electron gives rise to e.p.r. The odd isotopes of Mo (^{95}Mo and ^{97}Mo) have a nuclear spin of 5/2, so that the e.p.r. spectrum provides a great deal of information. Owing to the hyperfine interaction the odd isotopes will account for six lines while the even isotopes, which account for 75% of the natural abundance, give one strong e.p.r. line. In NH_4Cl because there are three magnetically inequivalent complexes along $\langle 100 \rangle$, $\langle 010 \rangle$, and $\langle 001 \rangle$ one observes a more complicated spectrum. Such experiments provide conclusive evidence for the orientation of the molecular centre.[297] Figure 5 shows a typical e.p.r. spectrum for MoO^{3+} in NH_4Cl. In

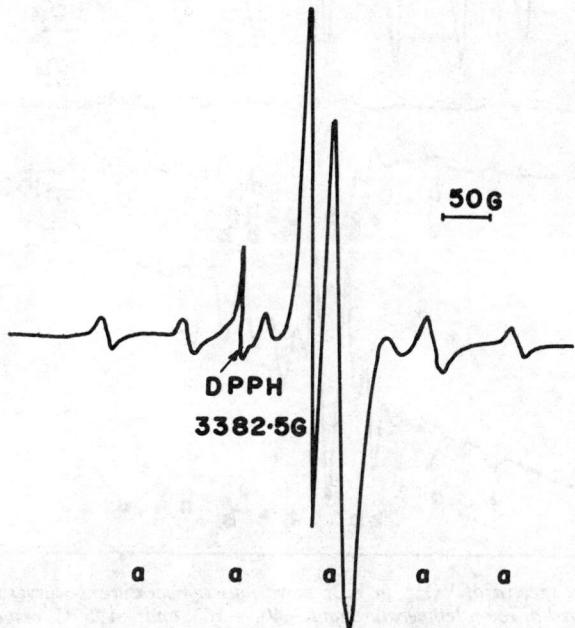

Figure 5 *Typical e.p.r. spectrum of* MoO^{3+} *in an* NH_4Cl *crystal at an arbitrary orientation*

NaCl Mo enters the lattice as an Mo^{3+} ion with a d^3 configuration. The spectrum obtained is angular independent, confirming that there is no preferential Mo—O direction.

In the visible region of the optical spectrum two bands are observed at 22 600 and 15 380 cm^{-1}. They correspond to the transitions $b_2 \to b_1^*$ and $b_2 \to e^*$, respectively.[297] In addition, two charge-transfer bands are observed at 28 980 and 32 360 cm^{-1}, corresponding to the transitions $b_2 \to e^b$ and $e \to e^b$. The e.p.r. parameters are correlated with the optical absorption results and the spin–orbit coupling constant of molybdenum is found to be $\lambda_M = 440$ cm^{-1}.

In alkali halides and NH_4Cl, VO_2^+ does not take any preferred orientation but is found to be rapidly tumbling inside the lattice.[298,299] The primary evidence for this interesting behaviour comes from e.p.r. studies in which an angle-independent spectrum is observed. In VO^{2+} the vanadium metal ion has a $3d^1$ outer configuration (V^{4+}). The vanadium nucleus has a nuclear spin of 7/2 and a characteristic spectrum with eight hyperfine lines is observed at room temperatures (Figure 6).

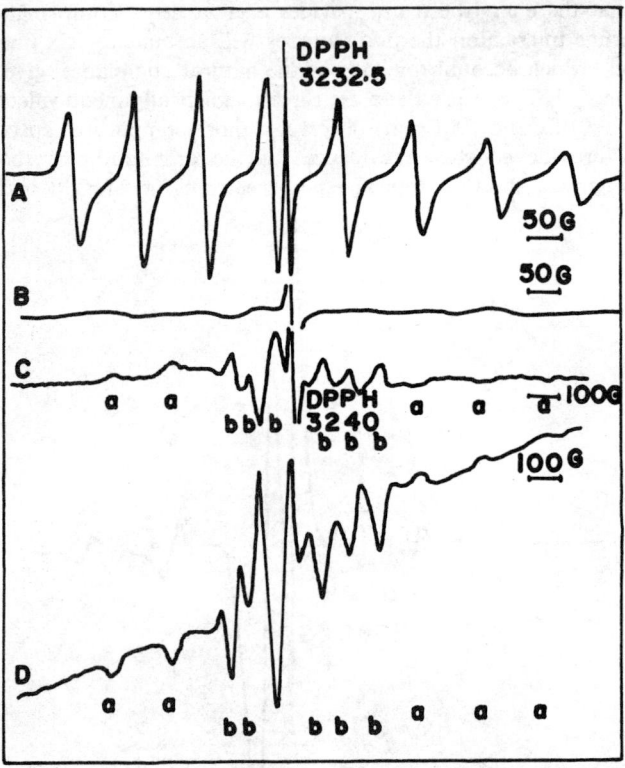

Figure 6 *E.p.r. spectra of VO^{2+} in KBr at various temperatures. Curves A—D are the spectra recorded at room temperature and -40, -160, and $-196\,°C$, respectively. Lines marked a are parallel components, and lines marked b are perpendicular components* (Reproduced by permission from *J. Mag. Resonance*, 1974, **16**, 199)

As the temperature is lowered the free tumbling motion freezes and a spectrum characteristic of preferred orientation (with a $g_{\parallel}$ and a $g_{\perp}$) is observed. The temperature at which the free tumbling motion freezes can be identified approximately by studying the temperature variation of the e.p.r. spectra. Table 20 summarizes data obtained for different alkali halides.

In view of limitations on space, details of results for all the molecules investigated could not be presented here; however, Table 21 gives a list of the various species studied and the local symmetries observed.

Table 20 Spin-Hamiltonian parameters for VO^{2+} in alkali halides. The values of A_0, A, and B are in the units of 10^{-4} cm^{-1}

	Lattice constant/ Å	T_{iso}-T_{aniso}/ °C	Room-temperature spectrum	Liquid-nitrogen-temperature spectrum	Linewidth for $m_I = \frac{3}{2}$/G
NaCl[a]	5.62	−160	$g_0 = 1.97 \pm 0.001$ $A_0 = 104.0 \pm 1.0$	$g_\parallel = 1.925 \pm 0.002$ $g_\perp = 1.996 \pm 0.002$ $A = 176.8 \pm 2.0$ $B = 64.2 \pm 2.0$	20
KCl[a]	6.28	−160	$g_0 = 1.969 \pm 0.002$ $A_0 = 106.8 \pm 2.0$	$g_\parallel = 1.932 \pm 0.002$ $g_\perp = 1.988 \pm 0.002$ $A = 186.0 \pm 3.0$ $B = 70.4 \pm 5.0$	16.9
KBr[b]	6.58	−35	$g_0 = 1.965 \pm 0.001$ $A_0 = 108.8 \pm 1.0$	$g_\parallel = 1.933 \pm 0.002$ $g_\perp = 1.986 \pm 0.002$ $A = 190.2 \pm 1.0$ $B = 61.8 \pm 2.0$	16.3
KI[b]	7.06	−40	$g_0 = 1.932 \pm 0.001$ $A_0 = 109.5 \pm 1.0$	$g_\parallel = 1.932 \pm 0.002$ $g_\perp = 1.983 \pm 0.002$ $A = 194.5 \pm 1.0$ $B = 60.2 \pm 2.0$	10

[a] A. V. Jagannadham and P. Venkateswarlu, *Proc. Indian Acad. Sci.*, 1969, **69**, 307; [b] S. Radhakrishna, B. V. R. Chowdari, and A. K. Viswanath, *J. Mag. Resonance*, 1974, **16**, 199.

8 Impurity Ions with the S^2 Configuration

When alkali halide crystals with the rock-salt structure are doped with impurity ions having the ground-state configuration s^2, four new absorption bands appear in the optical absorption spectrum (Hilsch [300,301]). These bands are designated A, B, C, D in order of increasing energy. The first three bands are due to excitation of the substitutional impurity ion (Seitz [302]); the D band is caused by a perturbed exciton transition (Inohara [303]). An energy-level diagram showing the effects of spin–orbit coupling, crystal field, and electron–lattice interactions is shown in Figure 7. The A band is due to the $^1A_{1g}$–$^3T_{1u}$ transition mode allowed by spin–orbit coupling. The B band is associated with the transitions $^1A_{1g}$–^3E_u, $^1A_{1g}$–$^3T_{2u}$, made allowed by vibronic mixing of B with the A and C states. The C band is due to the allowed transition $^1A_{1g}$–$^1T_{1u}$. The bands are all broad, indicating strong electron–lattice interaction. In some systems, *e.g.* KBr:Tl$^+$,[304] the A, B, and C bands *appear* to be single although they are not Gaussian in shape and display a temperature-dependent asymmetry. An extensive study by Fukuda [305] showed that this is not the general rule and that the majority of systems display a temperature-sensitive fine structure. Although most experimental work has been done using host crystals with the NaCl structures, Pb^{2+} doped CsCl, CsBr, and

[300] R. Hilsch, *Z. Physik.*, 1927, **44**, 860.
[301] R. Hilsch, *Proc. Phys. Soc.*, 1937, **49**, 40.
[302] F. Seitz, *J. Chem. Phys.*, 1938, **6**, 150.
[303] K. Inohara, *Sci. Light* (*Tokyo*), 1965, **14**, 92.
[304] P. W. M. Jacobs and S. A. Thorsley, *Crystal Lattice Defects*, 1974, **5**, 51.
[305] A. Fukuda, *Sci. Light* (*Tokyo*), 1964, **13**, 64.

Table 21 Observed symmetry and orientation of molecular impurities in ionic crystals. Subscripts s and b indicate stretching and bending modes, respectively

Molecular impurities investigated	Vibration modes and their activity	Point group symmetry of free molecules	Observed symmetry inside crystal	Orientation of molecular ions	Reference
N_2^-, O_2^-, S_2^-, Se_2^-	ν_s (R)	$D_{\infty h}$	$D_{\infty h}$	$\langle 110 \rangle$ in fcc	a–l
OH^-, OD^-, CN^-, SH^- SSe$^-$, SeH$^-$	ν_s (i.r., R)	$C_{\infty v}$	$C_{\infty v}$	$\langle 100 \rangle$ in fcc $\langle 111 \rangle$ in bcc	m–z
N_3^-, BO_2^-	ν_{1s} (R) ν_{2b} (i.r.) ν_{3s} (i.r.)	$D_{\infty h}$	$D_{\infty h}$	$\langle 111 \rangle$ in fcc	aa–cc
NCO^-, NCS^-	ν_{1s} (i.r., R) ν_{2s} (i.r., R) ν_{3s} (i.r., R)	$C_{\infty v}$	$C_{\infty v}$	$\langle 111 \rangle$ in fcc $\langle 100 \rangle$ in bcc	dd, ee
NO_2^-, NH_2^-, ND_2^-, S_3^-, H_2O	A_1–ν_{1s} (i.r., R) A_1–ν_{2b} (i.r., R) B_2–ν_{3s} (i.r., R)	C_{2v}	C_{2v}	two-fold axis in $\langle 110 \rangle$ direction in fcc	ff–gg
ClO_3^-, IO_3^-, SeO_3^-	A_1–ν_{1s} (i.r., R) A_1–ν_{2b} (i.r., R) E–ν_{3s} (i.r., R) E–ν_{4b} (i.r., R)	C_{3v}		three-fold axis along $\langle 111 \rangle$ direction for fcc	hh, ii
NO_3^-, CO_3^{2-}, BO_3^{2-}	A'–ν_{1s} (R) A''–ν_{2b} (i.r.) E'–ν_{3s} (i.r., R) E'–ν_{4b} (i.r., R)	D_{3h}	D_{3h}, C_{3v}, C_s	plane of ion is perpendicular to $\langle 111 \rangle$ direction in fcc	jj–ll

270   Surface and Defect Properties of Solids

| NH$_4^+$, ND$_4^+$, BD$_4^-$, BF$_4^-$, ClO$_4^-$, MnO$_4^-$, SO$_4^{2-}$, SO$_4^{3-}$, CrO$_4^{2-}$, MoO$_4^{2-}$, BeF$_4^{2-}$, PO$_4^{3-}$ | T_d | $A_1-\nu_{1s}$ (R)
$E-\nu_{2b}$ (R)
$T_2-\nu_{3s}$ (R, i.r.)
$T_2-\nu_{4b}$ (R, i.r.) | T_d, C_s, C_{2v}
(D_{2d} in bcc lattices) | $ii, mm-ww$ metal ion occupies cation site, and tetrahedral bonds are developed along ⟨111⟩ for fcc |

[a] W. Kanzig and M. H. Cohen, *Phys. Rev. Letters*, 1959, **3**, 509; [b] J. R. Morton, *J. Chem. Phys.*, 1965, **43**, 3418; [c] L. E. Vannotti and J. R. Morton, *Phys. Rev.*, 1967, **161**, 282; [d] J. R. Morton, *J. Phys. Chem.*, 1967, **71**, 89; [e] L. E. Vannotti and J. R. Morton, *J. Chem. Phys.*, 1967, **47**, 4210; [f] W. Holzer, W. F. Murphy, and H. J. Bernstein, *J. Mol. Spectroscopy*, 1969, **32**, 13; [g] J. Rolfe, F. R. Lipsett, and W. J. King, *Phys. Rev.*, 1961, **123**, 447; [h] J. H. Schulman and R. D. Kirk, *Solid State Comm.*, 1964, **2**, 105; [i] J. Rolfe, *J. Chem. Phys.*, 1968, **49**, 4193; [j] M. Ikezawa and J. Rolfe, *J. Chem. Phys.*, 1973, **58**, 2024; [k] G. J. Vella and J. Rolfe, *J. Chem. Phys.*, 1974, **61**, 41; [l] M. Höhne and M. Stasiw, *Phys. Stat. Sol.*, 1967, **20**, 657, 667; [m] V. Narayanamurti and R. O. Pohl, *Rev. Mod. Phys.*, 1970, **42**, 201; [n] S. Kapphan, *J. Phys. Chem. Solids*, 1974, **35**, 621; [o] F. Bridges, *CRC Crit. Rev. Solid State Sci.*, 1975, **5**, 1; [p] M. V. Klein, *Phys. Rev.*, 1961, **122**, 1393; [q] K. Brugger and W. P. Mason, *Phys. Rev. Letters*, 1961, **7**, 270; [r] S. Kapphan and F. Lüty, *J. Phys. Chem. Solids*, 1973, **34**, 969; [s] R. V. Jiminez and F. Lüty, *Phys. Stat. Sol.*, 1972, **52**, K27; [t] F. Bridges, *Solid State Comm.*, 1973, **13**, 1877; [u] J. G. Peascoe and M. V. Klein, *J. Chem. Phys.*, 1973, **59**, 2394; [v] R. D. Kirby, A. E. Hughes, and A. J. Sievers, *Phys. Rev. (B)*, 1970, **2**, 481; [w] W. Kelly and F. Bridges, *Bull. Amer. Phys. Soc.*, 1974, **19**, 651; [x] J. G. Peascoe, W. R. Fenner, and M. V. Klein, *J. Chem. Phys.*, 1974, **60**, 4208; [y] G. Zibold and F. Lüty, *J. Nonmetals*, 1972, **1**, 1, 17; [z] H. U. Beyeler, *Bull. Amer. Phys. Soc.*, 1974, **19**, 328; [aa] J. I. Bryant and G. Turrel, *J. Chem. Phys.*, 1962, **37**, 1069; [bb] W. C. Price, W. F. Sherman, and G. R. Wilkinson, *Spectrochim. Acta*, 1960, **16**, 663; [cc] H. W. Morgan and P. A. Staats, *J. Appl. Phys.*, 1962, **33**, 364; [dd] J. C. Decius, J. L. Jacobson, W. F. Sherman, and G. R. Wilkinson, *J. Chem. Phys.*, 1965, **43**, 2180; [ee] M. A. Cundill and W. F. Sherman, *Phys. Rev.*, 1968, **168**, 1007; [ff] A. R. Evans and D. B. Fitchen, *Phys. Rev. (B)*, 1970, **2**, 1074; [gg] C. Ruhenbeck, *Z. Phys.*, 1967, **207**, 446; [hh] G. N. Krynauw and C. J. H. Schutte, *Spectrochim. Acta*, 1965, **21**, 1947; [ii] V. P. Dem'yanenko, Yu. P. Tsyashchenko, and E. M. Verlan, *Phys. Stat. Sol. (B)*, 1971, **48**, 737; [jj] K. Kato and J. Rolfe, *J. Chem. Phys.*, 1967, **47**, 1901; [kk] S. Radhakrishna and A. M. Karguppikar, *J. Brenna*, 10, 895; [ll] S. C. Jain, A. V. R. Warrier, and H. K. Sehgal, *J. Phys. (C)*, 1973, **6**, 189; [mm] J. C. Decius, E. H. Coker, and G. L. Brenna, *Spectrochim. Acta*, 1963, **19**, 1281; [nn] T. I. Maksimova, *Phys. Rev. (B)*, 1971, **4**, 1382; [oo] D. N. Mirlin and I. I. Reshina, *Sov. Phys. Solid. State*, 1968, **10**, 895; [pp] S. Radhakrishna, *Phys. Rev. (B)*, 1971, **4**, 1382; [qq] S. C. Jain, A. V. R. Warrier, and S. K. Agarwal, *Chem. Phys. Letters*, 1972, **14**, 211; [J. Agarwal, NBS–Data Book, NSRDS–NBS Publication No. 52, July, 1974; [tt] P. Manzelli and G. Todder, *J. Chem. Phys.*, 1969, **51**, 1484; [uu] W. C. Price, *Phys. Chem. Solids*, 1973, **34**, 209; [rr] S. Radhakrishna and G. R. Wilkinson, *Proc. Roy. Soc.*, 1960, **A255**, 5; [vv] E. H. Coker and D. E. Hofer, *J. Chem. Phys.*, 1968, **48**, 2713; [ww] H. Bonadeo and E. Silberman, *J. Mol. Spectroscopy*, 1969, **32**, 214.

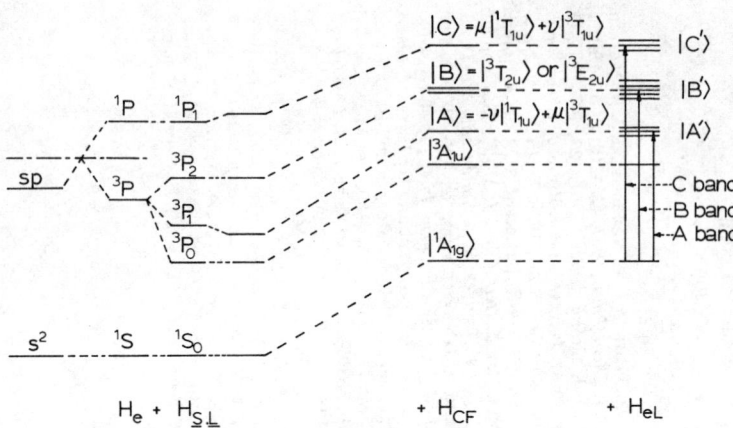

Figure 7 *Energy-level diagram for an impurity ion with the s^2 ground-state configuration in a cubic crystal, showing the effects of spin–orbit coupling ($H_{S.L}$), the static crystal field (H_{CF}), and the dynamical electron–lattice interaction (H_{eL})*

CsI have been investigated by Radhakrishna and Pande.[306] Bi^{3+} [307] and Sb^{3+} [308] also have the s^2 configuration and give similar A, B, and C bands. Toyozawa and Inoue [309] explained the triplet structure of the C band and the doublet structure of the A band, with its temperature-dependent asymmetry, in terms of the dynamical Jahn–Teller effect. Using the Condon and semi-classical approximations they calculated the theoretical lineshape of the A and C bands using a single adjustable parameter, the coupling constant c, which describes the coupling between the triply degenerate electronic T_{1u} state and the lattice vibrational modes of T_{2g} symmetry. The model of Toyozawa and Inoue has been used by Cho [310] in a numerical calculation of lineshapes for transitions from orbital singlets to triplet states in cubic symmetry, using various combinations of the numerical values of the coupling constants (a, b, and c) which represent the strength of the interaction of the electronic T_{1u} state with vibrational modes of A_{1g}, E_g, and T_{2g} symmetry. However, the effects of the mixing of the triplet state with other states by spin–orbit coupling and electron–lattice interactions were not considered. These effects have been accounted for by Jacobs and Oyama,[311–314] who calculated the lineshapes of the A and C bands in $KBr:In^+$ and $KBr:Sn^{2+}$ and compared the results with experimental data for these systems, which were typical of other In^+ and Sn^{2+} doped alkali halides.[311–314] The shape of the A and C bands at low temperatures was well accounted for by the model which allowed for spin–orbit coupling, vibrational coupling of the A, B, and C states, and the dynamical Jahn–Teller effect in the A′

[306] S. Radhakrishna and K. P. Pande, *Phys. Rev.* (B), 1973, **7**, 424.
[307] R. Reisfeld and A. Honigbaum, *J. Chem. Phys.*, 1968, **48**, 5565.
[308] S. Radhakrishna and R. S. Setty, *Phys. Rev.* (B), 1976, **14**, 969.
[309] Y. Toyozawa and M. Inoue, *J. Phys. Soc. Japan*, 1966, **21**, 1663.
[310] K. Cho, *J. Phys. Soc. Japan*, 1968, **25**, 1372.
[311] P. W. M. Jacobs and K. Oyama, *J. Phys.* (C), 1975, **8**, 851.
[312] P. W. M. Jacobs and K. Oyama, *J. Phys.* (C), 1975, **8**, 865.
[313] K. Oyama-Gannon and P. W. M. Jacobs, *J. Phys. Chem. Solids*, 1975, **36**, 1375.
[314] K. Oyama-Gannon and P. W. M. Jacobs, *J. Phys. Chem. Solids*, 1975, **36**, 1383.

and C′ (perturbed A and C) states. While the principal factor governing the lineshape is the coupling with modes of T_{2g} symmetry, the agreement with the experimental lineshape was much improved by convolution of the calculated lineshape with a Gaussian function which represented the dynamical interaction with modes of A_{1g} symmetry (Figure 8).

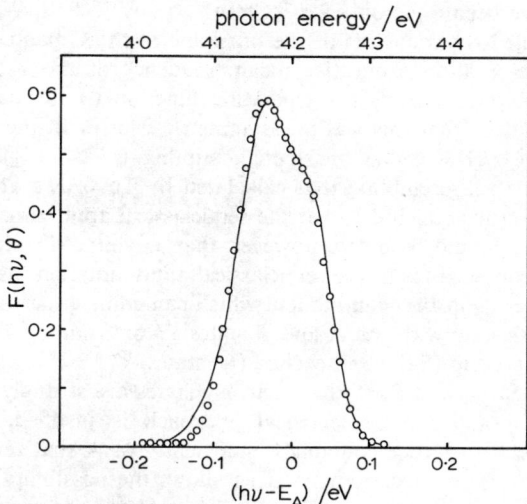

Figure 8 *Comparison of calculated and experimental lineshape of the A absorption band in* $KBr:In^+$ *at 18 K. Circles show experimental points. The continuous line is the calculated lineshape for* $c^2 = 1.69$ eV *and* $a/c = 0.4$
(Reproduced by permission from *J. Phys.* (C), 1975, **8**, 851)

For Sn^{2+} it is impossible to account for the lineshape in terms of the dynamical Jahn–Teller effect alone [313,314] and there is in addition a static splitting of the A' state which is due to a lowering in the symmetry of the crystal field caused by a nearby vacancy. There is clear evidence for this static splitting in the temperature-dependence of the second moment of the A and C bands and in the temperature-dependence of peak separation of the two components of the A band, or the three components of the C band.[313,314] The peak splitting is described by equation (4), in which ΔE_1 represents the static splitting. For

$$\Delta E(T) = \Delta E_1 + \Delta E_2 [\coth \hbar\omega_{\text{eff}}/2kT]^{\frac{1}{2}} \qquad (4)$$

$KBr:In^+$ and $KCl:In^+$ ΔE_1 is zero.[311]

Another way of obtaining information about absorption bands is by calculating the moments of the shape function instead of the lineshape. Numerical values of the parameters in the Hamiltonian can then be determined by comparing the theoretical expressions for the moments with experimental data. The moments method has been applied to the s^2 phosphors by Honma.[315-319] If the semiclassical

[315] A. Honma, *Sci. Light* (*Tokyo*), 1967, **16**, 212.
[316] A. Honma, *J. Phys. Soc. Japan*, 1968, **24**, 1082.
[317] A. Honma, *Sci. Light* (*Tokyo*), 1972, **21**, 119.
[318] A. Honma, *Sci. Light* (*Tokyo*), 1973, **22**, 1.
[319] A. Honma, *J. Phys. Soc. Japan*, 1973, **35**, 1115.

approximation is adopted then the coupling constants a^2 and $2b^2 + 3c^2$ can be evaluated. However, there is an inconsistency between the values of coupling constants derived from the moments method and from the lineshape.[317,320] If moments from dichroic spectra are used as well then individual values of the coupling constants a, b, and c for singlet and triplet states can be found. However, this seems to have been done only for KCl:In$^+$ so far.[319] Sati, Wang, and Inoue have also calculated the moments of the lineshape of the C band in terms of the coupling constant c and an effective mean frequency of the T_{2g} modes. They then used an arbitrary (Gaussian) generating function to calculate the C band lineshape and showed that this was an asymmetric triplet. In the limits of either high temperature ($kT \gg \hbar\omega_{\text{eff}}$) or strong coupling ($c^2 \gg \hbar\omega_{\text{eff}}$) this lineshape becomes symmetrical, resembling that calculated by Toyozawa and Inoue. Sati, Wang, and Inoue thus concluded that the semiclassical approximation is justified in these limits. It should be noted, however, that a symmetric triplet structure is not a necessary consequence of the semiclassical approximation: asymmetry arises from quadratic terms in the Hamiltonian which can come either from vibrational coupling of the C state with the A and B states [312] or from a difference in force constants of the ground ($^1A_{1g}$) and excited (C) states.[304,321,322]

For In$^+$, Tl$^+$, Sn^{2+}, and Pb^{2+} the electronic states are strongly coupled to the lattice vibrations, and the semiclassical approach is justified. However, for KCl:Au$^-$ the electron–lattice coupling is sufficiently weak that zero-phonon lines appear in the absorption spectrum.[323] This allows the possibility of determining the coupling parameters directly both from the broad band and from the zero-phonon line if measurements are made of the optical absorption of a crystal in a magnetic field and under uniaxial stress. The Huang–Rhys factor deduced from the broadband moments is 3.8 ± 0.4 for the A_{1g} modes in KCl:Au$^-$. For the E_g and T_{2g} modes it is 0.72. The Jahn–Teller energies for the non-cubic modes are equal and are a factor of 14 smaller than the spin–orbit coupling energy. The Huang–Rhys factors for the cubic and non-cubic modes could therefore also be deduced from the zero-phonon line data, with results in excellent agreement with those from the broad-band data. Accidental degeneracy between E and T_2 modes, which holds both for KCl:Au$^-$ and for the F^+ centre in CaO,[324] simplifies the theoretical analysis.[325,326]

The vibronic Hamiltonian for a triply degenerate electronic state coupled to modes of T_{2g} symmetry has been diagonalized [327] in an approximation that neglects quadratic interactions and spin–orbit coupling. The resulting lineshapes show a triplet structure and are very asymmetric, the asymmetry becoming less pronounced as the temperature or the coupling constant is increased. While the *a priori* nature of this approach is appealing, its usefulness for heavy-metal phosphors is limited by the importance in these systems of spin–orbit coupling, quadratic interactions,[314]

[320] A. Fukuda, *J. Phys. Soc. Japan*, 1969, **27**, 96.
[321] A. Honma, *Sci. Light (Tokyo)*, 1969, **18**, 33.
[322] P. W. M. Jacobs and L. M. Parsons, *Crystal Lattice Defects*, 1972, **3**, 155.
[323] D. Lemoyne, J. Duran, M. Billardon, and Le Si Dang, *Phys. Rev. (B)*, 1976, **14**, 747.
[324] R. Romestain and Y. Merle d'Aubigne, *Phys. Rev. (B)*, 1971, **4**, 4611.
[325] M. C. M. O'Brien, *J. Phys. (C)*, 1971, **4**, 2524.
[326] M. C. M. O'Brien, *J. Phys. (C)*, 1972, **5**, 2045.
[327] R. Englman, M. Caner, and S. Toaff, *J. Phys. Soc. Japan*, 1970, **29**, 306.

and coupling with A_{1g} [311–314,323] and E_g modes.[323] In the strong-coupling limit a semiclassical formulation [309,311–314] is adequate, whereas for weak coupling the moments method [323] is preferable.

Nasu and Kojima [328] have used the independent ordering approximation [329] in performing a quantum mechanical calculation of the lineshape for an A_{1g}–T_{1u} transition. For large coupling constants it is possible to neglect the time-dependence of the phonon correlation function, and in this approximation they recover the semiclassical adiabatic approximation thus confirming its usefulness in strong coupling situations.

Individual values of the coupling constants a, b, and c may not be obtained from lineshape measurements alone on unperturbed crystals. (For example, the analysis of Jacobs and Oyama [311–314] determines a and c having neglected coupling to E_g modes on the grounds that this does not induce splitting.) Hence there is considerable interest in measurements on crystals subjected to uniaxial stress or a magnetic field. The recently published work of Lemoyne et al.[323] has been alluded to above; stress-induced dichroism in the A, B, and C bands of Tl^+ and In^+ doped potassium halides has been measured by Dultz et al.[330,331] and analysed theoretically by Dultz [331] and by Lemos and Krolik.[332] The moments of dichroic spectra induced by an applied magnetic field or uniaxial stress have been calculated by Honma,[333] who has also developed a theory of the MCD lineshape.[317,319,334–336] However, in this theory the effects of the A and C bands on the B band MCD are treated by perturbation theory, which may not be a satisfactory approximation, and an alternative numerical approach [337] deserves consideration. Experimental measurements of MCD in s^2 phosphors have been published by Onaka, Mabuci, and Yoshikawa [338,339] ($KBr:Tl^+$, $KI:Tl^+$, $KBr:Pb^{2+}$, $KCl:Pb^{2+}$, $KCl:In^+$, $KBr:In^+$, $KI:In^+$), by Grasso et al.[340–342] ($KBr:Tl^+$ and $KBr:In^+$), and by Jacobs et al.[343] ($KBr:In^+$). Although the experimental lineshapes bear a qualitative resemblance to those calculated by Cho [337] and by Honma and Ooaku,[326] only for $KCl:In^+$ has a detailed comparison between experiment and theory been performed so far.[336]

Understandably, most of the emphasis in the s^2-phosphor literature is on the more prominent A and C bands. However, a survey of the properties of the vibration-induced B band due to In^+, Sn^{2+}, and Tl^+ in a number of alkali halides

[328] K. Nasu and T. Kojima, *Prog. Theor. Phys.*, 1974, **51**, 26.
[329] B. G. Vekhter, Yu. E. Perlin, V. Z. Polinger, Yu. B. Rosenfeld, and B. S. Tsukerblat, *Crystal Lattice Defects*, 1972, **3**, 69.
[330] D. Bimberg, W. Dultz, and W. Gebhardt, *Phys. Stat. Sol.*, 1969, **31**, 661.
[331] W. Dultz, *Phys. Stat. Sol. (B)*, 1971, **48**, 571.
[332] A. Lemos and C. Krolik, *Phys. Rev. (B)*, 1973, **7**, 1608.
[333] A. Honma, *Sci. Light (Tokyo)*, 1972, **21**, 119.
[334] A. Honma, *Sci. Light (Tokyo)*, 1975, **24**, 33.
[335] A. Honma, *Sci. Light (Tokyo)*, 1975, **24**, 57.
[336] A. Honma and S. Ooaku, *J. Phys. Soc. Japan*, 1976, **41**, 152.
[337] K. Cho, *J. Phys. Soc. Japan*, 1969, **27**, 646.
[338] R. Onaka, T. Mabuchi, and A. Yoshikawa, *J. Phys. Soc. Japan*, 1967, **23**, 1036.
[339] A. Yoshikawa, T. Mabuchi, and R. Onaka, *J. Phys. Soc. Japan*, 1968, **24**, 1405.
[340] V. Grasso, P. Perillo, and G. Vermiglio, *Solid State Comm.*, 1972, **11**, 563.
[341] V. Grasso, P. Perillo, and G. Vermiglio, *Nuovo Cimento*, 1973, **13B**, 42.
[342] C. Bagnato, V. Grasso, P. Perillo, and G. Vermiglio, *Nuovo Cimento*, 1975, **26B**, 263.
[343] P. M. W. Jacobs, M. J. Stillman, K. Oyama-Gannon, and D. Simkin, *Chem. Phys. Letters*, 1976, **42**, 530.

has been published by Tsuboi et al.[344] Particular emphasis has been given to the fine structure of the B band and in at least two systems, $KBr:Sn^{2+}$ and $KI:Sn^{2+}$, five components of the B band have been observed at low temperatures.[345] Semiclassical calculations of the lineshape have been published by Honma, Ooaku, and Mabuci [346] and by Matsushima and Fukuda.[347] The latter authors diagonalized the 12×12 matrix which arises from the vibrational interaction of the $^1A_{1u}$, $^3T_{1u}$, $^3T_{2u}$, ^3E_u, and $^1T_{1u}$ states numerically and employed Cho's Monte Carlo integration method [310] in the semiclassical calculation of the lineshape. Their calculated lineshape bears a striking qualitative resemblance to that observed for the B band in $KCl:In^+$.

There are also several papers [348—353] in the literature that deal with ions like Tl^0 or Sn^+ that have trapped an electron, or been reduced chemically, so that they have the s^2p ground-state configuration. Little work has been done until recently [354,355] on the effect of pressure on the structure of the A and C bands. Most early work dealt with the peak position of the A band. For In^+ in NaCl KCl, KBr, and KI at room temperature the A and C band splittings both become larger with increasing pressure, whereas for Sn^{2+} the splittings remain constant.[354] For $KI:In^+$ there is a discontinuous change in the A band splitting at the phase transition to the CsCl structure, presumably indicating a smaller coupling to the T_{2g} modes. For $NaCl:Sn^{2+}$ at 185 K the A band splitting decreases smoothly with increasing pressure,[355] instead of increasing as the semiclassical theory would imply [309] and as occurs for In^+ at room temperature. With $KCl:Sn^{2+}$ the separation of the A band components also decreases smoothly with pressure but shows hardly any change at the phase transition to the CsCl structure. These differences between the behaviour of In^+ and Sn^{2+} are presumably connected with differences in the local strains around the defect.

Emission.—Excitation of an impurity s^2 ion by light absorbed in the A, B, or C absorption bands produces luminescence. For $KI:Tl^+$ Edgerton and Teegarden [356,357] found five emission bands associated with transitions from the A, B, and C excited states to the ground state. Whereas the absorption spectra are determined by the Jahn–Teller splitting caused by small distortions from O_h symmetry (i.e. near $Q = 0$) emission occurs from the *relaxed* excited states. Early attempts [358,359] to explain the characteristics of the emission spectra of s^2 phosphors

[344] T. Tsuboi, Y. Nakai, K. Oyama, and P. W. M. Jacobs, *Phys. Rev.* (B), 1973, **8**, 1698.
[345] T. Tsuboi, K. Oyama, and P. W. M. Jacobs, *J. Phys.* (C), 1974, **7**, 221.
[346] A. Honma, S. Ooaku, and T. Mabuchi, *J. Phys. Soc. Japan*, 1974, **36**, 1708.
[347] A. Matsushima and A. Fukuda, *Phys. Stat. Sol.* (B), 1974, **66**, 663.
[348] T. Tsuboi and P. W. M. Jacobs, *J. Phys.* (C), 1973, **6**, L109.
[349] S. B. S. Sastry, V. Viswanathan, and C. Ramasastry, *J. Phys. Soc. Japan*, 1973, **35**, 508.
[350] S. Radhakrishna and A. M. Karguppikar, *Phys. Stat. Sol.* (B), 1974, **61**, 687.
[351] C. J. Delbecq, R. Hartford, D. Schoemaker, and P. H. Yuster, *Phys. Stat. Sol.* (B), 1975, **71**, K81.
[352] V. Osminin and S. Zazubovich, *Phys. Stat. Sol.* (B), 1975, **71**, 435.
[353] C. J. Delbecq, R. Hartford, D. Schoemaker, and P. H. Yuster, *Phys. Rev.* (B), 1976, **13**, 3631.
[354] S. Masunaga and E. Matsuyama, *J. Phys. Soc. Japan*, 1973, **34**, 1234.
[355] S. Masunaga and T. Abe, *J. Phys. Soc. Japan*, 1974, **37**, 1540.
[356] R. Edgerton and K. Teegarden, *Phys. Rev.*, 1963, **129**, 169.
[357] R. Edgerton and K. Teegarden, *Phys. Rev.*, 1964, **136**, A1091.
[358] P. D. Johnson and F. E. Williams, *J. Chem. Phys.*, 1952, **20**, 124.
[359] F. E. Williams and P. D. Johnson, *Phys. Rev.*, 1959, **113**, 97.

were based on a one-dimensioual configuration co-ordinate model. Kamimura and Sugano [360] introduced the concept of a multidimensional configuration co-ordinate model to allow for interactions of the A, B, and C states with normal modes of A_{1g}, E_g and T_{2g} symmetry. They showed that two kinds of 'X' minimum occurred on the adiabatic potential energy surface, thus accounting for the appearance of two emission bands (e.g. at 305 and 475 nm in KCl:Tl$^+$) following excitation in the A band (which peaks at 249 nm in KCl:Tl$^+$). Such minima can arise from either tetragonal or trigonal distortions of the nearest neighbour octahedral configuration, but a tetragonal distortion is necessary to explain the polarization of the 305 nm emission in KCl:Tl$^+$. The appearance of two emission bands following A band excitation occurs commonly in these phosphors: because of the identification of the high-energy emission band with three equivalent tetragonal minima it is designated the A_T band while the one occurring at longer wavelengths, called the A_X band, is associated with six equivalent minima of another kind.[361] Until recently their nature had not been identified for certain. An excellent account of the situation in 1972 has been given by Fukuda.[362] At low temperatures the A_T band is the prominent one [361] but as the temperature increases the relative intensity of the A_T and A_X emission bands changes in a complicated manner (Figure 9 and references 361 and 363).

Since vibrational modes of E_g and T_{2g} symmetry lower the symmetry of the nearest-neighbour octahedron from O_h to D_{4h} and D_{3d}, respectively, we should

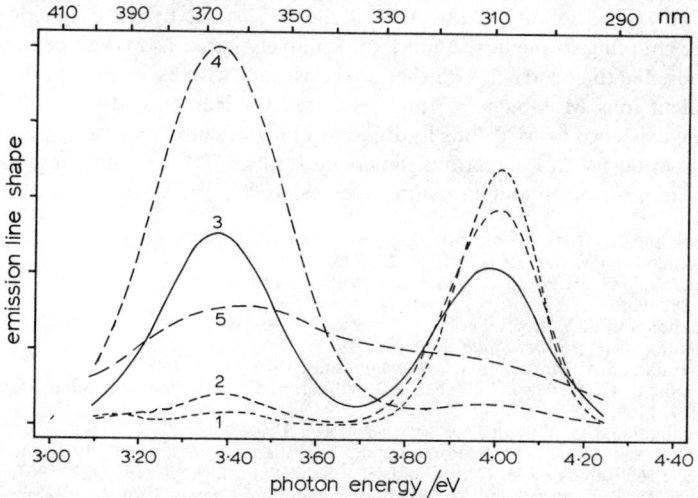

Figure 9 *Relative emission lineshape* (I/I_{max}) *as a function of photon energy for* KBr:Tl$^+$. 1, 15.5 K; 2, 57 K; 3, 83 K; 4, 107 K; 5, 296 K
(Reproduced by permission from *J. Luminescence*, 1974, **8**, 391)

---

[360] H. Kamimura and S. Sugano, *J. Phys. Soc. Japan*, 1959, **14**, 1612.
[361] A. Fukuda, *Phys. Rev. (B)*, 1970, **1**, 4161.
[362] A. Fukuda, in 'Physics of Impurity Centres in Crystals', Academy of Sciences of Estonian S.S.R., Tallinin, 1972, p. 505.
[363] P. W. M. Jacobs and S. A. Thorsley, *J. Luminescence*, 1974, **8**, 391.

expect the appearance of two kinds of minimum on the adiabatic potential energy surfaces (henceforth APS) from the tetragonal and trigonal distortions. However, when Öpik and Pryce [364] investigated the a_{1g} and t_{1u} excited states using the Russell–Saunders approximation for spin–orbit coupling, they found only one kind of minimum. If the spin–orbit coupling is very small then the A, C, and $^3A_{1u}$ APS display minima which are tetragonal if $|b| > |c|$ but are trigonal if $|b| < |c|$. The existence of minima of only one kind for the Russell–Saunders limit in the E_g sub-space was confirmed by Ranfagni [365,366] who also showed that two kinds of minimum appear for intermediate coupling. Although this modification of the Öpik and Pryce model leads to qualitative agreement with the temperature-dependence of the A_T and A_X emission intensities, the model is not quantitatively accurate [363] nor is it in harmony with the effects of a magnetic field [367,368] or hydrostatic pressure [369] on the A band emission. However, if quadratic interactions are included [370–373] the A_T and A_X emission bands can be correlated with three tetragonal minima and four trigonal minima on the AP. Each of the A_T and A_X relaxed excited states consist of two nearly degenerate sub-levels ($T_{1u} \rightarrow E_u + A_{2u}$) each with an associated trap level (A_{1u}). Evidence for the trap level comes principally from lifetime studies.[374] Since tetragonal distortions do not mix T_{1u} and A_{1u} states the transition $^3A_{1u} \rightarrow {}^1A_{1g}$ is still forbidden in the absence of a magnetic field, although there is some radiative probability due to lattice vibrations. In a magnetic field the radiative lifetime of the $^3A_{1u} \rightarrow {}^1A_{1g}$ transition decreases by almost two orders of magnitude in the range 0—4.2 T, owing to mixing of the $A(^3T_{1u})$ and $^3A_{1u}$ states by the magnetic field.[367] When the coupling to the lattice modes is relatively weak, as in Ag^- centres, it has proved possible to construct APS that are consistent with experimental data.[375,376] For bivalent ions Mössbauer [377] and polarized luminescence studies [377–379] have shown the existence of M^{2+} ions in different kinds of complex and aggregate.

At high impurity concentrations dimer centres like $(Tl^+)_2$ occur; these have been identified in absorption and emission experiments.[380–386] With bivalent impurities

[364] U. Öpik and M. H. L. Pryce, *Proc. Roy. Soc.*, 1957, **A238**, 425.
[365] A. Ranfagni, *Phys. Rev. Letters*, 1972, **28**, 743.
[366] A. Ranfagni, G. P. Pazzi, P. Fabeni, G. Villiani, and M. P. Fontana, *Phys. Rev. Letters*, 1972, **28**, 1035.
[367] A. Fukuda and P. Yuster, *Phys. Rev. Letters*, 1972, **28**, 1032.
[368] A. Fukuda, *J. Phys. Soc. Japan*, 1976, **40**, 776.
[369] A. Fukuda and A. Matsushima, *J. Luminescence*, 1976, **12/13**, 139.
[370] A. Ranfagni, G. P. Pazzi, P. Fabeni, G. Viliani, and M. P. Fontana, *Solid State Comm.*, 1974, **14**, 1169.
[371] A. Ranfagni and G. Viliani, *Phys. Rev. (B)*, 1974, **9**, 4448.
[372] M. Bacci, A. Ranfagni, M. P. Fontana, and G. Viliani, *Phys. Rev. (B)*, 1975, **11**, 3052.
[373] M. Bacci, A. Ranfagni, M. Cetica, and G. Viliani, *Phys. Rev. (B)*, 1975, **12**, 5907.
[374] S. Benci, M. P. Fontana, and M. Manfredi, *Solid State Comm.*, 1976, **18**, 1423.
[375] K. Kojima, S. Shimanuki, and T. Kojima, *J. Phys. Soc. Japan*, 1972, **33**, 1076.
[376] S. Shimanuki, *J. Phys. Soc. Japan*, 1973, **35**, 1680.
[377] E. Realo and S. Zazubovich, *Phys. Stat. Sol. (B)*, 1973, **57**, 69.
[378] W. C. Collins and J. H. Crawford, *Solid State Comm.*, 1971, **9**, 853.
[379] W. C. Collins and J. H. Crawford, *Phys. Rev. (B)*, 1972, **5**, 633.
[380] A. Yoshikawa, H. Takezoe, I. Maruyama, and R. Onaka, *J. Phys. Soc. Japan*, 1972, **32**, 472.
[381] A. Yoshikawa, H. Takezoe, and R. Onaka, *J. Phys. Soc. Japan*, 1972, **33**, 1632.
[382] S. Kinno, S. Kunieda, H. Takezoe, and R. Onaka, *J. Phys. Soc. Japan*, 1975, **38**, 290.
[383] T. Tsuboi and P. W. M. Jacobs, *J. Luminescence*, 1975, **11**, 227.
[384] A. Ermoshkin, R. Evarestov, R. Gindina, A. Maaroos, V. Osminin, and S. Zazubovich, *Phys. Stat. Sol. (B)*, 1975, **70**, 749.

like Pb^{2+} the dimer probably contains cation vacancies as well,[387] and higher aggregates [379] may be formed. Ionized dimer centres have also been observed.[388,389]

We acknowledge gratefully the considerable assistance received from Brenda Parker in data compilation and from Jacquie Taylor in the preparation of this work. J. Corish and S. Radhakrishna thank the Chemical Physics Centre at the University of Western Ontario for the award of Visiting Fellowships.

[385] H. Takezoe and R. Onaka, *J. Phys. Soc. Japan*, 1975, **38**, 804.
[386] H. Takezoe and R. Onaka, *J. Phys. Soc. Japan*, 1975, **38**, 810.
[387] T. Tsuboi and P. W. M. Jacobs, *Phys. Stat. Sol. (B)*, 1974, **62**, K59.
[388] T Tsuboi, Y. Matsusaka, and Y. Nakai, *J. Phys. Soc. Japan*, 1972, **33**, 1725.
[389] C. J. Delbecq, E. Hutchinson, and P. H. Yuster, *J. Phys. Soc. Japan*, 1974, **36**, 913.

6
Assessment of Crystal Perfection by X-Ray Topography

BY B. K. TANNER

1 Introduction

Despite the fact that high-resolution X-ray topography was developed some 18 years ago the technique has not yet achieved widespread use. Even now many scientists not working directly in materials science are totally unfamiliar with this very important analytical tool. X-Ray *diffraction* topography is concerned with a direct mapping of the lattice plane topography within a crystal. The technique probes lattice distortions and is not primarily a technique for studying surface contours. Surface features do often appear on X-ray topographs, but their presence is of secondary importance to the images of lattice distortions. The bulk of X-ray topographic studies have been concerned with imaging the long-range lattice distortion associated with dislocations, and most of the applications to date are concerned with revealing dislocation configurations in crystals.

At the most elementary level we can see how dislocations can be imaged in the following way. A perfect crystal with lattice spacing d will produce a strong diffracted beam from a monochromatic X-ray beam of wavelength λ only when the orientation is such that the Bragg relation (1) applies, where θ_B is the angle

$$\lambda = 2d \sin \theta_B \qquad (1)$$

between lattice planes and the incident beam direction. Around a defect such as a dislocation the lattice planes are distorted or misoriented, and clearly the Bragg relation cannot apply simultaneously to the perfect and imperfect regions. Consequently there is a local variation in the intensity of the diffracted beam, providing the image of the defect. The interpretation of this image is not altogether straightforward and it is often necessary to consider in detail the propagation of the X-rays in the distorted crystal, a problem which has still not been satisfactorily solved.

X-Ray topography has very many similarities with transmission electron microscopy of crystals and is often compared to it rather unfavourably, particularly by electron microscopists. True, due to its high strain sensitivity, X-ray topography has a much poorer spatial resolution than electron microscopy. Dislocation images in X-ray topographs are several micrometres wide whereas electron microscope images are a few hundred Ångstroms in width. Because of the wide images, in order to reveal individual dislocation images by X-ray topography, very low dislocation density crystals are required. However, in transmission electron microscopy very thin areas of specimen must be prepared, and these small volumes of crystal are examined with high magnification. Electron microscopy therefore needs high dislocation densities or there is very little chance of finding a dislocation in the field of view. Thus electron microscopy examines small areas of very thin crystals with high spatial resolution and is complemented by X-ray topography which examines large areas of relatively thick crystals with a poorer spatial resolution. Nowadays, crystals of many different materials can be grown with high crystalline perfection, and X-ray topography has an important role to play as an assessment tool for the crystal grower. As we will see in a later section, use of synchrotron radiation reduces the otherwise rather long exposure times to a level comparable with those in electron microscopy, and dynamic experiments are now possible.

This article gives a brief description of the more important X-ray topographic techniques, some ideas about image interpretation, and a review of some applications which are likely to be of interest to those with a training in chemistry. This inevitably means some very important areas of application get little or no mention, for example, magnetic and ferroelectric domain studies. More wide ranging accounts of applications, details of experimental technique and the basic theoretical background necessary for detailed interpretation of image contrast can be found in a recent monograph [1] and several review articles.[2-7]

2 Techniques of X-Ray Topography

Laboratory X-Radiation Techniques.—Three techniques are commonly used with conventional X-ray generators: (i) the Berg–Barrett method;[8,9] (ii) the Lang method;[10,11] (iii) the Double Crystal method.[12,13]

Berg–Barrett Topography. The essential features of the method are shown schematically in Figure 1. Using an extended source, the (single) crystal is set to diffract the characteristic radiation from a chosen set of lattice planes. The crystal is oriented either by eye using a known crystallographically oriented feature on

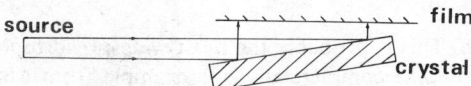

Figure 1 *The Berg–Barrett method*

---

[1] B. K. Tanner, 'X-ray Diffraction Topography', Pergamon Press, Oxford, 1976.
[2] A. R. Lang, in 'Modern Diffraction and Imaging Techniques in Materials Science', ed. S. Amelinckx *et al.*, North Holland, Amsterdam, 1970, p. 407.
[3] A. Authier, *Adv. X-ray Analysis*, 1969, **9**, 9.
[4] A. Authier, in ref. 2, p. 481.
[5] R. W. Armstrong and C. C. Wu, in 'Microstructural Analysis, Tools and Techniques', ed. J. L. McCall and W. M. Mueller, Plenum, New York, 1973, p. 169.
[6] B. J. Isherwood and C. A. Wallace, *Phys. in Technology*, 1975, **5**, 244.
[7] A. R. Lang, in 'Crystal Growth: An Introduction', ed. P. Hartman, North Holland, Amsterdam, 1973, p. 444.
[8] W. F. Berg, *Naturwiss.*, 1931, **19**, 391.
[9] C. S. Barrett, *Trans. AIME*, 1945, **161**, 15.
[10] A. R. Lang, *J. Appl. Phys.*, 1958, **29**, 597.
[11] A. R. Lang, *Acta Cryst.*, 1959, **12**, 249.
[12] U. Bonse and E. Kappler, *Z. Naturforsch.*, 1958, **13a**, 348.
[13] W. Bond and J. Andrus, *Amer. Mineralogist*, 1952, **37**, 622.

the specimen or by prior orientation using the back reflection Laue technique. A fine-grained photographic plate is used to record the diffracted beam. In principle the apparatus is extremely simple as no moving parts are required, and due to the extended source the angular range over which the crystal diffracts the characteristic line is large making precise angular positioning unnecessary. The diffracted beam can be detected on a fluorescent screen in a darkened room, but use of a photomultiplier and scintillator is preferable. There can be a considerable radiation hazard when setting up and anyone thinking of using the Berg–Barrett method would be well advised to construct a slightly more complex, but considerably safer, enclosed camera such as described by Liang and Pope.[14]

X-Ray topography in the laboratory is always bedevilled by the multiplicity of characteristic X-ray lines. Particularly troublesome is the closely spaced $K\alpha_1$–$K\alpha_2$ doublet, and in all topographic techniques some ingenuity is necessary to avoid producing double images of defects on the photographic plate. Conventionally in the Berg–Barrett method the plate is placed extremely close to the specimen. This gives high resolution as well as eliminating double images. Although the crystal diffracts both the $K\alpha_1$ and $K\alpha_2$ lines and these lines yield diffracted beams travelling in slightly different directions, the separation is negligibly small if the specimen to plate distance is small. Some care is necessary in cutting the crystal and selection of diffracting planes if high-resolution topographs of individual dislocations are required.

An important feature of the reflection Berg–Barrett method is that the X-rays only penetrate a very small distance (typically micrometres) into the crystal. One effectively examines a very thin slice of the crystal, and consequently reflection techniques can be used on crystals of much higher dislocation densities than can transmission techniques. In transmission, the overlapping of images from dislocations at different depths in the crystal limits the useful dislocation density to less than about 10^8 m^{-2} compared with about 10^{10} m^{-2} in reflection. When beginning a programme of crystal assessment, surface reflection techniques should be employed first of all.

The Lang Technique. This is probably the most widespread topographic technique and cameras are available commercially, for example from Marconi–Elliott Ltd. Simultaneous diffraction of the $K\alpha_1$ and $K\alpha_2$ lines is avoided by use of a fine-focus generator and collimating slit restricting the horizontal angular divergence of the X-ray beam to less than the angular separation of the $K\alpha_1$ and $K\alpha_2$ diffracting peaks. Only one line is then diffracted by the crystal at any angular setting. The difficulty, obvious from Figure 2, is that only a small region of the crystal is examined, and to view the whole crystal, the specimen and film must be traversed across the ribbon X-ray beam. It is the traversing mechanism which causes most trouble in the constructing of a Lang camera as the crystal, when traversing, must not twist by more than about 30 seconds of arc due to the high collimation of the beam. Further, the movement must be uniform or the resulting micrograph will show stripes across the image. A photograph of the Lang camera marketed by Marconi–Elliott is shown in Figure 3.

The traverse (or projection) topograph [12] may give an image of the whole crystal but it corresponds to an integration in space of many stationary topographs.

[14] S. J. Liang and D. P. Pope, *Rev. Sci. Instr.*, 1973, **44**, 956.

Assessment of Crystal Perfection by X-Ray Topography

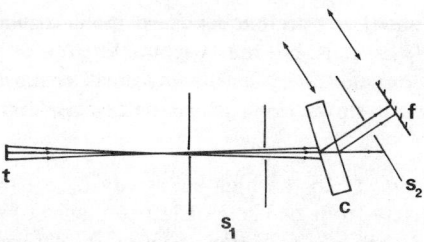

Figure 2 The Lang method. X-ray source t, collimating slits s_1, diffracted beam slits s_2, crystal c, and film f

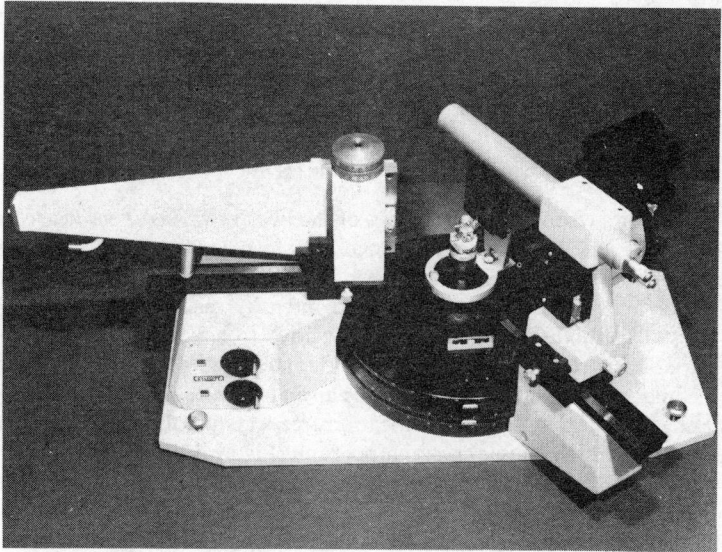

Figure 3 The Marconi–Elliott Lang camera. (Courtesy Marconi–Elliott Ltd.)

Stationary topographs of a small region of the crystal taken with a very narrow incident beam can provide detailed information on the direction of energy flow in the crystal and are known as section topographs.[11] A necessary prerequisite is that the beam width is less than the width of the fan of X-ray wavefields emerging at the exit surface of the crystal (see Figure 10).

When great care is taken to place the photographic plate within 10 mm of the crystal a geometrical resolution approaching 1 μm can be achieved. Several factors conspire to limit X-ray topographic resolution to at best 1 μm; geometrical factors, the grain size of the nuclear emulsions used for recording, the spectral linewidth of the X-ray lines, and the fundamental widths of dislocation images themselves. Great experimental care is required if this limit is to be reached, and details of experimental procedure can be found in references previously cited.[1,2] It is worth noting that no magnification is possible when taking a topograph and enlargement, typically 50 ×, of the nuclear plate image is performed on a microscope.

The Berg–Barrett and Lang techniques have many applications in studies of low dislocation density crystals, but they become less useful as the crystal quality improves. Neither technique is very sensitive to small changes in lattice parameter whose spatial variation is slow. For such materials, the double crystal technique should be used.

The Double-crystal Method. In the high sensitivity $(+-)$ parallel setting X-rays are diffracted successively from two sets of lattice planes of equal spacing first in the hkl reflection and then $\bar{h}\bar{k}\bar{l}$ reflection (Figure 4). Diffraction for the second

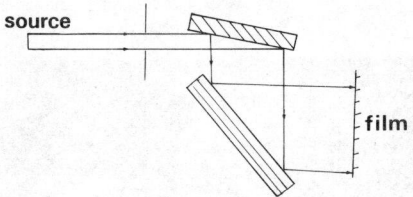

Figure 4 *The $(+-)$ parallel double crystal arrangement*

crystal only occurs when the lattice planes of the two crystals are very nearly parallel and the rocking curve, the angular range in which a strong diffracted beam is obtained, can be very narrow – typically a few seconds of arc. Local misorientations greater than this will result in *no* diffracted intensity reaching the plate, and the double crystal method can be extremely sensitive to small distortion even if the spatial variation is slow.[15] An example of the increased sensitivity of the double crystal technique over the Lang technique is given in Figure 5.

The Lang topograph of this dislocation-free crystal of silicon is featureless [Figure 5(a)], but in the double crystal topograph [Figure 5(b)] growth striations are clearly visible. Growth striations arise from small changes in lattice parameter occurring during crystal growth, generally due to fluctuations in the temperature of the crystal at the solid–liquid interface. In studies of dislocation-free low-oxygen float-zone-grown silicon (LOPEX) lattice parameter variations throughout the crystal of 2.5 parts in 10^7 were found with a spatial periodicity of 100 μm. Silicon crystals claimed free from carbon (PERFEX) were found to have lattice parameter variations of only a few parts in 10^8.[16] For identical lattice parameters in the two crystals the $(+-)$ parallel setting is non-dispersive in the sense that at any angular setting all wavelengths are diffracted. If a slight difference is made in the lattice parameters, only a band of wavelengths diffracts at any one setting and the angular sensitivity drops. However, use of very high Bragg angles and white radiation enables relative lattice parameter measurements to be made with quite high precision even when two different materials are used for the specimen and reference crystals.[17]

Use of two crystals also permits the observation of X-ray Moiré fringes. These behave similarly to optical Moiré fringes and both rotational and dilation Moiré patterns can be observed. The lattice planes of the two crystals act as the grids

---

[15] U. Bonse, 'Direct Observation of Imperfections in Crystals', ed. J. B. Newkirk and T. Wernick, Wiley, 1962, p. 431.
[16] P. J. E. Aldred and M. Hart, *Proc. Roy. Soc.*, 1973, **A332**, 223.
[17] A. Okazaki and M. Kawaminami, *Jap. J. Appl. Phys.*, 1973, **12**, 783.

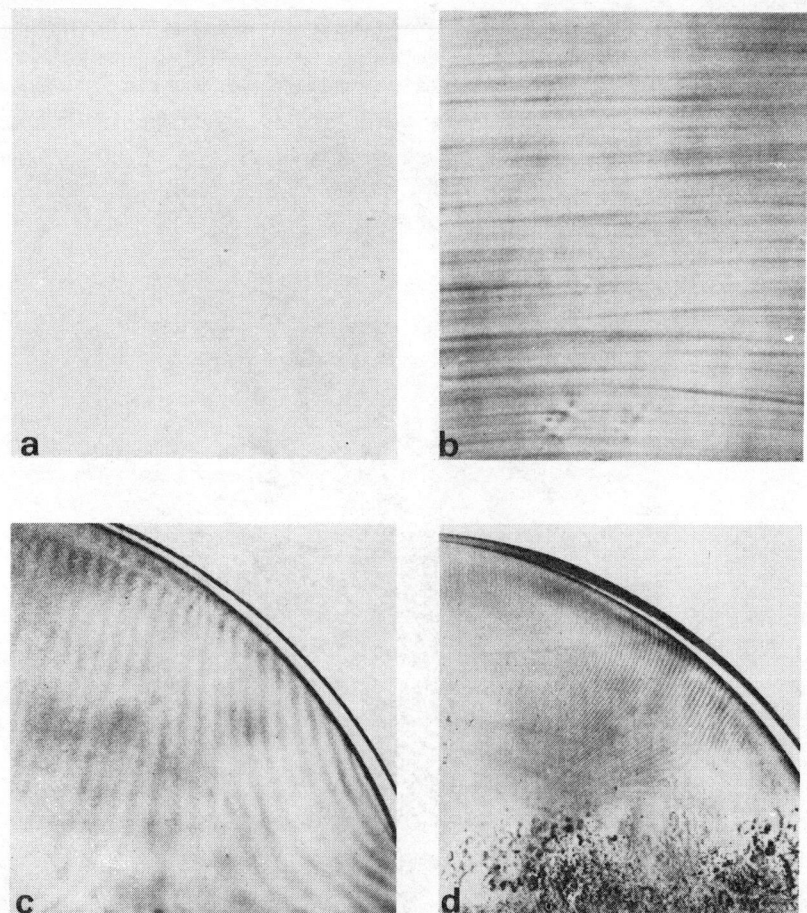

Figure 5 Comparison of the sensitivity of various techniques to lattice distortion in a highly perfect silicon crystal. (a) Lang topograph, 220 reflection, (b) double crystal topograph, 880 reflection, (c) Moiré pattern from an interferometer cut from the crystal, (d) contraction of Moiré fringes on oxidation. (Courtesy Dr. A. D. Milne)

and the Moiré spacing, D, is given by equation (2), where g_1 and g_2 are the

$$D^{-1} = |g_1 - g_2| \qquad (2)$$

reciprocal lattice vectors of the two crystals. A Moiré spacing of 1 mm corresponds to a lattice spacing difference of one part in 10^7. Bradler and Lang developed [18] the 'Moiré sandwich' in which the two crystals were separate and the second crystal was placed close to the first and oriented using a suitable jig to bring the reciprocal lattice points almost into coincidence. Figure 6(a) shows a traverse topograph of synthetic quartz taken using the Lang technique and Figure 6(b) shows the equivalent Moiré topograph. For the latter topograph the crystal was stuck to a

[18] J. Bradler and A. R. Lang, *Acta Cryst.*, 1968, **A24**, 246.

Figure 6 (Above) *Lang topograph of a synthetic quartz crystal.* (Below) *Moiré topograph of the same crystal. Field width* 1 mm.
(Reproduced by permission from A. R. Lang in 'Modern Diffraction and Imaging Techniques in Materials Science', North Holland, Amsterdam, 1970, p. 407

second (oriented) quartz crystal to produce the Moiré fringes. There is a discontinuity in the fringes at a dislocation which outcrops at the inner surfaces, between the two crystals. The number of new fringes equals $g \cdot b$, where g is the reciprocal lattice vector and b is the Burgers vector. Thus Moiré topography provides a useful means of determining the magnitude of the Burgers vector.

Moiré topographs are difficult to take in this manner, due to the problem of

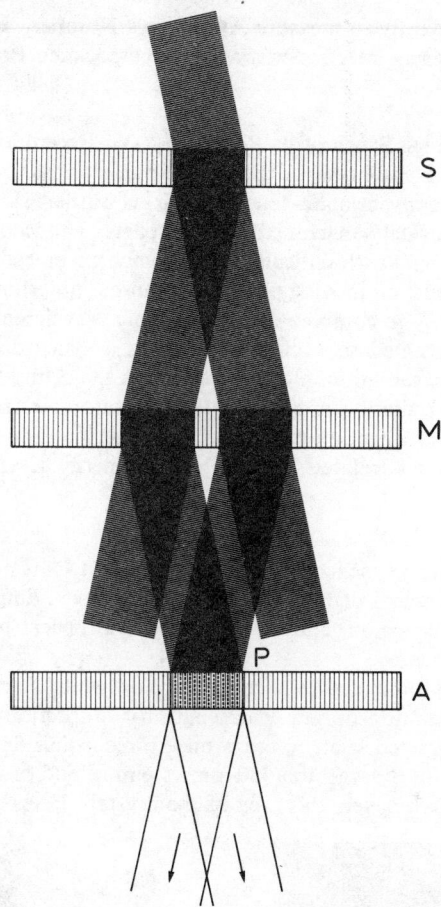

Figure 7 *The X-ray interferometer. Beam splitter*, S; *mirror*, M; *analyser*, A

bringing the reciprocal lattice points into coincidence and also retaining stability on the Ångstrom scale. Usually Moiré topographs are taken on monolithic interferometers [19] in which the lattice continuity is guaranteed by cutting the device from one single crystal using diamond loaded milling tools. The Moiré is formed between the beam splitter and analyser (Figure 7), the other section acting as a mirror to recombine the beams. Small rotations or distortions in the analyser produce Moiré patterns and, for example, interferometric studies of the defects and lattice distortions associated with ion implantation have been made.[20] Unfortunately, although it has high strain sensitivity, the interferometer does not have high spatial resolution [Figure 5(c) and (d)]. When both are required, the double crystal technique must be used. The X-ray interferometer has recently

[19] U. Bonse and M. Hart, *Appl. Phys. Letters*, 1965, **7**, 99.
[20] L. Gerward, G. Christiansen, and A. Lindegaard Anderson, *Phys., Letters*, 1972, **39A**, 63.

been used rather elegantly to measure Avogadro's Number. Hart [21] has given an excellent review of X-ray interferometry and the associated Bragg reflection X-ray optics.

X-Ray Topography using Synchrotron Radiation.—The recent availability of intense beams of synchrotron radiation in the 0.5—2 Å region has revolutionized X-ray topography. With conventional generators it is rather slow, requires critical adjustment and accurately machined moving parts, and can be combined with other measurements only with difficulty. With synchrotron radiation exposures are of the order of seconds, no moving parts are required, no critical adjustments need be made, and it is easy to combine the topographic experiment with another.

An electron constrained to a circular orbit in a synchrotron or storage ring emits electromagnetic radiation, highly polarized in the orbit plane. In the extreme relativistic limit the radiation appears in the laboratory frame as a narrow cone peaked in the forward direction tangential to the orbit. The angular divergence, θ, in the X-ray region is related to the electron energy E and rest mass m_0 by equation (3).

$$\theta \simeq m_0 c^2 / E \qquad (3)$$

For a 5 GeV electron synchrotron such as NINA at Daresbury Laboratory this is 10^{-4} rad. The spectrum of NINA is a continuum extending through the visible and u.v. to the X-ray region, and the flux is several orders of magnitude greater than conventional sources.

When a crystal is illuminated by such a source of 'white' X-radiation each crystal lattice plane selects an appropriate wavelength for diffraction and the well known Laue pattern of diffracted spots appears on the recording film. The synchrotron radiation beam is both of large area (30 mm × 6 mm) and of small divergence and the resulting Laue spot covers this area of the crystal. Each 'spot' is a topograph

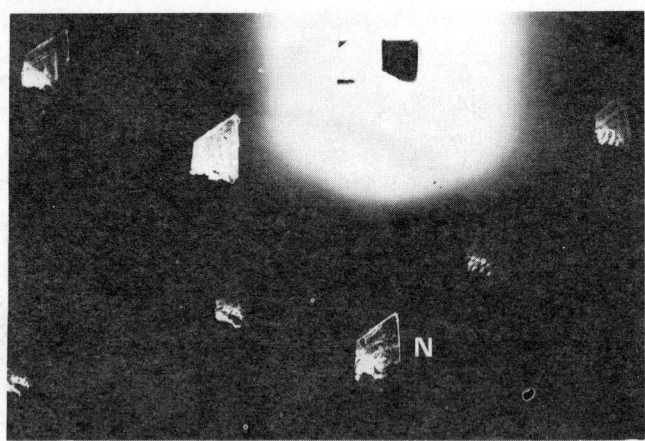

Figure 8 *X-Ray synchrotron topographs of a natural fluorite crystal. Reflection N is reproduced enlarged in Figure 17*

[21] M. Hart, *Reports Prog. Phys.*, 1971, **34**, 435.

of the crystal within which crystal defects such as dislocations and stacking faults can be observed (Figure 8). Taking synchrotron topographs is literally as easy as taking Laue photographs in the laboratory. For a specimen to plate distance as large as 100 mm a geometrical resolution of 1 μm can be obtained [22] and, as seen in Figures 9(a) and (b) the images of dislocations in transmission synchrotron topographs look very similar to images in a conventional Lang topograph.[23]

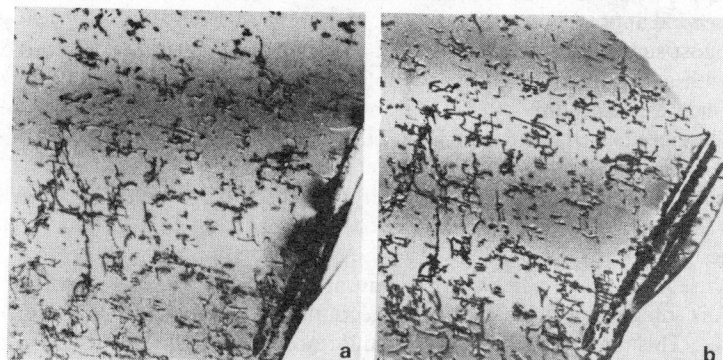

Figure 9 *Comparison of dislocation images in a wedge-shaped silicon crystal.* (a) *Lang topograph with* MoKa$_1$ *radiation.* (b) *Synchrotron topograph.* (220) *reflecting planes, field width* 2.5 mm

Many new experiments can be performed with synchrotron radiation, not only because of the great reduction in exposure time due to the high photon flux, but also because geometrical constraints imposed on the Lang technique are relaxed. No traversing mechanism is required and the specimen to plate distance can be greatly increased without catastrophic loss of resolution. Experiments such as studies of antiferromagnetic domain wall motion at temperatures down to 4.2 K and in fields of up to 1.5 Tesla become relatively easy to perform,[24] whereas they are impossible on a Lang camera using a laboratory X-ray source. Step-by-step experiments are straightforward when exposures are of the order of seconds, and in the next few years important experiments to study magnetic and electric domain motion, oxidation, and the early stages of plastic deformation are planned.

Direct Viewing of X-Ray Topographs.—Nuclear emulsion plates are very slow although giving excellent resolution, and considerable effort has been directed towards eliminating the initial photographic recording and viewing an X-ray topograph directly. Clearly this is most urgent now that with synchrotron radiation exposures on nuclear plates are of the order of seconds. While relatively rapid step-by-step experiments can be made using a simple cassette to change plates automatically,[24] truly dynamic experiments are desirable. With synchrotron

---

[22] M. Hart, *J. Appl. Cryst.*, 1975, **8**, 436.
[23] B. K. Tanner, D. Midgley, and M. Safa, *J. Appl. Cryst.*, 1977, **10**, 312.
[24] B. K. Tanner, M. Safa, D. Midgley, and J. Bordas, *J. Magnetism and Magnetic Materials*, 1976, **1**, 337.

radiation we anticipate that topographs could be recorded at about the frame speed of a normal television camera.

Two approaches have been made to the problem. In the first the X-rays are detected by a thin fine-grained or single-crystal phosphor and the resulting optical image is intensified by a high-gain image intensifier prior to viewing with a standard closed circuit TV monitor. There are variants in design, but all rely on conversion of the X-ray image into an optical one.[25,26]

The second approach is to convert the X-ray image directly into electrical signals. In the most successful system developed to date Chikawa has used a video camera fitted with a beryllium window and PbO photocathode within which the X-rays excite photoelectrons directly.[27] This system, which has a resolution of better than 15 μm has been used with a 50 kW rotating anode X-ray generator to observe dislocation movements during plastic deformation and crystal growth.[28] Similar direct conversion systems have been built using a silicon diode array camera fitted with a beryllium window. A review of developments up to 1971 has been given by Green.[29]

As all systems are at some point quantum limited it is vital for good resolution and short integration times that the detection of X-rays should be as efficient as possible. This implies either a thick fluorescent screen in an X-ray-to-optical conversion device or a thick photocathode on a direct conversion device. In the first case this itself yields poor resolution due to spreading of the image within the screen, and in the second type of device diffusion of the photo-induced carriers reduces the resolution. Clearly some compromise detector thickness has to be reached and accordingly a compromise resolution. In the quantum limited situation, resolution can be improved by increasing either the X-ray flux or the integration time, and it will depend on the experiment whether fast, but low resolution, topographs are desired in preference to relatively modest speed but higher resolution. Most X-ray topography groups do not have direct viewing systems but with the move from laboratory to the big machines producing synchrotron radiation, the development of systems will certainly accelerate.

3 Contrast on X-Ray Topographs

Perfect Crystal Phenomena – Pendellösung.—Intuitively, one expects that a traverse topograph, taken with the Lang technique in transmission, of a dislocation-free crystal should show a uniform intensity across the crystal. This is indeed found for a parallel-sided crystal [Figure 5(a)], but it is not so for a wedge-shaped crystal. In some of the earliest topographic experiments [30] fringes delineating contours of equal thickness were observed. These fringes, known as 'Pendellösung' fringes, had been predicted by Ewald [31] as early as 1916, and were the first evidence that dynamical diffraction processes were occurring in real crystals. It was quickly

---

[25] E. S. Meieran, J. K. Landre, and S. O'Hara, *Appl. Phys. Letters*, 1969, **14**, 368.
[26] A. R. Lang and K. Reifsnider, *Appl. Phys. Letters*, 1969, **15**, 268.
[27] J.-I. Chikawa and I. Fujimoto, *Appl. Phys. Letters*, 1968, **13**, 387.
[28] J.-I. Chikawa, *J. Crystal Growth*, 1974, **24/25**, 61.
[29] R. E. Green, jun., *Adv. X-ray Analysis*, 1971, **14**, 311.
[30] N. Kato and A. R. Lang, *Acta Cryst.*, 1959, **12**, 787.
[31] P. P. Ewald, *Ann. Phys.*, 1916, **49**, 38, 117.

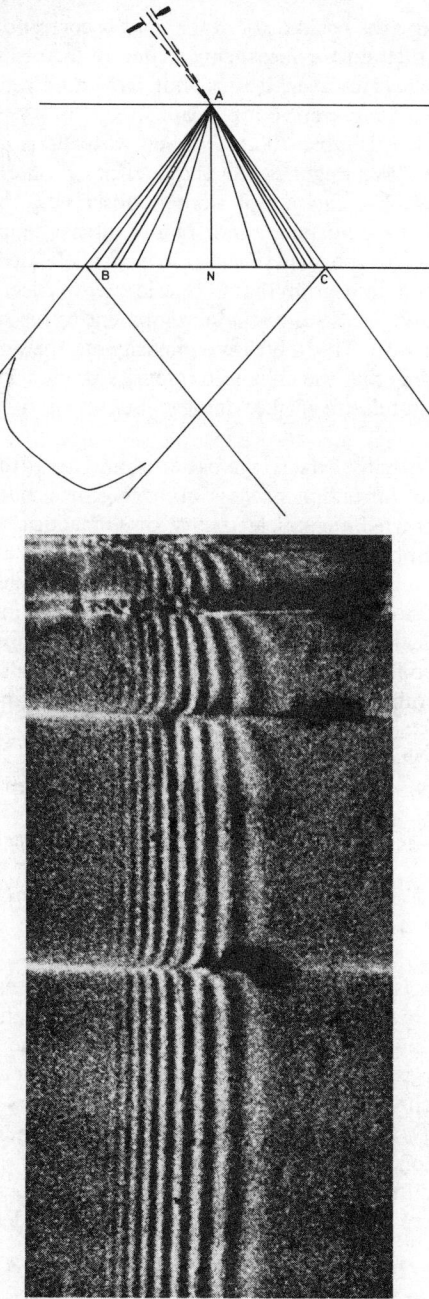

Figure 10 (Above) *The section topograph (schematic)*. (Below) *Spherical wave Pendellösung fringes in a parallel-sided crystal of silicon*
(Reproduced by permission from *Discuss. Faraday Soc.*, 1964, **38**, 292)

recognized that, because the periodicity of the fringes corresponds to one extinction distance, here was a method for measuring structure factors without the need for intensity measurements. However, it turns out that more reliable structure factor values can be obtained from section topographs.

The section topograph [Figure 10(a)] is taken with the crystal stationary and a narrow incident beam. We might expect the intensity profile on the photographic plate to be rectangular but this is not what is observed. Under low absorption conditions, the edges of the image are much more intense than the centre, and also fringes are observed across the image even from a parallel-sided plate [Figure 10(b)]. The fringes were the first indication that a spherical wave description of the incident wave was necessary and led Kato to the development of his spherical wave theory of X-ray diffraction.[32,33] The increased intensity at the margins of the image showed for the first time that the enhanced intensity previously seen at the edges of some Laue spots was not due to surface damage, but was in fact a dynamical diffraction effect.

To understand the origin of the fringe patterns, and later, the contrast of defects, we must consider the diffraction process occurring in a thick, effectively perfect crystal. The well-known kinematical theory of diffraction is inadequate in this situation as a very important inherent assumption is that the amplitudes of all diffracted beams are at all times small compared with the incident beam amplitude. This works well for small crystals, or crystals broken up into mosaic blocks by defects, but in a thick, highly perfect crystal there is an appreciable probability of scattering not just from incident to diffracted beams but also vice versa. In the dynamical theory of diffraction this interaction of incident and diffracted beams is taken into account.

Basically, the problem is to solve Maxwell's equations in a medium with a periodic susceptibility. One finds that solutions of the form (4) satisfy, provided

$$D = \exp i\omega t \sum_g D_g \exp(-2\pi i K_g \cdot r) \qquad (4)$$

that the wavevectors K_g are related by the Laue equation (5). We see that solutions

$$K_g = K_0 + g \qquad (5)$$

inside the crystal consist of a superposition of plane waves each corresponding to a diffracted wave. Now, in principle, we could study the behaviour of these component plane waves but this can prove troublesome as the plane waves are not normal modes. Energy is interchanged between the diffracted waves as the X-rays propagate through the crystal and the amplitudes vary with depth. It is this interchange of energy which leads to the Pendellösung interference. We should study the properties of the *total wavefield*.

Usually in X-ray diffraction only one reciprocal lattice vector contributes to the wavefield and we have a wavefield (neglecting time dependence) of the form (6).

$$D = D_0 \exp(-2\pi i K_0 \cdot r) + D_g \exp(-2\pi i K_g \cdot r) \qquad (6)$$

---

[32] N. Kato, in 'Crystallography and Crystal Perfection', ed. G. N. Ramachandran, Academic Press, New York, 1963, p. 153.
[33] N. Kato, in 'X-ray Diffraction', ed. L. V. Azaroff, McGraw Hill, New York, 1974, p. 295.

Substitution of this equation back into Maxwell's equations yields a relation between K_g and K_0 and an expression for the amplitude ratio D_g/D_0 in terms of K_g or K_0. The relation between K_g and K_0, when plotted graphically, is known as the dispersion surface. Far from a Bragg reflection there is only one wave excited in the crystal and the wavevectors lie on spheres about the reciprocal lattice points 0 or G of radius $|K_0| = nk$, where k is the vacuum wavevector and n the refractive index. Close to a Bragg reflection the crystal potential raises the degeneracy, and the dispersion surface has two branches for each polarization corresponding to D_g/D_0 positive or negative (*i.e.* the component plane waves either in phase or phase shifted by π). A section through the dispersion surface near to Bragg reflection is shown in Figure 11. Each point (tie point) on the dispersion surface uniquely defines one

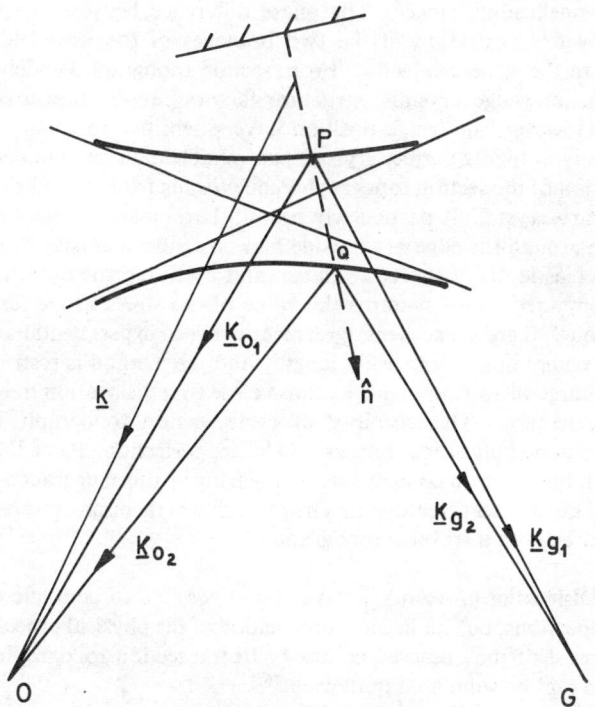

Figure 11 *The dispersion surface construction. Boundary conditions for the Laue geometry are shown*

wavefield within the crystal and it can be shown that the direction of propagation of any wavefield is determined by the normal to the dispersion surface at its corresponding tie point. Across any boundary, the normal component of the wavevector must be continuous and it follows that the wavefield excited within the crystal by an incident plane wave can be determined by drawing the normal to the crystal surface through the tip of the wavevector in vacuo. This has been done for the two tie points P and Q in Figure 11. An incident plane wave excites two wavefields, and we see that their direction of propagation is not generally in the same direction. If the incident plane wave is laterally limited in extent it should be possible to separate the two wavefields. This was first done experimentally by Authier[34] and confirmed that wavefields were not mathematical fictions but had physical reality.

However, it is only under very special circumstances that it is possible to describe the incident wave as a plane wave because usually the divergence of the incident beam is large compared with the perfect crystal reflecting range. In taking a section topograph (Figure 10) all tie points across the dispersion surface are excited coherently and the whole triangle ABC is filled with wavefields. In any direction, two wavefields propagate – one from branch 1 and the other from branch 2. Their associated wavevectors differ and they interfere. As we go along the base BC the phase difference between the arriving wavefields differs and we observe the section topograph Pendellösung fringes. The phase difference between the wavefields is determined by the separation of the two branches of the wavefield and this is proportional to the structure factor. From section topograph Pendellösung fringe measurements in wedge crystals, structure factors can be measured with high accuracy.[35] However, the fringe position is very sensitive to small strains in the crystal and only in highly perfect crystals can reliable data be obtained.

The distortion of the section topograph Pendellösung fringes can be used to study small strains in a crystal. A particularly beautiful example has been in the study of the distortion around the edge of an oxide film on a silicon crystal.[36] Here, a good model of the elastic distortion is available, and by fitting the experimental fringe pattern to computed fringe patterns the value of the stress in the film was determined. Although there is excellent agreement between experimental and simulated patterns, the computing is somewhat lengthy and the method is restricted to small distortions. Large distortions, such as those close to a dislocation line, destroy the Pendellösung pattern. The visibility of clear section topograph Pendellösung fringes is a very good indicator of high crystal lattice perfection. Patel[37] has elegantly shown how, in high oxygen content silicon, the fringe pattern degraded as precipitation occurred during heat treatment. The precipitates themselves were too small to be revealed directly in a traverse topograph.

Contrast of Dislocations.—It has not yet been possible to compute the complete images of dislocations, but sufficient is understood of the physical processes involved to interpret most of the observed contrast. In the section topograph, dislocation images can in fact be simulated quite well.[38]

Let us consider, as in Figure 12, the processes occurring when a dislocation line cuts through the fan of wavefields set up when taking a section topograph. (The traverse topograph is made up of a superposition of section topographs and so it is straightforward to generalize from the section to the traverse topograph.) In the vicinity of D the dislocation line cuts the direct beam AC which contains those X-rays which are not diffracted by the perfect crystal. Only X-rays within a narrow angular range $\Delta\theta_{\frac{1}{2}}$ (typically 10^{-5} rad) are diffracted by the perfect crystal and the

[34] A. Authier, *Bull. Soc. franç Mineral Crist.*, 1961, **84**, 51.
[35] N. Kato, *Acta Cryst.*, 1969, **A25**, 119.
[36] Y. Ando, J. R. Patel, and N. Kato, *J. Appl. Phys.*, 1973, **44**, 4405.
[37] J. R. Patel, *J. Appl. Phys.*, 1973, **44**, 3903.
[38] Y. Epelboin, *J. Appl. Cryst.*, 1974, **7**, 372.

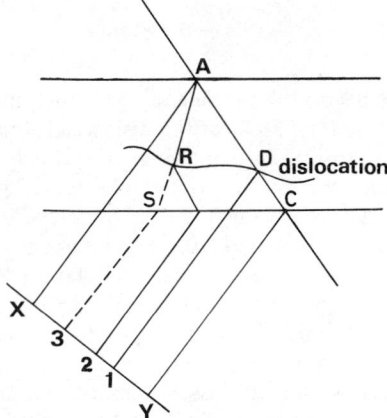

Figure 12 *Dislocation image formation in X-ray topographs.* 1, *direct image*, 2, *intermediary image*, 3, *dynamical image*

bulk of the intensity in the beam (of divergence typically 5×10^{-4} rad) is thus not diffracted. However, around the dislocation line, the lattice is distorted or misoriented and some of the X-rays not diffracted by the perfect crystal are diffracted by the strained region. In the small misoriented volume surrounding the dislocation the scattering is essentially kinematic and these X-rays do not suffer primary extinction as they pass through the perfect crystal. There is thus a net increase in intensity at the point 1 corresponding to this 'direct' or 'kinematical' image of the dislocation. The direct image begins to be formed when the effective misorientation around the dislocation exceeds roughly $\Delta\theta_{\frac{1}{2}}$, the full width at half maximum of the perfect crystal reflecting range.

When g is parallel to b, the misorientation $\delta(\Delta\theta)$ at a distance r from the core of a screw dislocation is approximately given by equation (7). The perfect crystal

$$\delta(\Delta\theta) = \frac{b}{2\pi r} \tag{7}$$

reflecting range in the symmetrical Laue condition is given by equation (8), where

$$\Delta\theta_{\frac{1}{2}} = 2/g\xi_g$$
$$= 2r_e \lambda^2 C(F_g F_{\bar{g}})^{\frac{1}{2}}/\pi V_c \sin 2\theta_B \tag{8}$$

ξ_g is the extinction distance related to the structure factor F_g, r_e is the classical electron radius, V_c the volume of the unit cell and $C = 1$ or $\cos 2\theta_B$ for σ and π polarization, respectively.

We see that the direct image width ($V_s = 2r$) is given by equation (9). For an

$$\frac{2}{g\xi_g} = \frac{b}{\pi V_s}, \quad \text{or generally,} \quad V_s = \frac{\xi_g}{2\pi}|\,g \cdot b\,| \tag{9}$$

edge dislocation, the corresponding relation is equation (10).[2,39] Typical X-ray

[39] A. R. Lang and M. Polcarova, *Proc. Roy. Soc.*, 1965, **A285**, 297.

$$V_E = \frac{0.88}{\pi}\xi_g|\,g\cdot b\,| \tag{10}$$

extinction distances are of tens of micrometres, so direct image widths are usually upwards of a few micrometres. Most of the dislocation images on traverse topographs taken under low absorption conditions are direct images and appear as regions of increased intensity on the topographs (see Figure 9a for example). Experimentally measured image widths are in quite reasonable agreement with values predicted using equations (9) and (10). Measurement of the image width on a topograph provides a means of checking the magnitude of the Burgers vector b. We note that the direct image contrast arises from the long-range strain field, and defects such as small dislocation loops do not give rise to strong direct images. They thus often pass undetected on traverse topographs.

Under certain conditions, the dislocation images become invisible and this provides, as in electron microscopy, an invaluable means of determining the direction of the Burgers vector. The effective misorientation of any dislocation is given by equation (11),[3] where $\partial/\partial s_g$ is the partial derivative in the diffracted beam direc-

$$\delta(\Delta\theta) = (\lambda/\sin 2\theta_B)[\partial/\partial s_g(g\cdot u)] \tag{11}$$

tion and u is the atomic displacement around the dislocation. For a general dislocation we can write equation (12), where b_e is the edge component of Burgers vector

$$u(r,\phi) = \frac{1}{2\pi}\left\{b\phi + b_e\frac{\sin 2\phi}{4(1-\nu)} + b\times l\left[\frac{(1-2\nu)\ln|r|}{2(1-\nu)} + \frac{\cos 2\phi}{4(1-\nu)}\right]\right\} \tag{12}$$

and ν is Poisson's ratio. For a pure screw dislocation the effective misorientation is zero when $g\cdot b = 0$ and for a pure edge dislocation it is zero when both $g\cdot b = 0$ and $g\cdot b \times l = 0$. In reflections satisfying these conditions, the dislocation is invisible. By finding two reflections with non-parallel diffracting vectors in which the dislocation is invisible one has an unambiguous determination of the direction of the Burgers vector b. Mixed orientation dislocations never completely disappear but often the second and third terms in equation (12) are small compared with the first and in reflections where $g\cdot b = 0$, only weak residual contrast appears.

The (black) direct image enables the direction and, in favourable cases, the magnitude of the Burgers vector to be determined. Unfortunately the direct image is insensitive to the sense of the Burgers vector and it is often only with difficulty that this information can be obtained.

The direct image is not the only image formed. Consider the two wavefields propagating along AR in Figure 12. On crossing the dislocation the dynamical diffraction conditions break down and the wavefields de-couple into their plane wave components. On re-entering perfect material below the dislocation the original wavefields excite new wavefields travelling in new directions. In the direction AS there is a loss of intensity and the dislocation casts a shadow which appears as a light region at 3. This is known as the 'dynamical image' and is the light banana-shaped image in Figure 12. It tends to be rather diffuse and spread out within the region XY on the plate but in certain conditions it can be observed on

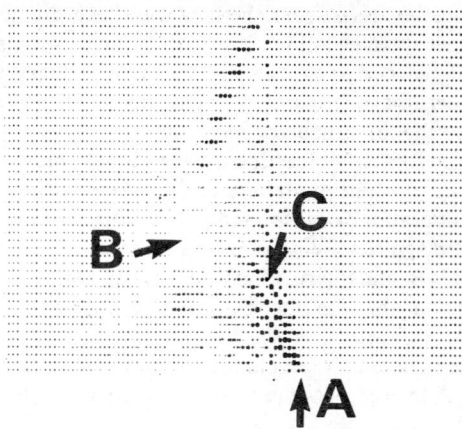

Figure 13 *Simulated image of a dislocation in a section topograph.* (Courtesy Dr. Y. Epelboin)

traverse topographs. One is in relatively low absorption conditions for particular dislocation geometries and the white images seen in Figure 9 are 'dynamical' images. The other situation where the dynamical image is important is under high absorption conditions.

Here, effectively all X-rays not undergoing dynamical diffraction are absorbed and hence no direct image is formed. However, to the surprise of the early investigators the dynamically diffracted X-rays undergo 'anomalous transmission'. Referring back to equation (6) we see that the total intensity in the wavefield is given by equation (13). The wavefield is spatially modulated with a periodicity

$$I = D^* \cdot D = D_0^2 + D_g^2 + 2D_0 D_g C \cos(2\pi g \cdot r) \tag{13}$$

equal to the Bragg plane spacing. On one branch of the dispersion surface the wavefield has nodes at the atom planes, on the other, antinodes. As absorption occurs *via* the photoelectric effect that wavefield with maximum intensity at the atomic planes is preferentially absorbed, the other is preferentially transmitted. One wavefield therefore experiences a very low effective absorption, and in highly perfect crystals quite intense transmitted beams are found where a reduction in intensity by a factor of 10^{13} is expected.

However, the anomalous transmission or Borrmann effect occurs only in perfect crystals, and the presence of a dislocation destroys the transmission. Again, the dislocation casts a shadow, and under high absorption conditions the dynamical image can have quite high spatial resolution if the dislocation is near the exit surface of the crystal. Destruction of anomalous transmission was one of the earliest methods by which dislocations were directly observed [40] and continues to be used as a useful means of studying thick, highly perfect crystals in the as-grown state.[41]

---

[40] G. Borrmann, W. Hartwig, and H. Irmler, *Z. Naturforsch.*, 1958, **13a**, 423.
[41] B. K. Tanner and S. H. Smith, *J. Crystal Growth*, 1975, **28**, 77.

A third type of image arises from the new wavefields created below the dislocation line. These new wavefields interfere with original wavefields not cutting the dislocation line and give rise to a series of interference fringes which lie between the dynamical and direct image (Figure 13). This image formed by the newly created wavefields is termed the 'intermediary image'.

Using a generalized form of diffraction theory it is possible to simulate the image of a dislocation on a high-speed digital computer. As the direct image is effectively excluded from consideration because it is highly localized (A in Figure 13) most of the section topograph image represents the dynamical and intermediary images. The simulated dynamical (B) and intermediary images (C) agree well with the experimental images. It turns out that the intermediary image is very sensitive both to sense and magnitude of the Burgers vector.[38] Comparison of the experimental micrograph with several simulated images enables the identification to be made. Unfortunately, it has not yet proved possible to simulate images in traverse topographs.

Contrast of Stacking Faults.—Rather less work has been done on stacking faults using X-ray topography, probably because stacking faults sufficiently large to be analysed by the technique are of high energy and therefore rather rare. Contrast arises from stacking faults due to the fault effectively introducing a phase shift in all the wavefields crossing it. At a stacking fault a translation of the lattice occurs by a vector R. On crossing the fault the wavefields decouple into their plane wave components. They re-enter perfect material immediately and create new wavefields. Just as any plane wave travelling in a vacuum excites two new wavefields (Figure 11), each component wave excites two wavefields. If we consider a point X on the stacking fault (Figure 14a) we find that two wavefields arrive and on crossing the fault excite new wavefields, two of which propagate on along AX, the other two propagate in a new direction XP. At a point P on the exit surface we have interference effects between the new waves propagating along XP, between original type waves arriving along AP which were not deviated on crossing the fault, and between the two sets, *i.e.* three interference terms. When an inclined fault cuts the whole fan of excited wavefields in a section topograph a rather delightful hour-glass image is

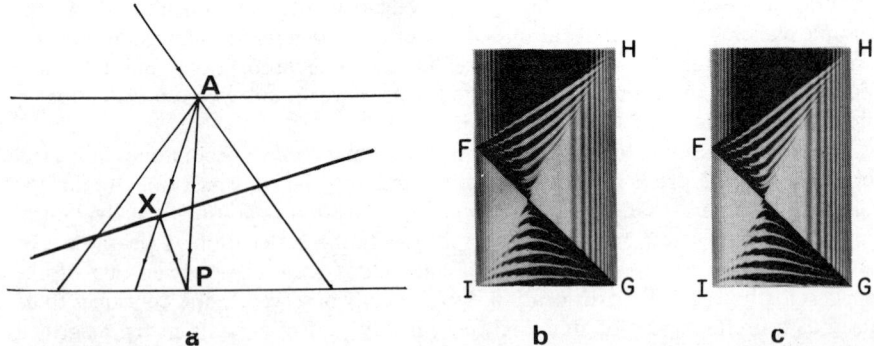

Figure 14 (a) *Paths of wavefields in a crystal containing a stacking fault.* (b) *Computer simulation of an intrinsic fault in a section topograph.* (c) *Simulation of an extrinsic fault under the same conditions.* (Courtesy Dr. J. R. Patel)

formed (Figures 14b and c). The fringe systems simulated on a computer (such as is shown here) are in excellent agreement with experimental images.[4,42,43]

In traverse topographs the fringe systems are not as pronounced but under suitable conditions can be observed. The integration on going from the section to the traverse topograph blurs the fringe with subsequent loss of information.

Stacking faults can be analysed in much the same way as dislocations. The magnitude of the fault vector can be determined by studying in which reflections the fault becomes invisible. In Figure 15a, for example, stacking faults parallel to the specimen surface are visible whereas in Figure 15b they disappear. The fault introduces a phase shift of $\alpha = 2\pi \, \mathbf{g} \cdot \mathbf{R}$ and when $\mathbf{g} \cdot \mathbf{R} = 0, 1, 2$ *etc.* the effective shift is zero and the fault invisible. Here the faults are invisible when $\mathbf{g} = \langle 11\bar{2}0 \rangle$

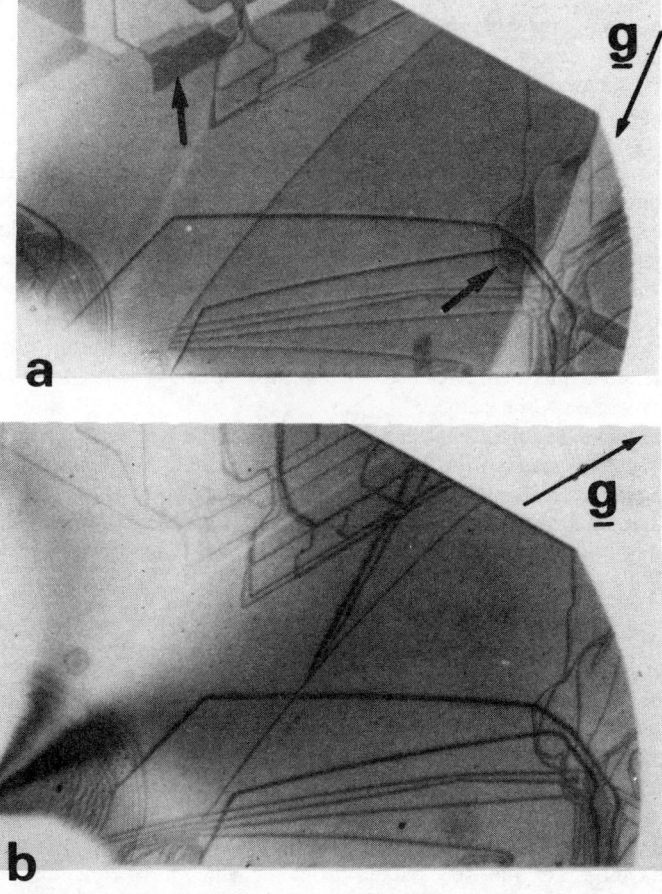

Figure 15 *Stacking fault images in a traverse topograph of a crystal of* SnS_2. (a) $\mathbf{g} = [\bar{1}100]$, (b) $\mathbf{g} = [1\bar{2}10]$. *Field width* 2.8 mm

[42] A. Authier, *Phys. Stat. Sol.*, 1968, **27**, 77.
[43] J. R. Patel and A. Authier, *J. Appl. Phys.*, 1975, **46**, 118.

and we deduce $R = \frac{1}{3} \langle 1\bar{1}00 \rangle$.[44] The actual direction of any fault vector can often be best determined by the Burgers vector of the bounding partial dislocations (D). It turns out however that the type of stacking fault, whether intrinsic of extrinsic is often of great importance. To determine this we must determine the sense of R and here one should go to the section topograph. The interference fringes such as pictured in Figure 14b are seen to be asymmetrical in contrast, there being a dark fringe at the bottom and a light fringe at the top. The sense of the last fringe depends on the sense of the phase shift $g \cdot R$, and the sign of the last fringe is opposite between intrinsic and extrinsic faults (Figures 14b and c). Although the nature of a stacking fault can, under certain conditions, be determined from the contrast of the last fringe in the traverse topograph,[45] the result can be ambiguous and the section topograph should be used whenever possible.[43]

Contrast of Twins and Subgrains.—While lamellae twins can give rise to contrast effects similar to those from stacking faults,[46] twins and subgrains are usually observed in X-ray topographs by a rather different mode. All the contrast effects discussed previously can be termed 'extinction contrast', those discussed here 'orientation contrast'. Across a twin or subgrain boundary, the lattice planes are misoriented. If the Lang technique is used, when the misorientation is greater than about 1 minute of arc only one twin or subgrain will be diffracted at any one setting. By measuring the angular separation between the two peaks in the rocking curve, the misorientation can be measured. In a Berg–Barrett or synchrotron topograph, where we have either an extended characteristic line source or a spectral continuum, simultaneous diffraction does take place. However, the directions of the diffracted beams from the two regions differ and either overlap or separation of the beams occurs with subsequent enhancement or loss of contrast. The subgrain boundaries show up as white or black lines.[22]

As with stacking faults, twins can be characterized by observation of the reflections in which the image of the boundary is absent, and one can conclude that there the Bragg planes are continuous. The nature of the twin can then be elucidated. Such studies are particularly important on 90° ferroelectric and ferromagnetic domains, both of which behave as twin boundaries in X-ray topographs.

4 Applications to Crystal Defect Studies

Crystal Growth.—From the earliest stages of its development it was clear that X-ray topography was ideally suited to studies of dislocation configurations in as-grown crystals. The technique is capable of mapping large volumes of material and has the advantage (at least in transmission) of revealing the dislocation configuration inside the crystal, unlike, for example, etch pit observations. In many cases it has proved possible to examine a whole crystal in the as-grown state without there being any need for cutting or polishing.

There have been two loosely related problems which have conspired to slow the rate of application of X-ray topography. First, in order to resolve dislocation

---

[44] H. P. B. Rimmington, A. A. Balchin, and B. K. Tanner, *J. Crystal Growth*, 1972, **15**, 51.
[45] A. R. Lang, *Z. Naturforsch.*, 1972, **27a**, 461.
[46] A. Authier, A. D. Milne, and M. Sauvage, *Phys. Stat. Sol.*, 1968, **26**, 469.

images, highly perfect crystals are needed and in the early days relatively few materials met this criterion. Earliest studies were on synthetic melt grown semiconductor and oxide crystals and natural crystals. Only aluminium amongst the metals was of sufficiently low dislocation density for serious X-ray topographic study.[47] Secondly, there has been a tendency for the crystal grower and X-ray topographer to work independently rather than as a team. Topographers have characterized crystals but the feed-back of information to the crystal grower was poor and the crystal growers continued for a long time to develop their art along traditional lines. However, during the 1960s crystal quality greatly improved and the characterization studies built up a considerable body of data on dislocation configurations. Some ideas as to the origin of the observed configurations began to emerge, and with the emphasis in crystal growth swinging from art to science, subsequent progress, particularly in the 1970s has been rapid.

In the latter half of the 1960s some very important observations were made on solution grown crystals. It was discovered that very low dislocation density crystals could be grown from aqueous solution using very unsophisticated apparatus. Provided that care was taken to avoid temperature or compositional fluctuations, pure materials were used and growth was slow, low dislocation density crystals could be grown, if not in the back kitchen, on any laboratory bench. The observations on cubic crystals, namely NaCl[48] and hexamethylene tetramine[49] showed rather similar configurations such as sketched in Figure 16a. Dislocations were predominantly generated at the seed and ran outwards in straight lines. They tended to bunch and run normal to the growth faces with the result that quite large volumes of crystal free from dislocations appeared as growth continued (Figure 16b). Only occasionally were dislocations nucleated during the growth, usually at inclusions of the mother liquor, and ran in pairs outwards to the growth face. It had generally been assumed that such crystals were good examples of 'mosaic crystals' in which kinematical diffraction theory was valid and these first results came as something of a surprise to crystallographers. Somewhat similar configuration of dislocations nucleated at the seed and running outwards in straight lines were also observed in natural diamonds.[50]

It was, however, the work of Klapper on organic crystals grown from aqueous solution which led to an understanding of the significance of these configurations. Klapper found that in elastically anisotropic materials dislocations again ran outward from the seed in straight lines but that the directions of the lines were in many cases non-crystallographic and made large angles with the normals to the growth faces. He postulated that, in order to minimize the free energy of the crystal, dislocation lines preferred to run in such directions as to minimize their elastic line energy per unit growth length. This is equivalent to the condition that the dislocations run in a direction in which the force on them by the free surface is zero. In a number of studies the elastic line energy of dislocations with various Burgers vectors was computed using anistropic elasticity theory. Using X-ray topography, the Burgers vectors and line directions of dislocations in a number of organic

---

[47] A. R. Lang, *J. Appl. Phys.*, 1959, **30**, 1748.
[48] S. Ikeno, H. Maruyama, and N. Kato, *J. Crystal Growth*, 1968, **3/4**, 683.
[49] R. A. Duckett and A. R. Lang, *J. Crystal Growth*, 1973, **18**, 135.
[50] A. R. Lang, *J. Crystal Growth*, 1974, **24/25**, 108.

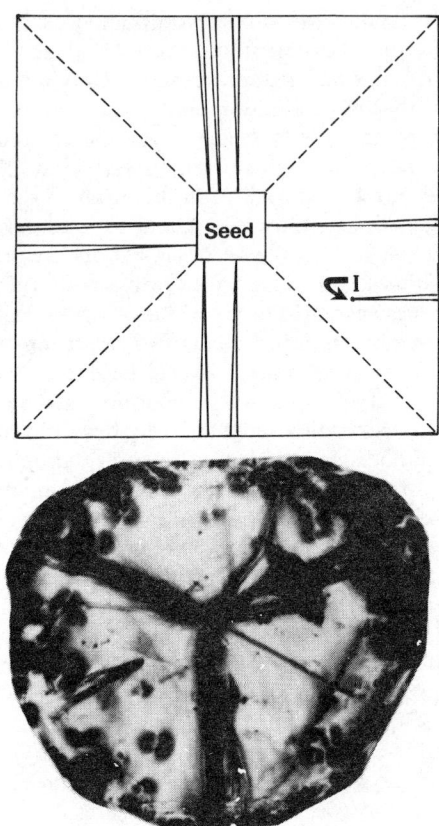

Figure 16 (Above) *Schematic diagram of the dislocation configuration in a cubic crystal grown from aqueous solution. Inclusion I nucleates dislocations.* (Below) *Traverse topograph of a potash alum crystal 3 mm wide*
(Reproduced by permission from *Phil. Mag.*, 1969, **19**, 7)

crystals were examined. The observed and calculated line directions were in excellent agreement in most cases [51,52] In crystals where the Peierls energy was high some discrepancies were found as the theory assumed continuum elasticity theory and neglected variations of dislocation core energy with direction [51] An interesting aspect of the theory is that it accounts for the abrupt 'refraction' of dislocation lines quite commonly observed when dislocations cross a boundary between regions of crystal growing on different faces. The line direction changes in order to minimize the elastic line energy per unit growth length for the new growth direction.

The theory assumed even greater significance when it was found that the line directions of dislocations in hydrothermally grown quartz could be explained using the minimum energy criterion. Further, excellent agreement was found for

[51] H. Klapper and H. Kuppers, *Acta Cryst.*, 1973, **A29**, 495.
[52] H. Klapper, *Z. Naturforsch.*, 1973, **28a**, 614.

natural crystals of calcite, topaz, and fluorite.[53] Very recently, studies of crystals of $KNiF_3$ and $KCoF_3$ grown from the flux at high temperatures have shown that the same theory can be used successfully to predict dislocation line directions in these crystals.[54] The basic concept seems to underlie the whole of solution growth despite the apparent disparity between regimes. This work must have important consequences in crystal growth both theoretically and experimentally in the future.

Where a substantial amount of dislocation climb takes place, dislocations do not lie in straight lines. Climb occurs by addition or subtraction of vacancies or interstitials to the dislocation core and results in the dislocation leaving its slip plane. In the early studies of aluminium, long rows of coaxial dislocation loops were observed which were shown by Burgers vector analysis to be prismatic.[55] The authors concluded that the loops formed first by vacancy condensation onto screw dislocations, resulting in the screw dislocation climbing into a helix and followed by an interaction of the helices with themselves or other defects to form the prismatic loops. It was subsequently shown that the dislocation density in strain-anneal grown aluminium was a direct function of the cooling rate.[56] Later experiments [57] have, however, cast doubts on the mechanism of formation of the prismatic loops. There is no doubt that the bulk of the configurations can be described by dislocation climb, and probably the most spectacular climb effects occur in unpublished work by G'Sell (see ref. 1). Dislocations in cadmium were observed to climb into spirals on cooling, and on subsequent heating, the spirals unwound. Recently, several groups have developed heating stages for Lang cameras which enable such heating experiments to be performed and this has greatly increased the possibilities for topographic studies.

Metallurgy and Plastic Deformation.—Despite the great effort which has gone into attempting to understand the very complex processes occurring during plastic deformation, primarily by use of transmission electron microscopy, there are several areas where very fundamental questions remain unanswered. One of these is in the very early stages of plastic deformation and concerns the initial nucleation and multiplication of dislocations. It is a problem which, at first sight, seems ideal for X-ray topography but is in fact fraught with experimental difficulties. First, by definition, metallurgists are concerned with metals and only with extreme care has it proved possible to produce low dislocation density single crystals of metals such as copper, zinc, cadmium, or aluminium. Secondly, it is even more difficult to mount the crystals, now in the form of thin foils typically 100 μm thick, in a suitable straining jig without accidentally introducing dislocations. All the metals mentioned are extremely soft and stress concentrations at the grips of the straining stage can be as much a problem as stress concentrations at tweezers. A third problem is that the dislocation density rises very rapidly as deformation proceeds and only a very small part of the stress–strain curve can be sampled. Further, with conventional generators, experiments take a long time. The number of systematic straining

---

[53] Y. Epelboin, A. Zarka, and H. Klapper, *J. Crystal Growth*, 1973, **20**, 103.
[54] M. Safa, B. K. Tanner, H. Klapper, and B. M. Wanklyn, *Phil. Mag.*, 1977, **35**, 811.
[55] A. Authier, C. B. Rogers, and A. R. Lang, *Phil. Mag.*, 1965, **12**, 547.
[56] E. Nes and B. Nøst, *Phil. Mag.*, 1966, **13**, 855.
[57] C. G'Sell, B. Baudelet, and G. Champier, *J. Crystal Growth*, 1972, **13/14**, 252.

studies is rather small and the nature of the dislocation sources remains a mystery in most cases. It has, however, been shown in copper that grown-in dislocations do not act as sources of dislocations but that dislocations are nucleated at the edges of the crystal, though the sources were not identified.[58] More recent experiments have confirmed that the slip bands thicken during deformation by cross slip and that the predominant type of dislocation was determined by the specimen geometry rather than the Schmidt factor.

In aluminium, on the other hand, it has been shown that grown-in dislocations can act as dislocation sources.[59] Under an applied stress, the end segments of dislocations outcropping at the specimen surface were induced to break away. Subsequently these dislocations extended to the edge of the crystal and acted as surface sources at which dislocation multiplication took place.[60]

Another area in which X-ray topography can be useful to the metallurgist is in dislocation velocity measurements. Unlike etch pitting techniques, the whole length of the dislocation is observed and changes of shape can be studied and any dependence of velocity on Burgers vector determined. To date the technique has not been applied to metals and the most reliable data have come from high-temperature studies of silicon.[61]

Mineralogy.—During crystal growth, small changes in temperature or composition of the liquid can cause small variations in lattice parameter in the crystal, presumably due to changes in the amount of impurity incorporated in the lattice. These changes

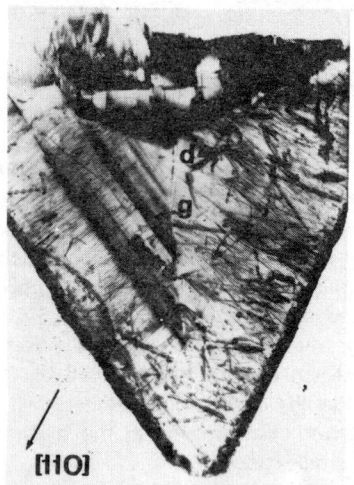

Figure 17 X-Ray synchrotron topograph of a natural fluorite crystal. Note many dislocations d nucleated at inclusions and running almost normal to the growth faces. Only one set of growth bands g is visible in this reflection. Field width 7 mm

[58] F. W. Young, jun. and F. A. Sherrill, *J. Appl. Phys.*, 1971, **42**, 230.
[59] O. Rustad and O. Lohne, *Phys. Stat. Sol. (A)*, 1971, **6**, 153.
[60] O. Lohne and O. Rustad, *Phil. Mag.*, 1972, **25**, 529.
[61] A. George, C. Escaravage, G. Champier, and W. Schröter, *Phys. Stat. Sol. (B)*, 1972, **53**, 483.

in lattice parameter can often be detected by X-ray topography in the same way that stacking faults can be visualized. Sometimes, if the lattice parameter change is very great, direct image contrast can also be observed due to unaccommodated strains present in that region.

Several interesting studies have been made relating growth banding to impurity content. In fluorite, a good correlation has been found between bands of colour and growth bands.[62] The colour centres are believed to arise from yttrium impurities and the correlation with some kind of fluctuation in the growth conditions has strengthened this idea. A correlation between bands of birefringence and growth bands has been found in fluorite (Figure 17). In the crystal shown it has been found that the major growth bands also correlate with edges of the coloured regions and that the growth sectors show quite strong birefringence which shows that the lattice is distorted from cubic symmetry by impurity inclusion. Again, the birefringence bands and the growth bands coincide confirming that all effects have a common origin. The dislocation bundles revealed in the X-ray topographs can also be identified in the birefringence micrograph. The growth bands provide a record of the growth history of the crystal, and this has important implications in mineralogical studies where data on the past morphology of a crystal can be of value in attempting to determine the growth conditions of the crystal. There is much potential in careful section by section studies through natural crystals, from which the detailed shape of the crystal throughout its growth can be reconstructed. Either the crystal can be cut and projection topographs taken, as in topaz,[63, 64] or section topographs can be used to make effective sections through the whole crystal. In studies on diamond,[50] the latter technique is preferable!

Control of Electronic Devices.—One of the basic applications of X-ray topography in the electronics industry has been the characterization of crystals from which devices are subsequently fabricated. This X-ray topographic analysis of synthetic quartz crystals prior to the cutting of transducers has become common. It turns out that transducers containing a growth sector boundary perform badly. Similarly, Lang and double crystal topographs have been used to monitor the gadolinium gallium garnet substrates and the epitaxial films on which magnetic bubble devices are produced. The presence of dislocations and major lattice parameter fluctuations in growth bands can cause non-uniform propagation of the magnetic bubble domains and impair device performance. In silicon devices, however, X-ray topography can also be used to monitor the total production process.

Integrated circuits are manufactured on a single 2 in diameter disc of silicon by successive diffusions through oxide masks into which the pattern of the circuit has been inscribed using photo-masking techniques. During the diffusion processes to produce successive n and p type regions the crystal lattice is distorted due to the mismatch in the ionic radii of the silicon and impurity atoms. When the dopant concentration becomes too large, dislocations can be generated which often impair device performance. For example, a correlation was found between the leakage current in diodes and the presence of dislocations extending through the junction

---

[62] G. Calas and A. Zarka, *Bull. Soc. franç Mineral Crist.*, 1973, **96**, 274.
[63] C. Giacovazzo, E. Scandale, and A. Zarka, *J. Appl. Cryst.*, 1975, **8**, 315.
[64] M. Isogami and I. Sunagawa, *Amer. Mineralogist*, 1975, **60**, 889.

Figure 18 *Traverse topograph of a processed silicon slice. Note the dislocations for example at* A. (Courtesy Dr. A. D. Milne)

region.[65] The devices themselves are visible on the topographs due to the strains present at the edges of the oxide windows through which the diffusion takes place (Figure 18). Thus the presence of dislocations in any component can, at least in principle, be detected. Several useful features have been revealed by the topographic

[65] J. M. Fairfield and G. H. Schwuttke, *J. Electrochem. Soc.*, 1966, **113**, 1229.

studies. At one time it was the practice to scribe the slice number on its back surface with a diamond knife. It turned out that during high temperature processing, when silicon becomes plastic, the stress in the scratches was relieved by plastic deformation. Slip bands were observed in the X-ray topographs extending right across the crystals and device yield maps showed corresponding poor regions. The practice ceased.

Although routine on-line monitoring of every slice has been attempted, a big problem is that the packing density of devices is so high that there is a large amount of contrast due to the strained regions at the component edges (Figure 18). It then becomes rather difficult to detect dislocations such as those existing at A in the micrograph. However, if the technique has limitations in actual on-line monitoring of production slices, it is invaluable in the development stage and enables control parameters to be optimized to give maximum yields. Thanks to such studies, 'swirl' patterns due to vacancy complex clusters have been eliminated from silicon vidicons.[66]

Oxidation Studies.—During oxidation it has been found that changes take place in the dislocation configurations in crystals. In X-ray topographic studies of cadmium and zinc [67,68] growth by climb of large dislocation loops and dipoles has been observed. Burgers vector analysis conclusively ruled out the possibility of these dislocations resulting from plastic deformation. Extensive climb of screw dislocations was also found, dislocations having been observed to climb at rates of up to 1 μm per day. Oxidation studies of tin [69] showed that long edge dislocations appeared over a period of several days. The sense of the Burgers vector was determined and it was shown that it corresponded to a sheet of vacancies injected from the surface. It is presumed that the oxide–metal interface acts as a source of vacancies which diffuse into the crystal as oxidation proceeds. Thus, dislocations are created during oxidation. The subject has not been explored in any depth and no studies of chemical attack other than oxidation have been performed. Here is an area where X-ray topography has shown its usefulness and an area in which the chemist may have more than a passing interest.

[66] A. J. R. de Kock, *Philips Tech. Rev.*, 1974, **34**, 244.
[67] C. G'Sell and G. Champier, *Phil. Mag.*, 1975, **32**, 283.
[68] B. Roessler and S. J. Burns, *Phys. Stat. Sol. (A)*, 1974, **24**, 285.
[69] R. Fiedler and A. R. Lang, *J. Mater. Sci.*, 1972, **7**, 531.

7
The Plasticity of Highly Plastic Molecular Crystals

BY R. M. HOOPER AND J. N. SHERWOOD

1 Introduction

For the most part, molecular crystals are close-packed solids of low symmetry in which asymmetrical molecules occupy well-defined positions on lattice sites. They are usually brittle or of limited plasticity and melt with a high entropy of fusion (ΔS_f) characteristic of the large change from order to disorder on melting (*e.g.* naphthalene $\Delta S_f = 55$ J mol^{-1} K^{-1}). In many cases motional and structural transformations occur in the solid state. These lead to a lowering of order, but in general contribute little to the total entropy. The entropy of fusion remains relatively high (*e.g.* benzene, rotation around particular axes, $\Delta S_f = 36$ J mol^{-1} K^{-1}). The resulting melts usually have a long temperature range of existence.

In contrast, solids which comprise approximately spherical molecules show a markedly different behaviour. At low temperatures they form crystals with lattices of low symmetry which are similar in general characteristics to normal organic solids. As the temperature is increased the solids undergo one or more crystallographic transformations to yield lattices of higher symmetry (f.c.c., b.c.c., or h.c.p.) in which the molecules execute a rapid, endospherical, re-orientational motion. In this phase the solids are highly plastic. In extreme cases they flow under gravity. The entropies of transition to these phases are high, but the orientationally disordered crystals melt with a very low entropy of fusion ($\Delta S_f = R$—$2.5R$). The melts have a relatively short range of existence. This considerable distinction in behaviour from what were regarded as the more normal organic solids led early workers to speculate that these materials in this state were mesophases of a type parallel to the Liquid Crystalline Mesophases. It was proposed that they could represent a three-dimensionally ordered liquid-like state. This speculation led to the acceptance of these materials as a unique phase of matter and as a consequence of their very high plasticity, they were named *Plastic Crystals*. It is now realized that such a designation is rather limited in that it implies that other molecular crystals are non-plastic. This is certainly not the case and it has recently been proposed that the term *Orientationally Disordered Solids* would be a more realistic description. More detailed accounts of the molecular properties and phase relationships of these solids will be found in refs 1 to 4 (which also give further references).

[1] J. Timmermans *et al.*, *J. Phys. and Chem. Solids*, 1961, **18**, pp. 1—92.
[2] L. A. K. Staveley, *Ann. Revs. Phys. Chem.*, 1961, **13**, 351.
[3] J. G. Aston, in 'Physics and Chemistry of the Organic Solid State', ed. D. Fox, M. M. Lakes, and A. Weissberger, Wiley, New York, 1963, Vol. 1, p. 543.
[4] G. Smith, in 'Advances in Liquids Crystals', ed. G. H. Brown, Academic Press, New York, 1975, Vol. 1, p. 189.

Although the name is inappropriate, these solids are unique among molecular crystals. It is of interest therefore to define their mechanical properties and to assess whether or not plastic deformation occurs predominantly by solid-like or liquid-like mechanisms.

The initial direct examination of the plasticity of these materials relative to other organic solids was carried out by Michels and first reported in 1948.[5] His experiments were conceived assuming that the materials had a liquid-like character and involved the extrusion of a number of organic solids through a small (mm size) orifice. His observations are summarized in Table 1 which records the pressures

Table 1 *Flow properties of organic crystals* [5]

Materials	$\Delta S_f/R$	Crystalline form		$P/\text{kg cm}^{-2}$	
		I[a]	II[c]	I	II
CBr_4	1.3	C[b]	M	247	1500
C_2Cl_6	2.2	C[b]	T	250	3600
d-Camphor	1.4	C[b]	R	247	454
d-Borneol	2.1	—	—	240	864
Naphthalene	6.7	M	—	2240	—
Hexamethylbenzene	5.6	T	—	2800	—
Dotriacontane	28	O	—	1200	—
Stearic acid	19.5	M	—	1120	—

[a] Form I is that which exists immediately below the melting point; [b] Plastically crystalline places;
[c] C = cubic, M = monoclinic, R = rhombohedral, T = triclinic, O = orthorhombic.

required to extrude the various phases of the solids. As will be seen, the highly plastic crystals were found to be 2—10 times more plastic than both the corresponding low temperature phases and other non-rotator phase solids. Michels made no mechanistic assessment of the process, but he comments in his paper that the appearance of all the extruded materials was more consistent with the deformation of crystalline solids than of liquid-like materials. Consequently a solid-state approach to the problem should be more appropriate. More recent studies have been made on this basis.

For the most part the plastic crystals are low melting (< 373 K) and descriptions of their high plasticity refer either to flow under gravity near the melting point or to the ease with which they may be deformed by bending or compression at lower temperatures. In solid state terms these experiments could be classified roughly as high temperature/low stress and low temperature/high stress deformation experiments respectively.

2 High Temperature/Low Stress Deformation – Self-diffusion Control

Studies of the deformation of ultra-pure single crystals under constant stress at high temperatures yield creep curves characteristic of crystalline solids (Figure 1). The initial rapid deformation which occurs on loading (Region A ≡ primary creep) gives way after a short period of time to a long linear region (B ≡ secondary or steady-state creep). For many crystalline solids it has been found that the stress (σ)

[5] A. Michels, *Bull. Soc. chim. belges*, 1948, **57**, 575.

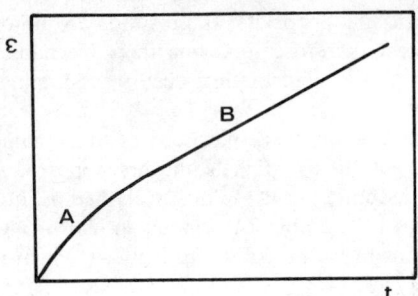

Figure 1 *Typical plastic deformation curve for a crystal deformed in compression*

and temperature (T) dependence of the creep rate ($\dot{\varepsilon}$) is best represented by an empirical equation of the form [6]

$$\dot{\varepsilon} = \frac{A}{T}\left(\frac{\sigma}{\mu}\right)^n \exp\left(\frac{-E_c}{RT}\right) \qquad (1)$$

where E_c is the activation energy for the process, μ is the bulk modulus of the solid and A and n are, respectively, structural and mechanistic constants.

Figures 2(a) and 2(b) show the influence of stress and temperature on the deforma-

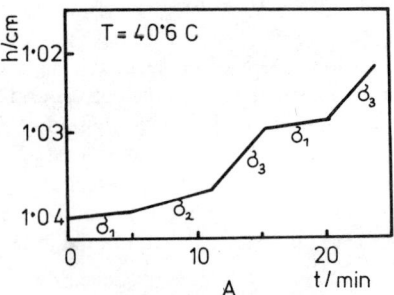
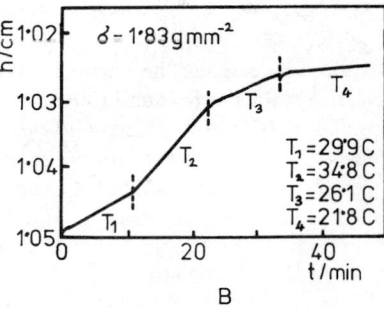

Figure 2 *Examples of the influence of varying stress and temperature on the deformation of (A) succinonitrile (B) dl-camphene (h is the crystal height)*

tion rate of succinonitrile [7] and *dl*-camphene,[8] and Figure 3 the rationalization of similar data for phosphorus [9] in terms of equation (1). Values of n and E_c derived from deformation experiments on a wide range of materials are given in Table 2. We note that with one exception (P_4) all values of n lie in the range 4—6. This immediately contradicts the propositions of a simple liquid-like behaviour since liquids should undergo Newtonian flow and yield $n = 1$.

Values of the observed order are more characteristic of the deformation processes which occur in crystalline solids under similar relative conditions.[6] It has been

[6] J. E. Dorn, *J. Mech. Phys. Solids*, 1955, **3**, 85.
[7] H. M. Hawthorne and J. N. Sherwood, *Trans. Faraday Soc.*, 1970, **66**, 1792.
[8] H. M. Hawthorne and J. N. Sherwood, *Trans. Faraday Soc.*, 1970, **66**, 1799.
[9] E. M. Hampton, P. McKay, and J. N. Sherwood, *Phil. Mag.*, 1974, **30**, 853.

Table 2 Values of the stress exponent (n) and activation energy (E_c/kJ mol^{-1}) for high-temperature plastic deformation in organic crystals

Materials	$\Delta S_f/R$	n	E_c	L_s
Plastic crystals				
Pivalic acid [10]	10.8	4.9	94	59
Cyclohexane [11]	1.0	4.9	112	46
Phosphorus [9]	1.0	7.0	126	59
dl-Camphene [12]	1.2	5.0	102	51
Hexamethyldisilane [13]	1.3	4.5	77	37
Carbon tetrabromide (I) [14]	1.3	5.2	103	47
d-Camphor	1.4	5.0	202	80
Succinonitrile [7]	1.5	4.6	57	70
Tetra(fluoromethylmethane) [15]	1.7	4.8	88	43
Perfluorocyclohexane [15]	2.3	5.2	72	36
1-Bromoadamantane [14]	—	5.2	133	—
Hexamethylethane [16]	2.4	5.6	86	44
Adamantane [17]	2.5	4.4	151	63
Normal crystals				
Carbon tetrabromide (II) [14]	—	4.4	107	54
Naphthalene [18]	6.7	5.1	142	75
Biphenyl [19]	6.4	5.3	130	74

proposed that the dominant deformation mechanism will involve the climb of dislocations,[19] a process whereby vacancies migrate through the lattice and accumulate at the dislocation cores. Thus the dislocation climbs out of the lattice and the counter diffusion of molecules provides the necessary mass transport. Theoretical analysis of this mechanism leads to values of $n = 4.5$ [19] in adequate agreement with the observed spread of values in Table 2. A further consequence of this mechanism is that, since vacancy migration is the rate-determining step in the process, the activation energy for plastic deformation should be equivalent to that for self-diffusion (E_t). The equivalence between the two sets of parameters for those systems for which both types of measurements have been made is shown in Table 3. Recalling the general observation that the activation energy for radiotracer self-diffusion in pure molecular solids is approximately double the lattice energy (L_s) of the solid [22] extends the equivalence to all pure solids quoted in Table 2.*

---

* Succinonitrile is unique in this respect in that $E_t = L_s$.[7] As will be seen in Table 3, in spite of this distinction $E_c = E_t$ for this solid also.

[10] H. M. Hawthorne and J. N. Sherwood, *Trans. Faraday Soc.*, 1970, **66**, 1783.
[11] E. M. Hampton and J. N. Sherwood, *J.C.S. Faraday I*, 1976, **72**, 2398.
[12] N. T. Corke, N. C. Lockhart, R. S Narang, and J. N. Sherwood, *Mol. Crystals Liquid Crystals* in the press.
[13] P. Salthouse and J. N. Sherwood, *J.C.S. Faraday I*, in the press.
[14] N. C. Lockhart, Ph.D. thesis, Strathclyde University, 1971.
[15] P. Bladon, N. C. Lockhart, and J. N. Sherwood, *Mol. Crystals Liquid Crystals*, 1973, **19**, 315.
[16] N. C. Lockhart and J. N. Sherwood, *Faraday Symposium Chem. Soc.*, 1972, **6**, 57.
[17] H. Resing, N. T. Corke, and J. N. Sherwood, *Phys. Rev. Letters*, 1965, **20**, 1227.
[18] N. T. Corke and J. N. Sherwood, *J. Materials Sci.*, 1971, **6**, 68.
[19] J. Weertman, *J. Appl. Phys.*, 1957, **28**, 362.
[20] P. McKay and J. N. Sherwood, *J.C.S. Faraday I*, 1975, **71**, 2331.
[21] E. M. Hampton, N. C. Lockhart, and J. N. Sherwood, *Chem. Phys. Letters*, 1971, **23**, 191.
[22] J. N. Sherwood, in this series, 1972, Vol. 2, p. 250.

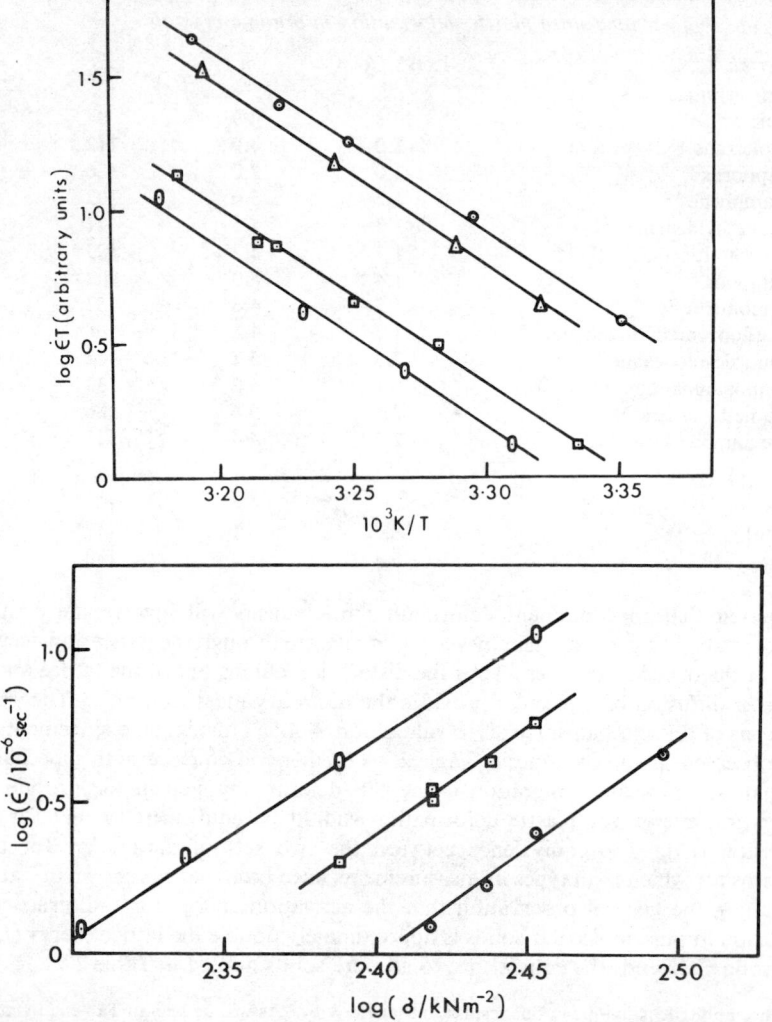

Figure 3 *Typical plots of* $\log \dot{\varepsilon}'T$ *vs.* T^{-1} *and* $\log \dot{\varepsilon}'$ *vs.* $\log \sigma$ *for the plastic deformation of phosphorus* (P_4)

The dominance of self-diffusion in the deformation mechanism is further confirmed by parallel measurements of the pressure dependence of the two processes. Accepting self-diffusion control leads to the re-statement of equation (1) as

$$\dot{\varepsilon} = A' \left(\frac{\sigma}{\mu}\right)^{4.5} D \tag{2}$$

The self-diffusion coefficient D can be expressed as

$$D = \gamma a^2 \nu \exp(-\Delta G_t/RT) \tag{3}$$

where γ, a, and ν are, respectively, a geometrical factor, the molecule jump distance and the lattice vibrational frequency. ΔG_t is the free energy for self-diffusion. We can thus define an activation volume V^+

$$V^+ = [\delta(\Delta G_t)/\delta P]_T = -RT[\delta(\ln D)/\delta P]_T = -RT[\delta(\ln \dot\varepsilon/\delta P]_T \quad (4)$$

This assumes that other terms in equation (3) are pressure independent within the experimental error. This is in fact the case.[20]

Table 3 *Comparison of activation parameters for high-temperature plastic deformation with those for radiotracer self-diffusion*

Plastic crystals	E_c/kJ mol^{-1}	E_t/kJ mol^{-1}	Molar volume[a]	
			V_c^+	V_t^+
Pivalic acid [10,20]	93.6	91	1.2	1.2
Cyclohexane [20]	112	92	1.0	1.0
Phosphorus [9]	126	114	1.3—1.7[b]	1.4
dl-Camphene [12,20]	102	96	1.0	—
Hexamethyldisilane [13]	77	78	1.2	—
Succinonitrile [7]	57	57	—	—
Hexamethylethane [16]	86	86	1.5—1.9[b]	1.7—1.8[b]
Adamantane [13,17,21]	151	138	1.2	1.2
Normal crystals				
Naphthalene [18]	142	179	—	1.2
Biphenyl [18]	130	169	—	—

[a] The subscripts refer to plastic deformation (c) and tracer (t) experiments; [b] Varies with temperature.

Values determined by experiment are quoted in Table 3. The agreement is again excellent, particularly noteworthy is the parallel temperature dependence of V^+ for hexamethylethane. That the self-diffusion occurs by vacancy migration has been proved by isotope-mass-effect experiments.[22]

With such excellent agreement, it remains only to confirm that the absolute

Table 4 *Comparison of self-diffusion coefficients (D_m) with the stress (σ_m) required to yield a deformation rate of $\dot\varepsilon = 10^{-6}$ s^{-1} both at the melting point, for several organic solids*

	D_m/m^2 s^{-1}	σ/kN m^{-2}	μ/10^{9}N m^{-2}
Plastic crystals			
Cyclohexane (f.c.c.)	10^{-13}	10	0.26
Pivalic acid (f.c.c.)	10^{-13}	70	0.64
Phosphorus	10^{-13}	80	1.8
dl-Camphene	10^{-12}	12	0.29
Carbon tetrabromide (I)	—	50	—
Adamantane	10^{-13}	300	2.5
Normal crystals			
Carbon tetrabromide (II)	—	600	—
Naphthalene	10^{-15}	1000	3.0
Biphenyl	10^{-15}	7000	2.0

difference in plasticity reflects the difference in self-diffusion coefficients between the highly plastic phases and normal crystals or phases.

The relevant data for a number of crystals are collected in Table 4. Approximate values of the self-diffusion coefficients and the stress [23] required to yield a deformation rate of 10^{-6} s^{-1} are both extrapolated to the melting point. For carbon tetrabromide the figures refer to the transition point between the two phases.

We note that the highly plastic crystals are 2—12 times more plastic than the normal crystalline phases. This is in good general agreement with Michels' observations (Table 1). Comparing, for example, adamantane with biphenyl and naphthalene we see the difference between the normal and highly plastic solids can be ascribed to the different rates of self-diffusion. From equation (2), for constant $\dot{\varepsilon}$ $(\sigma/\mu)^{4.5} \propto D$. The ratio of self-diffusion coefficients, $\approx 10^2$ is in good agreement with the ratio of $(\sigma/\mu)^{4.5}$: adamantane/biphenyl, 123, adamantane/naphthalene, 95. This agreement extends to the other solids within the errors of experiment and the reliability of the values of the bulk modulus. It is interesting to note one major distinction from the data of Michels. Whereas he found all of the plastic crystals which he examined to be equivalently plastic, we find a variation. In general this parallels the entropy of fusion (cf. Table 2) – the more disordered solids being the more plastic. For the range of highly plastic materials the variation in D is small, and isotope mass experiments prove that there is little or no variation in mechanism. For example, for the two extremes adamantane [13] and cyclohexane,[11] $D_m \approx 10^{-13}$ m^2 s^{-1} and the isotope-mass factor $E_{ab} = 0.78 \pm 0.02$, equivalent to the theoretical value for vacancy self-diffusion. The variation can be approximately rationalized by normalizing with respect to μ [equation (1)]. It must therefore be associated with a variation of some structural property of the solids or more specifically of the migrating dislocation (also related to μ, see below), which, should become more obvious from direct studies of dislocation motion. There is no doubt however that high-temperature plastic deformation occurs by a characteristically solid-state process.

3 Low Temperature/High Stress Deformation – Dislocation Slip

With increasing stress and decreasing temperature the mechanism of deformation changes. This is reflected by a decrease in the values of n and E_c [equation (1)].

Table 5 *The influence of decreasing temperature/increasing stress on the plastic deformation of highly plastic crystals*

	T/K	σ/kN m^{-2}	n	E/kN mol^{-1}
Cyclohexane [11]				
	270—279	1— 3	5	112
	270—279	4—10	3	52
Adamantane [23]				
	473—543	≈ 300	5	151
	373—453	≈ 700	3	75
	293—373	≈ 1000	2	42

Melting points: cyclohexane = 299 K; adamantane = 543 K

[23] B. S. Shah and J. N. Sherwood, *Trans. Faraday Soc.*, 1971, **67**, 1200, and unpublished work.

Table 5 shows the results of typical experiments on the two f.c.c. solids cyclohexane [11] and adamantane.[23] For adamantane, the successive values of n and E_c are constant within the temperature ranges indicated. The very minor increases and low values of the stress required to cause the change in cyclohexane are particularly noteworthy. These variations in n and E_c are compatible with a change from a diffusion-controlled mechanism to slip-controlled mechanisms.

Much exploratory work has been carried out on dislocations in the less plastic organic solids.[24–28] In contrast, the identification and characterization of dislocation slip processes in the present solids is limited to a very few observations on the f.c.c. solids d-camphor [29] and adamantane [23] and the b.c.c. solid succinonitrile.[29] Tensile tests at room temperature reveal that all solids exhibit a small elastic deformation giving way to plastic flow at yield stresses of 200—250 kN m^{-2}, 40—50 kN m^{-2} and 20—30 kN m^{-2} respectively. The association of such easy deformation with dislocation motion has been demonstrated by microscopic examination of deformed and etched samples. Light deformation of the {111} faces of both d-camphor [29] and adamantane [23] yields the characteristic distribution of slip traces (*e.g.* Figure 4a). Subsequent etching of adamantane yielded equivalent distributions of etch-pits

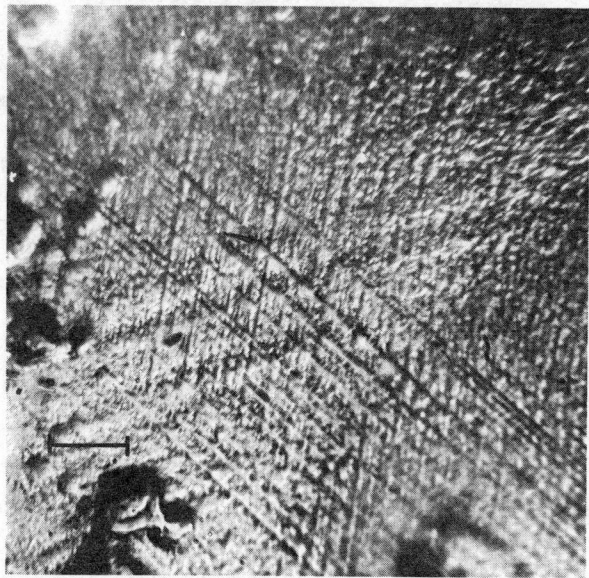

Figure 4a *Slip traces on the {111} face of an adamantane crystal following deformation by compression. The lines lie in the corresponding ⟨110⟩ directions. The scale line represents 25 μm*

[24] J. M. Thomas, *Phil. Trans. Roy. Soc.*, 1974, **277**, 251; *Adv. Catalysis*, 1969, **19**, 202.
[25] J. M. Thomas and J. O. Williams, *Progr. Solid State Chemistry*, 1971, **6**, 271.
[26] J. N. Sherwood, *J.C.S. Faraday I*, 1976, **72**, 2872.
[27] J. N. Sherwood, E. Hampton, R. M. Hooper, B. S. Shah, B. Escaig, and J. Di Persio, *Phil. Mag.*, 1974, **29**, 743.
[28] P. M. Robinson and H. G. Scott, *Acta Metallurgica*, 1967, **15**, 1581; *Mol. Crystals Liquid Crystals*, 1970, **11**, 13.
[29] G. J. Ogilvie and P. M. Robinson, *Mol. Crystals Liquid Crystals*, 1971, **12**, 379.

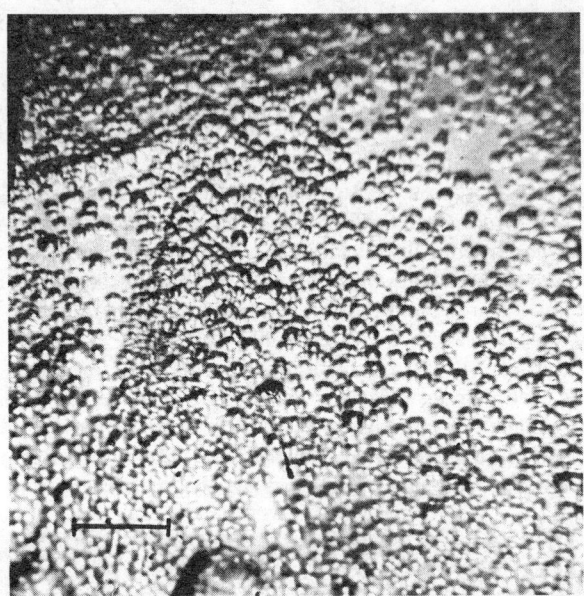

Figure 4b *The same area as Figure 4a after etching. The scale line represents* 25 μm

(Figure 4b). From this it can be reasonably concluded that deformation of these two solids occurs by the slip of $\{111\}\langle10\bar{1}\rangle$ dislocations, generally characteristic of f.c.c. solids. The observation of similar slip-patterns on the softer plastic crystals cyclohexane [11] and camphene [12] has also been reported. With these last materials it is less easy to confirm that any subsequent etch-patterns truly reflect dislocation etching.

The ready ease of dislocation multiplication and motion is perhaps best exemplified at present by the extreme difficulty experienced in the preparation and maintenance of large (cm³) single crystals of these materials.[30] Thus for the least plastic solid adamantane, growth from the melt yields a crystal of high dislocation content

Table 6 *Influence of growth methods growth temperature and stress on the perfection of adamantane crystals* [30]

	T/K	Dislocation count /cm^{-2}	Mosaic spread /seconds of arc.
Melt growth	543	10^8	1000
Vapour growth	383	10^7	200
	323	10^7	60
Solution growth	300	0	50
+1 g load	300	10^{5a}	200
+10 g load	300	10^{5a}	1000

^{a} Limit of resolution of X-ray topography.

[30] R. M. Hooper, B. S. Shah, and J. N. Sherwood, to be published.

and very high mosaic spread (Table 6). Better results can be obtained by low-temperature growth from the vapour phase or solution. In the latter case, volumes of crystal almost devoid of dislocations are generated in the outer parts of the growing crystal. The dislocations can be resolved by X-ray topography. The slightest mishandling of the crystal results in the ready multiplication of dislocations which, with consequential polygonization, leads to the rapid recrystallization of the solid. With such problems with the least plastic of these materials, it is hardly surprising that attempts to grow good single crystals of the more plastic solids yield samples of poor quality, *e.g.* in cyclohexane and perfluorocyclohexane mosaic spreads of *ca.* 0.5°. The problem is exacerbated by the presence of any impurity, which will inevitably segregate to cause lattice stresses and hence increase the dislocation content.

In spite of the very sparse quantitative information relating to deformation under these conditions, there is no doubt once again that the mode of deformation and the observed changes are basically characteristic of the changes known to accompany or to result from the slip of dislocations. There is obviously a great need for a more detailed quantitative study of these phenomena in these materials for two reasons. First, these solids make ideal model substrates for examining standard deformation processes at lower temperatures and at lower stresses than are normally usable with metals. Secondly, there is a need to define the average physical state of these systems. The interpretation of measurements of the molecular properties of these and other solids are inevitably referred to the perfect crystal. Whereas this may be a good or acceptable approximation for some materials, it may not be so for the present solids. In the absence of the necessary detailed information it is perhaps not unreasonable to speculate both on the relationship of the properties of dislocations in these materials relative to other solids and on the relative degree of their likely imperfection. We hope that this may stimulate the further interest necessary for better evaluation and consideration of their defect structure.

Evaluation of Dislocation Energies.—Dislocations have their origins in the growing crystal as a consequence of

(i) condensation of excess vacancies
(ii) impurity incorporation
(iii) thermal and mechanical stresses

The concentration and mobilities of lattice vacancies in these crystals are similar to those found in metals of similar structure.[31] Impurity precipitation of dissimilarly shaped impurities will lead to considerable lattice strain. These materials have low thermal conductances and high thermal expansivities. Thus the rapid freezing of semi-pure material could result in considerable internal strain. If this were relieved by dislocation generation then highly defective specimens would result even if the energies of dislocations were similar to those in other solids. With more careful conditions of preparation the concentration of growth-induced dislocations could still be relatively high.

Subsequent multiplication of the 'grown in' dislocations can occur *via* Frank–

---

[31] J. Friedel, 'Dislocations', Pergamon, London, 1964, provides the essential background to this section.

Read mechanisms or cross slip, and their motion in the lattice is limited by the stresses needed to overcome the barriers to dislocation motion (Peierls–Nabarro). A detailed discussion of these processes and evaluation of absolute stresses required to cause dislocation multiplication and motion is not possible simply because the necessary physical data are imprecise. It is sufficient to note that, within the present accuracy, these stresses are proportional to μ, the bulk modulus of the solid. The stresses required to operate Frank–Read sources and to overcome the Peierls–Nabarro barriers are, respectively, $\sigma_{FR} \approx 10^{-4}\mu$ and $\sigma_{PN} \lesssim 10^{-4}\mu$.[31] Table 6 gives the available values of μ for the extreme examples of the present materials and for two typical metals. Obviously, even on this naive basis the stresses required for dislocation multiplication and migration in these solids will be 2—20 times lower than those for metals. Thus it is not surprising that even carefully prepared samples of materials like camphene and cyclohexane are highly defective. Casually prepared samples are likely to be only semi-crystalline.

The realization that dislocation motion is so easy in these solids immediately leads to speculation as to the possible thermal generation of these defects and their contribution to the total entropy of the system particularly near the melting point.

The free energy F of the dislocation can be expressed in the usual form

$$F = U - TS \tag{5}$$

U is the total energy of the dislocation and is regarded as the sum of the core U_c and elastic strain U_{el} energies. For other solids U_c is found to be 0.1 U_{el} and is usually ignored.

For a dislocation line $U_{el} \approx \mu b^3$ per molecule plane threaded by the dislocation. Values of U_{el} are tabulated in Table 7 for four extreme cubic organic crystals.

Table 7 Calculated values of the line energy for common dislocation slip systems

	$\mu/10^9$ N m^{-2}	$b/10^{-10}$ m	U_{el}/eV
Adamantane (f.c.c.)	2.5	6.7	4.7
Cyclohexane (f.c.c.)	0.25	6.2	0.38
Silver (f.c.c.)	3.0	2.9	4.6
Hexamine (b.c.c.)	5.0	6.9	7.4
Camphene (b.c.c.)	0.29	6.1	0.6
Iron (b.c.c.)	8.3	2.5	8.1

Values for silver and iron are included for comparison. We note that hexamethylenetetramine (hexamine) does not undergo a rotator-phase transformation but its molecular similarity to adamantane does make it an ideal extreme of the b.c.c. system. Experiments confirm that slip in adamantane [23] and hexamine [32] occurs by the motion of the characteristic $\{111\}$ $a/2$ $\langle101\rangle$ and $\{1\bar{1}0\}$ $a/2$ $\langle111\rangle$ dislocations respectively. Slip traces on deformed single crystals of cyclohexane [11] and camphene [12] can be similarly (but less definitely) interpreted. The calculated values refer to these slip systems.

The energies of dislocations in the harder organic crystals are similar in magnitude

[32] J. Di Persio and B. Escaig, *Cryst. Lattice Defects*, 1972, **3**, 55.

to those calculated for the two metals. Those for the two solids with low entropies of fusion are considerably lower.

It is customary to neglect the entropy term in equation (5) on the basis that U_{el} for most materials is relatively large. Whereas this is still true for adamantane and hexamine it is not necessarily so for the other pair of solids.

The configurational entropy S_c is evaluated from a consideration of the number of orientations which the dislocation line can take up in the solid. The calculation yields $S_c = k \ln 11 = 2.4 \, k$ per unit length. Also to be included are terms 'associated with changes in lattice vibrations on the inclusion of the defect, $S_v \approx k$ per unit length, and lattice distortion, $S_d \approx 2k$ per unit length, i.e. $S_{total} \approx 5.4 \, k$. Thus at room temperature we evaluate $F \approx 0.25$ eV and 0.4 eV per molecule plane threaded for cyclohexane and camphene, respectively.

Under normal circumstances we could not envisage dislocations of length $l < 10b$ and hence the concentrations $[N_d = \exp(-lF/bkT)]$ in thermal equilibrium would be extremely low (e.g. cyclohexane $N_d \approx 10^{-50}$ at the melting point). However, consideration of the possibility that motion could occur by the migration of partials ($U_{partial} \approx 0.25 \, U_{full}$ for f.c.c. crystals) coupled with the likelihood that dislocations of smaller size become more acceptable at or near the melting point brings us into a region where thermal generation of small ($l/b = 1$) partial dislocations could become significant ($N_d \approx 10^{-2}$), i.e. equivalent to or slightly greater than the point defect concentration. Such processes could then represent a major contribution to the breakdown of the lattice prior to melting. Similar conclusions have been reached following molecular dynamics calculations on the melting of argon.[33] These speculations rely heavily on the accuracy of the published carbons of μ. In view of the possibilities indicated it would be of interest to obtain more precise values from measurement on well-characterized material.

4 Conclusions

In conclusion we recall that the deformation of these materials occurs by characteristically solid-state processes. It involves the climb or glide of dislocations depending on the temperature and stress applied to the system. Their high plasticity arises from the greater ease of multiplication and motion of dislocations relative to other solid types. The least plastic (e.g. adamantane) have plasticities which would appear to be equivalent to the simple metals whilst those at the other extreme are approximately ten times more plastic. In spite of the very low dislocation line energies which can be calculated for the latter it would appear that the thermal generation of full dislocations is still unlikely. It is not unreasonable to suggest that as the melting point is approached, there is a probability that dislocation generation may be a major contribution to pre-melting mechanisms. Even in the absence of thermal generation the very low line energies suggest that unless considerable care is taken in the preparation and handling of the more plastic crystals, i.e. those of low entropy of fusion, an 'intrinsically' highly defective sample will result. In such solids, molecular dynamic processes could differ considerably from those expected of the perfect lattice.

[33] E. J. Jensen, W. Damgaard-Kristensen, and R. M. J. Cotterill, *Phil. Mag.*, 1973, **27**, 623.

8
The Ultrastructure of Minerals as Revealed by High Resolution Electron Microscopy

BY J. L. HUTCHISON, D. A. JEFFERSON AND J. M. THOMAS

1 Introduction

Electron microscopic studies of minerals were well advanced even by the late fifties thanks largely to the development of techniques, based on diffraction contrast, by Hirsch, Whelan, Bollmann, Howie, and others.[1–3] Such studies emerged as a natural growth from earlier ones, utilizing chiefly optical microscopy, which, though of considerable value in the elucidation of the nature and properties of line defects (dislocations), lacked the resolution required to reveal the fine structure of the various faults that can occur in crystalline materials in general and minerals in particular. From the early sixties onwards, considerable progress was achieved in the electron microscopic study of extended defects in minerals, especial emphasis having been placed (by Amelinckx and co-workers notably)[4–10] on layered solids since these could be conveniently prepared, by cleavage, as slivers thin enough for transmission electron microscopy (TEM). The identity of individual (perfect or partial) dislocations, of various kinds of small-angle boundaries (composed of aligned dislocations) and of stacking fault ribbons were established by TEM, advantage being taken of the $g \cdot b = 0$ disappearance condition.[11] Other properties,[12] too, could be determined by the appropriate application of TEM, and quantitative information relating to stacking fault energies, surface energies, and the magnitude of Poisson's ratio for certain layered minerals could also be extracted from electron micrographs. All such work was accomplished

[1] P. B. Hirsch, R. W. Horne, and M. J. Whelan, *Phil. Mag.*, 1956, **1**, 677.
[2] P. B. Hirsch, A. Howie, and M. J. Whelan, *Phil. Trans. Roy. Soc.*, 1960, **252**, 499.
[3] W. Bollmann, *Phys. Rev,*, 1956, **103**, 1588.
[4] S. Amelinckx and P. Delavignette, *Phil. Mag.*, 1960, **5**, 533.
[5] S. Amelinckx and P. Delavignette, *Nature*, 1960, **185**, 603.
[6] S. Amelinckx and P. Delavignette, *J. Appl. Phys.*, 1960, **31**, 2126.
[7] P. Delavignette and S. Amelinckx, *Phil. Mag.*, 1960, **5**, 729.
[8] S. Amelinckx and P. Delavignette, in 'Direct Observation of Imperfections in Crystals', ed. J. B. Newkirk and J. H. Wernick, Interscience Publication, 1962, p. 295.
[9] S. Amelinckx, R. Siems, and P. Delavignette, in 'Chemistry and Physics of Carbon', ed. P. L. Walker, jun., Dekker, New York, Vol. 1, p. 1.
[10] S. Amelinckx, 'The Direct Observation of Dislocations', Academic Press, New York and London, 1964.
[11] Most methods of establishing the Burgers vector, *b*, of a dislocation utilize this so-called disappearance condition. When the diffraction vector, *g*, is perpendicular to the Burgers vector, the dislocation goes out of contrast. Hence, when an image is recorded under a condition such that the Burgers vector of the dislocation is perpendicular to the operative diffracting vector, that dislocation will be invisible.
[12] For a comprehensive review of the application of conventional electron optical techniques (*i.e.* electron diffraction and relatively low-resolution studies) see the monograph 'Electron Optical Investigation of Clay Minerals', ed. J. A. Gard, The Mineralogical Society, London, 1971.

The Ultrastructure of Minerals

using magnifications generally in the range from ×10 000 to ×50 000. The ultimate limit of resolution of transmission electron microscopes was not, in general, harnessed.[13]

Since the early seventies a new era in electron microscopy has dawned; magnifications well beyond ×300 000 (corresponding to resolutions of a few Å on high-resolution electron micrographs) may now be routinely employed, provided appropriate minerals are selected for study. The past three years have, consequently, witnessed a rapid growth in our knowledge of the ultra-structural characteristics of minerals (see J. E. Chisholm, in Vol. 4, p. 126). Indeed, so dramatic has been the burgeoning that the information contained in a book, devoted, *inter alia*, to the application of high-resolution electron microscopy (HREM) to mineralogy [14] and published as recently as 1976 has, in many important respects, been supplanted by subsequent results. Some of these results were reported in the proceedings of two major international conferences.[15,16] Moreover, a selection of other highly relevant papers has appeared in recent months (up to February 1977), all of which bear crucially upon the topics discussed here.

Our approach in this survey will be selective and reflective rather than rigorous and comprehensive. We shall begin with a very brief account of the less widely known and not generally applicable technique of etch-decoration coupled to electron microscopy, and then proceed to discuss, in turn, the pyroxenes, the pyroxenoids, amphiboles, serpentine minerals, layered silicates (which are further subdivided), orthosilicates, graphite and its intercalates, sulphides, and fluorocarbonates. Partly because of shortage of space, but also in view of the wealth of information that is coming to light in regard to their internal structure, we shall not consider the feldspars, which it is hoped will form the subject of an entire review in a forthcoming volume of this series. Before proceeding to discuss each category of mineral in turn, however, it is profitable to recall some relevant landmarks that have served as pointers in the territory already covered.

One of the most interesting early assessments of the power of electron microscopy in chemistry was made in 1949 by Turkevich and Hillier [17] using an instrument representative of attainable resolution at that time. Working with magnifications in the range 25 000 to 125 000 and some twenty or so distinct kinds of naturally occurring, as well as some synthetic, materials (colloidal sols of aluminium hydroxide, iron hydroxide, vanadium pentoxide, and gold), they were able to assess in hitherto unequalled detail the morphology of individual crystallites: they were not equipped to resolve low index, or any other planes, in specimens of the above-named materials. Indeed, their micrograph of chrysotile asbestos, as well as that taken shortly afterwards by Noll [18] of fibres of the same mineral, was the first to reveal the previously suspected [19,20] 'hollow' pipe (see Section 6) in this material.

[13] J. E. Chisholm, in this series, 1975, Vol. 4, p. 126.
[14] 'Electron Microscopy in Mineralogy', ed. H. R. Wenk *et al.*, Springer-Verlag, Berlin, 1976.
[15] 'EMAG 1975', ed. J. A. Venables, Academic Press, New York and London, 1976.
[16] 8th International Congress on Electron Microscopy, Canberra, 1974.
[17] J. Turkevich and J. Hillier, *Analyt. Chem.*, 1949, **21**, 475.
[18] W. Noll and H. Kircher, *Naturwiss.*, 1950, **37**, 540; 1952, **39**, 158.
[19] P. H. Koch, H. Freytag, and M. Ardenne, *Glastech Ber.*, 1943, **21**, 249.
[20] W. Noll, *Kolloid Z.*, 1944, **107**, 181.

A major advance in HREM came with the studies of Yada,[21,22] also on chrysotile. These and later studies are described in detail in Section 6: suffice it to mention that cross-sections of the fibres were carefully prepared for electron microscopic examination, and that the microscope itself was operated under defocused conditions thereby 'tuning' the instrument so as to render readily visible the curled up sheets of the mineral.

The next noteworthy advance was made by Ban and Heidenreich,[23–25] who showed that modern high-resolution microscopes could cope with the imaging of graphitized carbon blacks (interlayer spacing 3.4 Å), an achievement subsequently widely exploited in the study of graphite,[26] graphite intercalates,[27,28] and heat-treated coals and carbons [29–33] (see Section 9). By 1973, Iijima and others had demonstrated the utility of HREM in mineralogy, and rapid progress since made is discussed at length in this review. The principal points to bear in mind in so far as the feasibility of recording interpretable images of minerals is concerned are that, first, the sample has to be appropriately thinned (to a few tens of Å). Whereas this may sometimes be achieved by cleavage, it can also be accomplished by polishing followed by ion-bombardment.* Secondly, the structure has to be sufficiently 'open' to yield adequate variation in projected charge density down the zone axis chosen for imaging. A further point concerns electron-interaction with the sample. Apart from requiring the specimens to be stable in the electron microscope (radiation doses associated with high-resolution modes are often in the Grad range), it is desirable for there to be little multiple scattering, which can always be minimized by using very thin samples. Occasionally, however, as will emerge later, multiple scattering may be turned to advantage, and boundaries between polytypes are rendered visible as a result of changes in the phase and amplitude of electron-waves at such planar faults.[34]

During the past five years the ultra-structural characteristics of a wide variety of minerals have been exposed by HREM, and our intention in this review is to present a coherent picture which faithfully reflects current knowledge regarding, *inter alia*, patterns of structural behaviour on the inter-relationship between local chemical composition and local structure. The theoretical foundations upon which

---

* Ion beam thinning is sometimes deleterious and the mineral under study is irreversibly damaged.

[21] K. Yada, *Acta Cryst.*, 1967, **23**, 704.
[22] K. Yada and K. Iishi, in ref. 16, Vol. 1, p. 494.
[23] R. D. Heidenreich, W. M. Hess, and L. L. Ban, *J. Appl. Cryst.*, 1968, **1**, 1.
[24] L. L. Ban and W. M. Hess, 'Proc. 26th Annual EMSA', ed. C. J. Arcneaux, Claitor's Publ. Div., Baton Rouge, 1968, p. 256.
[25] L. L. Ban, in this series, 1972, Vol. 1, p. 54.
[26] D. A. Jefferson, G. R. Millward, and J. M. Thomas, *Acta Cryst.*, 1976, **A32**, 823.
[27] E. L. Evans and J. M. Thomas, *J. Solid State Chem.*, 1975, **14**, 99.
[28] J. M. Thomas, G. R. Millward, N. C. Davies, and E. L. Evans, *J.C.S. Dalton*, 1976, 2443.
[29] G. R. Millward and J. M. Thomas, 'Proc. Fourth International Carbon and Graphite Conference, Society of Chemical Industry, London, 1976, p. 492.
[30] E. L. Evans, J. L. Jenkins, and J. M. Thomas, *Carbon*, 1972, **10**, 637.
[31] H. Marsh, D. Augustyn, C. Connford, D. Crawford and G. Hermon, 'Proc. 12th Biennial Conference on Carbon', Pittsburgh, 1975, p. 117.
[32] D. J. Johnson and D. Crawford, *J. Microscopy*, 1973, **98**, 313.
[33] G. R. Millward and D. A. Jefferson, in 'Chemistry and Physics of Carbon', ed. P. L. Walker, jun., and P. A. Thrower, Marcel Dekker, in press.
[34] E. S. Crawford, D. A. Jefferson, and J. M. Thomas, *Acta Cryst.*, in press.

the technique of HREM rests have been discussed both at length [35,36] and in outline [37–39] elsewhere, and the reader is advised to consult these articles. It would also be prudent to recall the way in which conventional TEM has uncovered the properties of planar discontinuities (antiphase boundaries, stacking faults, twins, *etc.*) in a number of minerals: an admirable account, illustrated with numerous examples (molybdenite, MoS_2; leucite, $KAlSi_2O_6$; teallite, $PbSnS_2$), has been produced [40] by Amelinckx and Van Landuyt.

2 The Etch-decoration Technique and its Use in the Detection of Stacking Faults

This technique is not generally applicable for the study of defects and equilibrium faults in minerals, but it is, in principle, capable of offering unique information for all those highly anisotropic (generally layered) minerals that yield volatile products during an appropriate chemical reaction. It relies on the fact that the microtopography of an etched crystal is governed by the ultrastructural irregularities of the sub-surface or bulk regions of that crystal. The essence of the technique is best introduced by reference to work on graphite and the detection of minute concentrations of vacancies in that solid. Oxidation, by O_2, expands each vacancy at the outermost sheet of sp^2-bonded carbon atoms, thereby generating a shallow (3.4 Å deep) hole the diameter of which is directly proportional to the oxidation time. When vacancies at the second and lower sheets are duly exposed, fresh holes are produced, and each hole on the nth layer is tangential to the progenitor hole on the $(n-1)$th layer (see Figure 1). If, following oxidation, atoms of gold are allowed to condense and migrate on the graphite surface (for precise conditions see refs. 41–43), all etched-out holes are 'decorated' by crystallites of gold, which are readily visible in a transmission electron microscope (Plate 1). The key point to note is that, as demonstrated earlier,[44,45] gold-decoration detects steps of monatomic height.

We now illustrate how the etch-decoration* procedure, as utilized by Bahl *et al.*,[46–48] affords information concerning stacking faults (as well as residual point defects) in the mineral molybdenite, MoS_2. First we note the idealized X-ray based crystal structure of the hexagonal mineral (Figure 2). Each Mo atom is

---

* The chemical reaction that gives rise to volatile products with molybdenite entails introduction of O_2 at low pressures in the temperature range 400 to 600 °C:

$$2MoS_2(s) + 7O_2(g) \rightarrow 2MoO_3(g) + 4SO_2(g)$$

[35] J. M. Cowley and S. Iijima, ref. 14, p. 123.
[36] J. S. Anderson and R. J. D. Tilley, in this series, 1974, Vol. 3, p. 1.
[37] P. R. Buseck and S. Iijima, *Amer. Mineral.*, 1974, **59**, 1.
[38] J. S. Anderson, *Chem. in Britain*, 1977, **13**, 182.
[39] J. M. Thomas, *Chem. in Britain*, 1977, **13**, 175.
[40] S. Amelinckx and J. van Landuyt, ref. 14, p. 68.
[41] G. R. Hennig, in 'Chemistry and Physics of Carbon', ed. P. L. Walker, jun., Dekker, New York, 1966, Vol. 2, p. 1.
[42] J. M. Thomas, E. L. Evans, and J. O. Williams, *Proc. Roy. Soc.*, 1972, **A331**, 417.
[43] E. L. Evans, Ph.D. thesis, University of Wales, Bangor, 1967.
[44] G. A. Bassett, *Phil. Mag.*, 1958, **3**, 1042.
[45] H. Betche, *Surface Sci.*, 1964, **3**, 33.
[46] O. P. Bahl, E. L. Evans, and J. M. Thomas, *Surface Sci.*, 1967, **8**, 473.
[47] O. P. Bahl, E. L. Evans, and J. M. Thomas, *Proc. Roy. Soc.*, 1968, **A306**, 53.
[48] E. L. Evans, O. P. Bahl, and J. M. Thomas, *Trans. Faraday Soc.*, 1968, **64**, 3354.

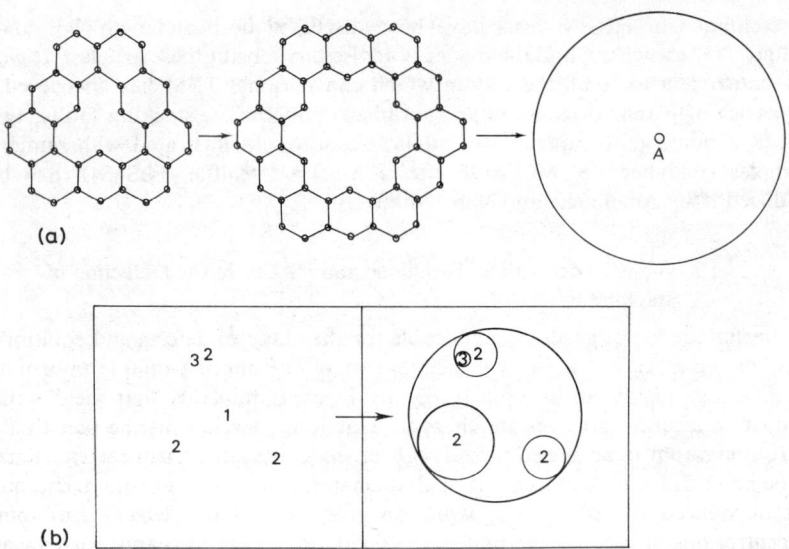

Figure 1 (a) *Schematic illustration of the 'development' by anisotropic oxidation of vacancies in the surface layer of graphite. The oxidizing conditions may be arranged so as to yield circular, rather than hexagonal, monolayer (3.4 Å) depressions. (b) Indigenous vacancies in the second, third, etc. layer are also, in due course, developed; the depression on a layer 'n' is always tangential to its 'parent' on layer $(n - 1)$. 1, 2, 3 refer to the siting of vacancies in the first, second, and third layers, respectively*

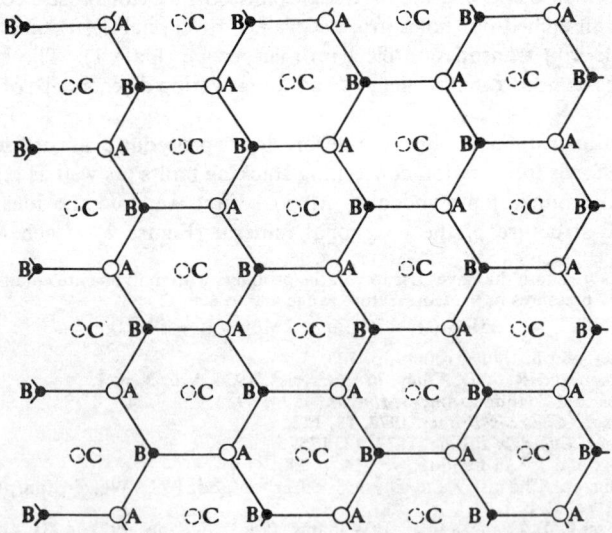

Figure 2 *Schematic diagram of the molybdenite structure. Molybdenum atoms (dark circles) are located at B sites, and the sulphur atoms (empty circles) at A sites which are above and below the plane in which the B sites lie. In hexagonal molybdenite the stacking sequence is $A_B B_A A_B$. In the rhombohedral variety, the C sites (dotted circles) become important, and the stacking sequence is $A_B A_C C_B B_A$*

situated at the centre of a trigonal prism of S atoms. If one S–Mo–S layer is denoted A_B, meaning that S atoms are at A sites and Mo atoms at B sites, then in the perfectly stacked hexagonal crystal the sequence of individual sheets is represented by $A_BB_AA_BB_A$. The perfect rhombohedral [49] variety of MoS_2 (which may be synthesized at high temperatures from the elements) would, on the other hand, be represented by $A_CC_BB_AA_CC_BB_A$. ... Suppose there are vacancies, consisting of one missing Mo atom and two missing S atoms, one each from above and below the plane of the Mo atoms, in each sheet. Such a defect in the outermost layer may be enlarged by oxidation, which would be greatly favoured along, but not perpendicular to, basal planes. A triangular depression possessing the requisite orientation would result from the stages schematized in Plate 2. Once the depression becomes large enough, vacancies on the second layer will be exposed. But owing to the $A_BB_AA_B$ ... sequencing characteristic of the hexagonal structure, the orientation of the triangular depression in the second layer will be turned through 60° with respect to that on the first (and third) layers (Plate 3). Such expectations are fully borne out by experiment (Plate 4), confirming that the hexagonal structure is indeed sustained up to the exterior surface of the vast majority of specimens of naturally recurring molybdenites studied by Bahl *et al.*

Suppose that stacking faults are introduced into the hexagonal structure. Even for the simple case where but one infraction to the regular sequencies occurs, we see (Table 1) that in principle, five distinct types of planar fault [parallel to (0001)] could arise. In two of these (types III and IV), no changes in orientation of the

Table 1 *Possible stacking faults for one infraction to the stacking order in hexagonal molybdenite (the orientation of triangular depressions is indicated)*

Normal sequence	Stacking faults				
	Type I	Type II	Type III	Type IV	Type V
A_B △	A_B △	A_B △	A_B △	A_B △	A_B △
B_A ▽	B_A ▽	B_A ▽	B_A ▽	B_A ▽	B_A ▽
A_B △	A_C ▽	C_B ▽	C_A △	B_C △	B_A ▽
B_A ▽	C_A △	B_C △	A_C ▽	C_B ▽	A_B △

triangular depressions, beyond the expected 60° rotation associated with the normal sequence, would occur. The type V stacking fault, since it involves a structure in which S atoms in successive layers are directly above one another, is unlikely. The other two stacking faults (types I and II) are an intrinsic feature of the known rhombohedral modification of MoS_2 and would therefore be expected to occur, intergrow, in intimate topological contact, within the hexagonal matrix. Electron micrographs (Plate 5) provide proof that we do indeed have the stacking sequence $A_BB_AA_C$ ... or $A_BB_AC_B$ ... interspersed within the regular $A_BB_AA_BB_A$ ... hexagonal solid. Confirmation of the correctness of the general interpretation is obtained from similar etch-decoration experiments on the parent rhombohedral (synthetic) MoS_2, where the triangular depression on adjacent sheets should be of the same orientation (Figure 3) and indeed are seen to be so (Plate 6).

[49] R. E. Bell and R. E. Herfert, *J. Amer. Chem. Soc.*, 1957, **19**, 3351.

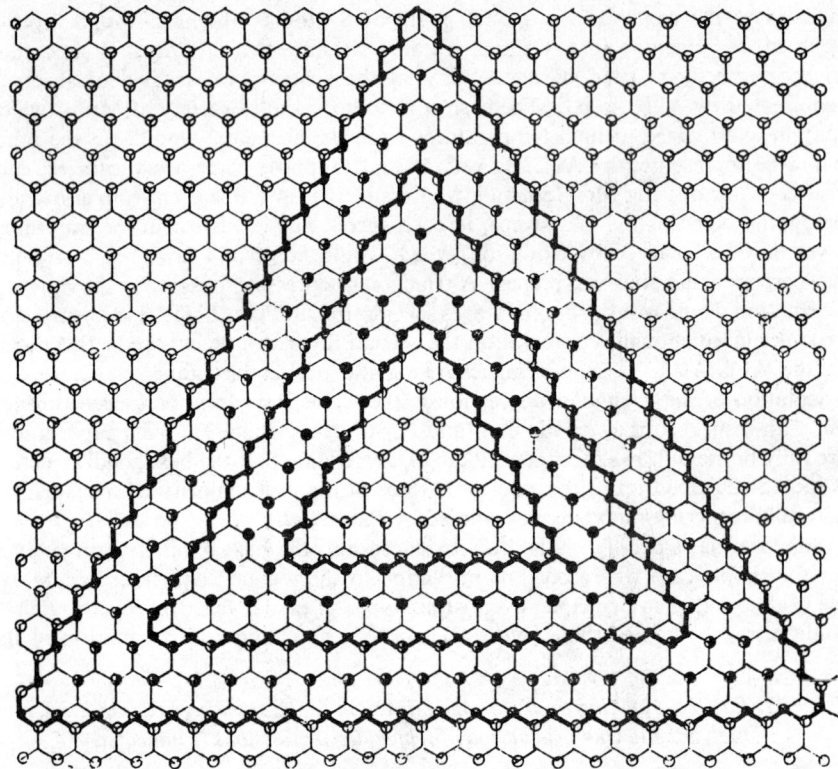

Figure 3 *Schematic illustration of the depressions formed as successive layer planes of rhombohedral MoS_2 are exposed leading to terraced triangular pits. Empty circles denote positions of sulphur atoms in the first and fourth layers, half-shaded circles the positions of sulphur atoms in the second layer, and dark circles the positions of sulphur atoms in the third layer. Each circle denotes two sulphur atoms, one above and one below the plane of the molybdenum atoms*

It should, by appropriate selection of etchant, be possible to extend this technique to the study of other layered minerals, especially the sheet silicates (micas, chlorites, and vermiculites). In view of its laborious nature and now that HREM can be used [34, 50—52] directly (see later) to assess the stacking disorder of such minerals (and others), the technique is not likely to be as widely applicable as that of direct imaging.

3 Pyroxene Minerals

The pyroxene family of minerals constitutes one of the most abundant components of the Earth's crust, being found in nearly all igneous rocks and a great number of

[50] D. A. Jefferson and J. M. Thomas, *J.C.S. Faraday II*, 1974, **70**, 1691.
[51] D. A. Jefferson and J. M. Thomas, *Mater. Res. Bull.*, 1975, **10**, 761.
[52] S. Iijima and P. R. Buseck, *Acta Cryst. (A)*, 1977 (in press).

Table 2 *Some of the more common members of the pyroxene mineral group*

Enstatite	$MgSiO_3$
Ferrosilite	$FeSiO_3$
Diopside	$CaMgSi_2O_6$
Hedenbergite	$CaFeSi_2O_6$
Johannsenite	$CaMnSi_2O_6$
Aegirine	$NaFe^{3+}Si_2O_6$
Spodumeme	$LiAlSi_2O_6$
Jadeite	$NaAlSi_2O_6$
Augite	$(Ca,Mg,Fe^{2+}Al)_2(Si,Al)_2O_6$
Pigeonite	$(Ca,Mg,Fe^{2+})(Mg,Fe^{2+})Si_2O_6$
Omphacite	$(Ca,Na)(Mg,Fe^{2+},Fe^{3+},Al)Si_2O_6$
Fassaite	$Ca(Mg,Fe^{2+},Fe^{3+},Al)(Si,Al)_2O_6$

metamorphic assemblages. The general formula of pyroxenes is ABO_3, with considerable scope for chemical substitution: the general formula can be written as $W_{1-P}(X,Y)_{1+P}Z_2O_6$ (W = Ca or Na, X = Mg, Fe^{2+}, Ni, Mn, or Li, Y = Al, Fe^{3+}, Cr^{3+}, or Ti^{4+}, Z = Si^{4+} or Al^{3+}).[53] Some of the more common compositional members are shown in Table 2. By far the most common group of pyroxenes occurs in the ternary system $CaSiO_3$–$MgSiO_3$–$FeSiO_3$, and in the quadrilateral formed by the compositions $MgSiO_3$–$CaMgSi_2O_6$–$CaFeSi_2O_6$–$FeSiO_3$. The more calcium-rich members of the ternary system adopt an alternative to the pyroxene structure, namely the pyroxenoid form, which is discussed later. Investigations of the more common pyroxenes may be grouped into two categories: those focused on compositions approaching ideal $MgSiO_3$, the enstatites, and those concentrating on exsolution structures in the system $(Ca,Mg,Fe)SiO_3$–$(Mg,Fe)SiO_3$. The latter category can be further subdivided into those which involve a monoclinic pyroxene as the host mineral and those which initiate from an orthorhombic structure; these groups form a convenient means of discussing defect structures in the pyroxenes.

Structures.—Four types of distinct structure are found in the more common pyroxenes, three of these being characteristic of enstatite ($MgSiO_3$) and the $MgSiO_3$–$FeSiO_3$ binary system, while the fourth is normally associated with the calcium-rich members of the $CaMgSi_2O_6$–$CaFeSi_2O_6$ system. All pyroxenes are based on the $(SiO_3)_n$ chain with a two-tetrahedra repeat, as shown in Plate 7(a). The chain can either be fully extended, or, if structural consideration necessitates, the tetrahedra in it can be twisted and the chain slightly compressed. The chains are arranged in pairs, with vertices pointing in opposite directions, and are linked together by metal cations which, at least approximately, are octahedrally co-ordinated by the oxygens of the chains. Considerable adjustments to cation co-ordination are possible by movement of the chains with respect to one another, a factor that accounts for the wide variety of chemical substitution in natural pyroxenes.

Ortho-enstatite and the orthopyroxenes are, in general, the low-temperature stable forms of pyroxenes lying in or near to the $MgSiO_3$–$FeSiO_3$ side of the ternary system. The structure of ortho-enstatite[54] is shown in Plate 7(b) and it can be

---

[53] H. H. Hess, *Amer. Mineral.*, 1949, **34**, 621.

seen that there are two distinct types of chain, with an offset between successive pairs of chains in the [100] direction. Two types of site exist for the metal cations: M_1, where the apices of tetrahedra point towards one another, and M_2, where the apices point away from each other. In the orthorhombic structure these are arranged such that the sequence of cations in the [100] direction is of the form AA BB AA BB ....[55] The space group of the orthorhombic cell is *Pbca*, and of the cation sites, M_2 is larger and is preferentially occupied by calcium if present. At higher temperatures, the orthorhombic *Pbca* structure is apparently replaced by a monoclinic form with space group $P2_1/c$ and only four chains within the unit cell.[56-58] This form has been termed clino-enstatite or pigeonite, if appreciable Fe^{2+} is present. Structural similarities to the orthorhombic *Pbca* form [Plate 7(c)] suggest that the latter is related to the monoclinic form by twinning: this will be discussed later. As in the *Pbca* structure, the chains are distorted from their ideal configuration and the sequence of cations in a direction perpendicular to the (100) planes is of the form AB AB AB .... Yet a third form of structure can occur if the iron content is less than 30%, this being proto-enstatite. In this orthorhombic *Pbcn*[59,60] structure [Plate 7(d)], the arrangement of the chains is somewhat different, in that they are fully extended, and the sequence of cations along the [100] direction is of the type AAAAAA .... Proto-enstatite is also a high-temperature form, and it has been suggested[61-63] that this is the true high-temperature form, while the ortho- and clino-enstatite structures are stable and metastable low-temperature variants, respectively. If calcium is assimilated into the structure in appreciable amounts, the considerable difference in size between Ca^{2+} and Mg^{2+} (or Fe^{2+}) necessitates a distinctive change in structure. In the calcium-rich pyroxenes, therefore, orthorhombic variants are not found, and the resulting monoclinic form has a different structure from that of clino-enstatite with a space group $C2/c$ rather than $P2_1/c$. In the structure of diopside,[64] there is only one unique type of chain, and the cation sites M_1 and M_2 differ considerably, with calcium favouring the larger M_2 site. All the pyroxenes in the system $CaMgSi_2O_6$–$CaFeSi_2O_6$, termed 'augites', possess this structure. Unit cell parameters also differ from those of the clino-enstatite or pigeonite structures, particularly in the β angle.

Enstatite Polymorphism.—Of the enstatite polymorphs, only the clino- and ortho-forms have been investigated in any detail, the possibility that these two forms are related by twinning on a unit cell scale having been advanced by several authors.[54,61,62,65,66] The clino-enstatite structure can be related to that of

---

[54] N. Morimoto and K. Koto. *Z. Krist.*, 1969, **129**, 65.
[55] N. Morimoto, *Carnegie Inst. Wash. Ann. Report Dir. Geophys. Lab.*, 1959, 197.
[56] N. Morimoto, *Proc. Japan Acad.*, 1956, **32**, 750.
[57] M. G. Bown and P. Gay, *Acta Cryst.*, 1957, **10**, 440.
[58] N. Morimoto, D. E. Applenan, and M. T. Evans, *Z. Krist.*, 1960, **114**, 120.
[59] J. V. Smith, *Acta Cryst.*, 1959, **12**, 515.
[60] J. R. Smyth, *Z. Krist.*, 1971, **134**, 262.
[61] W. L. Brown, N. Morimoto, and J. V. Smith, *J. Geol.*, 1961, **69**, 609.
[62] J. V. Smith, *Mineral Soc. Amer. Special Paper*, 1969, **2**, 3.
[63] J. R. Smyth, *Amer. Mineral*, 1974, **59**, 345.
[64] B. E. Warren and W. L. Bragg, *Z. Krist.*, 1928, **69**, 168.
[65] R. Sadanaga, E. P. Okamusa, and M. Takeda, *Mineral J.*, 1969, **6**, 110.
[66] R. S. Coe, *Cont. Min. Pet.*, 1970, **26**, 247.

ortho-enstatite by inserting a b-glide plane parallel to (100) in alternate pairs of chains, and, if such a mechanism is valid, the existence of clino-enstatite twinned on a microfine scale, as well as the ortho-form, should be possible. Streaking of spots parallel to x^* in the reciprocal lattices of both clino- and ortho-enstatite is common, and this would tend to confirm that such microfine twinning is indeed possible. Recently, however, twinning of this type has been confirmed directly by HREM.[37,67,68] As a change in periodicity occurs between the clino- and ortho-enstatites, the conversion of one into the other can be observed in either real or reciprocal space in projection down either [001] or [010]. In samples where the twins are of reasonable size (*ca.* 200 Å) the separate individuals can be observed in standard dark-field conditions. Alternatively, twins can be highlighted by slight tilting of the specimen in the bright-field, lattice imaging mode. One such micrograph, showing twins of several hundred Ångstroms width in projection down [010], is illustrated in Plate 8(a). A differently twinned structure is shown in Plate 8(b). Here the projection is the same but the dimensions of the twinned individuals are severely reduced, and in many regions, the twinning is on a unit cell scale, thereby producing small regions of the ortho-form. No attempt has been made to interpret such images in terms of actual structural features, the fringes merely indicating the extremities of the ortho- and clino-enstatite unit cells, although images have been reported [37] in which certain structural features are resolvable.

An alternative means of illustrating twinning in clino-enstatite is by producing a bright-field, tilted beam image such that the (002) fringes of one individual and the (20$\bar{2}$) fringes of the second are superimposed on the (100) fringes. Such an image is illustrated in Plate 9, together with the corresponding diffraction pattern indicating the relative positions of aperture and image-forming beams. In projection down [001], the structures of the clino- and ortho-enstatite forms are *identical*, as can be verified by construction of the appropriate ($hk0$) reciprocal lattice sections. Nevertheless, as a result of multiple diffraction, the larger periodicity of the ortho-form is sometimes evident in electron micrographs, as is the twinning (either isolated or on a unit cell scale), which necessitates a $b/2$ displacement. A typical micrograph (Plate 10) shows this very well along the line AA', with a second case, at BB' and CC', where two such twin planes are inserted some 9 Å apart, producing, in effect, one unit cell of the ortho-enstatite structure.

HREM thus provides direct evidence for unit cell twinning in clino-enstatite both as a structural defect in itself, and as a mechanism for the conversion of clino- into ortho-enstatite. One might argue that a feature such as this could be inferred from single-crystal X-ray diffraction data, where evidence exists (*i.e.* streaking of reflections with $l \neq 0$) for these features, but electron microscopic studies have revealed additional defects, which were apparently not indicated in X-ray diffraction patterns. One of the most striking of these is that apparent twin boundaries are by no means confined to planes parallel to (100). On the contrary, several micrographs have been obtained illustrating 'stepped' twin composition planes, one of which is illustrated in Plate 11(a). When viewed in projection down [010], the twin plane can here be seen to be inclined to the trace of (100) at a slight angle. No interpretation of the detailed features of the structure at the 'steps' so

[67] S. Iijima and P. R. Buseck, *Amer. Mineral.*, 1975, **60**, 758.
[68] P. R. Buseck and S. Iijima, *Amer. Mineral.*, 1975, **60**, 771.

produced is yet available. Another type of non-planar twin interface is shown in Plate 11(b). In this case, viewed down [001], the twin plane can be seen to change direction radically, being parallel at one point to the trace of (010). The suggestion has also been advanced that, apart from twinning, antiphase boundaries exist, where the register of the two types of chain in the structure, which normally alternate, is upset. Buseck and Iijima [68] considered the possibility that apparent twin interfaces such as those in Plate 11(b) are in reality anti-phase boundaries, and conclude that the magnitude of the 'steps' in the interface (multiples of *two* chain spacings) favour the twinning model, although, under ideal kinematic conditions the twinned structure would be indistinguishable from the untwinned variety, whereas a structure with anti-phase domains would not. The exact nature and interpretation of micrographs of this type are therefore somewhat uncertain.

Since twinning can occur on such a microfine scale in clino-enstatite, and because ortho-enstatite may be regarded as clino-enstatite twinned every 9 Å, the possibility of other polytypes, with twin planes slightly farther apart, arises. Such arrangements have been observed, some being shown in Plate 12. Using the criterion that a given arrangement could be classed as a unique polytype if three or more successive repeat units could be observed, Buseck and Iijima [68] described structures with 27, 36, and two with 54 Å repeat distances in the x-direction. These structures correspond to twin planes which are respectively 18, 27, 45, and alternately 27 and 18 Å apart. In no cases, however, were such structures observed over comparatively large (*i.e.* several hundred Ångstroms) regions of any one crystal.

Other features observed in HREM of enstatite include the exact nature of parting faults and cleavage cracks, which are related to twinning planes. No evidence to suggest the presence of the high-temperature phase, proto-enstatite, has been yet found, although it could be argued that with an a-axis of 9 Å this species, if present, would be indistinguishable from the clino-enstatite. However, direct resolution of the (002) and (20$\bar{2}$) fringes in the twinned clino-enstatite ruled out its presence in most cases.

Orthopyroxenes.—The general term 'orthopyroxenes' is usually applied to systems which lie close to the solid solution series $MgSiO_3$–$FeSiO_3$, predominantly at the $MgSiO_3$ end, but also contain some calcium. A solubility gap exists between pyroxenes of this composition and the augites, which have up to 50% of the magnesium replaced by calcium, and consequently most orthopyroxenes with appreciable calcium content will display some form of augite exsolution in the orthopyroxene host, depending upon the conditions of cooling. Exsolution in systems of this type usually occurs topotactically, with the orthorhombic *Pbca* orthopyroxene sharing (100) planes with the monoclinic $C2/c$ augite phase. This has been demonstrated optically [69] and by single-crystal X-ray diffraction methods,[70, 71] and recent electron-microscopic studies have confirmed this, and, in addition, cast new light on the finer points of the exsolution mechanism. In particular, the Ca:Mg ratios of the exsolved lamellae visible in low-resolution electron-microscopic studies have been determined using analytical electron

---

[69] A. Poldervaart and H. H. Hess, *J. Geol.*, 1951, **59**, 472.
[70] M. G. Bown and P. Gay, *Amer. Mineral.*, 1959, **44**, 592.

microscopy,[72] and agree with the accepted mechanism. In an orthopyroxene from the Sillwater complex (Montana), Champness and Lorimer [73,74] observed relatively narrow augite lamellae (ca. 0.5 μm in width) parallel to the (100) planes of the matrix, as was expected, but also noted considerable finer detail in the host regions between the augite lamellae. In particular, regions of the host close to the augite lamellae were almost completely precipitate-free [Plate 13(a)], whereas in the central regions between lamellae distinctive Guinier–Preston (GP) zones were observed. Thin, apparently coherent lamellae (AA′) were also noted. At higher resolution,[74] the GP zones can be studied directly [Plate 13(c)]. High-resolution micrographs of this type reveal that the zones are only 18 Å thick, with a diameter ranging, in the Stillwater sample, from 50 to 2000 Å. In a similar sample from the Bushveld complex, the thin, coherent (100) lamellae were much more readily observed, high-resolution micrographs [Plate 13(b)] revealing that these were monoclinic in nature, but the 9 Å (100) spacing observed was characteristic of the $P2_1/c$ monoclinic form rather than the $C2/c$ equilibrium augite phase. Multiple diffraction in the $C2/c$ structure could not give rise to fringes of this type, and analytical electron microscopic studies have shown that these precipitates, like the equilibrium augite lamellae, are also richer in calcium. Again, like the augite lamellae, these are surrounded by a precipitate-free zone. It therefore seems apparent that the exsolution process in orthopyroxenes produces three distinct types of precipitate, namely the main $C2/c$ augite lamellae, finer lamellae with the $P2_1/c$ space-group, and homogeneously nucleated GP zones.

One distinct difference between the principal augite lamellae and the finer, $P2_1/c$ lamellae in exsolved orthopyroxenes lies in the nature of their interface with the host orthopyroxenes. Whereas the finer, primitive lamellae [Plate 13(b)] appear to be entirely coherent with the matrix, very reminiscent of the twinning in clino-enstatite, the equilibrium augite phase possesses growth ledges at the matrix–lamellae interface some 18—72 Å in height, and it is believed that these ledges are associated with the process by which thickening of the lamellae occurs. HREM can reveal the nature of these ledges, one micrograph from a study by Vander Sande and Kohlstedt [75] being illustrated in Plate 14. Here the 18.3 Å periodicity of the host orthopyroxene can clearly be seen, with some subsidiary fringe structure due to the inclusion of higher order beams in the objective aperture, the larger periodicity stopping abruptly at the interface where the 4.6 Å (200) fringes of the $C2/c$ augite phase begin. The growth ledge (at BC) can clearly be seen, as can the terminating (200) augite fringes indicating the presence of dislocations. Motion of the ledges parallel to the augite lamellae can then be envisaged, thereby thickening the lamellae at the expense of the orthorhombic host material.

Clinopyroxenes.—These are generally classified as pyroxenes with a greater iron content than the orthopyroxenes discussed above. As with the orthopyroxenes, however, the chief characteristics of clinopyroxenes lies in their exsolution

[71] M. G. Bown and P. Gay, *Mineral. Mag.*, 1960, **32**, 379.
[72] G. W. Lorimer and P. E. Champness, *Amer. Mineral.*, 1973, **58**, 243.
[73] P. E. Champness and G. W. Lorimer, *J. Mater. Sci.*, 1975, **84**, 467.
[74] P. E. Champness and G. W. Lorimer, *Phil. Mag.*, 1974, **30**, 357.
[75] J. B. Vander Sande, and D. L. Kohlstedt, *Phil. Mag.*, 1974, **29**, 1041.

phenomena. Two types of exsolution can be visualized, these being exsolution of a monoclinic $C2/c$ augite phase from a $P2_1/c$ pigeonite, or *vice versa*, or exsolution between an augite and a *Pbca*-type orthopyroxene. One can also visualize one type of exsolution following another, the change of exsolution type occurring when the pigeonite monoclinic structure ceases to be stable and inverts to the orthopyroxene form. Where the lamellae formed are relatively coarse and easily characterized, optical studies [69] have shown that augite and pigeonite exsolve one another with a common (001) plane, unlike the case of orthopryroxene–augite exsolution. Single-crystal X-ray studies [71] have confirmed this general picture, although some diffusiveness of $h + k = 2n + 1$ reflections in pigeonite crystals has been interpreted in terms of microfine exsolution of augite.

Electron microscopic investigations of clinopyroxenes have been of comparatively recent origin, and, to date, have concentrated only on low-resolution diffraction contrast studies. Nevertheless, considerable light has been shed on the mechanism of exsolution in clinopyroxenes.[76–79] Owing to the close similarity between the augite and pigeonite structures, the principal mechanism appears to be not the nucleation and growth process of the orthopyroxenes, but a process of spinodal decomposition. Electron microscopic investigations in basalts have revealed structures consisting of coherent modulations parallel to both (001) and (100), the former having by far the greater amplitude, however, and it would appear that the second type of structure is unable to coarsen beyond modulations with a periodicity > 200 Å. In reciprocal space, the exsolution process can be noted by a region of diffuse scattering around $h + k = 2n$ reflections in the $P2_1/c$ pigeonite diffraction pattern, which gradually alter until distinct satellite maxima appear.

Spinodal decomposition, however, appears to be a mechanism associated chiefly with high cooling rates. At lower rates the traditional nucleation and growth process appears to take over.[80] However, specimens have also been found which are thought to result from intermediate cooling rates.[81,82] In such cases heterogeneously nucleated (001) lamellae have been clearly observed with a modulated type of structure between them. Also significant was the instance of (100) lamellae (although of very small dimensions) nucleated at ledges in the main (001) lamellae.

As yet, no exhaustive high-resolution lattice imaging studies on clinopyroxenes have been undertaken. Such studies, however, may prove extremely valuable in elucidating the exact nature of the spinodal decomposition process, in order to establish whether the structure modulation is a true modulation, or merely a very fine scale nucleation process with interfaces of extremely low intrinsic energy.

---

[76] J. M. Christie, J. S. Lally, A. H. Heuer, R. M. Fisher, D. T. Griggs, and S. V. Radcliffe, *Proc. 2nd Lunar Sci. Conf.* (*Geochim. Cosmochim. Acta, Suppl. 2*), 1971, **1**, 69.
[77] P. E. Champness and G. W. Lorimer, *Cont. Min. Pet.*, 1971, **33**, 171.
[78] P. E. Champness and G. W. Lorimer, 'Proceedings of the Fifth International Materials Symposium', Berkeley, 1972, p. 1245.
[79] J. S. Lally, R. M. Fisher, J. M. Christie, D. T. Griggs, A. H. Heuer, G. L. Nord, and S. V. Radcliffe, *Proc. 3rd Lunar Sci. Conf.* (*Geochim. Cosmochim. Acta, Suppl. 1*), 1972, 401.
[80] P. A. Copely, P. E. Champness, and G. W. Lorimer, *J. Petroleum*, 1974, **15**, 41.
[81] P. A. Copely, Ph.D. thesis, *Manchester University*, 1973.
[82] G. L. Nord, J. S. Lally, A. H. Heuer, J. M. Christie, S. V. Radcliffe, D. T. Griggs and R. M. Fisher, *Proc. 4th Lunar. Sci. Conf.* (*Geochim. Cosmochim. Acta, Suppl. 4*), 1973, **4**, 953.

4 Pyroxenoids

The pyroxenoids are a family of minerals with the general formula $MSiO_3$, where M can be Ca^{2+}, Mn^{2+}, Fe^{2+}, and possibly also Mg^{2+}. In the pyroxenoid series the structures are based upon similar principles to those of the pyroxenes, but the 'chain repeat' is larger, being three SiO_2 tetrahedra in wollastonite, pectolite, and bustamite, five in rhodonite, and seven in pyroxmangite. The various possible types of chain are illustrated in Plate 15. On the basis of chemical evidence, distinction between the pyroxenoid minerals is somewhat difficult. Wollastonite is commonly regarded as having the composition $CaSiO_3$, but considerable amounts of iron can be substituted for calcium, and possibly some managanese also. Pectolite appears to have a unique composition, namely $Ca_2NaH(SiO_3)_3$, with some form of hydrogen bonding between the (SiO_3) chains, but bustamite can only be written with a general formula $(Ca,Mn,Fe)SiO_3$. This latter formula can also be applied to both rhodonite and pyroxmangite, and possible structural relationships are discussed below. Virtually all work in the pyroxenoid series has been concentrated on wollastonite, with no electron-optical studies on the other members, which will therefore be discussed only briefly.

Wollastonite.—Three structures corresponding to the composition $CaSiO_3$ have been reported. One of these, pseudowollastonite,[83] has been shown to have a ring structure isomorphous with $SrGeO_3$, and as this cannot be classified as a true pyroxenoid structure it will not be discussed further. The other two forms of wollastonite both adopt the characteristic three-tetrahedra chain, one with a triclinic structure,[84] and one with a monoclinic arrangement,[85] the structure of the triclinic form, projected onto the (100) planes and along the [001] axis, being shown in Plate 16. The monoclinic form bears a close similarity to the triclinic structure, which suggests that the two variants are related by a polytypic transformation. If an 'idealized' structural view of wollastonite is taken, it can be described in terms of an orthogonal 'unit', with $2_1/c$ symmetry [outlined in Plate 16(a)], and these units will superimpose exactly in [010] and [001] directions to form a structural 'slab' or layer. In a direction perpendicular to the slabs, however, packing considerations prohibit any direct superposition of successive units, and two possible stacking offsets or displacement vectors exist, these being either $+0.25b - 0.11c$ or $-0.25b - 0.11c$. Regular repetition of either one of these will then yield the triclinic structure, whereas regular alternation of both vectors will produce the monoclinic form. Slight distortions of the actual structure away from the idealized 'unit' which are actually observed will then probably arise as the structure adopts itself to the particular mode of stacking employed. Twinning of wollastonite about [010], which is commonly observed, can also be explained in terms of a change in displacements, where a crystal which has been adopting one displacement suddenly changes to using the alternative one. Owing to the $2_1/c$ symmetry of the idealized 'unit', this will be equivalent to rotation about [010].

In reciprocal space, differences between the two polytypes are confined to rows with $k = 2n + 1$, other rows being identical in both forms. On the $k = 2n + 1$

[83] J. W. Jeffery and L. Heller, *Acta Cryst.*, 1953, **6**, 807.
[84] Kh. S. Mamedov, and N. Belov, *Doklady Akad. Sci. U.S.S.R.*, 1956, **107**, 463.
[85] F. J. Trojer, *Z. Krist.*, 1968, **127**, 291.

rows, however, the monoclinic structure is readily discernible by the presence of twice as many spots as those shown by the triclinic form. Furthermore, the monoclinic structure can also be distinguished from the twinned triclinic structure, which also gives a doubling of spots on these rows, by the fact that spots corresponding to the monoclinic form are shifted by 0.25 a^* (triclinic) away from the positions of triclinic spots.[86] In his structure determination of the monoclinic form, Trojer reported [85] observing a faint doubling of spots on $k = 2n$ rows, but this has not been confirmed by later workers.

Planar Defects in Wollastonite. Samples of wollastonite are frequently observed to give diffraction patterns characteristic of neither monoclinic nor triclinic forms, showing instead, rows of spots with $k = 2n + 1$ which are streaked out parallel to x^*. This was first interpreted by Jeffery [87] as resulting from a situation of a disordered intergrowth of both forms, where the sequence of displacements was irregular throughout the crystal. Further single-crystal X-ray studies by Wenk [88] showed similar features, and also indicated apparently more complex polytypes; in particular a four-layer triclinic structure was described, but this has been subsequently explained [86] as an intergrowth of monoclinic and *twinned* triclinic forms. In this latter study a sample of Californian wollastonite showing considerable disorder was examined in some detail by X-ray single-crystal analysis, and it was concluded that considerable short-range order was present. The complexities of the 'average Patterson function' [89] required for this type of analysis were so great that no determination of the relative sizes of triclinic and monoclinic regions within the crystal was possible.

Electron-microscopic lattice imaging methods provide the obvious means of studying a disorder of this type, but several problems of image interpretation exist, arising principally from the fact that the heavier atoms in the structure (Ca,Si) which, in fact, superimpose in a projection along [001], are too close together to allow a simple interpretation of images in terms of the projected charge density approximation (PCD),[90] unlike the situation that obtains for the case of tungsten–niobium oxides.[91] However, where whole unit cells are displaced, as occurs in a polytypic disorder, simple lattice fringe images can with profit be employed. (This is not, in general, the case if structural changes within the unit cells are being investigated.) In wollastonite, therefore, a resolution of only 7 Å is adequate, and there is the added advantage that dark-field methods can be employed, consequently increasing image contrast. Images of this type have been obtained for several disordered wollastonites;[51,92,93] typical micrographs are shown in Plate 17. The simplest type of dark-field image, formed with an aperture around a pair of $k = 2n$ spots [Plate 17(b)] gives rise ideally to a series of regular fringes corresponding to crystallographic planes unaffected by the disorder. Consequently the presence of irregularities such as that at AA' is surprising, and although these

[86] D. A. Jefferson and M. G. Bown, *Nature (Phys. Sci.)*, 1973, **245**, 43.
[87] J. W. Jeffery, *Acta Cryst.*, 1953, **6**, 821.
[88] H. R. Wenk, *Cont. Min. Pet.*, 1969, **22**, 238.
[89] W. Cochran and E. R. Howells, *Acta Cryst.*, 1954, **7**, 412.
[90] D. F. Lynch, A. F. Moodie, and M. A. O'Keefe, *Acta Cryst.*, 1975, **A31**, 300.
[91] J. G. Allpress, E. A. Hewatt, A. F. Moodie, and J. V. Sanders, *Acta Cryst.*, 1972, **A28**, 528.
[92] D. A. Jefferson and J. M. Thomas, *Acta Cryst.*, 1975, **A31**, S295.
[93] W. F. Muller and H. R. Wenk, *Acta Cryst.*, 1975, **A31**, S294.

effects could arise through multiple scattering, it seems very likely that they signify some hitherto unsuspected structural modification. If the same aperture is placed around a portion of the $k = 2n + 1$ streak [Plate 17(c)] a one-dimensional (systematics) lattice image is again generated, with fringes of irregular spacing, delineating monoclinic and triclinic regions, as at BB' and CC', respectively, by the different spacings observed in these regions. In addition, regions also appear which show a greater periodicity than the 15.4 Å of the monoclinic form, as at DD', and these can be interpreted either as true superstructures, or as regions of the alternative twinned triclinic form. This ambiguity can be resolved by employing two-dimensional images of the type shown in Plate 17(d), where the region DD' can be identified as the alternative triclinic form. Using this type of image, therefore, the entire pattern of disorder across a crystal can be elucidated and compared with X-ray single-crystal data, and the concept of interspersing one type of displacement into a crystal containing principally the other displacement (*i.e.* a $b/2$ stacking fault) and thereby producing small regions of the monoclinic form can be verified directly. Similar results can also be observed using bright-field microscopy. Plate 18 shows a typical image from a wollastonite from New York State,[94] where again the polytypes present and the disorder can be clearly observed.

Linear Defects and other Anomalous Features in Wollastonite. In addition to confirming the stacking disorders inferred from X-ray data, lattice imaging techniques have also revealed defects of a type hitherto unsuspected. Plate 19(a) shows a dark-field, one-dimensional image of Californian wollastonite, with a region of 7.7 Å fringes ostensibly terminating within a crystal. Plate 19(b) shows a similar effect in a different sample, this time in bright-field, which shows the arrangement more clearly. In such a defect one triclinic form actually terminates within the other, producing, in effect, an edge dislocation with Bergers vector $b = \frac{1}{2}[010]$. Phakey and Liddell [95] have also reported defects of this type; its precise structure remains somewhat enigmatic, as current electron microscopic resolution is insufficient for recording the exact atomic arrangement. The model utilizing short lengths of pyroxene chain inserted into the wollastonite chain,[96] illustrated in Plate 19(c), would preserve continuity across the defect: short lengths of a rhodonite-type chain could equally well be used. A further type of line defect has also been reported,[97] a micrograph being reproduced in Plate 20. In this case it appears that an extra 'slab' of structure has been inserted only incompletely into the crystal, producing a classical 'edge' dislocation, although no apparent change in structural type occurs away from the dislocation core. Once again the structural perturbations at the centre of such a defect remain undetermined, although more disruption could be expected than with the preceding one.

Apart from the defects described above, several other puzzling electron microscopic contrast features have been observed in wollastonites. One of the most distinctive is that typified by Plate 21(a). Here an abrupt change in contrast occurs along EE', with, however, no alteration in the stacking sequence, although the

[94] J. L. Hutchison and A. C. McLaren, *Cont. Min. Pet.*, 1976, **55**, 303.
[95] P. P. Phakey and N. A. Liddell, 1975, in preparation.
[96] D. A. Jefferson and J. M. Thomas, in 'Developments in Electron Microscopy and Analysis', ed. J. A. Venables, Academic Press, London, 1975, p. 275.
[97] J. Morales, D. A. Jefferson, J. L. Hutchison, and J. M. Thomas, *J.C.S. Dalton*, 1977 (in press).

contrast variation appears to extend completely over the crystal. A proposed explanation lies in the suggestion of Wenk *et al.*[98] that (100) reflection twinning, coupled with translation at the interface, may occur, a schematic illustration of this, together with a diagram of the normal triclinic structure, being shown in Plate 21(b). Although the two arrangements look very similar in projection, the direction in which the tetrahedra point changes across the interface of the reflection twin, as the $-0.11c$ component of either displacement vector will be reversed. Twinning of this type has been observed in X-ray diffraction patterns of wollastonite but on a macroscopic scale only, and it would appear that this type of twin could only be investigated by producing wollastonite samples oriented such that the electron-beam is parallel to [010], a problem which should be resolvable using the technique of ion-thinning.

Other Pyroxenoids.—Of the other pyroxenoid minerals, both pectolite and bustamite have a very similar structure to wollastonite. Pectolite[99] possesses an almost identical arrangement, with a slight decrease in the a-dimension attributed to hydrogen-bonding between adjacent slabs. In bustamite,[100] the structure is again very similar, although with considerable amounts of iron and manganese present a superstructure is adopted, with a doubling of a and c dimensions. Polytypism has not been observed in pectolite, but electron diffraction studies of bustamite[93] have shown similar effects to those found in wollastonite. Rhodonite and pyroxmangite[101,102] both have triclinic structures, utilizing the longer-repeat pyroxenoid chains: no polytypism has been reported in either of these, and an electron-microscopic study of a Devon rhodonite by the present authors failed to reveal any type of stacking disorder.

Although it would seem likely that, as is the case in the pyroxene group, opportunities for distinct chemical compositional relationships between the various members of the pyroxenoid family should exist, our present knowledge of this system is somewhat vague. Early experimental studies by Voos[103] showed a complete solid solubility between wollastonite, bustamite, and rhodonite, but this is at variance with the known optical properties of these minerals, and a possible explanation for this apparent paradox[104] may lie in the suggestion that solid solubility is complete at higher temperatures, with exsolution occurring as the temperature is lowered. This has been confirmed by single-crystal X-ray studies.[105] As regards rhodonite and pyroxmangite, co-existing samples of both have been found,[106,107] but doubts have been expressed as to the equilibrium nature of these assemblages, although it remains possible that the two structures are polymorphs of the same mineral.

[98] H. R. Wenk, W. F. Müller, N. A. Liddell, and P. P. Phakey, in ref. 14, p. 331.
[99] M. J. Buerger, *Z. Krist.*, 1956, **108**, 248.
[100] M. J. Buerger, *Proc. Nat. Acad. Sci. U.S.A.*, 1956, **42**, 113.
[101] F. Liebau, W. Hilmer, and G. Lindemann, *Acta Cryst.*, 1959, **12**, 182.
[102] F. Liebau, *Acta Cryst.*, 1959, **12**, 177.
[103] E. Voos, *Z. anorg. Chem.*, 1935, **222**, 201.
[104] F. Liebau, M. Sprung, and E. Thilo, *Z. anorg. Chem.*, 1958, **297**, 213.
[105] L. S. Dent Glasser and F. P. Glasser, *Acta Cryst.*, 1961, **14**, 818.
[106] C. E. Tilley, *Amer. Mineral.*, 1937, **22**, 720.
[107] M. C. Burrell, Ph.D. thesis, Harvard University, 1942.

5 Amphiboles

The amphiboles constitute another class of silicates, in which the SiO_4 tetrahedra are linked in double chains rather than the single chains found in pyroxenes. The recognition by Schaller[108] that tremolite had the formula $Ca_2Mg_5Si_8O_{22}(OH)_2$, with hydroxyl ions an essential component, was followed by Warren's X-ray structure determination[109] and demonstration that several amphiboles of widely varying composition had very similar structures.[110] The wide range in composition has given rise to a plethora of varietal names, the family name 'amphibole' (Greek 'amphibolos' = ambiguous)[111] being most appropriate.

The double silicate chains are the essential feature of all amphiboles, having the composition $(SiO_4O_{11})_n$. The repeat distance along them is about 5.3 Å – the c parameter for the unit cell. These double chains are sandwiched between planes of cations, as shown in Figure 4: M_1, M_2, and M_3 sites contain Mg, Fe, Al, etc.;

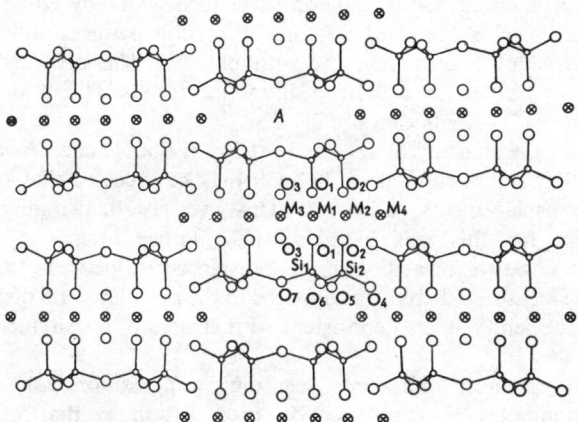

Figure 4 *Projection of clinoamphibole on* (001), *showing double* $[Si_8O_{22}]$ *chains end-on, and positions of* M_1, M_2, M_3, M_4, *and 'A' cations*
(Reproduced by permission from *Acta Cryst.*, 1960, **13**, 291)

M_4 sites contain Ca, Na, K, etc., and there may be an additional A-site cation, such as Ca, Na, etc. In most amphiboles the stacking produces a monoclinic cell, as in tremolite, space group $C2/m$, $a = 9.84$, $b = 18.05$, $c = 5.27$ Å, $\beta = 104.7°$. For amphiboles with low (Ca, Na, K) content the stacking may result in an orthorhombic cell, as an anthophyllite[112] $(Mg,Fe^{2+})_7[Si_8O_{22}](OH,F)_2$ with $a = 18.6$, $b = 18$, $c = 5.3$ Å; space group *Pnma*. The compositional limits for ortho-amphiboles have been studied from a point of view of crystal chemistry by Whittaker.[113]

---

[108] W. T. Schaller, *U.S. Geol. Surv.*, 1916, Bull. 610.
[109] B. E. Warren, *Z. Krist.*, 1930, **72**, 42.
[110] B. E. Warren, *Z. Krist.*, 1930, **72**, 493.
[111] C. Haüy, Traite de Mineralogie (Paris), 1801.
[112] B. E. Warren and D. I. Modell, *Z. Krist.*, 1930, **75**, 161.
[113] E. J. W. Whittaker, *Acta Cryst.*, 1960, **13**, 291.

A number of the amphiboles became commercially important as asbestos minerals, finding a wide range of uses from Roman funeral cloths to reinforcing fibres. Since electron microscopy has been restricted mainly to fibrous varieties they will be dealt with in detail. These include tremolite, grunerite $(Fe^{2+})_4(Fe^{2+},Mg)_3[Si_8O_{22}](OH,F)_2$ (as the fibrous form it is known as amosite), anthophyllite, and riebeckite, $Na_2Fe_3^{2+}Fe_2^{3+}[Si_8O_{22}](OH,F)_2$ (known in fibrous form as crocidolite).

X-Ray studies of fibrous amphiboles are not numerous.[114-118] In a study of amosite[116] it was suggested that there could be some disorder in the stacking of the silicate chains. The main contributions of HREM have been in the elucidation of this and other kinds of disorder, in fibrous amphiboles.

In his study of amosite and crocidolite Chisholm[119] attempted to explain some apparently anomalous electron diffraction results in terms of frequent planar defects on (010) and microtwinning on (100). Although the evidence for these effects was sparse, confirmation came in subsequent work by Hutchison, Irusteta and Whittaker,[120] who described electron diffraction patterns and lattice images for amosite, crocidolite, tremolite, and anthophyllite. This provided explanations for the apparently anomalous electron diffraction results[119] and also agreed with many of the earlier X-ray findings.[116]

As a result of the stacking of successive strips of double chains which gives rise to the monoclinic cell in clino-amphiboles, it may be expected that stacking faults, with opposite displacements, can occur. This gives rise to twinning on (100) and direct evidence for this was found in $(h0l)$ lattice images of amosite and crocidolite.[116] It was associated with a $c/2$ displacement between twin boundaries which could be measured directly (as opposed to the postulated $c/3$ displacement).[119] The $c/2$ displacement was also consistent with structural requirements for atomic matching.

The crystal chemical arguments regarding composition ranges for ortho- and clino-amphiboles[113] can also be used to show that stacking faults and related microtwinning should be restricted to the ortho-amphiboles, grunerite, cummingtonite $(Mg,Fe^{2+})_7[Si_8O_{12}](OH,F)_2$, and clinoholmquistite $Li_2(Mg,Fe^{2+})_3(Al,Fe^{3+})_2[Si_8O_{22}](OH,F)_2$, *i.e.* those with small ions (Mg,Li and also Fe) in the M_4 sites. This was qualitatively consistent with the finding that the twins observed in crocidolite had a lower probability and a markedly different distribution from those in amosite. The crocidolite twins were suggested to have arisen possibly by deformation. (100) twins have also been observed in 'nephrite', a fibrous actinolite $Ca_2Fe_5[Si_8O_{22}](OH,F)_2$, as shown in Plate 22, but their occurrence is rare.[121]

The term 'Wadsley defect' was originally used to describe planar coherent intergrowths in crystallographic shear structures, with specific reference to complex

---

[114] J. C. Rabitt, *Amer. Mineral.*, 1948, **33**, 263.
[115] E. J. W. Whittaker, *Acta Cryst.*, 1949, **2**, 312.
[116] E. J. W. Whittaker, *Brit. J. Appl. Phys.*, 1950, **1**, 162.
[117] R. I. Garrod and C. S. Rann, *Acta Cryst.*, 1952, **5**, 285.
[118] L. W. Finger, *Carnegie Inst. Year Book*, 1970, **68**, 283.
[119] J. E. Chisholm, *J. Mater. Sci.*, 1973, **8**, 475.
[120] J. L. Hutchison, M. C. Irusteta, and E. J. W. Whittaker, *Acta Cryst.*, 1975, **A31**, 794.
[121] L. G. Mallinson, M.Sc. thesis, Aberystwyth, 1976.

non-stoicheiometric oxides.[122] Their existence in silicates was postulated by Chisholm[119] and the concept amplified in a subsequent review[13] in which structural relationships between pyroxenes (single chain), amphiboles (double chain), and talc (sheet) were derived in terms of a crystallographic shear model. Direct evidence for such defects was obtained in lattice images[120] from both $(0k0)$ and $(0kl)$ diffraction patterns. It was possible to identify diffraction patterns over a wide range of tilt angles and, by this means, to set the fibrils in any desired orientation. Planar faults were found on (010) in amosite, tremolite, and anthophyllite, and it was shown that they probably consisted of inserted triple-chain lamellae (corresponding to the loss of one cation and replacement of the oxygens by two hydroxyls). In anthophyllite these faults were seen to segregate into domains of ordered triple-chain structure.[123] More recently, a study of nephrite (actinolite) has revealed similar n-chain defects,[124] but in varying widths up to at least $n = 6$ (Plate 23). Such defects indicate severe chemical inhomogeneity and in the extreme cases, the fault lamellae may be regarded as intergrown layers of embryonic talc. Intergrowth of talc and anthophyllite, on a coarser scale, has been recorded but the evidence was based on low resolution micrographs[119] or electron diffraction.[125]

Buseck and Iijima[37] observed 'antiphase boundaries' in kaersutite, a titanium-containing amphibole. These boundaries were detected in ab sections and were parallel to (100). In addition, some (100) boundaries did not appear to be antiphase but the possible structure was not discussed in any detail.

Ion-beam thinning has been used by Alario-Franco et al.[126] to obtain thin sections of crocidolite, normal to the fibre c-axis. In addition to the fibre morphology, lattice images from these sections (Plate 24) revealed triple-chain Wadsley defects on (010) and also (110) planar faults with displacement vector of $\frac{1}{2}b$ [010], in the ab plane. The resolution of these images was not sufficient to allow a detailed analysis of the fault plane.

Exsolution in Amphiboles.—In 1923 a miscibility gap between hornblende and cummingtonite was postulated by Asklund,[127] and subsequently this and several other miscibility gaps have been extensively documented.[128] Exsolution of calcium-rich (e.g. hornblende) from calcium-poor (e.g. cummingtonite, grunerite) amphiboles and vice versa, was detected by optical methods and X-ray diffraction,[129,130] prior to its demonstration by electron microscopic studies.[14,131] Grunerite was found as fine (100) platelets between (100) and ($\bar{1}$01) hornblende lamellae. Growth

---

[122] A. D. Wadsley and S. Andersson, in 'Perspectives in Structural Chemistry', ed. J. D. Dunitz and J. A. Ibers, Wiley, New York, 1970, Vol. 3.
[123] F. Liebau and K. F. Hesse, *Acta Cryst.*, 1975, **A31**, S74.
[124] J. L. Hutchison, D. A. Jefferson, L. G. Mallison, and J. M. Thomas, *Mater. Res. Bull.*, 1976, **11**, 1557.
[125] J. L. Hutchison, M. C. Irusteta, and E. J. W. Whittaker, in ref. 16, Vol. 1, p. 492.
[126] M. Alario-Franco, J. L. Hutchison, D. A. Jefferson, and J. M. Thomas, *Nature*, 1977, **266**, 520.
[127] B. Asklund, *Sveriges Geol. Unders. Ansb*, 1923, **17** (6).
[128] C. Klein, *J. Petroleum*, 1968, **9**, 281; *Amer. Mineral.*, 1969, **54**, 212.
[129] P. Eskola, *Amer. Mineral.*, 1950, **35**, 729.
[130] M. Ross, J. J. Papike, and K. W. Shaw, *Mineral. Soc. Amer.*, Special Paper, 1969, **2**, 275.
[131] M. F. Gittos, G. W. Lorimer, and P. E. Champness, *J. Mater. Sci.*, 1974, **9**, 184.

ledges were observed in the precipitates and their existence accounted for the slight angular difference between these planes and the 'average' interfaces.[132] Similar steps were described in exsolved pyroxenes.[75] The variations in calcium content between the two intergrown phases was verified by X-ray microanalysis.

6 Serpentine Minerals

The serpentines have the idealized formula $Mg_3Si_2O_5(OH)_4$ and structually are built up by alternate stacking of two layers: (Si_2O_5) silica tetrahedra networks interleaved with octahedral brucite $Mg(OH)_2$ layers. Two-thirds of the hydroxyls at the base of each brucite layer are replaced by oxygens on the apices of (SiO_4) tetrahedra. Owing to an inherent dimensional mismatch between these two types of layer [the $(SiO_5)_n$ network having a smaller repeat distance] the serpentines display departures from the simplified, planar structure, which may be manifested in distortions within the layers, dimensional changes *via* substitution of some of the cation, or by curvature of the component sheets (Plate 25).

Serpentines have been classified into three main groups,[133] the classification being based upon the type, and degree, of distortion present. They are, chrysotile, antigorite, and lizardite and in this section we will show how (and to what extent) HREM has contributed to current understanding of these minerals. A recent review [12] describes early applications of electron diffraction and low-resolution electron microscopy up to 1970, and gives a wealth of data which will not be duplicated here.

Chrysotile.—This is a fibrous serpentine, commercially important as an 'asbestos mineral' with a high tensile strength [134] (*ca.* 31 000 kg cm^{-2}). The silica-brucite [17,18,135,136] layers are wound round the fibre axis (usually *a*). As hinted earlier in Section 1, electron micrographs showed fairly uniform fibres, often with a central light (electron transparent) stripe along their length; this was interpreted in terms of hollow tubular morphology. This led to a series of X-ray studies of various chrysotiles [137—141] which supported the concept of a cylindrical structure. Lattice imaging studies have convincingly shown the validity of these studies,[21,142,143] and in particular, micrographs of cross-section chrysotile fibres [143] present startlingly clear evidence for the hollow tubes. Examples of these are shown in Plate 26 and it can be seen that both spirally wound and concentric lattices are present. Multilayer spirals, with several composite layers wound together [Plate 26 (a)]supported one of the earlier X-ray based hypotheses.[137] Single-layer spirals (*i.e.* only one composite silica/brucite layer) were occasionally observed

[132] P. Robinson, H. W. Jeffe, M. Ross, and C. Klein, *Amer. Mineral.*, 1971, **56**, 909.
[133] E. J. W. Whittaker and J. Zussman, *Mineral. Mag.*, 1956, **31**, 107.
[134] A. A. Hodgson, 'Fibrous Silicates', Royal Institute of Chemistry, Lecture Series No. 4, 1965.
[135] T. F. Bates, L. B. Sand, and J. F. Mink, *Science*, 1950, **111**, 512.
[136] W. Noll and H. Kircher, *Neues. Jahrb. Min., Monatsh*, 1951, 219.
[137] H. Jagodzinski and G. Kunze, *Neues Jahrb. Min., Monatsh*, 1954, 95.
[138] E. J. W. Whittaker, *Acta Cryst.*, 1954, **7**, 827.
[139] E. J. W. Whittaker, *Acta Cryst.*, 1955, **8**, 261, 265, 726.
[140] E. J. W. Whittaker, *Acta Cryst.*, 1956, **9**, 855.
[141] E. J. W. Whittaker, *Acta Cryst.*, 1957, **10**, 149.
[142] J. L. Hutchison, Ph.D. thesis, Glasgow, 1970.
[143] K. Yada, *Acta Cryst.*, 1971, **A27**, 659.

[Plate 26(b)] in apparent conflict with Whittaker's argument [140] that such structures would be incompatible with the two-layer unit cell. In addition, perfectly concentric structures, as shown in Plate 26(c), occur and it was noted that their frequency varied widely with locality. It was suggested [143] that growth of these concentric structures could take place by means of a dislocation mechanism, as proposed for the multilayer spirals.[137]

There has been considerable controversy due to a number of density measurements which suggested [144-147] total void volumes of 1—15% in bulk specimens, compared with 10—20% as derived from the observed structures. Various reasons have been advanced to account for the obvious discrepancy, but none is entirely satisfactory. There is good evidence from micrographs such as Plate 27 that some of the fibres contain 'plugs' of amorphous material – presumably a silica gel.[148] However, such features as this are not sufficient to account for the difference between theoretical and observed densities. Whittaker [149] suggested that the void volume in bundles of fibres (ca. 25% for the dimensions used) was filled with fragmented curved laths, in the hollow centres and also between fibres. The micrographs show no evidence for this. Presumably, the outstanding discrepancy is due either to artefacts in specimen preparation, or to difficulties in obtaining accurate density values of sealed blocks of fibres.

Para-chrysotile, with the fibre axis parallel to the crystallographic b-axis, has been identified by lattice imaging, but its occurrence is rare.

Various growth patterns, ranging from stepped layers to 'cone-in-cone' formation [22] (Plate 28) have been found, all of which contribute to the understanding of possible growth mechanism. Yada also described various 'growth defects' in the cross-sectioned fibres, but it is not clear to what extent they may be artefacts due to the microtome sectioning process itself. Ion-beam thinning procedures have been recently applied to fibrous and 'splintery' serpentines [150] (as will be discussed below) and it appears that less mechanical damage is incurred, although an amorphous layer is created by the ion beam.

Antigorite.—This material has a platy or lath-like morphology, and is characterized by a large a parameter, usually ca. 43 Å. The b and c parameters are close to ideal 'serpentine' values. The large a parameter arises from corrugation of the layers curved around the b-axis. This is in contrast with the complete wrapping-round of the layers in chrysotile. Clearly, there is scope for variation in the 'wavelength' of these corrugations, as was shown in electron diffraction patterns, a values between 16.8 and 51 Å having been recorded.[151-153] Lattice imaging studies of antigorites are few; a Manchurian sample ('Yu-Yen Stone') yielded fringes

[144] F. L. Pundsack, *J. Phys. Chem.*, 1956, **60**, 361.
[145] G. L. Kalousek and L. E. Muttart, *Amer. Mineral.*, 1957, **42**, 1.
[146] F. L. Pundsack, *J. Phys. Chem.*, 1961, **65**, 30.
[147] C. W. Huggins and H. R. Shell, *Amer. Mineral.*, 1965, **50**, 1058.
[148] T. F. Bates and J. J. Comer, *Clays and Clay Minerals*,, 1959, **6**, 237.
[149] E. J. W. Whittaker, *Acta Cryst.*, 1966, **21**, 461.
[150] B. A. Cressey and J. Zussman, *Canad. Mineral.*, 1976, **14**, 307.
[151] J. Zussman, G. W. Brindley, and J. J. Comer, *Amer. Mineral.*, 1957, **42**, 133.
[152] J. A. Chapman and J. Zussman, *Acta Cryst.*, 1959, **12**, 550.
[153] G. Kunze, *Fortschr. Mineral.*, 1961, **39**, 206.

ca. 90—100 Å in spacing,[154] and 40 Å fringes were later recorded for another variety.[152] More recent work by one of the authors (D. A. J.,unpublished) shows spacings and also the 'serpentine' sublattice, as can be seen in Plate 29. Variations in fringe spacing [152,155] (within single crystals) indicate variations in the layer corrugations and may be related to chemical inhomogeneities. Although the sensitivity of these images to crystal thickness, orientation, and defocus was recognized in these early studies, their quantitative effects have only recently been clearly understood. Studies of antigorites using high-resolution techniques coupled with X-ray microanalysis of small, structurally characterized areas, will be a fruitful area for further work.

Lizardite.—Serpentines with platy cleavage, which are neither chrysotile nor antigorite, have been defined as lizardite. It approaches most closely to the ideal serpentine structure but the crystallinity is usually limited in extent. A single-crystal X-ray investigation of a lizardite suggested three distinct types of disorder:[156] bending of platelets, layers displaced by $\pm\frac{1}{3}b$, and layers rotated by 180°. Although there are several papers describing general morphological features in low-resolution micrographs,[151—157] HREM evidence is lacking apart from a single micrograph of a synthetic lizardite (Plate 30) which shows a mosaic structure (50—80 Å domains) with an abundance of dislocations. The possibility of various stacking arrangements has been suggested but this awaits confirmation either by means of careful electron diffraction/tilting experiments or by lattice imaging of lizardite along the layers.

'Polygonal Serpentine'.—Krstanović and Pavlović [158] identified a number of clinochrysotile specimens which had unusually sharp X-ray fibre diffraction patterns and called them 'Povlen-type clinochrysotile'. A detailed study using X-ray and electron diffraction [159] confirmed the existence of these structures and additionally revealed a 'Povlen-type orthochrysotile' in some samples. The detailed structure was not identified with certainty, but a possible model was put forward which was consistent with the diffraction data: part of the structure consisted of a normal cylindrical chrysotile lattice, but this was enclosed by an outer shell of polygonally arranged flat layers. Concurrent with this work was an electron microscopic study of ion-thinned serpentinites,[150] including samples containing 'Povlen-type' chrysotiles. Where thinning had produced fibre cross-sections, micrographs presented good evidence for the polygonal structures, as shown in Plate 31. The largest fibre is incomplete but the polygonal nature is clearly seen. Such fibres were found with and also without cylindical cores.

'Garnierites'.—This term has been applied to nickel-containing serpentines which, because of their fine-grained, inhomogeneous, and poorly ordered nature have not yielded much to X-ray studies. Most electron microscopic studies have been

---

[154] G. W. Brindley, J. J. Comer, R. Uyeda, and J. Zussman, *Acta Cryst.*, 1958, **11**, 99.
[155] Y. Kamiya, M. Nonoyama, and R. Uyeda, *J. Phys. Soc. Japan*, 1959, **14**, 1334.
[156] J. C. Rucklidge and J. Zussman, *Acta Cryst.*, 1965, **19**, 381.
[157] N. M. Popov and B. B. Zvyagin, *Izvest. Akad. Nauk S.S.S.R., Ser. fiz.*, 1958, **23**, 670.
[158] I. Krstanović and S. Pavlović, *Amer. Mineral.*, 1964, **49**, 1769.
[159] A. P. Middleton and E. J. W. Whittaker, *Canad. Mineral.*, 1976, **14**, 301.

morphological, revealing detail of the microstructure. Most garnierites crystallize as either '7 Å' (serpentine-like) or '10 Å' (talc-like) types, depending upon basal spacings. An electron optical study by Uyeda et al.[160] was successful in 'resolving' these spacings and showed, in addition to varying morphology, intergrowth of 7 Å and 10 Å layers. No correlation between Ni^{2+} content and morphology was noted, and it was concluded that Ni^{2+} replacement of Mg^{2+} had no obvious effect on the crystal morphology.

7 Silicates based on the Mica Type of Layer

Whereas only kaolinite and the serpentines are based on the 1:1 (T:O) aluminosilicate layer, several distinct groups of minerals contain the 2:1 (T:O:T) layer, *i.e.* a layer with silicon–oxygen tetrahedral components on either side of the basic octahedral component. The four principal groups based upon this structural type are (i) the mica group and related minerals (illite, vermiculite, montmorillonite, together with talc and pyrophyllite), (ii) chlorites, (iii) the rare mineral zussmanite, and (iv) stilpomelane. Idealized structures of these are illustrated in Plate 32. In almost all cases, some form of linkage occurs between individual structural layers; in the micas, for example, this is achieved by the presence of larger, interlayer cations, usually Na^+, K^+, or Ca^{2+}, with one such cation for every six-membered ring on the layer, although this figure is progressively reduced in the series mica, illite, vermiculite, montmorillonite, reaching zero in the case of talc and pyrophyllite, where layers are held together merely by residual van der Waals forces. In chlorites, unlike the micas, linkages between layers are effected by means of a brucite-type layer, hydrogen-bonding between these and the silicon–oxygen components of mica layers holding the whole structure together. In contrast to the chlorites, where interlayer bonding is weaker than in micas, the corresponding bonds in zussmanite and stilpnomelane are much stronger, owing to interlayer silicon–oxygen linkages. In all the minerals of this group the principal type of planar defect found is a stacking disorder arising from polytypism, although evidence has been forwarded [13,119] for a form of crystallographic shear occurring in talc. Apart from pure structural determinations, relatively few X-ray investigations of these layered silicates have been undertaken, partly because of the scarcity of good single crystals and partly because most minerals in this group show considerable disorder. Electron microscopy and diffraction, however, has proved particularly successful in elucidating the polytypism of these minerals, and, when coupled with improved specimen preparation techniques, have yielded a detailed picture of the disorder in many systems.

Mica Group.—Polytypism in micas arises in an exactly analogous manner to that in silicon carbide, namely, the presence of close-packed layers of atoms in the structure. Two close-packed layers of atoms with their interstitial cations (Plate 32a), form the octahedral component of the structural layer: consequently, although the presence of interlayer cations ensures that SiO_4 tetrahedra on the lower surface of one layer superimpose above those on the upper surface of the layer beneath, no such superposition is possible between tetrahedra on either side

[160] N. Uyeda, Pham Thi Hang, and G. W. Brindley, *Clays and Clay Minerals*, 1973, **21**, 41.

of one octahedral component. A number of offsets or displacement vectors are therefore possible (so far as subsequent tetrahedra are concerned) at the centre of each layer. In general, displacement vectors can be divided into two groups, those producing rhombohedral stacking of octahedra in successive layers and those giving a hexagonal arrangement, and for each octahedron within the unit cell, a possible displacement vector in each group will exist. Using single-crystal X-ray techniques, systematic surveys of polytypes in the mica group have been possible,[161,162] and the simplest possible structures characterized. Six displacement vectors are possible in micas, falling into two groups of three. Repeated usage of any one vector produces a one-layer monoclinic polytype (1M), whereas with the alternate use of two vectors from the same group a two-layer monoclinic form (2M$_1$) is produced, and a regular sequence of all three vectors in any one group produces a three-layer trigonal structure (3T). If vectors are used from both groups, alternate use of two vectors will produce either a two-layer monoclinic structure (2M$_2$), (structurally distinct from 2M$_1$) or a two-layer orthorhombic form (2O), while regular use of all six vectors give rise to a six-layer hexagonal (6H) polytype. With longer sequences of vectors, the complexity of the polytypes obtained can be very great, and more recent work [163] has shown that polytypes with c-spacings in the region of 200 Å are possible.

One important respect in which mica polytypism appears to differ from that in materials such as silicon carbide is in the dependence, at least partial, upon the degree of chemical substitution in the layers. The general formula of the mica group is AB$_{2-3}$(Al$_{0-2}$Si$_{2-4}$O$_{10}$)(OH,F)$_2$ (A = Na$^+$, K$^+$, or Ca^{2+}; B = Fe^{2+}, Fe^{3+}, Mg^{2+}, Li$^+$, or Al^{3+}). The possibilities of substitutions are therefore very great indeed, some idea of the different possible combinations being shown in Table 3, which lists the more common varieties. Certain of the simpler polytypes are not commonly observed, particularly those involving vectors from both groups (the 2M$_2$, 2O, and 6H structures), and this led Radoslovich [164,165] to conclude that ditrigonal distortions of the six-membered rings of SiO$_4$ tetrahedra in a

Table 3 *Some of the more common cationic substitutional variants of the basic mica structure*

Muscovite	KAl$_2$(AlSi$_3$O$_{10}$)(OH)$_2$
Paragonite	NaAl$_2$(AlSi$_3$O$_{10}$)(OH)$_2$
Phlogopite	KMg$_3$(AlSi$_3$O$_{10}$)(OH)$_2$
Biotite	K(Mg,Fe)(AlSi$_3$O$_{10}$)(OH)$_2$
Polylithionite	K(Li$_2$Al)(Si$_4$O$_{10}$)(OH)$_2$
Zinnwaldite	K(Li,Fe,Al)(AlSi$_3$O$_{10}$)(OH)$_2$
Pyrophyllite	Al$_2$(Si$_4$O$_{10}$)(OH)$_2$
Talc	Mg$_3$(Si$_4$O$_{10}$)(OH)$_2$
Margerite	CaAl$_2$(Al$_2$Si$_2$O$_{10}$)(OH)$_2$
Xanthophyllite	CaMg$_3$(Al$_2$Si$_2$O$_{10}$)(OH)$_2$

[161] S. B. Hendricks and M. E. Jefferson, *Amer. Mineral.*, 1939, **24**, 729.
[162] J. V. Smith and M. S. Yoder, *Mineral. Mag.*, 1956, **31**, 209.
[163] M. Ross, M. Takeda, and D. R. Wones, *Science (Washington)*, 1966, **515**, 191.
[164] E. W. Radoslovich, *Nature*, 1959, **183**, 253.
[165] E. W. Radoslovich, *Acta Cryst.*, 1960, **13**, 919.

determined structure of muscovite prevented a mixing of displacement vectors from the two groups defined above, and suggested that the $2M_2$, $2O$, and $6H$ polytypes would require a degree of chemical substitution such that those rings were very nearly hexagonal. Later workers [166-168] have discussed such structural control in more detail, and experimental studies on lithium micas such as polylithionite, where the rings do exhibit the higher symmetry, have provided confirmatory evidence. However, some apparently anomalous systems do exist, as shown by a recent structural investigation of a $2M_2$ mica.[169]

With the exception of the final example above (a structure determination by electron diffraction) all structural investigations of mica have employed single-crystal X-ray diffraction techniques. No detailed investigation of a disordered phase using this technique has been reported, and apart from certain studies of dislocations in mica using diffraction contrast methods,[8,170] little electron-optical work has been carried out on well crystalline micas, although numerous investigations of mica-based clay minerals have been reported, chiefly to obtain morphological information. The principal difficulty in any electron-optical investigation of polytypism in micas, either by electron diffraction or direct lattice imaging, lies in the presence of the pronounced basal cleavage of these minerals, which ensures that in the normal, fine-grained samples used, crystals are in precisely the wrong orientation for studying the layer stacking arrangement. Where flakes of micaceous minerals are extremely thin, their edges sometimes curl up, and at the extremities a lattice image of $(00l)$ planes may be obtained. This method has been used in the case of natural and organo-montmorillonites,[171,172] but its usefulness is somewhat limited. More encouraging results can be obtained from the technique of cutting petrological thin sections (ca. 30 μm thick) with the layer planes normal to the section, and subsequent argon-ion thinning. This was first done by Phakey et al.[173] in an electron diffraction study, but is also readily applicable to lattice imaging studies; recently, lattice images of the $(0kl)$ and $(00l)$ planes of muscovite have been so obtained.[174] Plate 33(a) shows a lattice image of mica so obtained, with $(00l)$ and (020) fringes clearly visible. The pronounced change in direction of the (020) fringes at the plane AA' indicates a change in polytype from $1M$ to $2M_1$, which can be determined from the theoretical angles ($100°$ and $95°$) between $(00l)$ and (020) fringes for the various simple polytypes. Polytypism and disorder in micas have also been the subject of an investigation by Buseck and Iijima [37,52] using resin-embedded microtome-cut sections. Initial images [17] were of the one-dimensional systematics type only, displaying fringes which corresponded to the triple aluminosilicate layer in mica, but owing to the fact that (in the absence of multiple diffraction) all mica polytypes give rise to an identical set of $(00l)$ diffraction maxima, no information regarding the polytypes present

[166] M. Franzini, *Cont. Min. Pet.*, 1969, **21**, 203.
[167] M. Takeda and C. W. Burnham, *Mineral. J. Japan*, 1969, **6**, 102.
[168] J. W. McCauley and R. E. Newnham, *Amer. Mineral.*, 1971, **56**, 1626.
[169] A. P. Zhoukhlistov, B. B. Zvyagin, S. V. Soboleva, and A. F. Fedotov, *Clays and Clay Minerals*, 1973, **21**, 465.
[170] S. Amelinckx and P. Delavignette, *J. Appl. Phys.*, 1961, **32**, 341.
[171] L. M. Barclay and D. W. Thompson, *Nature*, 1969, **222**, 263.
[172] E. Suito, M. Arakawa, and T. Yoshida, *Proc. Int. Clay. Conf., Tokyo*, 1969, **1**, 757.
[173] P. P. Phakey, C. D. Curtiss, and G. Oestel, *Clays and Clay Minerals*, 1972, **20**, 193.
[174] D. A. Jefferson and J. M. Thomas, *Acta Cryst.*, 1975, **A31**, S295.

could be obtained. In a more recent study,[52] however, two-dimensional images with the electron beam parallel to [100] have given clear information as to the polytypes present in the crystal [Plate 33(b)] and show also a good correspondence between the image intensity and the projected charge density in the structure. With projections down [010], [Plate 33(c)] microscope resolution was inadequate to form a two-dimensional image, but these authors noted that, owing to multiple scattering in the thicker parts of the crystals, information regarding the polytypes present could be readily observed even in the one-dimensional systematics case. This phenomenon has also been reported in stilpnomelane and chloritoid (see below). Muscovite, although providing adequate lattice images in the above cases, is by no means the best material for this purpose, tending to give very thin flakes which are usually twisted or distorted. The more brittle micas (margerite and xanthophyllite), however, with partial or complete replacement of potassium by calcium, may be expected to be more stable and give thicker flakes in the [001] direction, and in view of data obtained from zussmanite and stilpnomelane, would almost certainly yield adequate lattice images more readily.

Chlorite Group.—The chlorite minerals are more complex than micas in that, although they are made up of a mica-type layer, they contain in addition a brucite-type hydroxide layer. In simple terms, these brucite-like layers are then bonded to the mica or talc layers *via* hydrogen bonds; there are no other interlamellar ions. The general formula of chlorite minerals is then $A_6B_8O_{20}(OH)_4 \, C_6(OH)_{12}$, the former part being that of mica, without the interlayer cations, the latter that of brucite. In general, A and C can be Mg, Fe, Al, while B is Al, Si. Some of the better characterized chlorite compositions are listed in Table 4. Polytypism in chlorites is related to that of micas, but, owing to the replacement of interlayer cations by the brucite-layer, an offset exists between silicon–oxygen components on adjacent surfaces of neighbouring mica layers, unlike the situation that prevails for the micas proper. A maximum of six offsets of this type is possible, making, when coupled with the possible offsets between silicon–oxygen components on the mica layer, a total of 12 possibilities for every 'chlorite' layer. Consequently the number and complexity of possible polytypes is very great. An initial survey of polytypism in chlorites by Brindley, Oughton, and Robinson [175] considered polytypes with one-layer monoclinic, and one-, two- and three-layer triclinic structures. Single-crystal X-ray structure determinations have been undertaken on both of the one-layer forms,[176,177] and on ortho-hexagonal polytypes.[178] Electron-

Table 4 *Some of the more distinct compositional variations of the chlorite family of minerals*

Auesite	$MgAl_2(Al_2Si_2O_{10})(OH)_2 \, Mg_3(OH)_6$
Corundophyllite	$(Mg,Fe)Al_2(Al_2Si_2O_{10})(OH)_2(Mg,Fe)_3(OH)_6$
Clinochore	$Mg_2Al(AlSi_3O_{10})(OH)_2 \, Mg_3(OH)_6$
Talc-chlorite	$Mg_3(Si_4O_{10})(OH)_2 \, Mg_3(OH)_6$

[175] G. W. Brindley, B. M. Oughton, and K. Robinson, *Acta Cryst.*, 1950, **3**, 408.
[176] M. Steinfink, *Acta Cryst.*, 1958, **11**, 191.
[177] M. Steinfink, *Acta Cryst.*, 1958, **11**, 195.
[178] M. Shirozu and S. W. Bailey, *Amer. Mineral.*, 1965, **50**, 868.

optical investigations of chlorites have been comparatively few, mainly owing to the sensitivity of the material to the electron beam, and this beam-sensitivity would preclude the use of ion-thinning of polished thin sections to reveal stacking disorder directly, as has been possible with micas. However, Brown and Jackson [179] employed the alternative ultra-microtome sectioning method of preparing thin sections, and have produced several micrographs of the (00*l*) lattice fringes in chlorite, one of which is reproduced in Plate 34. Lattice images obtained in this way, however, have proved to be of the one-dimensional (systematics) variety only, and two-dimensional images needed to determine the stacking sequence directly have yet to be observed.

Zussmanite.—The rare mineral zussmanite bears many similarities, both structurally and compositionally, to the mica minerals, and has the experimental advantage of occurring in crystals of a size suitable for X-ray and electron-microscopic investigations, with a good basal cleavage, although not so pronounced as that of micas. The determined structure of zussmanite [180] shows distinct differences from the mica layout in that the six-membered rings of the silicon–oxygen component are completely isolated from one another, and are linked by separate three-membered rings, the latter also linking the six-membered rings to their counterparts on adjacent layers. As in micas, larger cations sit between the six-membered rings on neighbouring layers, but, owing to the fact that there are Si–O–Si linkages between layers, such interlayer cations are not essential to maintain direct superposition between adjacent silicon–oxygen networks. Possible displacement vectors, therefore, arise solely at the level of the octahedral components. In all, 26 possible offsets or displacement vectors occur, in two groups of 13, and of these only two will produce rhombohedral stacking arrangements. The remainder give rise to triclinic structures, and when repeating units of more than one displacement vector are considered, the complexity of the polytypes which can form is very great. In addition to the complicated stacking arrangements, the presence of isolated six-membered rings of SiO_4 tetrahedra necessitates a larger unit cell than that of mica, and this is reflected in the compositional formula, which is $XA_{13}B_{18}O_{42}(OH)_{14}$, with X being K^+, Na^+, or a vacancy, A being Fe^{2+}, Al^{3+}, Mg^{2+}, or Mn^{2+}, and B being Si or Al.

Polytypism and disorder in zussmanite have been the subjects of detailed single-crystal X-ray investigation [181] which showed that, apart from the accepted rhombohedral form of zussmanite, most crystals showed a tendency to form other triclinic polytypes. Two types of triclinic structure have been identified, the simpler being a one-layer arrangement analogous to the 1M mica polytype, and the other a two-layer triclinic form analogous to the $2M_1$ mica polytype. Only the latter, however, could be observed in single crystals with a thickness >0.1 mm, the former only occurring in disordered crystals, where the predominant mode of stacking was of the rhombohedral variety. As in the case of most micas, it appeared that only one group of displacement vectors could be used at any one time. An electron microprobe study, allied to single-crystal investigation, also revealed that

[179] J. L. Brown and M. L. Jackson, *Clays and Clay Minerals*, 1973, **21**, 1.
[180] A. Lopes-Vieica and J. Zussman, *Mineral. Mag.*, 1969, **37**, 49.
[181] D. A. Jefferson, *Amer. Mineral.*, 1976, **61**, 470.

disorder could be directly correlated with the degree of chemical substitution in this mineral, the disorder increasing as interlayer potassium cations were removed, with a corresponding removal of Al^{3+} from the silicon–oxygen network. A consequent increase in the manganese content of the octahedral component was also noted, and the triclinic polytypes were believed to represent an intermediate stage between a completely disordered system, with no potassium present, and a regular rhombohedral structure with its full complement of interlayer cations.

As with micas, an electron-optical study, either by diffraction or by lattice imaging, requires the use of ion-thinned sections cut perpendicular to the basal cleavage if the stacking sequence is to be observed directly. However, zussmanite is much more suitable for this method than mica, existing in larger flakes, and readily yields lattice images of suitable quality. A typical image so obtained [Plate 35(a)][50] shows a region of the principal, rhombohedral structure intergrown with a two-layer triclinic structure. The exact arrangement of the two polytypes is shown in Plate 35(b). Analysis of polytypism in zussmanite by lattice imaging methods has proved to be remarkably simple, and unlike the case of wollastonite (see Section 4), perfection in the plane of the layers appears to be complete, with no line or point defects having been reported.

Stilpnomelane.—The structure of stilpnomelane [182] is based upon the same principles as that of zussmanite, but with the six-membered rings being replaced by 'rafts' of seven six-membered rings, and the three-membered rings by pairs of six-membered rings, situated directly above one another, and with only trigonal symmetry. The unit cell is thus considerably larger than that of either mica or zussmanite in the plane of the layers, and the interlayer spacing is also increased. Stilpnomelanes are of very variable composition, and an approximate general formula is $X_n A_{48} B_{72} OL_{80}(OH)_{48}$, where X can represent larger cations (Na^+, K^+, etc.) although no clear sites exist for these in the structure, A represents Fe^{2+}, Mg^{2+}, Mn^{2+}, etc., and B represents Al^{3+}, Fe^{3+}, Si^{4+}, etc. Stilpnomelane is a relatively common mineral, but X-ray single-crystal investigations of its possible polytypism have been few, partly because of the difficulty in obtaining sufficiently large crystals, and partly because of, with such a variable composition, difficulty interpreting the results obtained. In the original structure determination six possible one-layer triclinic polytypes are described, in two symmetry-related groups, which are almost, but not quite, rhombohedral. Further polytypes are referred to by Smith and Frondel,[183] but no structural details are given, and in both studies the single crystals used were heavily disordered. In all, 16 one-layer structures are possible in stilpnomelane, and more recent X-ray single-crystal evidence [184] indicates that a mixing of displacements from the two principal groups is possible, unlike the case of zussmanite.

Whereas severe difficulties exist in obtaining suitable crystals for X-ray analysis, no such drawbacks are found with electron-microscopic investigation, providing that the technique of ion-thinning is used on appropriately oriented sections. A recent such study [34] has revealed further details of the complexity of polytypism

---

[182] R. A. Eggleton, *Mineral. Mag.*, 1972, **38**, 693.
[183] M. L. Smith and C. Frondel, *Mineral. Mag.*, 1968, **36**, 893.
[184] D. A. Jefferson, Ph.D. thesis, Cambridge University, 1973.

in stilpnomelanes. Three samples were investigated, from New Zealand, North Wales, and California, the first two being relatively normal in that they displayed only single, or twinned versions of the determined triclinic structure with varying degrees of disorder. The Californian sample, however, showed a great complexity of polytypes which included, in addition to the one-layer triclinic variety, a two-layer triclinic type, a two-layer trigonal form involving displacement vectors from both groups, a three-layer trigonal structure, and four-, five-, nine-, and 14-layer polytypes with as yet undetermined symmetry. Structures of this type could be characterized by electron diffraction methods, experimentally observed patterns of the five- and nine-layer polytypes being illustrated in Plate 36 (a) and (b), and for the simpler structures, comparison of experimental and calculated diffraction intensities enabled an exact definition of the sequence of displacement vectors in each polytype. High-resolution lattice imaging techniques also enabled structural features not revealed in the electron diffraction pattern to be recorded. Plate 36(c), for example, indicates three- and four-layer sequences interspersed in a crystal with a principally two-layer repeat: Plate 36(d) indicates a regular repetition of nine- and 14-layer structures. In some cases a true 21-layer sequence was also observed. As with the study of mica described above, the multiple scattering in thicker regions of the specimen could be used to great advantages to yield systematics images which showed the *true* periodicity of the structure in each region. Again, as in the case of both the mica minerals and zussmanite, no indication of stacking faults terminating within a crystal was noted.

The precise reasons for such a variety of polytypic forms in stilpnomelane remains obscure, although, as appears to be the case with zussmanite, it seems likely that their existence is due to a gradual progression from a severely disordered arrangement to a more ordered one. In view of the high degree of compositional variability in stilpnomelane, investigations of such variability with the degree of order would be extremely valuable. However, in view of the very small crystal size of stilpnomelane, ultramicroanalytical techniques are necessary, a facility which is only just becoming available.

8 Orthosilicates

Chloritoid.—This is a rather unusual orthosilicate, frequently grouped with the layer silicates owing to its morphological characteristics. The determined structure of a monoclinic chloritoid is shown in Figure 5 [185] with the principal differences being that the continuous sheets of six-membered rings of tetrahedra in mica are replaced by isolated SiO_4 tetrahedra, and that, instead of the large interlayer cations in mica, structural 'layers' in chloritoid are held together by a second octahedral component (L_2), which is arranged such that the tetrahedra share a face with a vacant octahedral site in this component. The principal octahedral component (L_1) plays exactly the same role in chloritoid as it does in mica, and some hydrogen-bonding between oxygens of L_2 and hydroxyls of L_1 also helps to hold adjacent structural layers together. The isolated nature of the SiO_4 tetrahedra, however, necessitates a different repeat unit in the plane of the layers from that of mica, and the differences are also manifested in the chemical formula: $(Mg,Fe)_2Al$

[185] G. W. Brindley and F. W. Harrison, *Acta Cryst.*, 1952, **5**, 698.

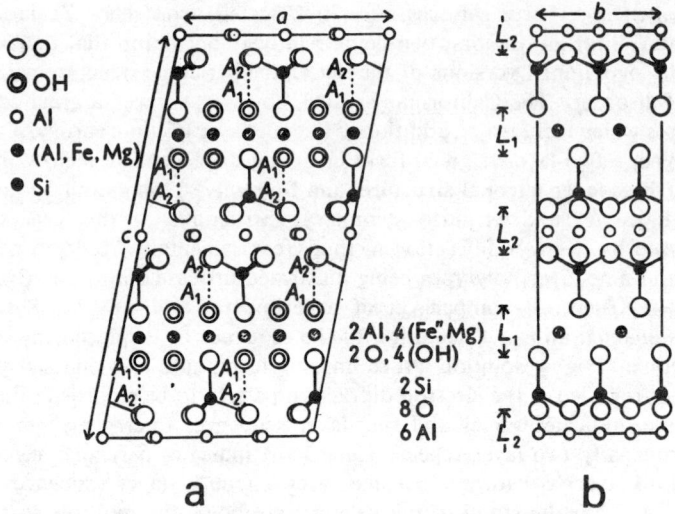

Figure 5 (a) *The structure of* $2M_2$ *chloritoid projected on* (010). (b) *The structure projected down* [100]

$(Al_3Si_2O_{10})(OH)_4$ which shows very different Si:Al or Si:(Mg,Fe) ratios from those of micas and the kandites.

Owing to the basic unit of one layer being the L_1 edge-sharing octahedral component, however, chloritoid exhibits strong similarities to the layer silicates, the chief of these being polytypism. X-Ray studies of chloritoid [186] have shown that, in addition to the monoclinic structure studied by Brindley and Harrison,[185] a triclinic form also exists. The latter, corresponding to the lower half of the monoclinic unit cell, is exactly analogous to the 1M mica polytype, despite the reduction in symmetry. The monoclinic form, which can be regarded as the triclinic structure twinned about [010] with composition planes at every layer, is a two-layer structure which is analogous not to the common $2M_1$ mica form but the $2M_2$ structure, owing to the fact that subsequent L_1 octahedral components in this form produce a hexagonal stacking of octahedra rather than a rhombohedral one. In accordance with current nomenclature, these polytypes are designated 1Tc and $2M_2$, respectively. Recent X-ray diffraction and electron microscopic investigations of chloritoid [187] have revealed further polytypes: in particular, a second monoclinic form has been identified by X-ray diffraction, this being equivalent to the triclinic form twinned about [100] rather than [010], which produces a $2M_1$ polytype analogous to the $2M_1$ mica structural variant. Detailed X-ray single-crystal studies have also revealed that disordered intergrowths of all three polytypes are of frequent occurrence, and this has been confirmed by HREM.

HREM studies of chloritoid bear many resemblances to those carried out on micas. One important difference, however, is the existence of a reasonably good {110} cleavage in chloritoid, with some 30% of all chloritoid flakes having this as

[186] L. B. Halferdahl, *Carnegie Inst. Wash. Ann. Report Dir. Geophys. Lab.*, 1956, 200.
[187] D. A. Jefferson and J. M. Thomas, 1977, in preparation.

the principal cleavage rather than the customary {001} cleavage. Consequently the more complicated methods of sample preparation, such as ion-thinning of oriented sections or microtome cutting, are unnecessary, and conventionally prepared specimens (crushed grains on holey carbon films) can be used. As for mica, the reciprocal lattice of chloritoid is pseudo-hexagonal, and two types of reciprocal lattice section are obtained, with rows parallel to c^* separated by distances corresponding to 4.73 and 2.74 Å. With current electron microscope goniometer stages, the latter is not generally resolvable unless tilted illumination is employed. Nevertheless, using the effect of multiple scattering, alterations in the periodicity of the structure in the [001] direction can be observed in a crystal in a one-dimensional (systematics) image, and this can highlight intergrowths between the triclinic (1Tc) polytypes and the monoclinic ($2M_1$ or $2M_2$) forms. However, where the two latter types are intergrown, both variants have the same (001) interplanar spacing, and can only be distinguished if two-dimensional images are found.

Although single-crystal X-ray diffraction studies indicated severely disordered 1Tc polytypes of chloritoid, electron microscopic investigation reveals areas showing a high degree of order [Plate 37(a)], and, surprisingly, some regions within ostensibly triclinic crystals showing a well-characterized monoclinic structure. The micrograph illustrated corresponds to a ($h0l$)-type reciprocal lattice section of the triclinic form, and the 9 Å periodicity of the triclinic structure extends over the entire crystal, with few stacking faults visible. In contrast, the monoclinic sample giving the micrograph illustrated in Plate 37(b) shows quite severe faulting. The electron beam in this case is parallel to the [100] direction in the $2M_2$ monoclinic structure, and the image is of the systematics type discussed above. Apart from the principal 18 Å periodicity, several regions are clearly shown which exhibit a 36 Å periodicity, and in some cases, a 54 Å variety. One region of particular interest (circled) shows a periodicity of 54 Å evident near the edge of the crystal, which terminates and becomes a normal 18 Å spacing with an adjacent 36 Å periodicity. In such a defect, as in the case of linear defects in wollastonite (see Section 4) considerable structural rearrangement must take place.

In the monoclinic example of Plate 37(b) the 'faults' appear to consist of regions where the overall periodicity is larger than the normal monoclinic one. However, in a second, basically monoclinic, sample, two types of fault can be distinguished [Plate 37(c)]. These consist of the 36 Å periodicities described above, with, in addition, some regions of a 9 Å periodicity. The image of Figure 42(c) is again of the systematics type, so that any interpretation of the faults requires some consideration of multiple scattering, but it appears likely that the first type of defect arises where the stacking sequence differs from that in the normal monoclinic structure, but is not, of itself, a regular sequence.

In the second type of defect, however, the sequence, while differing from that in the host monoclinic structure, appears to involve identical stacking of each layer, producing, in effect, a small region of the triclinic polytype. Similar faults to these can be readily observed in two-dimensional images. In particular, Figure 42(d) shows a region where the two monoclinic polytypes ($2M_1$ and $2M_2$) are intergrown, as shown by the abrupt change in fringe direction. Indeed, in this projection, which corresponds to the electron beam being parallel to the [1$\bar{1}$0] direction of the $2M_2$ form, the different monoclinic structures appear as reflection twins, with (001) composition plane. In the particular region shown, the right-hand twin boundary is characterized by a single unit cell strip (AA') of a triclinic form. An adjacent region also shows a triclinic strip (BB'), but in the latter case the orientation of the triclinic cell differs from that at AA'.

Mullite.—Mullite is a high-temperature aluminosilicate which, although rare in nature, is a common component in slags. It has a composition range from $1.5Al_2O_3,SiO_2$ to $2Al_2O_3,SiO_2$. X-Ray diffraction patterns reveal its close similarity to sillimanite (Plate 38) Al_2O_3,SiO_2. Diffuse scattering and occasional superstructure reflections are indicative of certain disorder. According to structure analyses of a mullite with composition $2Al_2O_3,SiO_2$[188,189] the structure may be derived from sillimanite by statistical random removal of approximately one fifth of the 'O_3' oxygen atoms, compensatory replacement of some Si^{4+} by Al^{3+}, and slight shift in certain cation positions to regain tetrahedral co-ordination with oxygen.

Thus it was shown[188] that mullite could be regarded as a non-stoicheiometric phase, intermediate in structure between sillimanite and andalusite.

Nakajima et al.[190] have studied a synthetic mullite of composition $1.86Al_2O_3,SiO_2$, which displays sharp superstructure X-ray reflections; ($h0l$) electron diffraction patterns show this to be a non-integral n-fold superstructure with $n = 3.4$ along the a^* axis. Lattice images taken close to the optimum defocus show an array of white spots interpreted as regions of low atomic density, in projection, as shown in Plate 39. This, it is argued, is due to oxygen vacancies and the associated displacements of surrounding cations. A modulation in contrast is evident about every third or fourth layer, which would correspond to the '$n = 3.4$' superlattice. As the defects were essentially *point* vacancies and not *linear* along the beam direction, the considerable variation in contrast of the white spots was taken to represent the probability of oxygen vacancies along b.

Lattice imaging studies of a variety of complex oxides[38] have demonstrated that oxygen atoms do not contribute significantly to the image contrast, which is dominated by contributions from the (heavy) cations. This may be less true in the case of an aluminosilicate such as mullite, as the cations are relatively light. However, it seems unlikely that the bright contrast is due entirely to oxygen vacancies, as proposed by Nakajima et al.[190] When *ab* and *ac* projections of sillimanite are considered, it is seen that the rows of tetrahedral cations parallel to (100) are alternately ca. 2.3 and 1.5 Å apart, neither distance being readily resolvable under the experimental conditions. Accordingly, in a micrograph of sillimanite[190] only intersecting (101) and ($\bar{1}01$) fringes were visible, there being no structural details. In the defect structure of mullite the 'flipping over' of certain tetrahedra caused by shifts of the cations (see Plate 38) opens out these cation rows to an estimated 4 Å. This distance is well within the intuitive projected charge density range and would certainly show up as bright contrast and build-up of such

---

[188] R. Sadanaga, M. Tokanami, and Y. Takéuchi, *Acta Cryst.*, 1962, **15**, 65.
[189] S. Ďurovič, *Kristallografiya*, 1962, **7**, 339.
[190] Y. Nakajima, N. Morimoto, and E. Watanabe, *Proc. Japan Acad.*, 1975, **51**, 173.

features along the *b*-axis (parallel to electron beam) would give corresponding contrast variations. Thus the vacancy probabilities can be assessed directly from the image, but only because of the displacements and not through any contributions of the oxygens to image contrast.

9 Graphite and its Intercalates

With the publication of the high-resolution micrographs of heat-treated carbon blacks [23,24] (by Ban and Heidenrich) and of heat-treated coals by Evans *et al.*[30] and Oberlin *et al.*,[191,192] in which the 3.4 Å spacing of turbostratic layered carbons were shown to be clearly resolved, interest soon focused on graphite itself, and especially on intercalates of graphite which are of great current chemical interest.

Although (*hkil*), as well as (000*l*), images of graphite may be obtained [27] with currently available high-resolution microscopes, such studies have been of less value for the elucidation of the ultrastructure of graphite itself, than for the assessment of order and disorder in quasi-crystalline carbons which, on progressive heat-treatment, increase their degree of crystalline perfection. We shall not pursue summaries of such studies here, since they have recently been fully discussed elsewhere.[26,33] It is, however, important to note that the phase-grating approximation breaks down (even qualitatively) when graphite samples are of thickness >50 Å. This fact has serious repercussions in the study of structural imperfections, and, unless great caution is exercised in the interpretation of lattice images of graphite carbons, gross errors regarding the presence or absence of individual (000*l*) planes in particular specimens may be made.[26]

So far as graphite intercalates themselves are concerned these tend to be air-sensitive and unstable *in vacuo*. For some systems, notably graphite–iron(III) chloride, so-called residue compounds are relatively stable,[28] and the same is true, but to a lesser extent, of the potassium–graphite system, which by electron diffraction, may readily be seen to display superlattice structures [Plate 40(a) and (b)]. When the samples are viewed along the layer-planes, high-resolution images reveal that, in the residue potassium–graphite compound, the intercalated K atoms are located in ordered sheets, but at random along the c^* direction (see Figure 8A of ref. 27). Likewise, as seen in Plate 41, the residue of the so-called first-stage intercalate (idealized composition C_7FeCl_3, in which a sheet of one octahedron thickness of $FeCl_3$ is inserted between every sheet of graphitic carbon) displays a random distribution of $FeCl_3$ along the c^* direction.

The electron micrograph shown in Plate 42 exhibits interesting features. If the light striations are taken to coincide with sheets of $FeCl_3$ and the less pronounced striations between them with the graphite layer planes, the specific identification of several distinct *n*th stages of the graphite–iron(III) chloride intercalates becomes possible. The *c* direction repeat unit in trigonal $FeCl_3$ ($R\bar{3}$, with $a = 6.06$ Å, $c = 17.40$ Å), consists of three layers of $FeCl_3$ octahedra, and the *c* direction repeat of 6.71 Å in graphite consists of two layers of carbon. Thus the *c* direction repeat distance in any *n*th stage graphite–$FeCl_3$ intercalate is given by

---

[191] A. Oberlin and G. Terriere, *J. Microscopie*, 1973, **18**, 247.
[192] A. Oberlin, G. Terriere, and J. L. Boulmier, *Tanso*, 1975, **80**, 29; **81**, 68.

$\left(6.71\frac{n}{2} + \frac{17.40}{3}\right)$ Å, so that first-, second-, third-, fourth-, fifth-, and sixth-stage intercalates should possess repeat distances of 9.2, 12.5, 15.9, 19.2, 22.6, and 25.9 Å, respectively. Plate 43 summarizes the sequence of the various stages of intercalation proposed from the evidence in Plate 38. The repeat distances measured from Plate 43 were 9.2, 12.6, 23.0, and 26.0 Å, corresponding to the first, second, fifth, and sixth stages, respectively. What is remarkable about this micrograph is that it provides direct evidence for the coexistence of chemically distinct (stoicheiometrically different) units of various-stage intercalates.

10 Sulphides

Many naturally occurring transition-metal chalcogenides are structurally related to the NiAs type, which is shown schematically in Plate 44. These include NiSb (breithuptite), NiS (millerite), and $Fe_{1-x}S$ (pyrrhotite).[193] The chalcogenide anions form a hexagonal-close-packed array (centres of the trigonal prisms in Plate 44) with the cations occupying the octahedral sites in the lattice. Of this series of minerals only two have been examined in any detail by HREM: pyrrhotite [194] and a synthetic vanadium sulphide VS_{1+x},[195] which is not, strictly speaking, a mineral.

It was early recognized that the variations in composition in natural pyrrhotites were due to cation vacancies,[196] the ordering of which gave rise to superstructures of the basic (NiAs) structure, which had a hexagonal cell* with $A = 3.45$ and $C = 5.75$ Å.[193] Five distinct superstructures have been found and classified by Morimoto et al.[197] as follows: $2C$ (FeS), the mineral troilite;[198] $6C$ ($Fe_{11}S_{12}$);[199] $11C$ ($Fe_{10}S_{11}$);[197] $5C$ (Fe_9S_{10});[200] and $4C$ (Fe_7S_8);[201] $6C$, $11C$, and $5C$ are special forms of 'intermediate pyrrhotite', or nC. In addition, $1C$ and $3C$ pyrrhotites, with variable composition, are found at high temperatures.[202,203]

4C Pyrrhotite.—The structure of $4C$ pyrrhotite (monoclinic, with $a = 11.88$, $b = 6.87$, $c = 5.70$ Å, $\beta = 90.47°$) has been accurately determined by X-ray work [201,204] and is shown in simplified form in Plate 45. For clarity the sulphur layers are omitted. The ordering of iron vacancies is obvious, the vacancy-containing layers exhibiting an ABCDABCD ... stacking along z. These layers

---

* To avoid confusion, we shall retain the use of capital letters for the **a** and **c** dimensions of the NiAs structure.

[193] A. Alsen, Geol. Foren. Forh, 1925, **47**, 18.
[194] H. Nakazawa, N. Morimoto, and E. Watanabe, Amer. Mineral., 1975, **60**, 359.
[195] H. Horiuchi, I. Kawada, M. Nakano-Oneda, K. Kato, Y. Matsui, F. Nagata, and M. Nakihara, Acta Cryst., 1976, **A32**, 558.
[196] G. Hägg and I. Sucksdorff, Z. phys. Chem., 1933, **22**, 444.
[197] N. Morimoto, H. Nakazawa, and E. Watanabe, Science, 1970, **168**, 964.
[198] L. G. Berry and R. H. Thomson, Geol. Soc. Amer. Mem, 1962, **85**, 61.
[199] M. E. Fleet and N. Macrae, Canad. Mineral., 1969, **9**, 699.
[200] R. H. Carpenter and G. A. Desborough, Amer. Mineral., 1964, **49**, 1350.
[201] E. F. Bertaut, Acta Cryst., 1953, **6**, 557.
[202] M. Corlett, Z. Krist., 1969, **126**, 124.
[203] N. Morimoto and H. Nakazawa, Science, 1968, **161**, 577.
[204] M. Tokanami, K. Nishiguchi, and N. Morimoto, Amer. Mineral., 1972, **57**, 1066.

are structurally identical, and vary only in their orientations relative to one another. Since optimum high-resolution lattice images show contrast related to projected charge density, it is instructive to consider various projections of the structure. Two of these are shown in Plate 46 along [010] and [110] of the monoclinic cell. They reveal two different projections of rows of Fe sites: those completely filled (black circles) and those only half occupied (squares). Nakazawa et al.[194] obtained lattice images of $4C$ pyrrhotite with the electron beam parallel to [110] (i.e. [100] of the hexagonal subcell). At optimum defocus these images showed an array of white dots (Plate 47) which were interpreted as projected half-filled rows of Fe sites, with the completely Fe-filled rows and the sulphur layers constituting the unresolved dark contrast features.

The work of Horiuchi et al.[195] on $VS_{1.155}$ (more correctly $VS_{0.865}$) supported this interpretation as, in addition to micrographs, they produced computed images which agreed with the observed contrast. These computations revealed the strong dynamical nature of these objects, however, and it was shown that the regions of 'intuitive' contrast occurred beyond the first extinction band (as can be seen in Plate 47), as well as at the extreme edge of the crystal. Thus the projected charge-density region was limited to a thickness of 60 Å, the more extensive area of thicker crystal providing most of the useful image. This phenomenon of 'intuitive' projected charge-density contrast reappearing in thick crystal has been previously noted for some complex oxide systems,[205] and it was seen to be important in some systems discussed above (e.g. Section 8 on chloritoid).

It can be seen in Plate 47 that the horizontal rows of white spots (normal to z) occur in two distinct sequences: those in which successive rows are displaced by $\frac{1}{2}$ repeat normal to z, and those in which this displacement takes place in every third layer. Referring to Plate 46, it can be seen that the former corresponds to the arrangement of half-filled cation rows along [010], the latter corresponding to the [110] projection. This was regarded by Nakazawa et al. as fine-scale rotation (60°) twinning around z with (001) as the composition plane. Twin lamellae varied in width from two to ten layers, consistent with the fact that twinning was invariably found in X-ray precession photographs of small 'single' crystals.[194] The twinning was associated with differences in vacancy ordering and left the anion network unaltered, the small deviation from orthogonality (β = 90.47°) not being detectable by lattice imaging.

'Out of step' boundaries were also observed in which the stacking of vacancy-containing layers altered within a single layer, from one orientation to the other, via regions in which the discrete white spots gave way to white streaks, signifying the superposition of two or more arrangements along the beam direction. An example of this is shown in Plate 48.

Intermediate (nC) Pyrrhotite.—This term covers the composition range $Fe_9S_{10}(5C)$—$Fe_{11}S_{12}(6C)$ and its members are characterized by their long c-axes, often a non-integral multiple of C_{NiAs}.[206] Initial lattice images [207] revealed considerable disorder in the vacancy distribution, compared with the $4C$ structure.

[205] P. L. Fejes, S. Iijima, and J. M. Cowley, *Acta Cryst.*, 1973, **A29**, 710.
[206] H. Nakazawa and N. Morimoto, *Mater. Res. Bull.*, 1971, **6**, 345.
[207] H. Nakazawa, N. Morimoto, and E. Watanabe, ref. 16, Vol. 1, p. 498.

Further studies [208] confirmed this and showed that the vacancies were not confined to every second Fe layer, but had a more random distribution. An example of an nC pyrrhotite ($c = 28.67$ Å, an average value $n = 5$) is shown in Plate 49, together with an image of $4C$ pyrrhotite for comparison. It was noted that ordered superstructures ($n = 4$, 6, or 11) were *not* found at the unit cell level, although they had been suggested previously as constituents of nC pyrrhotites.[200] On the other hand, a parallel study by Pierce and Buseck,[209,210] who used electron diffraction as well as lattice imaging (bright and dark field), revealed superstructure spacings of $4C$, $5C$, $7C$, $10C$, and $28C$, in one 'single crystal', as well as non-integral nC spacings with $4 \leqslant n \leqslant 5.5$. These findings reflect the degree of variation possible in samples from different localities. They treated the stacking variations, described by Nakazawa *et al.* as twinning,[194] as anti-phase domains and were able to simulate some of the electron diffraction patterns using the method of Minagawa [211] for calculating X-ray intensities from arrays of anti-phase domains. Their discussion would apply to an orthorhombic lattice, but the monoclinic angle (90.47° for the $4C$ sample [194]) is so close to 90° that an unambiguous interpretation of the images would not be possible.

nA **Pyrrhotite.**—The $3C$ high-temperature form [202] was subsequently shown [203] to have an additional superstructure along the x-axis, giving the cell parameters $a = nA$, $c = 3C$. It has been noted [207] that under electron irradiation, nC type pyrrhotite may transform into an nA superstructure, at temperatures > 220 °C. Nakazawa postulated a structure for this type, which was based on three domains related by a $\frac{1}{3}c$ glide along the x-axis.[212] Lattice images of irradiated nC pyrrhotite occasionally showed the development of these nA superstructures,[213] but their complexity was such that complete structures could not be derived from the images.

Smythite, Fe_3S_4.[214]—With $A_{hex} = 3.47$ and $c = 34.5$ Å, smythite may be regarded as a six-fold superlattice of NiAs cell. X-Ray studies suggest that every fourth 'Fe' layer is completely empty. No HREM observations of this mineral have been reported, but the possibilities for superlattice variability are obvious and a high-resolution study would be fruitful.

Troilite, FeS.[198]—Below 140 °C, stoicheiometric FeS has a $\sqrt{3}A$, $2C$ supercell;[215] annealing experiments (90—130 °C) carried out on a meteoritic troilite induced what was described as a phase separation: [216] electron microprobe analysis revealed small variation (*ca.* 2%) in iron content between the intergrown lamellae and the matrix and it was argued that the two phases lay on either side of the stoicheiometric composition; the separated lamellae, with a $2A$, $1C$ superstructure, were iron-deficient and the superstructure was presumed to arise from vacancy

[208] N. Morimoto, H. Nakazawa, and E. Watanabe, *Proc. Japan Acad.*, 1974, **50**, 765.
[209] L. Pierce and P. Buseck, *Science*, 1974, **186**, 1209.
[210] L. Pierce and P. Buseck, ref. 14, p. 137.
[211] T. Minagawa, *Acta Cryst.*, 1972, **A28**, 308.
[212] H. Nakazawa, *Dissertation (Osaka Univ.)*, 1968.
[213] H. Nakazawa, N. Morimoto, and E. Watanabe, ref. 14, p. 304.
[214] R. C. Erd and R. H. Richter, *Amer. Mineral.*, 1957, **42**, 309.
[215] E. F. Bertaut, *Bull. Soc. franç Mineral. Crist.*, 1956, **79**, 276.
[216] A. Putnis, *Phil. Mag.*, 1975, **31**, 689.

ordering; the remaining matrix was said to be iron-rich, with iron interstitials, and to retain the $\sqrt{3}A$, $2C$ supercell of low-temperature troilite.

The electron-diffraction patterns of the 'phase separated' intergrowths were similar to those observed in various of the pyrrhotites, and a possible reason could be that some reduction resulted in the formation of intergrown pyrrhotite phases of different composition, but both sub-stoicheiometric. Since it is suggested that the sulphur lattice is continuous, migration of iron (and presumably therefore iron vacancies) obviously takes place on annealing.

11 Fluorocarbonates

The structure of the rare-earth mineral bastnaesite, $LnFCO_3$, was determined by Donnay and Donnay,[217] and the related minerals synchistite, $LnFCO_3,CaCO_3$, and parisite, $2 LnFCO_3,CaCO_3$, have been analysed according to the same structural principles. Bastnaesite has a structure consisting of alternating (ionic) LnF layers and layers of CO_3 groups; synchistite has, in addition, a calcium layer sandwiched between CO_3 layers, giving the sequence:

$$\overbrace{/LnF/CO_3/Ca/CO_3}^{S}/LnF/CO_3/Ca/CO_3/\ldots$$

Caro[218] pointed out that the other two minerals in this series could be built up by varying the proportions of bastnaesite (B) and synchistite (S) blocks. Thus parisite corresponds to the sequence BS:

$$\overbrace{/LnF/CO_3/Ca/CO_3}^{B}\overbrace{/LnF/CO_3/LnF/CO_3/Ca/CO_3}^{S}/\ldots$$

and roentgenite to the sequence BS_2.

Van Landuyt and Amelinckx[219] succeeded in identifying several other sequences by lattice imaging methods. Using both bright and dark field (000*l*) lattice fringes they deduced a simple imaging 'code': 14 Å inter-fringe spacings corresponded to a compound BS block; 9 Å spacings represented an S block. In a subsequent paper[220] this 'code' was further simplified to 'a wide dark line corresponding to a B block, a narrow dark line to an S block'. Although not stated, presumably this applied only to the bright field images, and to special conditions of thickness and focus. By means of this 'code', Van Landuyt and Amelinckx were able to demonstrate coherent intergrowth between the various structures (described as 'syntaxy'[218]). An example of this is shown in Plate 50, in which small domains of parisite, roentgenite, and synchistite are randomly intergrown. The distinction was drawn between 'sequential faults' which left the overall composition unchanged, and 'compositional faults' which may be regarded as Wadsley defects – random, planar faults which altered the composition. An example of both types of fault is shown

---

[217] G. Donnay and J. D. H. Donnay, *Amer. Mineral.*, 1953, **38**, 932.
[218] P. E. Caro, *J. Solid State Chem.*, 1973, **6**, 396.
[219] J. van Landuyt and S. Amelinckx, *Amer. Mineral.*, 1975, **60**, 351.
[220] J. van Landuyt and S. Amelinckx, ref. 14, p. 68.
[221] J. S. Anderson and J. L. Hutchison, *Cont. Phys.*, 1975, **16**, 443.

in Plate 51. Three new ordered sequences were found: BS_4, B_3S_4, and B_3S_2. The designation was based on the observation of at least ten repeats in a minimum of three crystals. The validity of interpreting such features as 'ordered structures' is discussed, in relation to similar findings in hexagonal barium ferrites, by Anderson and Hutchison.[221]

12 Conclusion

This selective report has necessarily focused on a few groups of minerals and the examples chosen illustrate the ways in which electron microscopy, in particular high-resolution lattice imaging, has made a significant contribution to the chemical study of minerals. It has again to be emphasized that this technique has its limitations: many minerals are sensitive to the vacuum environment and, more seriously, to the electron beam. Some, like chrysotile, degrade slowly enough to allow lattice images to be obtained; others, like many of the zeolites, may not survive long enough to give the observer more than a fleeting glimpse of a diffraction pattern before crystallinity is lost. Furthermore, not all minerals have structures which yield readily interpretable projected-charge-density, and lattice imaging may not be applicable in these instances.

Despite these limitations, however, the electron microscope has emerged as a powerful tool for the study of defective minerals, and, with current instruments offering very high resolution (2—3 Å) as well as good specimen tilting facilities and the possibility of *in situ* X-ray microanalysis of small, structurally characterized areas, future developments should prove both exciting and challenging.

It is a pleasure to acknowledge the help of the following and their collaborators in providing many of the micrographs used in this report: Dr. J. L. Brown, Professor P. R. Buseck, Dr. P. E. Champness, Mrs. B. Cressey, Dr. S. Iijima, Professors M. L. Jackson, J. Van Landuyt, and N. Morimoto, Dr. J. B. Vander Sande, and Professor K. Yada. Dr. S. Iijima provided a copy of paper 52 prior to its publication.

Author Index

Abe, T., 276
Aboagye, J. K., 223, 225
Abrikosov, A. A., 9
Adams, D. L., 35, 40, 52, 176, 216
Adrianes, M. R., 83
Agarwal, G., 17
Agarwal, S. K., 263, 265, 271
Agrawal, V. K., 246, 253
Ailion, D. C., 220, 221
Ajayi, O. B., 235
Alario Franco, M., 339
Albrecht, H. E., 145
Alderman, D. W., 257
Aldred, P. J. E., 284
Alford, N., 180
Allen, C. A., 227, 231
Allen, F. G., 179
Allen, R. E., 65
Allen, R. T., 79
Allnatt, A. R., 222, 225, 243, 244
Allpress, J. G., 334
Alsen, A., 354
Amelinckx, S., 320, 323, 345, 357
Anders, L. W., 157
Anderson, A. B., 157
Anderson, J., 176, 204, 209, 212, 215
Anderson, J. R., 46
Anderson, J. S., 323, 357
Andersson, S., 339
Ando, K. J., 111
Ando, Y., 294
Andreev, A., 171
Andreev, A. M., 220, 228
Andrus, J., 281
Angenmuller, H., 196
Apai, G., 179, 200, 212
Apker, L., 185
Appelbaum, J., 205
Applenan, D. E., 328
Arakawa, E. T., 187
Arakawa, M., 345
Aramati, V. S., 169
Ardenne, M., 321
Armstrong, R.A., 177
Armstrong, R. D., 249
Armstrong, R. W., 281
Ash, L., 123
Ashcroft, N. W., 181, 182
Asklund, B., 339
Aston, J. G., 4, 31, 32, 308
Atkinson, S. J., 83
Atkinson, S. R., 212
Augustyn, D., 322
Authier, A., 281, 294, 299, 300, 303
Avery, J., 182
Avgul, N. N., 4
Axelmann, R. C., 127
Azaroff, L. V., 56
Azizov, U. V., 147

Bacci, M., 278
Bacigalupi, R. J., 4
Baetzold, R. C., 157, 158, 165
Bäverstam, U., 137
Bagchi, S. N., 39
Bagnato, C., 275
Bahl, O. P., 323
Bailey, S. W., 346
Bainbridge, J., 116
Baker, A. D., 213
Baker, B. G., 46, 102, 147
Baker, D., 213
Baker, J. M., 212
Balchin, A. A., 300
Ballhausen, C. J., 263
Ballu, Y., 196
Baltog, I., 229
Ban, L. L., 322
Bancroft, G. M., 127, 134
Baner, E., 169
Barash, Yu. S., 5
Barbaux, Y., 94
Barber, M., 180
Barclay, L. M., 345
Bardeen, J., 4, 91, 140, 150
Barford, B. D., 168
Barr, L. W., 219, 220, 228
Barrer, R. M., 4
Barrett, C. S., 281
Barsis, E., 225
Bartell, L. S., 157
Bassani, G. F., 184
Bassett, G. A., 323
Bates, T. F., 340, 341
Batra, A. P., 219, 220, 223, 227, 233
Batra, I. P., 157, 208
Baudelet, B., 303
Bauer, C. F., 225
Beaufils, J.-P. A., 94
Beaumont, J. H., 221
Becker, D., 220
Becker, E., 162, 212
Becker H., 196
Beder, E. C., 21, 78, 79
Beevers, C. A., 249
Bell, A. E., 92
Bell, B., 156, 164
Bell, R. E., 325
Belov, N., 333
Benci, S., 278
Bengtzelius, A., 221
Bénière, F., 220, 222, 225, 227, 228, 231, 234, 246
Bénière, M., 220, 222, 225, 227, 231
Bennett, A. J., 21, 140
Bennett, L. H., 114, 131
Berard, M. F., 220
Berg, W. F., 281
Berge, S., 147
Berger, H., 114
Berggren, K. F., 190
Berglund, C. N., 183

Bergmark, T., 85
Bernasconi, G. F., 162, 180
Bernasek, S. L., 167
Berndt, H., 175
Berndt. W., 162
Bernstein, H. J., 259, 271
Berry, L. G., 354
Bertaut, E. F., 354, 356
Berteit, P., 226
Besocke, K., 149, 150
Betche, H., 323
Bethe, H. A., 182
Bettman, C., 249
Beyeler, H. U., 253, 271
Beyeler, M., 226
Bienfait, M., 44, 65, 103
Bijvank, E. J., 233
Billardon, M., 274
Bimberg, D., 275
Birman, J. L., 209
Bishop, H. E., 101
Bladon, P., 311
Blakely, J. M., 176
Block, J. H., 147
Bloeck, W. L., 147
Bloembergen, N., 221
Blott, B. H., 145
Blumenstock, D., 257
Blyholder, G., 157, 161, 165 177
Bochnen, K. P., 203
Bogan, A., 81
Bohm. C., 137
Bokemeyer, H., 114
Bollmann, W., 225, 320
Bommel, H., 111
Bonadeo, H., 266, 271
Bonchev, Z. W., 114, 137
Bond, G. C., 1. 161
Bond, W., 281
Bonse, U., 281, 282, 287
Bonzel, H. P., 170
Bootsma, G. A., 100
Bordas, J., 289
Born, M., 202
Borrmann, G., 297
Boswarva, I. M., 235, 236, 245
Boudart, M., 112
Boulmier, J. L., 353
Bouwman, R., 91, 92
Bouznik, V. M., 252
Bowen, D. H., 242
Bowen, S. P., 253
Boyer, T. H., 13
Bowman, R. C., 243
Bown, M. G., 328, 330, 331, 334
Brady, J. J.. 145
Bradley, J. N., 248
Bradshaw, A. M., 179
Bragg, W. L., 328
Brauer, P., 245
Bredig, M. A., 251
Brenna, G. C., 263, 271

Author Index

Brennan, D., 43, 46
Brewer, D. F., 104
Bridges, F., 252, 257, 271
Briggs, A., 242
Brindley, G. W., 341, 343, 346, 349
Brink, E., 220
Brodén, G., 83
Brohult, S., 249
Bron, W. E., 257
Brown, F., 115
Brown, J. L., 347
Brown, N., 222, 225, 231
Brown, W. L., 328
Bruch, L. W., 65
Bruesch, P., 248
Bruggar, K., 271
Brumbach, S. B., 67
Brun, A., 231, 234
Brunauer, S., 1. 97
Brundle, C. R., 34, 83, 85, 162, 179, 212, 213
Bryant, J. I., 271
Buckingham, M. J., 185
Bührer, W., 248
Buerger, M. J., 336
Büttner, P., 162
Bullett, D. W., 161
Bulmer, R. S., 249
Burkstrand, J. M., 29, 83, 147
Burley, G., 249
Burnham, C. W., 345
Burns, S. J., 307
Burrell, M. C., 336
Burton, J. W., 113
Buseck, P. R., 323, 326, 329, 356
Butcher, P., 212
Buzek, F., 167, 169
Byer, N. E., 254, 257

Cabrera, N., 74
Calas, G., 305
Calderwood, J. H., 226, 231
Calia, V. S., 81
Callcott, T. A., 196
Callen, H. B., 13
Cambell, H. C., 114
Campos, V. B., 233
Capelletti, R., 233, 234
Cares, W. R., 112
Carley, A. F., 83, 156
Carneiro, K., 58
Caro, P. E., 357
Caroli, C., 182
Carpenter, R. H., 354
Carr, V. M., 252
Cartier, P. G., 168
Cashion, J. K., 184
Casimir, H. B. G., 9
Catlow, C. R. A., 222, 235, 237, 238, 239, 241, 242, 243, 244
Cederbaum, L., 179
Celli, V., 74
Cerisier, P., 243
Černý, S., 160, 167, 168, 169
Cessac, G. L., 265
Cetica, M., 278
Chadwick, A. V., 220, 221, 225, 237, 252
Champier, G., 303, 304, 307
Champness, P. E., 331, 332, 339

Chan, N.-H., 228, 231
Chandra, S., 225, 233, 253
Chaney, R. E., 231
Chang, C. C., 101
Chang, D. R., 19
Channing, D. A., 127
Chapman, J. A., 229, 231, 341
Charbrier, G., 185
Chaudhuri, S., 240
Chauvin, G., 196
Chemla, M., 220, 222, 225, 227, 228, 231
Chen, W. K., 228
Chesters, M. A., 44, 46, 47
Chikasaki, M., 246
Chikawa, J.-I., 290
Chinn, M. P., 44, 54
Chisholm, J. E., 321, 338
Cho, K., 272, 275
Choi, S., 250
Chon, H., 31, 32
Chow, H., 74
Chowdari, B. V. R., 266, 269
Chowdhury, S., 236
Christensen, N. E., 184, 202
Christiansen, G., 287
Christie, J. M., 332
Christmann, K., 168, 169, 170
Chrzanowski, E., 147
Chuikov, B. A., 169
Chutjian, A., 42
Clark, A., 163
Clark, B. C., 66
Clark, D. T., 85
Clarke, J. A., 212
Clavenna, L. R., 169
Cinti, R., 196
Citrin, P. H., 36
Cochran, W., 334
Cochrane, G., 225
Coe, R. S., 328
Coey, J. M. D., 111
Coghlan, W. A., 220
Cohen, M. H., 243, 258, 259, 271
Cohen, P. I., 44, 50, 65
Coker, E. H., 263, 266, 271
Collins, A. J., 225
Collins, D. M., 34, 179
Collins, R. L., 114, 131
Collins, W. C., 278
Comer, J. J., 341, 342
Comes, R., 250
Compaan, K., 248
Comrie, R. M., 163
Connford, C., 322
Conrad, H., 168, 175, 177
Constabaris, G., 4, 111
Cook, J. S., 228, 229
Coomes, E. A., 147
Cooper, R. L., 19
Cooper, T. L., 29
Copely, P. A., 332
Corish, J., 218, 222, 238, 239, 244
Corke, N. T., 311
Corlett, M., 354
Coslett, V. E., 122
Cotterill, R. M. J., 235, 320
Cotts, R. M., 257
Coulomb, J. P., 44, 65, 103
Cowley, J. M., 323, 355
Cranshaw, T. E., 123
Crawford, D., 322

Crawford, E. D., 322
Crawford, E. S., 322
Crawford, J. H., 218, 228, 278
Cressey, B. A., 341
Creswell, D. J., 104
Crissman, J., 233
Crouser, L. C., 92
Crowell, A. D., 1, 3, 4, 43
Cundill, M. A., 271
Curtiss, C. D., 345
Cutler, P. H., 142
Cyrot-Lackmann, F., 162, 185, 203
Czyzewski, J. J., 162

Dabiri, A. E., 169
Dagneaux, D., 196
Damgaard-Kristensen, W., 320
Danemar, A., 243
Dansas, P., 231, 234
Dash, J. G., 33, 57, 58
Daunt, J. G., 31
Davenport, J. W., 210
Davidson, E. R., 158
Davies, B., 16
Davies, B. R., 120
Davies, J. A., 115
Davies, N. C., 322
Davies, R. O., 4
Davis, J. C., 142
Davis, L. E., 101
Davison, S. J., 181
Davisson, C. J., 40
Dawson, K., 219
Deans, II. A., 163
Deb, B. M., 164
de Boer, J. H., 3, 21
Decius, J. C., 271
Degras, D. A., 169
Dehmer, J. L., 210
de Kock, A. J. R., 307
Delavignette, P., 320, 345
Delbecq, C. J., 276, 279
Delchar, T. A., 147,176
Delgass, W. N., 85, 112
Delong, W. T., 114
Demuth, J. E., 34, 83, 162, 175, 178
Dem'Yanenko, V. P., 265, 266, 271
Den Hartog, H. W., 233
Dent Glasser, L. S., 336
Deportes, C., 251
Derrington, C. E., 251, 252
Desborough, G. A., 354
Deschamps, L., 250
Desjonqueres, M. C., 203
de Souza, M., 253, 257
Desplat, J. L., 176
Desjonguères, M. C., 162
Dettmann, K., 223
Devonshire, A. F., 20, 74, 79, 253
De Wames, R. E., 63
de Wette, F. W., 65
Díaz Góngora, A., 257
Diamond, J. B., 157
Dick, B. G., 253
Dickens, P. G., 250
Dickerson, S. M., 127
Dickey, J. M., 44, 101
Dickinson, T., 249

Author Index

Dienes, G. J., 239, 242, 243
Dierk, E. A., 91
Dieterly, D. K., 112
Dietz, E., 179, 196
Dill, D., 210
Diller, K. M., 222, 237, 238, 239
Dillon, J. A., 145
Di Persio, J., 315, 318
Di Salvo, F. J., 180, 189
Dobrezow, L. N., 142
Dobson, C. M., 113
Domcke, W., 179
Domke, M., 168
Donaldson, E. E., 168
Donnay, G., 357
Donnay, J. D. H., 357
Dorn, J. E., 310
Dovesi, R., 4
Dowden, D. A., 161
Downing, H. L., 220, 222
Doyen, G., 170
Drechsler, M., 168
Dresser, M. J., 88, 147
Dreyfus, R. W., 257
Drummond, J. E., 19
Drwiega, M., 131
Dryden, J. S., 228, 229
Ducastelle, F., 185
Duckett, R. A., 301
Dudley, G. J., 218, 251
Dudnik, D. F., 257
Duke, C. B., 21, 63, 90, 140, 201
Dultz, W., 275
Duran, J., 274
Durović, S., 352
Dutta, J., 229
Duval, X., 99
Dweydari, A. W., 148
Dworecki, Z., 92
Dworkin, A. S., 251
Dzyaloshinskii, I. E., 9, 10

Eastman, D. E., 34, 83, 145, 178, 179, 181, 182, 183, 184, 212, 215
Eastment, R. M., 148
Ebert, L. B., 225
Ebiko, H., 127
Eckardt, D., 114
Eckold, G., 248
Edgerton, R., 276
Edwards, D. E., 29
Egelhoff, W. F., 83, 180, 197
Eggleton, R. A., 348
Ehrenreich, H., 200
Ehrlich, G., 4, 149, 160, 176
Eijkelenkamp, A. J., 225
Einstein, T. L., 205
Eisberg, R. M., 180
Eisenstadt, M., 225
Ekdahl, T., 137
Ekkerman, V. M., 223
Eley, D. D., 168
Elliot, R. J., 197
Endriz, J. G., 181, 187
Engel, O., 169
Eugel, T., 3, 92, 147
Engelhardt, H. A., 176
Engelke, B., 91
Englman, R., 274
Epelboin, Y., 294, 303
Erd, R. C., 356

Erhardt, J. J., 67, 72
Erickson, N. E., 88
Erlich, G., 92, 97
Ermoshkin, A., 278
Ertl, G., 168, 169, 170, 175, 177
Escaig, B., 315, 318
Escaravage, C., 304
Eskola, P., 339
Esterman, I., 66
Estle, T. L., 257
Estrup, P. J., 40, 170, 175, 176
Etsell, T. H., 251
Evans, A. R., 262, 271
Evans, B. E., 92
Evans, E. L., 322, 323
Evans, M. T., 328
Evarestov, R., 278
Ewald, P. P., 290

Fabeni, P., 278
Fadley, C. S., 85, 180
Fahlman, A., 85
Fain, S. C., jun., 44, 54, 158
Fairfield, J. M., 306
Falicov, L. M., 156
Fan, H. Y., 183
Farge, Y., 225
Farnsworth, H. E., 40
Farrell, H. H., 44, 62, 101
Fassaert, D. J. M., 157, 171
Faux, I. D., 242
Feder, R., 196
Fedorus, A. G., 145
Fedotov, A. F., 345
Feher, G., 255
Fehrs, D. L., 145
Feibelman, P. W., 181, 182
Feit, M. D., 221
Fejes, P. L., 355
Feldman, V., 149
Fener, P., 79
Fenger, J., 117
Fenner, W. R., 256, 271
Fermi, F., 233
Ferrell, R., 204
Feuerbacher, B., 34, 83, 162, 179, 182, 184, 194, 196
Fiedler, R., 307
Fielder, W. L., 251
Figueroa, D. R., 220, 221, 225 229, 231, 252
Finger, L. W., 338
Fiory, A. T., 257
Fisch, G., 168
Fischer, C. R., 242
Fischer, F., 233
Fisher, R. A., 31, 32
Fisher, R. M., 332
Fitchen, D. B., 262, 271
Fitton, B., 83, 179, 194, 196
Fleet, M. E., 354
Flengas, S. N., 251
Fletcher, N. H., 225
Fletcher, R., 235
Flinn, P. A., 114
Flodstrom, S. A., 187, 197
Flygare, W. H., 235
Flynn, C. P., 235
Fong, F. K., 244
Fontana, M. P., 278
Ford, R. L., 244
Fordyce, J. S., 251
Forester, D. W., 135
Forestier, M., 251

Fortsmann, F., 162, 205
Fort, T., jun., 99
Foster, D. L., 233
Frankel, D. R., 74
Frankel, J., 226
Franklin, A. D., 233
Frantsesson, A. V., 257
Franzini, M., 345
Fredericks, W. J., 219, 227, 231
Freeman, A. G., 117, 137
Freeman, D. L., 4, 104
Freeman, J. H., 134
Freidt, J. M., 137
Freytag, H., 321
Friauf, R. J., 222, 223, 225, 234
Friedel, J., 317
Frisch, R., 66
Frondel, C., 348
Fryer, G. M., 223
Fuchs, R., 17, 181
Fueki, K., 220
Fuggle, J. C., 179
Fujimoto, I., 290
Fujioke, M., 137
Fukuda, A., 269, 274, 276, 277, 278
Fulde, P., 203
Fuller, R. G., 219, 222
Fumi, F. G., 237
Funke, K., 248

Gadzuk, J. W., 162, 182, 204, 205
Gaffari, M., 250
Gager, H. M., 111, 112
Gainotti, A., 234
Gal'chinetskii, L. P., 223
Gallon, T. E., 101
Gardner, W., 179
Garg, H. C., 246
Garner, F. M., 147
Garnes, Y. G., 257
Garrod, R. I., 338
Garrone, E., 4
Garten, R. L., 112
Gartland, P. O., 147, 196, 197
Gaune, P., 243
Gavriliuk, V. M., 145
Gay, P., 328, 330, 331
Gebhardt, W., 275
Geller, S., 249
George, A., 304
George, T. H., 149
Gerhardt, U., 179, 196
Gerlach, E., 15
Germer, L. H., 40, 176, 216
Gerward, L., 287
Gessell, T. F., 187
Ghita, C., 229
Ghosh, A. K., 240
Giacovazzo, C., 305
Gibb, T. C., 107
Gibbs, G. B., 239
Gilbey, D. M., 79
Gillerlain, J. D., 79
Gindina, R., 278
Ginzburg, V. L., 5
Giraud, E., 147
Gittos, M. F., 339
Glasser, F. P., 336
Glenon, B. M., 11
Glick, M., 249

Author Index

Gobeli, G. W., 179
Goldanski, V. I., 112
Goldstein, M., 243
Golodets, G. J., 173
Golopentia, D., 229
Gomer, R., 3, 32, 92, 147, 149, 170, 205
Gomez, M., 253
Góngora, A., 255
Gonser, U., 121
Good, G. A., 134
Good, R. H., 92
Goodman, F. O., 4, 20, 67, 74, 79
Goodwin, R. P., 113
Gorbatyi, N. A., 147
Gordon, R. G., 31
Gor'kov, L. P., 9
Gorlich, P., 225
Gosar, P., 254
Goto, Y., 104
Goudonnet, J. P., 185
Govindarajan, J., 246
Graham, M. J., 46, 127
Graham, R. L., 115
Grant, W. B., 265
Granzer, F., 229
Grasso, V., 275
Gregg, S. J., 46
Green, R. E., jun., 290
Greene, P. D., 248
Greenwood, N. N., 107
Grepstasl, J. K., 197
Griggs, D. T., 332
Grimes, N. W., 252
Grimley, T. B., 162, 180, 205, 207
Grimmer, D. P., 104
Grňo, J., 231
Grunze, M., 169
Gruverman, I. J., 112
G'Sell, C., 303, 307
Gudat, W., 215
Guiliano, E. S., 197
Guillot, C., 175, 177, 196
Guinier, A., 56
Guirgea, M., 229
Gundry, P. M., 4, 32, 92
Gunnarson, O., 205
Gurman, S. J., 181, 183, 185, 197
Gurney, R. W., 205, 239
Gustafsson, T., 215
Guy, I. D., 212
Gyftopoulos, E. P., 91
Gyorffy, B. I., 196, 197

Haas, W. C., 250
Hackenberg, R., 248
Hägg, G., 354
Härtel, H., 253
Hagen, D. E., 4
Hagen, D. J., 168
Hagenmuller, P., 250
Hagrman, D. L., 255
Hagström, S. B. M., 180, 197
Hagstrum, H. D., 35, 83, 156, 162, 212
Halferdahl, L. B., 350
Halff, A. F., 225
Halsey, G. D., jun., 4
Halverson, F., 259
Hamann, D. R., 36, 205
Hampton, E. M., 310, 311, 315

Hamrin, K., 85
Han, H. R., 169
Hansen, R. S., 157
Hansson, G. V., 197
Harding, B. C., 221
Hardy, J. R., 236
Haridoss, S., 233
Harley, R. T., 251
Harris, J., 17, 22
Harris, L. B., 234
Harrison, J. P., 257
Hart, H. R., 254
Hart, M., 284, 287, 288, 289
Hartford, R., 276
Hartmanová, M., 231
Hartwig, W., 297
Hartzell, R. A., 114
Hatcher, R. D., 242, 243
Haüy, C., 337
Hauk, W., 225
Haven, Y., 225, 247, 248
Hawthorne, H. M., 310, 311
Haydock. R., 185, 187, 205
Hayes, W., 218, 251
Hayns, M. R., 241
Hedin, L., 182, 204
Hedman, J., 85
Heidenreich, R. D., 42, 322
Heine, V., 154
Heinike, W., 257
Heinisch, H. L., 245
Heinmann, P., 179, 196
Heinrichs, J., 16, 17
Heist, R. H., 244
Heller, L., 332
Heller, W. R., 253
Hellwig, S., 147
Henderson, B., 242
Hendricks, S. B., 344
Heneisen, J. D., 220
Hennig, G. R., 323
Herbst, J. F., 83
Herendeen, R. A., 255, 257
Herfert, R. E., 325
Herman, R., 66
Hermann, H., 91
Hermon, G., 322
Herring, C., 90, 140
Hess, H. H., 327, 330
Hess, W. M., 322
Hesse, K. F., 339
Heuer, A. H., 332
Hewatt, E. A., 334
Heyne, H., 4, 92, 149
Hickernell, D. C., 104
Hickmott, T. W., 92
Hillier, J., 321
Hilmer, W., 336
Hilsch. R., 269
Himpsel, F. J., 196
Hinrichs, C. H., 92
Hirsch, P. B., 320
Hisatake, K., 137
Hjelinberg, H., 205
Hladik. J., 218
Hobbs, L. W., 218
Hobgood, H., 31
Hobson, J. P., 19, 33, 67
Hobson, M. C., 111, 112
Hodges, C. H., 154
Hodges, L., 200
Hodgson, A. A., 340
Höhne, M., 271
Hofer, D. E., 266, 271

Hoffmann, R., 157
Hohenberg, P., 23
Hohne, M., 261
Holland, U., 257
Holscher, A. A., 147
Holt, A. C., 237
Holzer, W., 259, 271
Homan, C. G., 226
Honigbaum, A., 272
Honma, A., 273, 274, 275, 276
Hoodless, I. M., 220, 225
Hooper, A., 246
Hooper, R. M., 315, 316
Hopkins, B. J., 91, 145, 147, 175, 176, 177
Hor, A. M., 234
Horiuchi, H., 354
Horne, R. W., 320
Hosemann, R. H., 39
Hoshino, H., 225, 233
Houston, J. E., 90, 101
Howells, E. R., 334
Howie, A., 320
Hsieh, C. H., 225
Huang, S. S., 12
Hudda, F. G., 92, 149
Hudson, D. E., 147
Huen, T., 187
Huggins, C. W., 341
Huggins, R. A., 225, 235, 250
Hughes, A. E., 255, 257, 271
Hughes, H. L., 179
Hughes, T. R., 85
Hughey, L. R., 180
Huntington, H. B., 161, 185
Husa, D. L., 104
Hussain, M., 44, 46
Hutchins, B. A., 162
Hutchinson, E., 279
Hutchison, J. L., 335, 338, 339, 341, 357

Ibach, H., 90
Idawi, I., 181
Ignatiev, A., 44, 49, 63, 65
Iijima, S., 323, 326, 329, 355
Iishi, K., 322
Ikeda, T., 227
Ikeno, S., 301
Ikeya, M., 243
Ikezawa, M., 260, 261, 271
Ikonnikov, D. S., 175
Ilver, L., 196
Imangulova, N. I., 147
Inohara, K., 269
Inoue, M., 272
Ionov, N. I., 142
Ipatova, I. P., 58
Irmler, H., 297
Irusteta, M. C., 338, 339
Ischuk, V. A., 176
Isherwood, B. J., 281
Isirikyan, A. A., 4
Isogami, M., 305
Isozumi, Y., 117
Itoh, H., 157, 171
Itoh, N., 243
Ives, H. E., 179

Jackson, B. J. H., 225
Jackson, J. D., 10
Jackson, M. L., 347
Jacobi, K., 212

Author Index

Jacobs, P. W. M., 218, 221, 222, 225, 231, 233, 234, 238, 239, 244, 246, 269, 272, 274, 275, 276, 277, 278, 279
Jacobson, J. L., 271
Jähnig, G., 168
Jaffee, R. I., 218
Jagannadham, A. V., 269
Jagodzinski, H., 341
Jain, S. C., 263, 265, 266, 271
Janak, J. F., 182, 183, 184
Jaswal, S. S., 245
Jeffe, H. W., 340
Jefferson, D. A., 322, 326, 334, 335, 339, 345, 347, 348, 350
Jefferson, M. E., 344
Jeffery, J. W., 332, 334
Jenkins, J. L., 322
Jensen, E. J., 319
Jepsen, D. W., 162
Jimenez, R. V., 253, 271
Johansson, G., 85
Johnson, B. B., 147
Johnson, D. J., 322
Johnson, K. H., 157, 161, 177
Johnson, P. D., 276
Johnsrud, A. L., 179
Johnston, H. S., 163
Jones, A. A., 180
Jones, A. V., 44, 49
Jones, E. R., 65
Jones, R. O., 22, 181, 182
Jonson, M., 30
Jordanov, A., 114, 137
Joshi, A., 101
Jost, A., 248
Jost, W., 239, 241, 248
Jowett, C. W., 91
Joyce, D. O., 127
Joyner, R. W., 156, 178, 212
Jung, P., 221
Juretschke, H. J., 140

Kader, I., 117
Kadyrov, P. M., 147
Känzig, W., 254
Kalennikova, T. A., 226
Kalkstein, D., 185
Kalousek, G. L., 341
Kambe, K., 205
Kamimura, H., 277
Kaminsky, M., 20
Kamiya, Y., 342
Kane, E. O., 179, 183
Kankeleit, E., 114
Kanomata, I., 147, 169
Kanzig, W., 257, 258, 259, 271
Kaplow, R., 56
Kapphan, S., 253, 254, 257, 271
Kappler, E., 281
Kar, N., 204
Karguppikar, A. M., 262, 271, 276
Karl, R., jun., 4
Karlsson, S. E., 85
Katane, R., 117
Kato, K., 262, 271
Kato, N., 56, 290, 292, 294, 301, 354
Katz, E., 225
Kautz, H. E., 251
Kawada, I., 354
Kawaminami, M., 284

Kebbekus, E. R., 43
Keck, J. C., 80
Keenan, A. G., 243
Keeton, S. C., 241, 245
Kellerman, E., 113, 128
Kelly, M. J., 185, 205
Kelly, W., 271
Kemeny, P. C., 202
Kennedy, J. H., 249
Kern, R., 100
Kessler, A., 226, 229
Kestner, N. R., 19
Keune, W., 129
Khanna, R. K., 265
Kichon, A., 199
Kigaura, M., 137
Kikuchi, R., 250
Kim, Y. S., 31
King, D. A., 167
King, R. D., 242
King, W. J., 260, 271
Kingery, W. D., 243
Kinno, S., 278
Kirby, R. D., 255, 257, 271
Kircher, H., 321, 340
Kirk, C. F., 4, 92, 149
Kirk, R. D., 260, 271
Kiselev, A. V., 4
Kiser, R. W., 165
Kisliuk, P., 169
Kittel, C., 56, 156
Kjems, J. K., 57, 58
Klapper, H., 302, 303
Kleiman, G. G., 4, 17, 83
Klein, C., 339, 340
Klein, M. V., 256, 271
Kleinman, L., 17
Klemperer, D. F., 92
Kliewer, K. L., 17, 181
Klimenko, E. V., 145
Knapp, A. G., 92
Knapp, J. A., 204
Kneringer, G., 170
Knop, K., 257
Knop, W., 254
Knor, Z., 160, 161, 162
Knudson, S. K., 157
Koch, J., 169, 170
Koch, P. H., 321
Kölbel, H., 162
Kohlstedt, D. L., 331
Kohn, W., 4, 9, 23, 25, 90, 140, 161, 182
Kohrt, C., 170
Kojima, K., 278
Kojima, T., 275, 278
Kokott, Ch., 233
Koshkin, V. M., 223
Kostopoulos, D., 225
Koster, G. F., 202
Koto, K., 328
Kowalski, J., 131
Koyama, R., 180, 183
Krakowski, R. A., 115
Kramer, H. M., 44, 103
Kramp, D., 225
Krause, J. K., 231
Krause, K. L., 227
Kraut, E. A., 63
Kravchenko, V. B., 257
Krolik, C., 275
Krolikowski, W. F., 183
Krstanović, I., 342
Krumhansl, J. A., 197, 253

Krylov, O. V., 112
Krynauw, G. N., 271
Ku, R., 170
Kuchynka, K., 160
Küppers, D., 168
Küppers, H., 302
Küppers, J., 90, 170, 177
Kuhn, U., 252
Kulik, V. N., 223
Kummer, J. T., 246, 249
Kundig, W., 111
Kunieda, S., 278
Kunze, G., 341
Kupryazhkin, A. Ya., 231
Kupihara, S., 220
Kventsel, G. F., 173
Kvist, A., 221

Lagally, M. G., 40
Laine, J., 243
Lally, J. S., 332
Lambert, M., 225
Lambert, W. H., 163
Landau, L. D., 9, 80
Lander, J. J., 40, 43, 44
Landman, U., 4, 17, 40, 52, 63, 90, 97, 105
Landre, J. K., 290
Lang, A. R., 281, 285, 290, 295, 301, 303, 307
Lang, N. D., 9, 21, 25, 90, 140, 164, 200, 205
Langbein, D., 5
Lansiart, S., 226
Lapeyre, G. P., 204, 209, 212, 215
Lapoujoulade, J., 168, 170
Laramore, G. E., 90
Laredo, E., 229, 231
Larher, Y., 99
Laskar, A. L., 220, 225, 227, 233
Latham, D., 196
Latta, E. E., 168, 175, 177
Law, B., 212
Lawless, K. R., 169
Lazareth, O., 242, 243
Lazarski, A., 132
Lazarus, D., 221, 226, 227
Leal Ferreira, G. F., 233
Leath, P. L., 197
Lébl, M., 231
Lecante, J., 177, 196
Le Cars, Y., 250
Le Claire, A. D., 228
Lederer-Rozenblatt, D., 182
Lee, D. I., 117
Lee, T. I., 145
Lefelhocz, J. F., 112
Leibfried, G., 223
Leidheiser, H., jun., 113, 128
Lemos, A., 275
Lemoyne, D., 274
Lennard-Jones, J. E., 4, 20, 74
Le Si Dang, 274
Levine, J. D., 91, 181
Levinson, L. M., 239
Levy, R. A., 114
Li, C. H., 81
Li, M. S., 253, 257
Liang, S. J., 282
Liddell, N. A., 335, 336
Lidiard, A. B., 235, 242, 243, 244, 245, 251

Liebau, F., 336, 339
Liebmann, S. P., 197
Liebsch, A., 162, 204
Liehr, A. D., 263
Lifschitz, E. M., 6, 9, 10
Liljequist, D., 137
Lilley, E., 229, 231, 245
Lin, S. F., 83
Lindau, I., 34, 83, 178, 179, 180
Lindberg, B., 85
Lindegaard Anderson, A., 287
Lindemann, G., 336
Linder, A., 251
Lindgren, I., 85
Lindquist, R. H., 111
Linnett, J. W., 83, 163, 180, 197, 212
Lipmann, B. A., 180
Lipsett, F. R., 260, 271
List, T., 226
Littleton, M. J., 234
Liu, S. F., 178
Llabador, Y., 137
Lloyd, D. R., 180, 191, 196, 212, 213
Lockhart, N. C., 311
Loftus, E., 243
Logan, H., 31
Logan, R. M., 20, 67, 80
Lohne, O., 304
Lombardo, G., 252, 257
London, F., 5
Lopes-Vieica, A., 347
Lorimer, G. W., 331, 332, 339
Lu, K. Y., 179
Lucas, A. A., 29
Lüty, F., 252, 253, 254, 255, 257, 271
Lundén, A., 221
Lundquist, B. I., 182, 204
Lundquist, S., 204
Lundy, T. S., 220
Luszczyskii, K., 104
Luther, A., 263
Lygina, I. A., 4
Lynch, D. F., 334
Lynch, D. W., 236
Lyo, S., 205

Maaroos, A., 278
Mabuchi, T., 275, 276
McCammon, R. D., 31, 32
McCauley, J. W., 345
McClure, J. D., 81
McCombie, C. W., 235
McElhiney, G., 44, 47
McFeely, F. R., 179, 200
McGeehin, P., 246
Machida, H., 227, 231
McKay, P., 310, 311
Mackie, W., 145
McKinney, J. T., 65
Mackinson, R. E. B., 185
McLachlan, A. D., 4
McLane, S. B., 162
McLaren, A. C., 335
McLean, R. K., 187, 201
McMenamin, J. C., 83
McNichol, B. D., 225
McRae, E. G., 35, 40
Macrae, N., 354
McTague, J. P., 58
Madden, H. H., 43, 170

Madey, T. E., 4, 88, 97, 101, 162, 168, 175, 176, 177
Madhukar, A., 156, 164
Mahan, G. D., 182
Mahnig, N., 168
Maire, G. L. C., 147
Maksimov, Yu. V., 112
Maksimova, T. L., 263, 271
Malathi, N., 112
Mallison, L. G., 338, 339
Mamedov, Kh. S., 333
Manfredi, M., 278
Manson, R., 74
Manzelli, P., 265, 271
Maradudin, A. A., 17, 58, 66
Marcus, P. M., 162
Margenau, H., 4, 5, 19
Margolis, L. Ya., 112
Mariani, E., 231
Marsh, H., 322
Martin ,G., 221
Martin, J. W., 235
Martino, F., 190
Maruyama, H., 301
Maruyama, J., 278
Masel R. I., 67, 72
Mason, R., 212
Mason, W. P., 271
Masri, P., 44, 65
Masunaga, S., 276
Mathews, L. D., 169
Matsui, Y., 354
Matsusaka, Y., 279
Matsushima, A., 276, 278
Matsuyama, E., 276
Matsuzawa, T., 113
Mattheiss, L. F., 190
Matzke, H., 221
Mavroyannis, C., 4
Maydell-Ondrusz, E., 132
Mayer, H., 183
Mayer, J. E., 236
Mayne, J. E. O., 127
Mazak, R. A., 114, 131
Medvedev, U. K., 145
Mee, C. B., 148
Meessen, A., 183
Meieran, E. S., 290
Meijer, D. T., 147
Meisal, W., 114
Menzel, D., 101, 157, 176, 177 179
Merle D'Aubigne, Y., 274
Merrill, R. P., 30, 32, 67, 69, 72, 163
Messiah, A., 182
Messmer, R. P., 157, 161, 177
Methfessel, S., 187
Meyer, F., 100, 101
Meyers, J. A., 74
Michels, A., 309
Middleton, A. P., 342
Midgley, D., 289
Mignolet, J. C. P., 3, 91
Mikheieva, E. V., 145, 148
Mikušík, P., 162
Milford, F. J., 4
Miliotis, D., 231
Miller, P. J., 265
Miller, R. B., 115
Miller, W. H., 72
Mills, D. L., 17, 66
Millward, G. R., 322
Milne, A. D., 300

Mimkes, J., 219
Minagawa, T., 356
Mink, J. F., 340
Minkov, B. I., 223
Minkova, A., 114, 137
Mirlin, D. N., 263, 271
Misra, K. D., 225, 233
Mitchell, J. L., 221
Mitchell, K., 185
Modell, D. I., 337
Mössbauer, R. L., 106
Montroll, E. W., 58, 97
Moodie, A. F., 334
Moodie, K. S., 234
Moore, J., 249
Morabito, J. M., jun., 32, 44, 100
Morales, J., 335
Morgan, H. W., 271
Moriarty, M. C., 257
Morimoto, N., 328, 352, 354, 355, 356
Morimoto, T., 228
Morrison, F. W., 349
Morrison, J., 43, 44
Morton, J. R., 258, 271
Morton, J. W., 233
Moruzzi, V. L., 182, 183
Moskvich, Yu. N., 252
Mothes, H., 225
Mott, N. F., 234, 239
Mourikos, S., 225
Müller, E. W., 92, 162
Mueller, F. M., 200
Müller, J., 149
Müller, M., 240
Müller, P., 225, 239
Mukaibo, T., 220
Muller, O., 251
Muller, R. H., 32, 44, 100
Muller, W. F., 334, 336
Mulliken, R., 95
Mullis, D., 233
Murakani, Y., 113
Murin, A. N., 220, 226, 228, 231
Murin, I. V., 220, 226, 228, 231
Murphy, W. F., 259, 271
Murthy, C. S. N., 226, 233, 240
Murti, Y. V. G. S., 223, 233, 240
Muttart, L. E., 341

Nabarro, F. R. N., 239
Nadler, C., 225, 231
Nagata, F., 354
Nagel, L. E., 252
Nakai, Y., 276, 279
Nakajima, Y., 352
Nakamura, F., 246
Nakano-Oneda, M., 354
Nakazawa, H., 354, 355, 356
Nakihara, M., 354
Narang, R. S., 311
Narayanamurti, V., 252, 257
Nasa, S., 113
Nasu, K., 275
Naumovets, A. G., 145
Navrotsky, A., 252
Nechitalo, A. E., 112
Neddemeyer, H., 179, 196
Neil ,K. S., 168
Nerenberg, M. A., 246
Nes, E., 303

Author Index

Netzer, F. P., 170
Neumann, H., 168
Neustadter, H. E., 4
Newnham, R. E., 345
Newns, D. M., 142, 205
Nichols, M. H., 90, 140
Nicolas, F., 228
Niehus, H., 169
Nieuwenhuys, B. E., 92, 94, 147
Nijboer, B. R. A., 9
Nikliborc, J., 92
Nilsson, P. O., 184, 196
Ninham, B. W., 16
Nishiguchi, K., 354
Nøst, B., 303
Noll, W., 321, 340
Nonoyama, M., 342
Nordberg, R., 85
Nordling, C., 85
Norgett, M. J., 222, 234, 235, 237, 238, 239, 240, 241, 242, 243, 245
Norris, C., 184, 196, 199
Norton, R. P., 168, 169
Novaco, A. D., 4, 57
Nowick. A. S., 225, 233, 253

Oberlin, A., 353
O'Brien, M. C. M., 274
Öpik, U., 278
Oestel, G., 345
Ogilvie, G. J., 315
Oguri, T., 147, 169, 176
O'Hara, S., 290
Ohlsen, W. D., 255
Ohring, M., 120
Oishi, Y., 243
Okada, I., 228
Okamusa, E. P., 328
Okazaki, A., 284
O'Keefe, M. A., 334
O'Keeffe, M., 247, 251, 252
Okuno, E., 233, 234
Oliver, J. P., 43
Olpin, A. R., 179
Oman, R. A., 81, 145
Onaka, R., 275, 278, 279
Ong, S. H., 225, 233
Onodera, H., 127
Ooaku, S., 275, 276
Orr, W. J. C., 4
Osminin, V., 276, 278
Oswald, R., 120
Osswald, R., 255, 257
Oughton, B. M., 346
Overbeck, J. T. G., 9
Ovtushinnikov, A. P., 145
Owens, B. B., 248
Oyama, K., 272, 276
Oyama-Gannon, K., 272, 275

Pace, E. L., 4
Page, P. G., 212
Page, P. J., 83
Paigne, J., 175, 196
Palmberg, P. W., 44, 45, 46, 101, 170
Pande, K. P., 233, 263, 271, 272
Pandey, G. K., 243, 253
Panitz, J. A., 162
Pansare, A. K., 220, 225
Pantelis, P., 222
Papike, J. J., 339

Papp, H., 44, 47
Park, D. S., 225, 233
Park, R. L., 43, 90
Parsegian, V. A., 16
Parsons, L. M., 274
Passell, L., 57, 58
Patankar, A. V., 220, 225
Patel, J. R., 294, 299
Patigny, J., 94
Pattanyak, D. N., 17
Pavlíková, M., 231
Pavlovic, S., 342
Pauling, L., 163
Paulson, R. H., 205
Pazzi, G. P., 278
Pearson, R. G., 167
Peascoe, J. G., 256, 271
Pederson, L. B., 235
Penchina, C. M., 199
Pender, R. R., 175, 176
Pendry, J. B., 40, 181, 182, 183, 187, 197, 201
Penn, D., 147
Penn, D. R., 205
Perdereau, M., 101
Peressini, P. P., 257
Perillo, P., 275
Perlin, Yu. E., 275
Perlman, M. M., 228, 229
Perry, D. L., 83, 180, 197, 212
Pershits, Ya, N., 225, 226
Peters, C., 249
Peterson, N. L., 228
Petkov, D., 171
Petroff, Y., 196
Peuckert, W., 140
Pfister, G., 254, 257
Phakey, P. P., 335, 336, 345
Pham Thi Hang, 343
Pianetta, P., 34, 83, 178, 179
Pierce, L., 356
Pierotti, R. A., 4
Pignet, T., 169
Pinchaux, R., 196
Pines, D., 29
Piott, J. E., 104
Piper, T. C., 63
Pirc, R., 254
Pisani, C., 4, 205
Pitaevskii, L. P., 11
Placzek, G., 56
Plummer, E. W., 83, 143, 147, 182, 196, 213, 215
Pohl, R. O., 252, 257
Polák, K., 231
Poícarova, M., 295
Polder, D., 9
Poldervaart, A., 330
Polinger, V. Z., 275
Pollack, G. L., 51
Pollard, W. G., 4
Pollock, F., 31
Ponec, V., 160
Pooley, D., 263
Pope, D. P., 282
Popov, N. M., 342
Portens, J. O., 63
Portnyagin, V. I., 220, 228, 231
Poshkus, D. P., 4
Pound, R. V., 221
Powell, M. J. D., 235
Prasad, P. S., 223
Pratt, P. L., 225, 226
Price, D. M., 221

Price, W. C., 266, 271
Pringle, J. P. S., 115
Pritchard, A. M., 113
Pritchard, J., 44. 46, 47, 91
Propst, F. M., 29, 63
Prosen, E. S. R., 4
Protopopov, O. D., 145, 148
Pryce, M. H. L., 278
Ptushinskii, Yu. G., 169
Puff, H., 183
Pundsack, F. L., 341
Purcell, E. M., 221
Puri, S. P., 112
Putnam, F. A., 99
Putnis, A., 357

Quentel, G., 100
Quinn, C. M., 180, 181, 191, 196, 197
Quinn, J. J., 204

Rabbitt, J. C., 338
Radcliffe, S. V., 332
Radhakrishna, S., 223, 231, 233, 262, 263, 265, 266, 269, 271, 272, 276
Radoslovich. E. W., 344
Radzilowski, R. H., 249
Raether, H., 90
Rahman, A., 65
Raistrick. I. D., 235
Ramasastry, C., 225, 231, 276
Ramdas, S., 240, 244
Ranfagni, A., 278
Rann, C. S., 338
Rao, C. N. R., 236, 240, 244
Rawlings, P. K., 150
Realo, E., 278
Reddy, V., 255
Ree, F. H., 237
Regnier, J., 99
Reifsnider, K., 290
Reilly, M. H., 222
Reimenn, K. J., 114
Reis, H., 150
Reisfeld, R., 272
Renken, C. J., 114
Reshinaa, I. I., 263, 271
Resing, H., 311
Reuter, G. E. H., 17
Reymann, G. A., 254
Reynolds, F. L., 147
Rhodin, T. N., 35, 40, 44, 49, 63, 65, 83, 97, 143, 147, 162, 175
Ricca, F., 4
Rice, M. J., 248, 250
Rich, F. J., 226
Richards, M. G., 104
Richards, P. J., 169
Richardson, N. V., 180, 191, 196, 212
Richter, R. H., 356
Rickard, J. M., 100
Rickert, H., 247
Ridgeway, P., 127
Rimbey, P., 181
Rimmington, H. P. B., 300
Ringström, B., 137
Ritchie, R. H., 17
Rivan, R., 175, 177
Rivière, J. C., 92, 101, 145
Robaux, O., 157, 208
Robert, G., 251

Roberts, M. W., 83, 156, 178, 179, 212
Roberts, R. H., 44, 47
Roberts, S., 254
Robinson, K., 346
Robinson, P., 340
Robinson, P. M., 315
Rösch, N., 157
Roessler, B., 307
Rössler, K., 223
Roetti, C., 4
Roev, L. M., 171
Rogers, C. B., 303
Rokbani, R., 231
Rolfe, J., 225, 233, 259, 260, 261, 262, 271
Rollefson, B. J., 104
Roloff, H. F., 179, 196
Rolt, J., 104
Romestain, R., 274
Rootsaert, W. J. M., 92
Rosenfeld, Yu. B., 275
Ross, M., 4, 339, 340, 344
Ross, M. A., 249
Ross, S., 43
Ross, T. A., 241
Rossel, J., 225, 231
Roth, W. L., 248, 250, 251
Roulet, B., 182
Rovida, G., 169
Rowe, J. E., 90, 183, 191
Roy, D., 236, 240
Royce, B. S. H., 243
Rucklidge, J. C., 342
Ruggeri, S., 197
Ruhenbeck, C., 262, 271
Rusch, W., 257
Rushbrook-Williams, S., 219, 220
Rushworth, A. J., 251
Rustad, O., 304
Ryan, J. F., 251
Rye, R. R., 168

Sabharwal, K. S., 219
Sachdev, M., 235
Sachs, R. G., 4
Sachtler, W. M. H., 91, 92, 94, 143, 147
Sack, H. S., 254, 257
Sadanaga, R., 328, 352
Safa, M., 289, 303
St James, D., 182
Saito, N., 228
Salpeter, E. E., 182
Salthouse, P., 311
Saltsburg, H., 67, 72
Samberg, M., 242
Sammells, A. F., 249
Sams, J. R., jun., 4, 33
Samson, J. A. R., 35
Samuelson, G. L., 220
Sand, L. B., 340
Sander, L. M., 243, 253
Sanders, J. V., 334
Sandstrom, D. R., 83, 177
Sargood, A. J., 91
Sastry, M. D., 266
Sastry, S. B. S., 276
Sato, H., 250
Sau, R., 30, 32, 69
Sauvage, M., 300
Sawicka, B., 131, 132, 138
Sawicki, J., 131, 132, 138

Sawyer, E. W., 227
Scandale, E., 305
Schaich, N. W., 182
Schaich, W. L., 181
Schaller, W. T., 337
Schamp, H. W., 225
Scheffler, M., 205
Schiff, L. I., 25
Schmidt, L. D., 168, 169
Schmit, J., 29
Schober, O., 168
Schoeffer, E. A., 114
Schoemaker, D., 276
Schoonman, J., 225
Schram, A. S., 94
Schram, K., 9
Schrieffer, J. R., 161, 164, 205
Schroeder, K., 223
Schröter, W., 304
Schubin, S., 183
Schulman, J. H., 260, 271
Schulte, F. K., 140
Schulz, K., 91
Schulze, P. D., 236
Schunck, J. P., 137
Schutte, C. J. H., 271
Schwuttke, G. H., 306
Scott, H. G., 315
Seeger, A., 239
Sehgal, H. K., 263, 266, 271
Seib, D. H., 199
Seidal, H., 257
Seitz, F., 269
Seltzer, M. S., 218
Sen, S. K., 236
Seraphin, B. O., 184
Sette-Camara, A., 129
Setty, R. S., 272
Seward, W. D., 255, 257
Sexton, B. A., 102
Shah, B. S., 314
Sham, L. J., 23, 182
Shaner, J. W., 255
Shapkina, Yu. S., 231
Sharma, B. D., 223, 231, 263
Sharma, M. N., 225, 233
Shaw, K. W., 339
Sheinberg, B. N., 145, 148
Shell, H. R., 341
Shepherd, I. W., 255, 257
Sherman, W. F., 266, 271
Sherrill, F. A., 304
Sherwood, J. N., 225, 310, 311, 315, 316
Shevchik, N. J., 199, 202
Shimanuki, S., 278
Shimoji, M., 225, 233
Shinjo, T., 112, 113
Shirley, D. A., 36, 164, 179, 200, 212
Shirozu, M., 346
Shopov, D., 171
Shore, H. B., 243, 253, 255, 257
Shukla, A. K., 236, 240, 243, 244
Shuppe, G. N., 145, 147, 148
Shurmann, R., 91
Sickafus, E. N., 90, 101
Sickhaus, W. L., 167
Siedal C. W., 112
Siegbahn, K., 85
Siegel, E., 131
Siems, R., 320
Sievers, A. J., 255, 257, 271

Silberman, E., 266, 271
Silsbee, R. H., 255, 257
Simkin, D., 275
Simmons, G. W., 113, 128
Simpson, J. H., 236
Sing, K. S. W., 46
Singer, J., 251
Singh, R., 263
Sittig, R., 257
Sivkov, V. P., 226
Sixou, P., 234
Sladký, P., 233
Slagsvold, B. J., 147, 196, 197
Slater, J. C., 22, 161, 202
Slichter, C. P., 221
Slifkin, L. M., 218, 219, 220, 223, 227, 229, 232
Smereka, T. P., 145
Smirnov, B. G., 147
Smith, D. L., 70
Smith, F. C., 161
Smith, G., 308
Smith, J. N., jun., 67, 72
Smith, J. R., 25, 140, 150, 161, 205
Smith, J. V., 328, 344
Smith, M. L., 348
Smith, M. W., 11
Smith, N. V., 179, 180, 183, 184, 185, 187, 190, 191, 204
Smith, R. J., 212, 215
Smith, S. H., 297
Smoluchowski, R., 91, 149, 225, 243
Smorodinova, M. I., 147
Smutek, M., 167, 169
Smyth, J. R., 328
Snaith, J. C., 92
Soboleva, S. V., 345
Somorjai, G. A., 32, 44, 62, 67, 100, 167
Sondheimer, E. H., 17
Sors, A. I., 245
Soven, P., 164, 185, 204
Spicer, W. E., 34, 83, 178, 179, 181, 183, 184, 187, 199
Spijkerman, J. J., 113, 114
Sprung, M., 336
Srinivasan, G., 30
Staats, P. A., 271
Stanek, J., 131, 138
Stasiw, M., 261, 271
Staveley, L. A. K., 308
Steele, B. C. H., 218, 251
Steele, R. B., 4
Steele, W. A., 2, 4, 33, 43
Steele, W. C., 221
Stefansson, V., 137
Steiger, R. F., 32, 44, 100
Stein, R. J., 162
Steinfink, M., 346
Steinkelberg, M., 179
Steinmann, W., 196
Steinrisser, F., 90
Stern, O., 66
Stickney, R. E., 67, 72, 80, 145, 169
Stiddard, M. H. B., 92
Stier, P. M., 149
Stillman, M. J., 275
Stocks, G. M., 196, 197
Stoebe, T. G., 225
Stohr, J., 179, 200, 212
Stoll, A. G., jun., 67, 70, 72

Author Index

Strachan, C., 20
Strässler, S., 248
Strange, J. H., 220, 221, 225, 233
Strauch, D., 253
Strayer, R. W., 145
Strock, L. W., 249
Strongin, M., 44, 101
Strozier, J. A., 182
Strutt, J. E., 229
Stuart, R. N., 183, 187
Sturm, K., 196
Sucksdorff, I., 354
Suetin, P. E., 231
Süptitz, P., 220
Sugano, S., 277
Suito, E., 345
Sultanov, V. M., 147
Sunagawa, I., 305
Sussmann, J. A., 253
Suzanne, J., 44, 65, 103
Suzdalev, I. P., 112
Suzuki, K., 229
Swalin, R. A., 221
Swaminathan, P. K., 244
Swanson, K. R., 113
Swanson, L. W., 92, 145
Swartzendruber, L. J., 114, 131
Sytaya, E. P., 147

Takada, T., 113
Takafuchi, M., 117
Takaishi, T., 104
Takamura, J., 246
Takayama, H., 203
Takeda, M., 328, 344, 345
Takeuchi, Y., 352
Takezoe, H., 278, 279
Tamaru, K., 167, 174
Tamm, I., 183
Tamm, P. W., 168, 169
Tan, Y. T., 225
Tanaka, K., 174
Tanner, B. K., 281, 289, 297, 300, 303
Tannhauser, D. S., 219
Taub, H., 57, 58
Taylor, A., 225
Taylor, N. J., 101
Taylor, T. N., 170, 175
Teegarden, K., 276
Teltow, J., 244
Temmerman, W., 196
Tendulkar, D. V., 72
Tennakoon, T. B., 112
Terriere, G., 353
Theimer, O., 245
Thery, J., 250
Thilo, E., 336
Thiry, R., 196
Thomas, H., 183
Thomas, M. W., 251
Thomas, R. N., 122
Thomas, J. M., 112, 117, 121, 127, 180, 315, 322, 323, 326, 334, 335, 339, 345, 350
Thompson, D. W., 345
Thompson, E. D., 74
Thompson, A. L., 104
Thomson, R. H., 354
Thomy, A., 99, 101
Thorpe, B. J., 205
Thorpe, R. K., 134
Thorsley, S. A., 269, 277

Tibbetts, G. G., 83
Tilley, C. E., 336
Tilley, R. J. D., 323
Tillmetz, K. D., 162
Timmermans, J., 308
Timusk, T., 261
Toaff, S., 274
Todd, C. J., 147
Todder, G., 265, 271
Toennies, J. P., 67
Tokanami, M., 352, 354
Tomášek, M., 162
Tompkins, F. C., 4, 32, 92, 239
Tong, S. Y., 40, 44, 63, 66
Toombs, G. A., 248
Topping, J., 91
Torijana, T., 137
Torrens, I. M., 81
Torrey, H. C., 221
Torrini, M., 205
Tosi, M. P., 237
Toyozawa, Y., 272
Tracy, J. C., 46, 83, 101, 170, 176
Traum, M. M., 180, 183, 185, 189, 191, 204
Tricker, M. J., 112, 117, 121, 123, 127, 134, 137
Trilling, L., 80
Trimm, D. L., 212
Trnovcová, V., 231
Trojer, F. J., 333
Tsang, Y. W., 156
Tskhakaya, V. K., 175
Tsong, T. T., 92
Tsuboi, T., 276, 278, 279
Tsuchida, A., 4, 74
Tsukerblat, B. S., 275
Tsyaschenko, Yu. P., 265, 266, 271
Tucker, C. W., 177
Turk, L. A., 161
Turkevich, J., 321
Turner, D. W., 213
Turner, R. G., 220, 225
Turrel, G., 271
Turtle, R. R., 196
Tykodi, R. J., 4

Ugo, R., 164
Ullermayer, L. S., 4
Unger, S., 228, 229
Unguris, J., 44, 50
Usami, S., 147, 175, 177
Uyeda, R., 342, 343

van der Avoird, A., 157, 171
Vander Sande, J. B., 331
Van der Voort, E., 242
van Gool W., 246, 248, 249
Van Hove, L., 56
Van Kampen, N. G., 9
van Landuyt, J., 323, 357
Vannotti, L. E., 258, 271
Van Oirschot, T. G. J., 92, 145
van Oostrom, A., 149
Van Reijen, L. L., 92
Van Sciver, W. J., 228, 231
van Someren, L., 147
Varotsos, P., 225, 231
Vasilos, T., 221
Vasko, N. P., 169
Veisman, V. L., 225
Vekhter, B. G., 275

Vella, G. J., 260, 271
Venkateswarlu, P., 266, 269
Verbeek, H., 157
Verlan, E. M., 265, 266, 271
Vermiglio, G., 275
Vernier, P. J., 185
Verwey, E. J. W., 9
Viliani, G., 278
Vineyard, G. H., 235
Viswanath, A. K., 266, 269
Viswanathan, V., 276
Viswanatha Reddy, K., 225, 231
Völter, J., 175
Volobjev, P. V., 231
Voos, E., 336
Vorburger, T. V., 83, 147, 162, 177
Voronov, V. N., 252
Vrakking, J. I., 101

Waclawski, B. J., 83, 162, 177, 196
Wadsley, A. D., 339
Wagner, H., 149, 150
Wagner, J. B., jun., 247
Wahl, J., 257
Wallace, C. A., 281
Wallden, L., 184
Wallis, R. F., 66
Wang, J. C., 250
Wanklyn, B. M., 303
Warren, B. E., 56, 328, 337
Warrier, A. V. R., 263, 265, 266, 271
Watanabe, E., 352, 354, 355, 356
Watanabe, H., 127
Watson, R. E., 203
Webb, M. B., 40, 44, 50, 65
Wedepohl, P. T., 237
Wedin, G., 204
Wedler, G., 168
Weertman, J., 311
Wehner, P. S., 179, 200, 212
Weinberg, W. H., 30, 67, 69, 163
Weis, M., 169
Weiss, G. H., 16, 58
Weiss, R. J., 56
Welch, D. O., 242, 243
Welton, T., 13
Wenk, H. R., 334, 336
Wentz, M., 229
Wertheim, G. K., 107
Whelan, M. J., 320
Whitam, 226, 231
White, R. E., 67, 72
White, W. W., 161
Whitefield, R. L., 145
Whitmore, D. H., 225
Whittaker, E. J. W., 337, 338, 339, 340, 341, 342
Whittingham, M. S., 246, 247, 250
Wiedersich, H., 249
Wiese, W. L., 11
Wigner, E., 91, 150
Wilkinson, G. R., 266, 271
Wilks, R. S., 243
Williams, A. R., 164, 182, 183, 184, 205
Williams, B., 175
Williams, C. B., 176

Williams, F. E., 276
Williams, G. P., 196, 199, 233
Williams, J. O., 315, 323
Williams, P. M., 83, 180, 196, 212
Williams, R. H., 180
Williams, R. S., 179, 212
Willis, B. T. M., 242
Willis, R. F., 34, 162, 182, 196
Wilmer, P. C., 176
Wilson, A. J. C., 56
Wilson, J. C., 263
Wilson, W. D., 241, 245
Winterbottom, A. P., 117, 121, 127, 137
Wohlfahrt, K., 114
Wojciechowski, K. F., 4, 145
Wolf, D., 221
Wolf, E., 17, 202
Wolf, H. C., 255, 257
Wolfram, T., 63
Wolken, G., jun., 74
Wones, D. R., 344
Wood, E. A., 43

Wood, J., 196
Woods, S., 212
Wooten, F., 183, 187
Workowski, C. J., 147
Wu, C. C., 281
Wuensch, B. J., 221
Wuttig, M., 219
Wylde, L. E., 220, 225

Yada, K., 322, 341
Yagnik, C. M., 114, 131
Yamamoto, H., 127
Yamazaki, H., 169
Yampolskii, Yu. P., 112
Yates, J. T., jun., 4, 88, 101, 162, 168, 175, 176
Yates, K., 180
Ying, S. C., 161
Yoder, M. S., 344
Yong, C. Y., 157
Yokose, S., 225, 233
Yoon, D. N., 226, 227
Yoshida, T., 345
Yoshikawa, A., 275, 278

Yound, R. A., 56
Young, A. C., 19
Young, D. A., 225
Young, D. M., 1, 3, 4, 43
Young, F. W., jun., 304
Young, K. F., 233
Yu, K. Y., 34, 83, 178
Yuen, P. S., 243
Yuster, P. H., 276, 278, 279

Zakis, J. R., 227
Zandberg, E. Ya., 142, 147
Zaremba, E., 4
Zarka, A., 303, 305
Zasuhka, V. A., 171
Zazubovich, S., 276, 278
Zeikats, V. P., 227
Zhoukhlistov, A. P., 345
Zibold, G., 255, 271
Zierold, K., 229
Zingerman, Ya. P., 176
Zussman, J., 340, 341, 342, 347
Zvyagin, B. B., 342, 345
Zykov, V. M., 175